BIOMATERIALS SCIENCE AND ENGINEERING

Edited by **Rosario Pignatello**

INTECHOPEN.COM

Biomaterials Science and Engineering
Edited by Rosario Pignatello

CBS, Edition **2016**

Published by InTech
Janeza Trdine 9, 51000 Rijeka, Croatia

Notice

Publishing Process Manager Mirna Cvijic

Technical Editor Teodora Smiljanic

Cover Designer InTech Design Team

Image Copyright Yannis Ntousiopoulos, 2011. Used under license from Shutterstock.com

Additional hard copies can be obtained from orders@intechopen.com

Biomaterials Science and Engineering, Edited by Rosario Pignatello
p. cm.
ISBN 978-953-307-609-6

Contents

Preface

Scientists who dedicate their research activity to biomaterials pass through the typical dichotomy that often characterizes the basic research. On one side is the wish of exploring new frontiers of chemistry, physics, biology, medicine, pharmaceutics and all other disciplines to which biomaterials can be applied. The constantly improving scientific knowledge would feed the freedom of attempting new strategies for producing materials with always tailored and improved characteristics. On the other side, one should one have a look to the different 'official' definitions given for biomaterials. It is evident how the restriction imposed by words would limit the fantasy and effectiveness of fundamental scientific research. Just as an example-biomaterials are defined as a 'nonviable material used in a medical device, intended to interact with biological systems' (Consensus Conference of the European Society for Biomaterials, 1986), or as 'any substance (other than a drug) or combination of substances, synthetic or natural in origin, which can be used (…) as a whole or as a part of a system which treats, augments, or replaces any tissue, organ, or function of the body (NIH), or even 'a systematically and pharmacologically inert substance designed for implantation within or incorporation with living systems' (Clemson University Advisory Board for Biomaterials). Essentially, the only common property is that a biomaterial would be different from a biological material, that is produced by a biological system. Clearly, none of the proposed definitions can succeed to cover the whole landscape of properties and applications of these peculiar compounds, but they can only enlighten a particular aspect of their potentials. A similar situation can be applied for nanomedicine – a research field with which the field of biomaterials actually often shares technologies and applications – and for which is the gap between 'official' definitions and the originality of published researches even larger.

These considerations have been one of the basis of the present editorial task, that will comprehend three volumes focused on the recent developments and applications of biomaterials. These books collect review articles, original researches and experimental reports from eminent experts from all over the word, who have been working in this scientific area for a long time. The chapters are covering the interdisciplinary arena which is necessary for an effective development and usage of biomaterials. Contributors were asked to give their personal and recent experience on biomaterials, regardless any specific limitation due to fit into one definition or the other. In our opinion, this will give readers a wider idea on the new and ongoing potentials of different synthetic and engineered macromolecular materials.

In the meantime, another editorial guidance was not to force the selection of papers concerning the market or clinical applications or biomaterial products. The aim of the book was to gather all results coming from very fundamental studies. Again, this will allow to gain a more general view of what and how the various biomaterials can do and work for, along with the methodologies necessary to design, develop and characterize them, without the restrictions necessarily imposed by industrial or profit concerns.

The chapters have been arranged to give readers an organized view of this research field. In particular, this book collects 22 chapters reporting recent researches on new materials, particularly dealing with their potential and different applications in biomedicine and clinics: from tissue engineering to polymeric scaffolds, from bone mimetic products to prostheses, up to the strategies to manage their interaction with living cells. The book is structured in sections covering different aspects: the first section on presents some reviews and articles on the production and/or chemical modification of novel biomaterials, along with their potential applications. The second section presents six articles exploring the use of engineered biomaterials for the treatment of bone and cartilage diseases, along with a further chapter on the mechanobiology of dental implantable devices. The third section contains five chapters and it is focused on recent studies on tissue engineering by cells and biomaterials. Finally, two conclusive chapters deal with the application of biomaterials to hearth and vascular prostheses.

I hope that you will find all these contributions interesting, and that you will be inspired from their reading to broaden your own research towards the exciting field of biomaterial development and applications.

Prof. Rosario Pignatello
Department of Pharmaceutical Sciences,
Faculty of Pharmacy,
University of Catania,
Italy

Part 1

Production, Modification and Biomedical Applications of Novel Materials

Biomaterials and Sol-Gel Process: A Methodology for the Preparation of Functional Materials

Eduardo J. Nassar et al*
*Universidade de Franca, Franca, Sao Paulo,
Brazil*

1. Introduction

There are many kinds of materials with different applications. In this context, biomaterials stand out because of their ability to remain in contact with tissues of the human body. Biomaterials comprise an exciting field that has been significantly and steadily developed over the last fifty years and encompasses aspects of medicine, biology, chemistry, and materials science. Biomaterials have been used for several applications, such as joint replacements, bone plates, bone cement, artificial ligaments and tendons, dental implants for tooth fixation, blood vessel prostheses, heart valves, artificial tissue, contact lenses, and breast implants [1]. In the future, biomaterials are expected to enhance the regeneration of natural tissues, thereby promoting the restoration of structural, functional, metabolic and biochemical behaviour as well as biomechanical performance [2]. The design of novel, inexpensive, biocompatible materials is crucial to the improvement of the living conditions and welfare of the population in view of the increasing number of people who need implants [3]. In this sense, it is necessary that the processes employed for biomaterials production are affordable, fast, and simple to carry out. Several methodologies have been utilized for the preparation of new bioactive, biocompatible materials with osteoconductivity, and osteoinductivity [4 - 13]. New biomaterials have been introduced since 1971. One example is Bioglass 45S5, which is able to bind to the bone through formation of a hydroxyapatite surface layer [14]. The sol-gel processes are now used to produce bioactive coatings, powders, and substrates that offer molecular control over the incorporation and biological behavior of proteins and cells and can be applied as implants and sensors [15 - 17]. In the literature there are several works on the use of the sol-gel process for production of biomaterials such as nanobioactive glass [18], porous bioactive glass [19], and bioactive glass [20 - 22], among others.

Hybrid inorganic-organic nanocomposites first appeared about 20 years ago. The sol-gel process was the technique whose conditions proved suitable for preparation of these materials and which provided nanoscale combinations of inorganic and organic composites

* Katia J. Ciuffi, Paulo S. Calefi, Lucas A. Rocha, Emerson H. De Faria, Marcio L. A. e Silva, Priscilla P. Luz, Lucimara C. Bandeira, Alexandre Cestari, Cristianine N. Fernandes
Universidade de Franca, Franca, Sao Paulo, Brazil

[23]. Natural bone is an inorganic-organic composite consisting mainly of nanohydroxyapatite and collagen fibers. Hybrid materials obtained by the sol-gel route combine the advantages of both organic and inorganic properties. Several kinds of organofunctional alkoxysilanes precursors have been studied for the production of silica nanoparticles. The sol-gel offers advantages such as the possibility of obtaining homogeneous hybrid materials under low temperature, thereby allowing for the incorporation of a variety of compounds [23 - 29].

The sol-gel process is based on the hydrolysis and condensation of metal or silicon alkoxides and is used to obtain a variety of high-purity inorganic oxides or hybrid inorganic-organic materials that are simple to prepare [30]. This process can be employed for the synthesis of functionalized silica with controlled particle size and shape [31 - 38].

Apart from the several applications mentioned in the first paragraph of this chapter, more recently, biomaterials have been utilized as drug delivery systems (DDSs). In this sense, polymers and biodegradable polymers emerge as potential materials, since they promote temporal and targeted drug release. Indeed, biomaterials have had an enormous impact on human health care. Applications include medical devices, diagnosis, sensors, tissue engineering, besides the aforementioned DDSs [39]. In the latter field, an ideal drug deliverer should be able to lead a biologically active molecule at the desired rate and for the desired duration to the desired target, so as to maintain the drug level in the body at optimum therapeutic concentrations with minimum fluctuation [1, 40]. The use of DDSs overcomes the problems related to conventional administration routes, such as oral and intravenous administration.

Several biomaterials have been applied as DDSs. This is because they are biocompatible and/or biodegradable, which allows for consecutive administrations. Hydroxyapatite-based materials, natural and synthetic polymers, silica, clays and other layered double hydroxides, and lipids are some examples of biomaterials that have been employed for the delivery of active molecules through the body. Liposomes, solid lipid nanoparticles, polymeric nano and microparticles, micelles, dendrimers, metallic nanoparticles, and nanoemulsion are currently utilized as DDSs.

Special attention has been given to DDSs comprised of biodegradable polymers and silica. In polymeric DDSs, the drugs are incorporated into a polymer matrix. Since biodegradable polymers are degraded to non-toxic substances, they do not have to be removed after implantation. So they have become attractive candidates for DDS applications. The rate of drug release from polymeric matrices depends on several parameters such as the nature of the polymer matrix, matrix geometry, drug properties, initial drug loading, and drug-matrix interaction. Moreover, the drugs can be effectively released by bioerosion of the matrices. [40]. Thus, both natural, frequently polysaccharides, and synthetic biodegradable polymers, usually aliphatic polyesters such as PLA, PGA, and their copolymer (PLGA), are the most extensively investigated biodegradable materials for drug delivery applications [1]. Inorganic materials, like silica, can offer the necessary properties for a nanoparticle to be applied as DDS, especially nontoxicity, biocompatibility, high stability, and a hydrophilic and porous structure. The drug release rate from the silica structures could be controlled by adjusting particle size and porous structure [41 - 45].

The sol-gel technology is also employed in the preparation of inorganic ceramic and glass materials. This technique was first used in the mid 1800s, when Ebelman and Graham carried out studies on silica gels [46]. Initially, the sol-gel process was utilized in the preparation of silicate from tetraethylorthosilicate (TEOS, $Si(OC_2H_5)_4$), which is mixed with water and a mutual solvent, to form a homogeneous solution. Recently, new reagents have appeared, so novel inorganic oxides and hybrid organic-inorganic materials can be synthesized using this methodology. Another process known as non-hydrolytic sol-gel has been developed by Acosta et al [47], who used the condensation reaction between a metallic or semi-metallic halide (M-X) and a metallic or semi-metallic alkoxide (M'-OR) to obtain an oxide (M-O-M'). The hydrolytic and non-hydrolytic sol-gel processes as well as their mechanisms are well discussed in the literature [46, 48, 30]. The sol-gel route is well-known for its simplicity and high rates. It is the most commonly employed technique for the synthesis of nanoparticles, and it involves the simultaneous hydrolysis and condensation reaction of the alkoxide or salt. The obtained materials have several particular features. The importance and advantages of nanoparticles have been scientifically demonstrated, and these particles have several industrial applications; e.g., in catalysis, pigments, biomaterials, phosphors, photonic devices, pharmaceuticals, and among others [36, 49 - 54].

In this chapter, we propose a brief review on materials prepared by the hydrolytic and non-hydrolytic sol-gel methodologies and their possible bioapplications.

2. Results and discussion

In the next topics, 2.1 and 2.2, the results and discussion about all the research developed in our laboratory using hydrolytic and non-hydrolytic methodologies in the synthesis of materials for bio applications such as glass ionomers, bioactive materials, coating on scaffolds obtained by rapid prototyping (RP), and materials for drug delivery are shown.

2.1 Preparation of biomaterials by the hydrolytic sol-gel process

In this topic the materials prepared by the hydrolytic sol-gel methodology and their characterization are described. We aimed to obtain materials whose properties would enable their application as biomaterials.

In a first work, materials containing Ca-P-Si were prepared by the sol-gel route by mixing TEOS, calcium alkoxide, and phosphoric acid [55]. The resulting materials were immersed in Simulated Body Fluid (SBF) [56], pH = 7.40, for 12 days. The sample was characterized before and after contact with SBF.

Transmission electron microscopy (TEM) can provide structural information about materials, such as particle shape, size, and crystallinity. Figures 1a, b, c, and d show TEM images of the Ca-P-Si matrix obtained by the sol-gel methodology before and after immersion in SBF.

The TEM images in Figure 1a reveal the formation of small particles with an average size of 20 nm. Electron diffraction gives evidence of an amorphous phase. The bright and dark fields in Figures 1b and c demonstrate that the materials contain crystalline and amorphous phases. Figure 1d displays the electron diffraction of the crystalline phase. The electron diffraction pattern shows planar distances of 2.86 Å and 1.88Å, which, according to Bragg's law, indicates that these distances correspond to $2\theta = 31.2°$ (211) and $48.6°$ (320). This peak can be ascribed to hydroxyapatite (JCPDS – 9-0432) [57]. The EDS spectra of the amorphous phase reveal large quantities of Si and O, indicating the presence of amorphous silicate. The crystalline phase, whose composition contained Ca and P, has been ascribed to hydroxyapatite crystallization.

In another work, samples with different Ca/P molar ratios were prepared by the sol-gel methodology, by mixing TEOS, calcium ethoxide, and phosphoric acid. The samples were analyzed before and after contact with SBF [50].

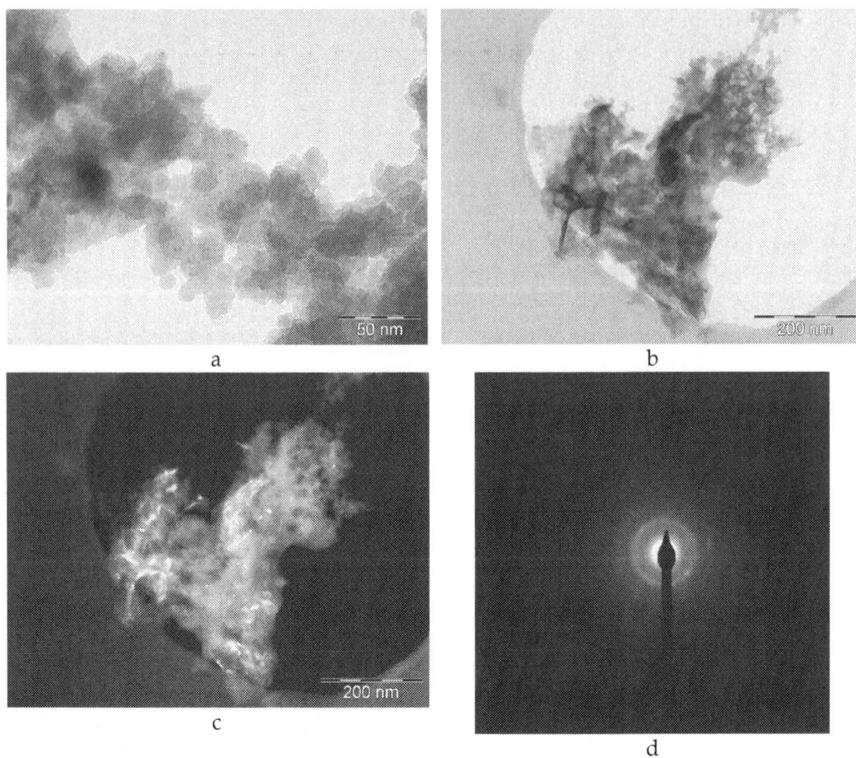

Fig. 1. TEM images of the Ca-P-Si matrix before (a) and after (b) immersion in SBF, bright (b) and dark field (c), and electron diffraction (d) of the sample´s crystalline phase after immersion in SBF.

Figures 2 and 3 depict the X-ray diffraction patterns for the samples prepared with different Ca/P molar ratios, before and after contact with the SBF solution, respectively.
The three starting powders present crystalline and amorphous phases, with well-defined diffraction peaks. The crystalline phase displays peaks at 2θ = 26.5, 32.5, 33.0, 49.2, and 53.1, which can be ascribed to hydroxyapatite (HA), whereas the peaks corresponding to calcium triphosphate (TCP-β) appear at 2θ = 26.5, 30.2, and 53.1. [58]. Several other peaks due to other phosphate silicates, such as $Ca_5(PO_4)_2SiO_4$ and $(Ca_2(SiO_4))_6(Ca_3(PO_4)_2)$, can also be observed.
The high percentages of phosphate ions (40%) in the present samples were crucial to the precipitation of crystalline phosphate nanoparticles in the silica matrix. However, according to literature reports, the double P=O bond favors phosphate phase formation in the silica network, thus increasing the tendency toward crystallization [59]. The peak broadening of the XRD reflection can be used to estimate crystallite size in a direction perpendicular to the crystallographic plane, using the Scherrer equation [60]. On the basis of the XRD data, the average crystallite size was calculated as being approximately 2 nm, which indicates the formation of calcium phosphate nanoparticles.

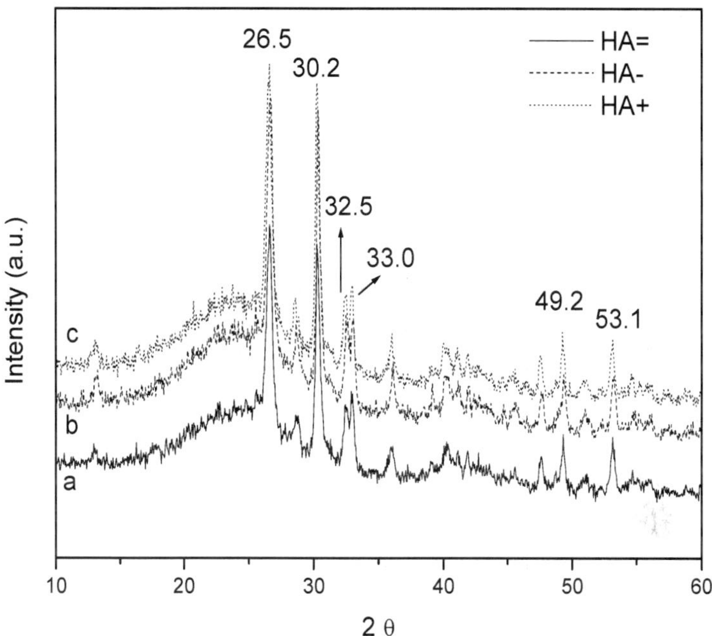

Fig. 2. X-ray diffraction of the samples (a) = HA, (b) HA-, and (c) HA+ before contact with SBF.

The technology based on RP is a processes employed for assemblage of materials in the powder, filament, liquid, or slide form, which in turn are stacked in successive thin layers until a three-dimensional structure is achieved. The process begins by designing a mold for the scaffold using computer-aided design (CAD) software. The mold can possess a branching network of shafts that will define the microchannels in the scaffold [61 - 65]. The layer-by-layer building approach allows for the preparation of highly complex structures that cannot be obtained by technologies based on material subtraction, which is the most frequently employed procedure nowadays. RP has several important applications in a number of areas, including aircrafts, automobiles, telecommunications, and medicine [66 - 68].

In our following works 3D piece prepared by RP on ABS and polyamide (nylon) was used, and the properties of this piece were modified by the sol-gel methodology. To this end, the sols were prepared by stirring TEOS and calcium alkoxide in ethanol. Two sols were synthesized, namely one containing phosphate ions (Si-Ca-P) and another without phosphate (Si-Ca) [69]. The sols were deposited onto ABS by using the dip-coating, as described in the literature [70, 71]. This technique consists in immersing a substrate directly into the prepared sols. The crystallization of phosphates was accomplished by immersing the samples into SBF for 15 days. The SBF treatment was performed in the static condition. The samples were then dried at 50°C and characterized.

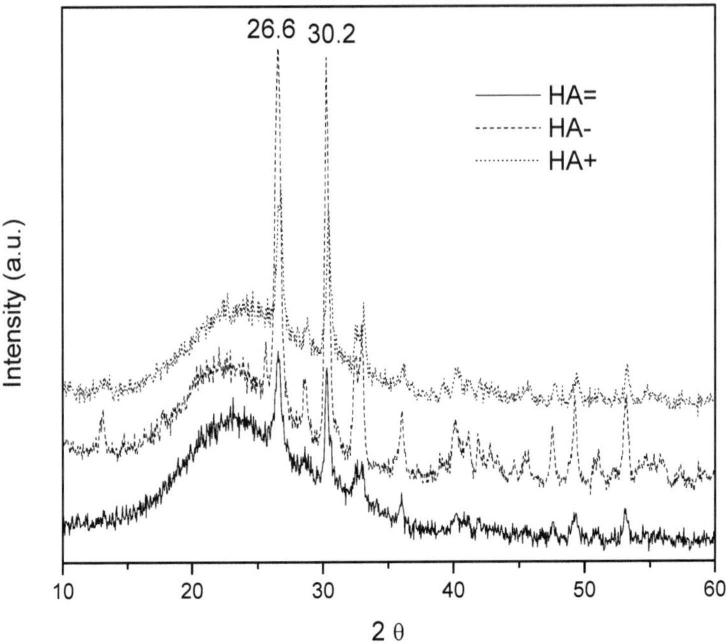

Fig. 3. X-ray diffraction of the samples (a) = HA, (b) HA-, and (c) HA+ after contact with SBF.

Figure 4 displays the XRD patterns obtained for the ABS substrate and for the AcP and AsP samples.
XRD analysis of the ABS substrate and of the samples coated by sol-gel revealed the presence of broad peaks between 10 and 30°, characteristic of amorphous materials. Figures 5 and 6 illustrate the XRD patterns of the samples AcP and AsP before and after contact with SBF, respectively.
After contact with SBF, the XRD patterns of the samples AcP and AsP clearly displayed peaks, which is evidence of initial crystallization. Peaks at 2θ = 10.6, 21.6, 31.6, and 45.2 were detected for the AcP sample after it was placed in SBF for 15 days, whereas the XRD pattern of the sample AsP displayed two peaks only, namely at 2θ = 31.6 and 44.9. The latter peaks correspond to a mixture of calcium phosphate silicates (JCPDS 21-0157; 11-0676). This observation is very important since it shows that the coating interacts with SBF to form a calcium phosphate, which is the main component of hydroxyapatite. Infrared spectroscopy of the ABS substrate presented peaks at 1077 and 465 cm-1, ascribed to the Si-O-Si vibration mode. These peaks were also verified in the spectrum of the sample AcP after contact with SBF, indicating that the silicate coating is still present on the ABS substrate. New peaks appeared at 3360 and 1653 cm-1, related to water vibrations, and at 610 and 550 cm-1, ascribed to the P-O vibration [72], thereby corroborating the observations from the X-ray

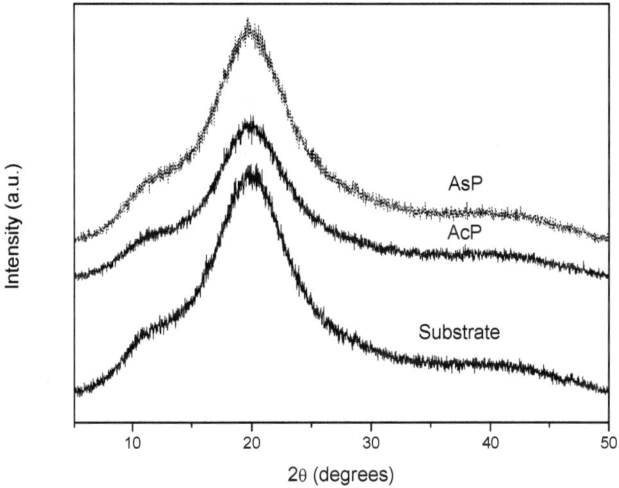

Fig. 4. XRD analysis of the ABS substrate and the ABS-coated samples AcP and AsP.

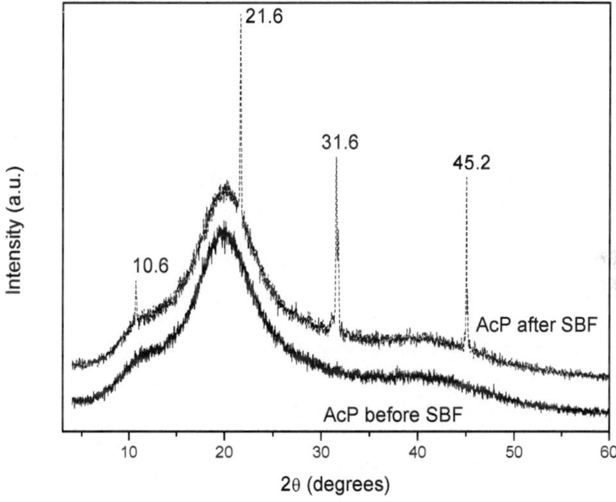

Fig. 5. XRD of the sample AcP before and after contact with SBF.

analysis and evidencing formation of the calcium phosphate silicate. The FTIR-ATR spectrum recorded for the sample AsP after contact with SBF displayed peaks characteristic of crystalline phosphate at 600 and 560 cm^{-1}, and carbonate hydroxyapatite, at 1451, 1408, and 874 cm^{-1} [72]. This suggests that these materials can be used for bioapplications.

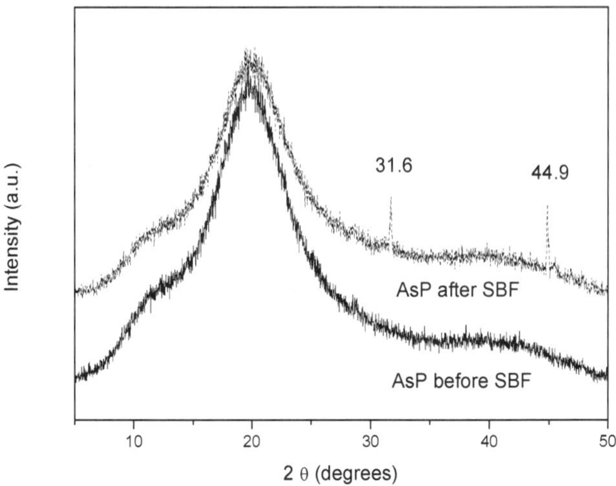

Fig. 6. XRD of the sample AsP before and after contact with SBF.

Changes in the properties of macroporous (pore size = 500μm) samples of the polymers polyamide 12 (nylon) and ABS, obtained by RP, were investigated herein. Sols containing silicon and calcium alkoxide, with or without phosphate anions, were deposited onto the polymers by the dip-coating technique [73]. The goal of this work was to coat the organic polymer materials with macroporous compounds and verify whether the resulting materials can be used as biomaterials. If the homogeneous composition of a coating can transform an organic polymer into a biocompatible material, then the latter can be applied in bone implant. RP promotes the building of pieces with different and complex forms. Figure 7 is a representation of the substrates based on the organic polymers polyamide 12 (nylon) and ABS prepared by RP, with a pore size of 500μm.

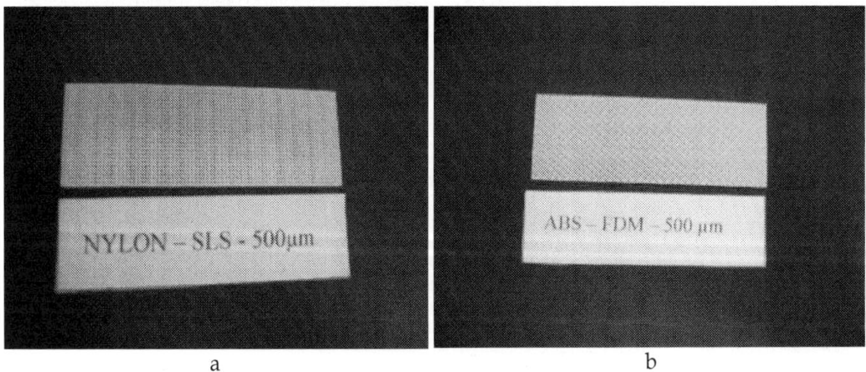

Fig. 7. Porous materials prepared by RP: (a) polyamide 12 and (b) ABS.

The SEM micrographs show that the polymers exhibit different surfaces. Polyamide 12 is rough, whereas ABS is smooth. Surface features can affect adherence of the coating to the polymer. Figure 8a, b, c, and d depict the SEM micrographs of polyamide 12 and ABS after coating with Si-Ca-P and Si-Ca, respectively.

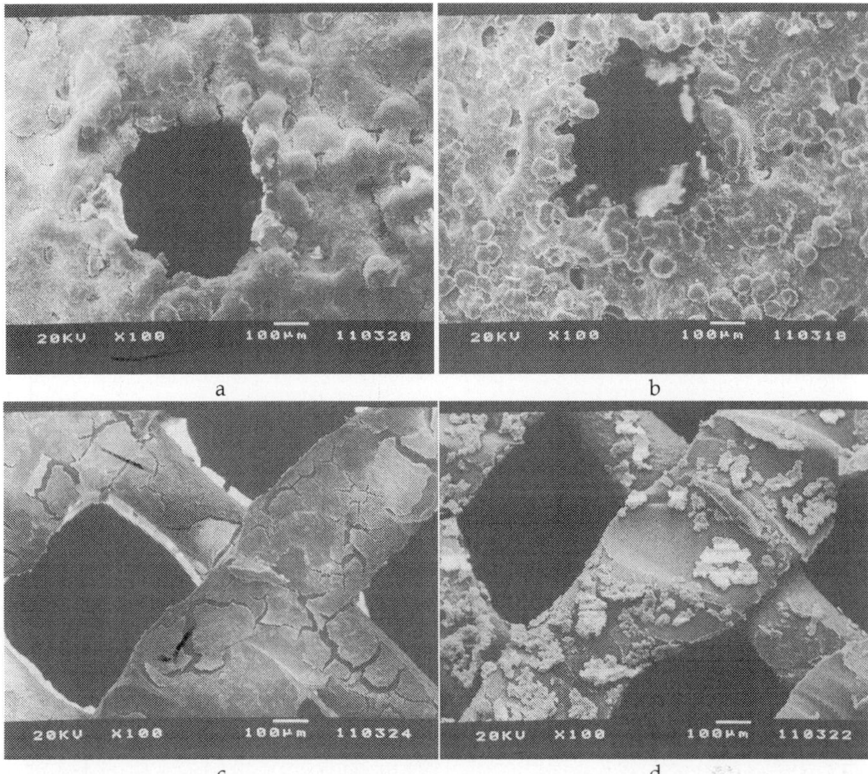

Fig. 8. SEM micrographs of the coated polymers: (a) polyamide 12 + Si-Ca-P, (b) polyamide 12 + Si-Ca, (c) ABS + Si-Ca-P, and (d) ABS + Si-Ca.

SEM furnishes information about the homogeneity, shape, size, and adherence of materials, which aids explanation about the change in the properties of the coated polyamide 12 and ABS. The polyamide 12 polymer has an initial pore size of 500 μm, which is reduced to 300 μm after coating and RP. This leads to the conclusion that the coating has a thickness of 200 μm. The coating prepared by combination of the sol-gel methodology with the dip-coating technique produces films with sizes in the nanometer range [74, 75]. In the present case, the compositions of both the sol and the polymer promote an increase in thickness. The thickness of the coating in polyamide 12 and ABS polymers influences the decomposition temperature, and thicker coatings lead to higher decomposition temperatures. This observation is corroborated by the SEM technique. The same coating was prepared on a 3D piece, represented in Figure 9.

Fig. 9. 3D piece of the polyamide structured by RP.

Bioactivity tests in SBF and fibroblast cell cultures were performed. There was growth of differentiated cells in the fibroblast culture for both functionalized polyamide and ABS. Figures 10 a, b, c, and d correspond to the results obtained with polyamide 12 without and with coating and ABS without and with coating after 4 days, respectively. It was possible to observe that the cells were not affected by the coating, showing that the materials present potential future applications for use in biomaterials.

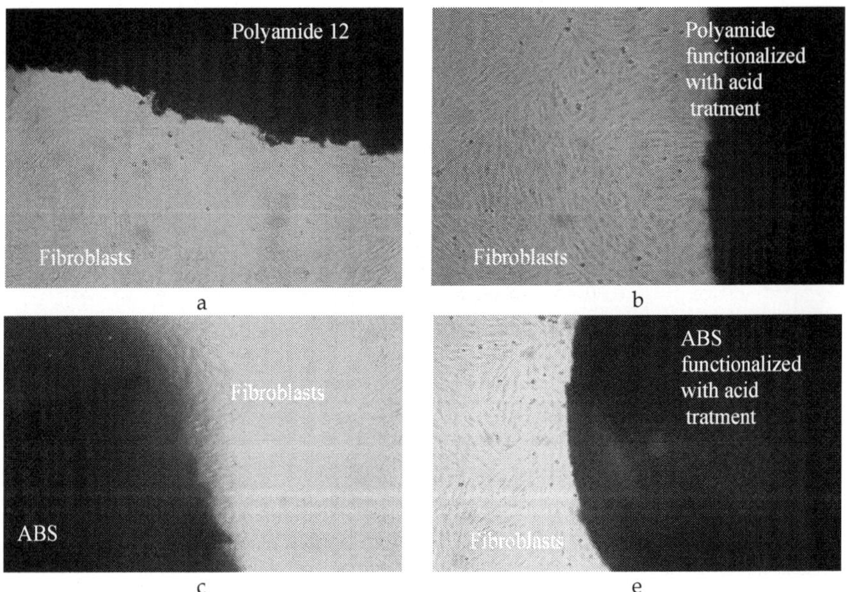

Fig. 10. Polyamide and ABS without and with coating after immersion into fibroblast cell culture for four days.

Formation of calcium phosphate on the polyamide substrate obtained by RP can be achieved by means of the sol-gel process. So a sol containing tetraethylorthosilicate, calcium alkoxide, phosphoric acid, and alginate was prepared. Figures 11 a and b depict the micrograph of the resulting substrate.

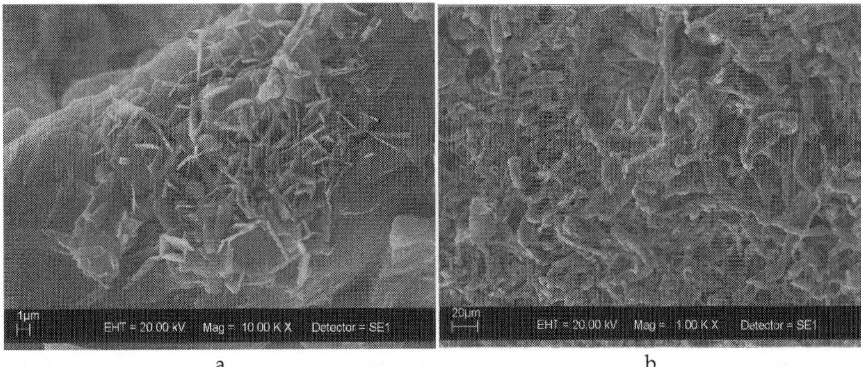

a   b

Fig. 11. Polyamide substrate functionalized with Si-Ca-P-alginate.

From Figure 11a it can be seen that calcium phosphate was formed on the alginate fibers. In Figure 11b, the alginate fibers and the red square corresponding to the calcium phosphate crystals can be observed. The EDS spectrum confirms the presence of the Si-Ca-P elements in the crystals. This material is a candidate for use in bone implant.

Silica nanospheres were prepared in order to test their use as DDS. In addition, a factorial design was developed, so that the influence of hydrolytic base-catalyzed sol-gel parameters and surfactant (Tween 80 or Pluronic F68) concentration over the nanoparticles morphology could be analyzed.

SEM characterization showed that the two surfactants employed herein provided very

a   b

Fig. 12. SEM images of the particles prepared by the hydrolytic sol-gel using (a) Pluronic F68 and (b) Tween 80 as surfactants.

different results. In general, the particles prepared using Tween 80 were more spherically shaped compared to those prepared with Pluronic F68. Figure 12 displays two examples of particles obtained from the factorial design, prepared under the same conditions, but with different surfactants.

The spherical and nanometric particles presented in Figure 12b meet the requirements of the morphology desired for a DDS. This morphology allows for different administration routes, including the intravenous route.

2.2 Preparation of biomaterials by the non-hydrolytic sol-gel process

The non-hydrolytic sol-gel process is another route for the production of materials for bioapplications such as dental and osseous substitutes. Here the preparation and characterization of the glass ionomer by this methodology is described. Calcium fluoroaluminosilicate glass containing phosphorus and sodium (Ca-FAlSi) consists of an inorganic polymeric network (mixed oxide) embedded in an aluminum and silicon matrix, comprising an amorphous structure. This material is currently employed in dentistry as restorative designated glass ionomer cement [76, 77]. Firstly, the calcium-fluoroaluminosilicate glass was prepared in oven-dried glassware and $AlCl_3$, $SiCl_4$, CaF_2, AlF_3, NaF, $AlPO_4$, and ethanol were reacted in reflux under argon atmosphere [53]. The ^{27}Al NMR results revealed the coordination of aluminum. Figure 13 shows the NMR spectrum of the calcium-fluoroaluminosilicate glass dried at 50°C and submitted to heat treatment at 1000°C.

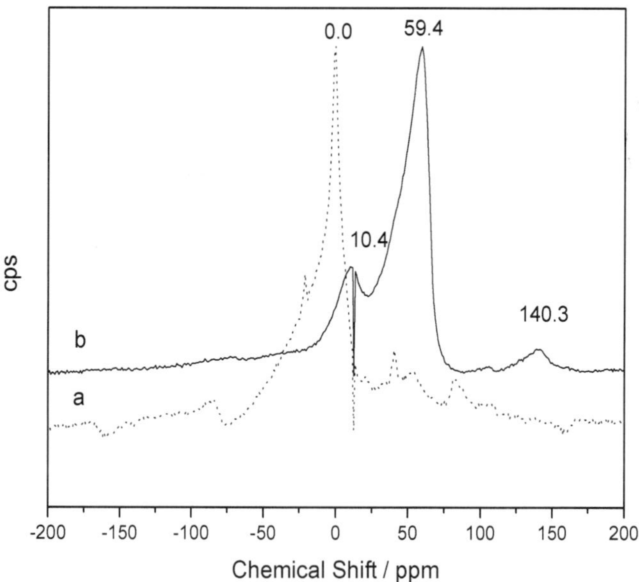

Fig. 13. ^{27}Al NMR spectrum of the sample dried at 50°C and treated at 1000°C.

The central transition (CT) frequency of the spectrum of a quadrupolar nucleus of half intereger spin, such as ^{27}Al (I = 5/2), depends on the orientation of each crystallite in the static magnetic field to the second order in the perturbation theory. The quadrupolar interaction between the nuclear electric quadrupole moment (eQ) and the electric field gradient of the nucleus (eq), arising from any lack of symmetry in the local electron distribution, is described by the quadrupolar coupling constant Cq (e^{2}qQ/h) and the symmetry parameter η. It should be noted that disordered materials such as glasses have a wide range of interatomic distances and, consequently, CT line broadening occurs due to the distribution of δ_{iso} and quadrupolar interactions [78]. After the material was heat-treated at 1000°C, a single peak corresponding to Al$^{(VI)}$ predominated at 0.0 ppm, indicating the structural change in the coordination state of aluminum. When Al atoms are in tetrahedral coordination Al$^{(IV)}$, their chemical shifts vary from 55 to 80 ppm. Chemical shifts in the range of -10 to 10 ppm correspond to coordinated octahedral Al$^{(VI)}$ [79, 80, 47]. The spectra of the two samples prepared here presented three peaks at 10.4, 59.4, and 140.1 ppm, which are characteristic of Al$^{(VI)}$, Al$^{(IV)}$, and spinning side bands [81], respectively. Although some authors have reported the presence of Al$^{(V)}$ atoms with chemical shifts at 20 ppm [82], this peak was not detected. The dominant species in the sample heat-treated at 50 °C corresponded to Al$^{(IV)}$. The chemical shifts between 50 and 60 ppm corresponding to Al$^{(IV)}$ depend on the Al/P molar ratio. Al$^{(IV)}$ has been found at 60 ppm in model glasses based on SiO$_2$Al$_2$O$_3$CaOCaF$_2$, and at about 50 ppm in glasses containing phosphate where the molar Al:P ratio was 2:1 [81]. In this study an Al/P molar ratio of approximately 10 was achieved, which is higher than that present in commercial glasses. In our case, this chemical shift was very difficult to observe because of the lower incidence of Al-O-P bonds. The ^{29}Si NMR results allow for analysis of the chemical environment around silicon atoms in silicates, where Si is bound to four oxygen atoms. The structure around Si can be represented by a tetrahedron whose corners link to other tetrahedra. The Q^n notation serves to describe the substitution pattern around a specific silicon atom, with Q representing a silicon atom surrounded by four oxygen atoms and n indicating the connectivity [83]. Figure 14 presents the NMR spectrum of the sample dried at 50°C.

The material displayed a peak at -100 ppm and a shoulder at -110 ppm, and the values in this range were attributed to Si atoms Q^4 and Q^4 or Q^3, respectively. Figure 15 illustrates the Q^3 and Q^4 structure.

As mentioned above, the chemical shift indicates the environment around the Si atoms in the glass. The commercial calcium-fluoroaluminosilicate glass presents a broad peak between -90 and -99 ppm [82]. On the basis of the results obtained here, our material exhibits a vitreous lattice. The number of nearest neighbor aluminum atoms is given in parentheses. Q^4 (3/4 Al) and Q^4(1/2) [78] are the structures represented in Figure 16.

The chemical shift ranges overlap, so the resonances in Fuji II cement (commercial glass) at -87, -92, -99, and -109 ppm may be due to Q^4 (3/4 Al), Q^4 (3 Al), Q^4 (1/2 Al), and Q^4 (0 Al), respectively [78]. In our case, the chemical shifts at -110 and 100 ppm may be due to the Si atoms Q^4 (1/2 Al) and Q^4 (0 Al), because of the molar ratio Al/Si < 1. Figure 17 illustrates the ^{29}Si NMR of the sample heat-treated at 1000°C for 4 hours.

In the present case, only one peak at -88 ppm was detected, which can be attributed to the presence of a Q^4 (3/4 Al) site in Si atoms due to the structural rearrangement of the

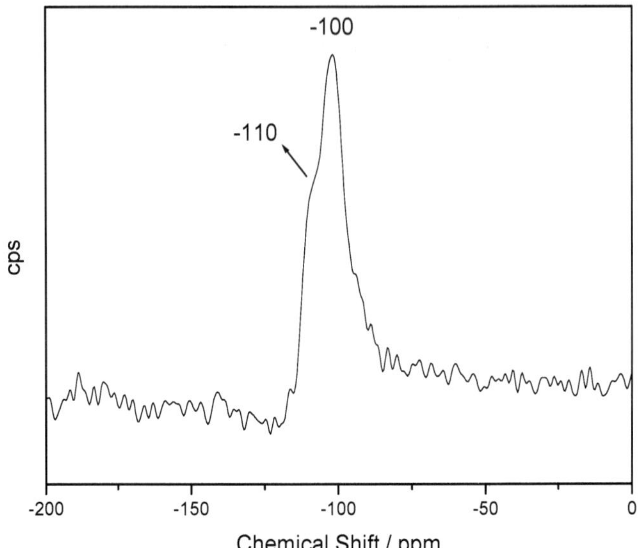

Fig. 14. ^{29}Si NMR spectrum of the sample dried at 50°C.

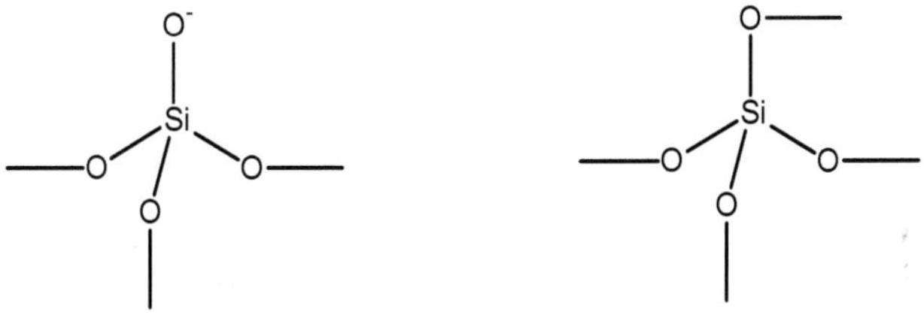

Q^3

Q^4

Fig. 15. Schematic representation of the Q^3 and Q^4 structures.

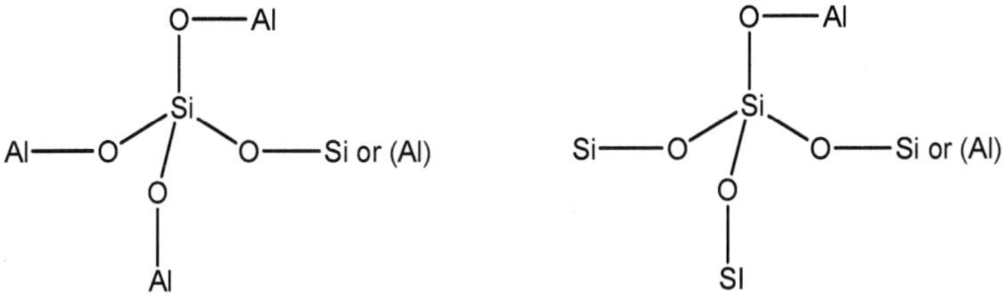

Q^4 (3/4 Al)

Q^4 (1/2 Al)

Fig. 16. Schematic representation of the Q^4 (3/4 Al) and Q^4 (1/2 Al) structure.

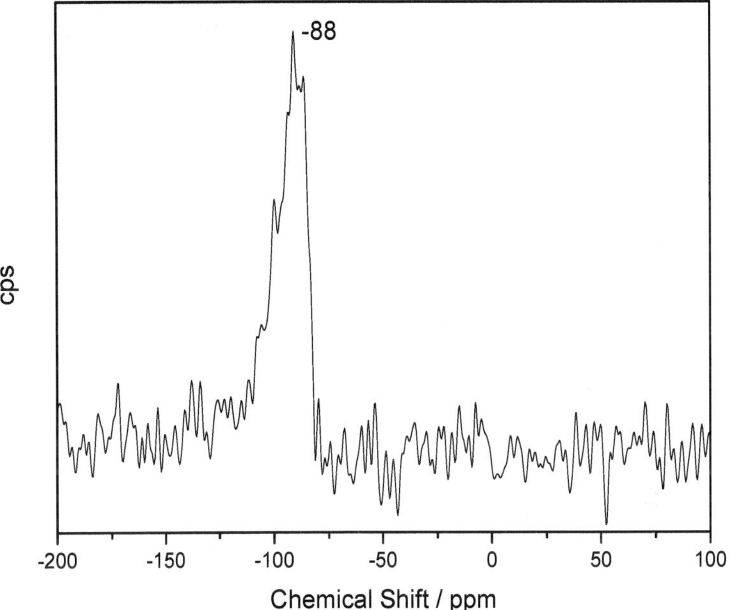

Fig. 17. ^{29}Si NMR spectrum of the sample treated at 1000°C.

aluminosilicate crystalline structures. This is consistent with our X-ray data. The nonhydrolytic sol-gel method has proved efficient for the preparation of materials with glass properties, as shown in this work. This process enables reaction control and the use of stoichiometric amounts of Al and Si reagents at low temperatures, near 110°C, thereby reducing production costs.

The preparation of aluminosilicate-based matrices by the nonhydrolytic sol-gel method, using varying concentrations of the glass components, especially the element phosphorus, has been accomplished by our group [84]. Figure 18 depicts the X-ray diffractograms for the samples A2, A3.3, and A4, all dried at 50°C.

An amorphous structure predominates in sample A2. The A3.3 material displays an amorphous structure with crystalline phases attributed to fluorapatite ($Ca_5(PO_4)_3OH$) and mullite ($3Al_2O_32SiO_2$), according to Gorman et al. Sample A4 also presents an amorphous phase and a crystalline phase, which is ascribed to mullite [85]. The X-ray diffraction analysis revealed the influence of the phosphorus concentration on the structural formation of the materials. The material prepared with the Wilson formulation has an amorphous structure, while the one prepared according to Hill displays an amorphous structure with a crystalline phase attributed to mullite. An increase in phosphorus in the Hill formulation allows the formation of an amorphous material with crystalline phases attributed to mullite and fluorapatite. The increase in phosphorus concentration affects the formation of the

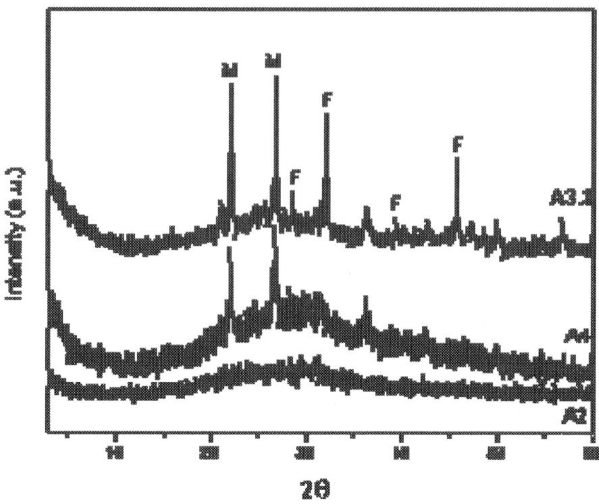

Fig. 18. X-ray diffractograms of samples A2, A3.3, and A4.

material's structure, since this increase leads to the formation of fluorapatite. The difference between the Wilson and Hill formulations [86, 87] also account for the formation of different materials: sample A2 material presents a structure devoid of crystallinity, while the structure of glass A4 contains a crystalline phase attributed to mullite. These are attractive factors in dental restorations. Calcium fluoroaluminosilicate glasses containing sodium and phosphorus are materials that can be employed in dentistry, as components of glass ionomer cement, and in medicine, as replacements for bone implants.

In this work, the preparation and characterization of matrices based on aluminossilicates obtained by the non-hydrolytic and hydrolytic sol-gel routes were investigated. Three different routes, namely the non-hydrolytic sol-gel route and the hydrolytic sol-gel route using either basic or acid catalysis, were employed in the preparation of three materials, namely IC1, IC2, and IC3. The obtained materials were characterized by different physical methods and antimicrobial activity tests. For evaluation of the antimicrobial activity, the materials were examined against the microorganisms *E. faecalis, S. salivarius, S. sobrinus, S. sanguinis, S. mutans, S. mitis,* and *L. casie,* using the double layer diffusion technique.

Scanning electron microscopy analysis coupled with energy dispersive spectroscopy (MEV/EDS) of the material IC1 (Figure 19) mapped the elemental distribution and showed that the silicon atoms are located in the same regions as the aluminum and oxygen atoms distributed throughout the particle surface. This indicates the possible formation of aluminosilicate, confirming the IR data. The presence of chlorine is evidence that this material contains a large amount of residual groups from the reagent $AlCl_3$. The calcium atoms are distributed over the IC1 matrix, but they appear as some clusters that coincide with the fluoride clusters, thereby indicating the presence of CaF_2 crystals, as shown by XRD.

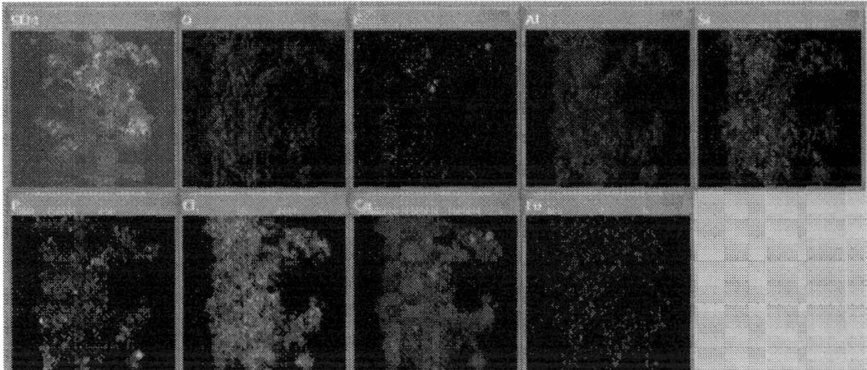

Fig. 19. Elemental distribution mapping of the material IC1 treated at 110 ºC.

For the material IC2, the MEV/EDS analysis (Figure 20) revealed that the silicon atoms are located in the same regions as the aluminum and oxygen atoms, pointing to the possible formation of aluminosilicate. The smaller amount of chlorine in relation to IC1 indicates that this material has fewer residues, probably originated from the catalyst (HCl). Also, it is important to bear in mind that in this case the precursor $AlCl_3$ was replaced with aluminum isopropoxide, which may be contributing to the lower amount of residual Cl groups. The phosphorus and calcium atoms are distributed over the matrix. This material contained no calcium and fluoride clusters, corroborating the XRD findings, which had not indicated the presence of CaF_2.

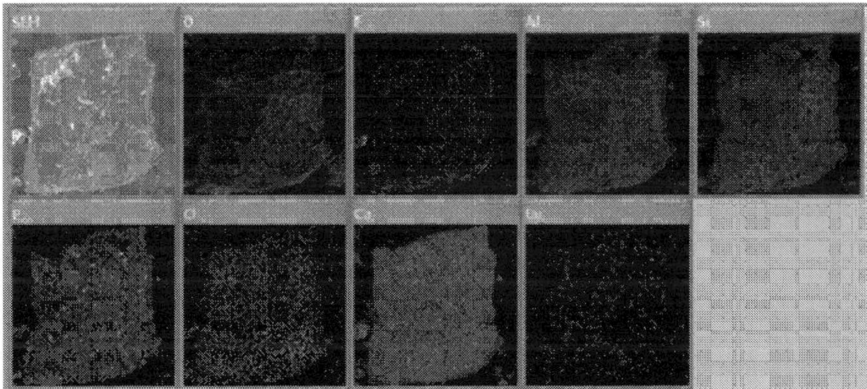

Fig. 20. Elemental distribution mapping of the material IC2 treated at 110 ºC.

In the case of the material IC3, the MEV/EDS (Figure 21) demonstrated that the silicon atoms are present in the same regions as the aluminum and oxygen atoms. The smaller amount of chlorine in relation to IC2 indicates that this material has few residual Cl. The calcium atoms are distributed in the matrix and appear as clusters in the same regions as the fluoride atoms, indicating the presence of CaF_2, as previously detected by XRD.

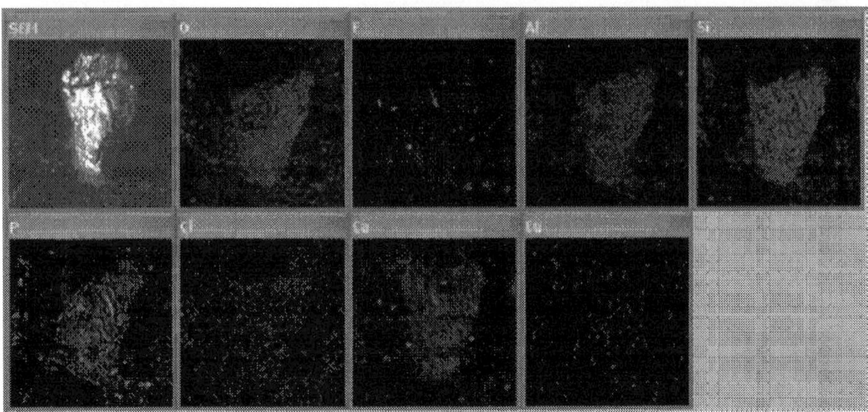

Fig. 21. Elemental distribution mapping of the material IC3.

The Ca-FAlSi used as a basis for glass ionomer almost always presents amorphous phase separation. For instance, materials obtained by the usual methods are composed by distinct regions with high Al and Si content as well as other regions containing Ca, P, and F. This fact undermines the homogeneity of the Ca-FAlSi material and impairs the setting reaction of the cement, because regions rich in Ca, P, and F are less reactive, which in turn promotes formation of inhomogeneous cement. This ruins the integrity of the restorative material and affects its physical and chemical resistance. In our case, there was no clear evidence of phase separation for the materials synthesized by the sol-gel methods, and the elements were well distributed on the surface of materials. The material IC2 presented high homogeneity compared to the other materials, since the elements were well distributed across the surface of the particles.

For evaluation of the antimicrobial activity of the materials prepared here, the dual layer diffusion technique was employed, using the microorganisms listed in Table 1.

The materials IC1 and IC2 presented antimicrobial activity similar to that of the commercial glass ionomer Ca-FAlSi, so they can be applied in the oral environment and avoid damage by microorganisms. In conclusion, the material IC2 can be used as a basis for glass ionomer cement and the fact that it is synthesized at lower temperature and leads to greater homogeneity compared with the Ca-FAlSi produced by industrial methods makes it advantageous over the commercially available materials.

The glass ionomer prepared in reference [84] was tested with respect to its biocompatible properties and compared to the materials obtained by the industrial methodology. Experimental and conventional GIC were analyzed in terms of morphology and of the morphometric reaction induced by the cement in the subcutaneous tissue of rats.

The methodology described in [88] was used for the biocompatible test. Table 2 lists the tested materials.

The experimental and conventional GIC powders were used to prepare the glass ionomer cement. The surface of the obtained materials was examined by scanning electron microscopy (Figures 22 a and b, respectively).

microorganism/ ATCC	Substance ( ) Extract ( ) Product (x)	Mean (mm) ± SD
E. faecalis ATCC 4082	IC 1	20.0
	IC 22x d	15.5 ± 2.1
	IC 3	0.0
	Gass ionomer (S.S. White Artigos Dentários LTDA)	22.0 ± 0.0
	Periorgard mouthwash (positive control)	13.0 ± 0.0
S. salivarius ATCC 25975	IC 1	18.0
	IC 22x d	12.0 ± 1.4
	IC 3	0.0
	Gass ionomer (S.S. White Artigos Dentários LTDA)	25.5 ± 0.7
	Periorgard mouthwash (positive control)	19.0 ± 0.0
S. sobrinuns ATCC 33478	IC 1	19.0
	IC 22x d	14.0 ± 1.4
	IC 3	0.0
	Gass ionomer (S.S. White Artigos Dentários LTDA)	9.5 ± 0.7
	Periorgard mouthwash (positive control)	21.0 ± 0.0
S. sanguinis ATCC 10556	IC 1	14.0
	IC 22x d	12.0 ± 2.8
	IC 3	0.0
	Gass ionomer (S.S. White Artigos Dentários LTDA)	10.5 ± 0.7
	Periorgard mouthwash (positive control)	18.0 ± 0.0
S. mutans ATCC 25175	IC 1	20.0
	IC 22x d	10.0 ± 2.8
	IC 3	0.0
	Gass ionomer (S.S. White Artigos Dentários LTDA)	10.0 ± 0.0
	Periorgard mouthwash (positive control)	21.5 ± 0.7
S. mitis ATCC 49456	IC 1	20.0
	IC 22x d	13.0 ± 2.8
	IC 3	0.0
	Gass ionomer (S.S. White Artigos Dentários LTDA)	14.0 ± 0.0
	Periorgard mouthwash (positive control)	18.5 ± 0.0
L. casei ATCC 11578	IC 1	19.0
	IC 22x d	16.0 ± 1.4
	IC 3	0.0
	Gass ionomer (S.S. White Artigos Dentários LTDA)	20.5 ± 0.7
	Periorgard mouthwash (positive control)	23.5 ± 0.7

Table 1. Antimicrobial activity tests for the materials prepared in this work.

GIC	Composition	Source
Experimental	Powder: Calcium fluoraluminiosilicates containing phosphorus and sodium Liquid: Tartaric acid, polyacrilic acid, distilled water	Sol-Gel Methodology
Conventional (Vidrion R)	Powder: Sodium fluorosilicates, calcium aluminium, barium sulphate, pigments. Liquid: Tartaric acid, polyacrilic acid, destilled water	SS White - Prima Dental Group, Gloucester, UK

Table 2. Tested materials.

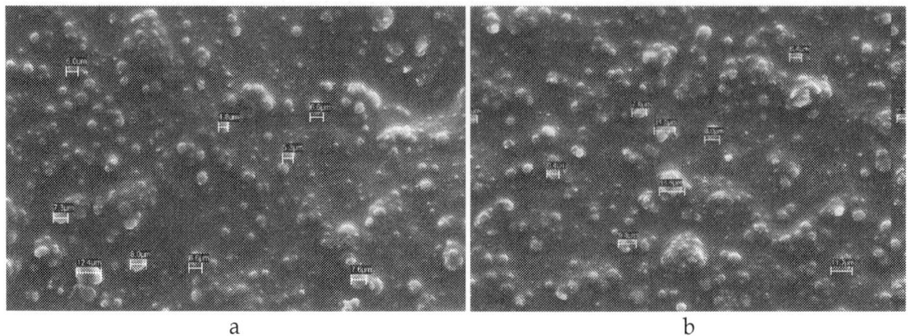

a   b

Fig. 22. SEM micrographs recorded for the glass ionomer cements. (a) experimental and (b) conventional

The micrographs revealed that the surface the conventional cement presents lower homogeneity, compared to the experimental cement. The better homogeneity of the experimental cement is due to the homogeneity of particle size, which promotes the same physical-chemistry properties all over the cement surface.

The biocompatible test was carried out on the basis of the response obtained with the tissue stimulated by the experimental cement and compared to that achieved with the conventional cement. These responses were analyzed by means of morphological and morphometric analyses of the reaction caused by these cements in the subcutaneous tissue of rats. Polyethylene tubes were obtained according to the methodology used by Campos-Pinto et al. [89]. To this end, an urethral catheter with an internal diameter of 0.8 mm was sectioned sequentially at 10 mm intervals. After sectioning, one of the tube ends was sealed with cyanoacrylate ester gel (Super Bonder, Aachen, Germany), to avoid extravasation of the material to be tested. The obtained tubes were placed in a metal box and autoclaved at 120°C for 20 min. [90].

The data obtained for all the histopathological events assessed in each period of study are presented in Table 3.

Histopathological Events	7 days			21 days			42 days		
	GGI	GGII	CCG	GGI	GGII	CCG	GGI	GGII	CCG
Polymorphonuclear	+++	+++	++	++	++	++	--	--	--
Mononuclear	+++	++++	++	++	++	++	--	--	--
Fibroblasts	+++	+++	++	++	++	++	++	++	++
Blood vessels	+++	++++	+++	+++	++	++	++	++	--
Macrophage	++	+++	++	--	++	--	--	--	--
Giant inflammatory cells	++	++	++	--	--	--	--	--	--

Score: (-) absent; (+) slight; (++) moderate; (+++) intense.
CG – control group; GI – experimental cement; GII – conventional cement.

Table 3. Summary of the data obtained for the histopathological events observed in each group at the different periods of study.

Figures 23, 24, and 25 show the biocompatible tests after 7, 21, and 42 days, respectively.

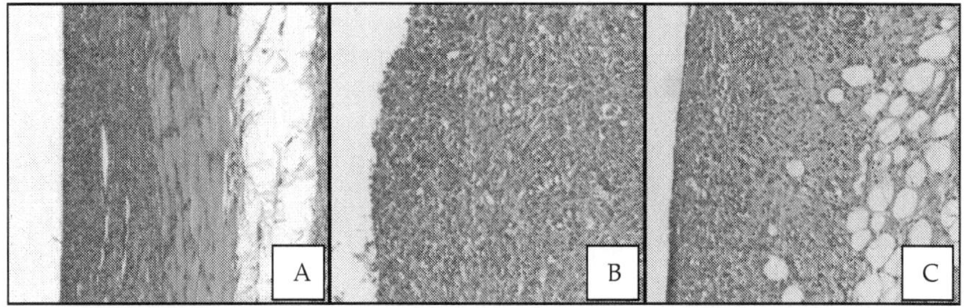

Fig. 23. Biocampatible tests after 7 days. Control group (A), conventional GIC (B), and experimental GIC (C) (100X. H.E.).

In the case of the control group (CG), connective tissue with delicate fibers, highly cellularized with fibroblasts and several blood vessels adjacent to all the analyzed faces was observed after 7 days (Figure 23A). A mild chronic inflammatory reaction was also detected. As for the experimental cement group (GI) (Experimental GIC), a layer of cellularized connective tissue with moderate chronic inflammatory reaction, predominantly formed by lymphocytes and blood vessels, was observed after 7 days. Small areas of necrosis were also noted (Fig. 23B). Few macrophages or foreign body multinucleated giant cells were seen. Concerning the conventional cement (GII) (Conventional GIC), an intense chronic inflammatory reaction, associated with hyperemic blood vessels and macrophages (Figure 23C) was detected after 7 days. Necrosis area was observed close to the dispersed material.

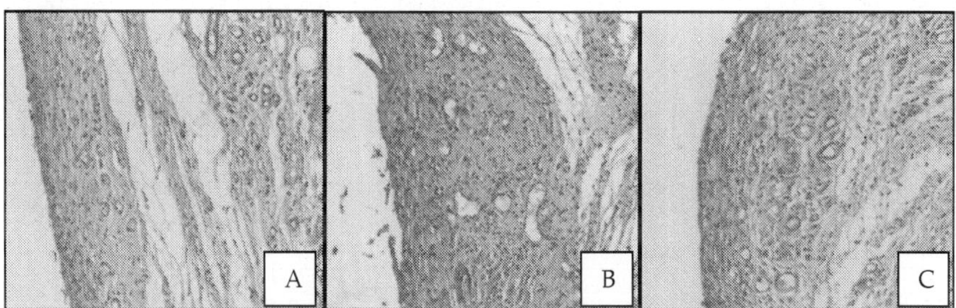

Fig. 24. Biocampatible tests after 21 days. Control group (A), conventional GIC (B), and experimental GIC (C) (100X. H.E.).

After 21 days, the CG group presented a mild chronic inflammatory reaction in this period (Figure 24A). As for GI (Experimental GIC), there was a mild chronic inflammatory reaction, with few lymphocytes and several fibroblasts, in the connective tissue adjacent to the open end of the tube. Foreign body multinucleated giant cells and macrophages were not observed (Figure 24B). In the case of GII (Conventional GIC), the connective tissue exhibited a moderate chronic inflammatory reaction. Phagocytic activity and rare necrosis areas were also observed (Figure 24C).

After 42 days, the CG presented no chronic inflammatory reaction (Figure 25A). Concerning the GI group (Experimental GIC), connective tissue with mild to absent chronic

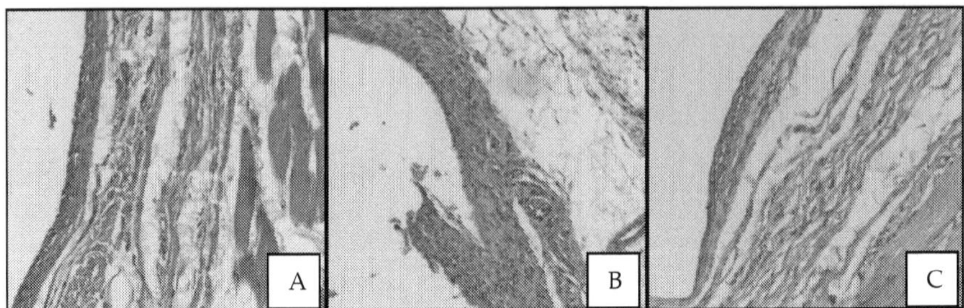

Fig. 25. Biocampatible tests after 42 days. Control group (A), conventional GIC (B), and experimental GIC (C) (100X. H.E.).

inflammatory reaction and residual dispersed cement was detected (Figure 25B), while necrosis areas and other changes were absent. As for GII (Conventional GIC), there was a mild to absent chronic inflammatory reaction, without foreign body giant cells or macrophages. Some dispersed material was observed, but no necrosis areas or degenerative changes were seen (Figure 25C).

In summary, both tested cements caused a mild to absent inflammatory response after 42 days, which is also acceptable from a biological standpoint.

3. Conclusion

Bioapplications that could provide better quality of life for humanity are desirable. It has been demonstrated here that a series of materials prepared by sol-gel methodology (hydrolytic and non-hydrolytic routes) present useful properties for application in biological medium. Indeed, it was possible to obtain multifunctional materials by this methodology. The combination of the different elements at molecular and atomic level affords potential candidates for a variety of applications. The very interesting results obtained by us in the present work indicate that this methodology can be applied for production of biomaterials with potential application in several fields such as medicine, dentistry, veterinary, engineering, chemistry, physics, and biology. Materials for use in the areas of bone implant, restorative tooth coating, diagnosis, membrane permeability, biosensor, scaffolding, and drug delivery can be prepared and transformed into biocompatible, bioactive, bioinducer materials with different properties and composition, since this methodology allows for production of materials with controlled stoichiometry and particle size.

4. Acknowledgment

The authors gratefully acknowledge the financial support of the Brazilian research funding agencies FAPESP, CNPq, and CAPES. Profa. Dra. Isabel Salvado from Departamento de Engenharia Cerâmica e do Vidro - CICECO - Universidade de Aveiro/Pt, Dr. Jorge V. L. Silva do Centro da Tecnologia da Informação Renato Archer (CTI), - Campinas - SP - Brazil, Profa. Dra. Shirley Nakagaki, Departamento de Química da Universidade Federal do Paraná – PR – Brazil, Profa. Dra. Fernanda de C. P. Pires de Souza, Departamento de Materiais Dentários da Universidade de São Paulo - Ribeirão Preto - SP - Brazil, and Prof. Dr. Carlos Henrique Martins da Universidade de Franca - Franca - SP - Brazil are also acknowledged.

5. References

[1] Nair, L. S., Laurencin, C. T. (2006). Polymers as Biomaterials for Tissue Engineering and Controlled Drug Delivery, *Adv Biochem Engin/Biotechnol*, Vol. 102, pp. 47–90.

[2] Hench, L. L. (1998), .Biomaterials: a forecast for the future.. *Biomaterials*, Vol. 19, No 16, pp. 1419-1423.

[3] Leeuwenburgh, S. C. G., Malda, J., Rouwkema, J., Kirkpatrick, C. J. (2008). Trends in biomaterials research: An analysis of the scientific programme of the World Biomaterials Congress 2008. *Biomaterials*, Vol. 29,pp. 3047-3052.

[4] Bonzani, I. C. , Adhikari, R., Houshyar, S., Mayadunne, R., Gunatillake, P., Stevens, M. M. (2007). Synthesis of two-component injectable polyurethanes for bone tissue engineering. *Biomaterials*, Vol. 28, No 3, pp. 423-433.

[5] Balani, K., Anderson, R., Laha, T., Andara, M., Tercero, J., Crumpler, E., Agarwal, A. (2007). Plasma-sprayed carbon nanotube reinforced hydroxyapatite coatings and their interaction with human osteoblasts in vitro. *Biomaterials*, Vol. 28, No 3, pp. 618-624.

[6] Chu, T. M. G., Warden, S. J., Turner, C. H., Stewart, R. L. (2007). Segmental bone regeneration using a load-bearing biodegradable carrier of bone morphogenetic protein-2. *Biomaterials*, Vol. 28, No. 3, pp. 459-467.

[7] Eglin, D., Maalheem, S., Livage, J., Coradin, T. (2006). In vitro apatite forming ability of type I collagen hydrogels containinrg bioactive glass and silica sol-gel particles. *J. Mat Scie. Mat In Medicine*, Vol. 17, No. 2, pp. 161-167.

[8] Skelton, K. L., Glenn, J. V., Clarke, S. A., Georgiou, G., Valappil, S. P., Knowles, J. C., Nazhat, S. N., Jordan, G. R. (2007). Effect of ternary phosphate-based glass compositions on osteoblast and osteoblast-like proliferation, differentiation and death in vitro. *Acta Biomaterialia*, Vol. 3, No. 4, pp. 563-572.

[9] Misra, S. K., Mohn, D., Brunner, T. J., Stark, W. J., Philip, S. E., Roy, I., Salih, V., Knowles, J. C., Boccaccini, A. R. (2008). Comparison of nanoscale and microscale bioactive glass on the properties of P(3HB)/Bioglass® composites. *Biomaterials*, Vol. 29, No. 12, pp. 1750-1761.

[10] Lee, H. J., Choi, H. W., Kim, K. J., Lee, S. C. (2006). Modification of Hydroxyapatite Nanosurfaces for Enhanced Colloidal Stability and Improved Interfacial Adhesion in Nanocomposites, *Chem. Mater.*, Vol. 18, pp. 5111-5118.

[11] Rámila, A., Padilla, S., Muñoz, B., Regí, M. V. (2002). A New Hydroxyapatite/Glass Biphasic Material: In Vitro Bioactivity. *Chem. Mater.*, Vol. 14, pp. 2439-2443.

[12] Doğan, O., Öner, M. (2006). Biomimetic Mineralization of Hydroxyapatite Crystals on the Copolymers of Vinylphosphonic Acid and 4-Vinilyimidazole. *Langmuir*, Vol. 22, pp. 9671-9675.

[13] Iwatsubo, T., Sumaru, K., Kanamori, T., Shinbo, T., Yamaguchi, T. (2006). Construction of a New Artificial Biomineralization System. *Biomacromolecules*, Vol. 7, pp. 95-100.

[14] Clupper, D. C., Hench, L. L. (2003). Crystallization kinetics of tape cast bioactive glass 45S5. *J. Non-Cryst. Solids*, Vol. 318,No. 1-2, pp. 43-48.

[15] Hench, L. L. (1997). Sol-gel materials for bioceramic applications. *Current Opinion in Solid State & Material Science*, Vol. 2, No 5, pp. 604-610.

[16] Pickup, D. M., Speight, R. J., Knowles, J. C., Smith, M. E., Newport, R. J. (2008). Sol–gel synthesis and structural characterisation of binary TiO_2-P_2O_5 glasses. *Mater. Research Bulletin*, Vol. 43, No. 2, pp. 333-342.

[17] Carla, D., Pickup, D. M., Knowles, J. C., Ahmed, I., Smith, M. E., Newport, R. J. (2007). A structural study of sol–gel and melt-quenched phosphate-based glasses. *J. Non-Cryst. Solids*, Vol. 353, No. 18-21, pp. 1759-1765.

[18] Saravanakumar, B., Rajkumar, M., Rajendran, V. (2011). Synthesis and characterisation of nanobioactive glass for biomedical applications, *Materials Letters*, Vol. 65, pp. 31–34.

[19] Lei, B., Chen, X., Wang, Y., Zhao, N., Miao, G., Li, Z., Lin, C. (2010). Fabrication of porous bioactive glass particles by one step sintering, *Materials Letters*, Vol. 64, pp. 2293–2295.

[20] Chen, Q.-Z., Li, Y., Jin, L.-Y., Quinn, J. M. W., Komesaroff, P. A. (2010). A new sol–gel process for producing Na_2O-containing bioactive glass ceramics, *Acta Biomaterialia*, Vol. 6, pp. 4143–4153.

[21] Kokubo, T., Kim, H.-M., Kawashita, M. (2003). Novel bioactive materials with different mechanical properties, *Biomaterials*, Vol. 24, pp. 2161–2175.

[22] Olmo, N., Martín, A. I., Salinas, A. J., Turnay, J., Vallet-Regí, M., Lizarbe, M. A. (2003). Bioactive sol–gel glasses with and without a hydroxycarbonate apatite layer as substrates for osteoblast cell adhesion and proliferation, *Biomaterials*, Vol. 24, pp. 3383–3393.

[23] Vrancken, K. C., Possemiers, K., Voort, P. V. D., Vansant, E. F. (1995). Surface modification of silica gels with aminoorganosilanes *Colloids Surf. A: Physicochem. Eng. Aspects*, Vol. 98, No. 3, pp. 235-241.

[24] Mark, J. E., Lee, C. Y. C., Bianconi, P. A., Hybrid Organic-Inorganic Composites. ISBN 9780841231481 (ACS Symp. Ser. 586, American Chemical Society Washington, DC, 1995).

[25] Cerveau, G., Corriu, R. J. P., Lepeytre, C., Mutin, P. H. (1998). Influence of the nature of the organic precursor on the textural and chemical properties of silsesquioxane materials *J. Mater. Chem.*, Vol. 12, No. 8, pp. 2707-2714.

[26] Corriu, R. (1998). A new trend in metal-alkoxide chemistry: the elaboration of monophasic organic-inorganic hybrid materials. *Polyhedron*, Vol. 17, No. 5-6, pp. 925-934 .

[27] Corriu, R. J. P., Leclerq, D. (1996) Recent Developments of Molecular Chemistry for sol-gel Processes. *Angew Chem Int Engl.*, Vol. 35 No. 13-14, pp. 1420-1436.

[28] Shea, K. J., Loy, D. A., Webster, O. (1992). Arylsilsesquioxane gels and related materials. New hybrids of organic and inorganic networks. *J Am Chem Soc.*, Vol. 114, No. 17, pp. 6700-6710.

[29] Jackson, C. L., Bauer, B. J., Nakatami, A. I., Barnes, J. (1996). Synthesis of Hybrid Organic–Inorganic Materials from Interpenetrating Polymer Network Chemistry. *Chem. Mater.*, Vol. 8, No. 3, pp. 727-733.

[30] Brinker, C. J.; Scherer, G. W. Sol-Gel Science, The Phys Chem. Sol-Gel Processing, ISBN 0121349705,Academic Press, San Diego, 1990.

[31] Nassar, E. J., Ciuffi, K. J., Ribeiro, S. J. L., Messaddeq, Y. (2003). Europium incorporated in the silica matrix obtained by sol-gel methodoly: Luminescent materials *Mater Research*, Vol. 6, No 4, pp. 557-562.

[32] Beari, F., Brand, M., Jenkner, P., Lehnert, R., Metternich, H. J., Monkiewicz, J., Siesler, H. W. (2001), Organofunctional alkoxysilanes in dilute aqueous solution: new accounts on the dynamic structural mutability. *J. Organo Chem.*, Vol. 625, No. 2, pp. 208-216.

[33] Nassar, E. J., Neri, C. R., Calefi, P. S., Serra, O. A. (1999). Functionalized silica synthesized by sol–gel process. *J Non-Cryst Solids*, Vol. 247, No 1-3, pp. 124-128.

[34] Stöber, W., Fink, A., Bohn, E. (1968). Controlled growth of monodisperse silica spheres in the micron size range. *J. Coll. Inter. Scie.*, Vol. 26, No. 1, pp. 62-69.

[35] Papacídero, A. T., Rocha, L. A., Caetano, B. L., Molina, E. F., Sacco, H. C., Nassar, E. J., Martinelli, Y., Mello, C., Nakagaki, S., Ciuffi, K. J. (2006). Preparation and characterization of spherical silica–porphyrin catalysts obtained by the sol–gel methodology. *Coll and Surfaces*, Vol. 275, No. 1-3, pp. 27-35.

[36] Nassar, E. J., Nassor, E. C. O., Ávila, L. R., Pereira, P. F. S., Cestari, A., Luz, L. M., Ciuffi, K. J., Calefi, P. S. (2007). Spherical hybrid silica particles modified by methacrylate groups. *J. Sol-Gel Scie. Techn.*, Vol. 43, No. 1, pp. 21-26.

[37] Ricci; G. P., Rocha; Z. N., Nakagaki; S., Castro; K. A. D. F., Crotti; A. E. M., Calefi; P. S., Nassar; E. J., Ciuffi, K. J. (2011). Iron-Alumina Materials Prepared by the Non-Hydrolytic Sol-Gel Route: Synthesis, Characterization and Application in Hydrocarbons Oxidation Using Hydrogen Peroxide as Oxidant, *Applied Catalysis A: General*, Vol. 389, No 1-2, pp. 147-154.

[38] Matos, M. G., Pereira, P. F. S., Calefi, P. S., Ciuffi, K. J., Nassar, E. J. (2009). Preparation of a GdCaAl$_3$O$_7$ Matrix by the non-hydrolytic sol-gel route. *Journal of Luminescence*, Vol. 129, pp. 1120-1124.

[39] Langer, R. (2000). Biomaterials in Drug Delivery and Tissue Engineering: One Laboratory's Experience. *Acc. Chem. Res.*, Vol. 33, No. 2, pp. 94-101

[40] Kumar, D. S., Banji, D., Madhavi, B., Bodanapu, V., Dondapati, S., Sri, A. P., (2009). Nanostructured porous silicon – a novel biomaterials for drug delivery. *International Journal of Pharmacy and Pharmaceutical Sciences*, Vol. 1, No 2, pp. 8-16

[41] Arruebo, M., Galán, M., Navascués, N., Téllez, C., Marquina, C., Ibarra, M. R., and Santamaria, J. (2006). Development of Magnetic Nanostructured Silica-Based Materials as Potential Vectors for Drug-Delivery Applications. *Chem. Mater.*, Vol. 18, No. 7, pp. 1911-1919

[42] Borak, B., Arkowski, J., Skrzypiec, M., Ziółkowski, P., Krajewska, B., Wawrzynska, M., Grotthus, B., Gliniak, H., Szelag, A., Mazurek, W., Biały, D., Maruszewsk, K., (2007). Behavior of silica particles introduced into an isolated rat heart as potential drug carriers. *Biomed. Mater.*, Vol. 2, No. 4, pp. 220-223.

[43] Vallet-Regí, M., Balas, F., Arcos, D. (2007). Mesoporous Materials for Drug Delivery *Angew. Chem. Int. Ed.*, Vol. 46, pp. 7548-7558.

[44] Yagüe,C., Moros, M., Grazú, V., Arruebo, M., Santamaría, J. (2008). Synthesis and stealthing study of bare and PEGylated silica micro- and nanoparticles as potential drug-delivery vectors., *Chemical Engineering Journal*, Vol. 137, No. 1, pp. 45-53.

[45] Yang, J., Lee, J., Kang, J., Lee, K., Suh, J-S., Yoon, H-G., Huh, Y-M., Haam, S. (2008). Hollow Silica Nanocontainers as Drug Delivery Vehicles. *Langmuir.*, Vol. 24, No. 7, pp. 3417-3421.

[46] Hench, L. L., West, J. K.(1990). The sol-gel process. *Chemical Reviews*, Vol. 90, No 1, pp. 33-72.

[47] Acosta, S., Corriu, R. J. P., Leclerq, D., Lefervre, P., Mutin, P. H., Vioux, A. (1994). Preparation of alumina gels by a non-hydrolytic sol-gel processing method *J. Non-Cryst. Solids*, Vol. 170, No.3 , 234-242.

[48] Wright, J. D., Sommerdijk, N. A. J. (2003). M. Sol-Gel Materials: Chemistry and Applications; ISBN 90-5699-326-7; Taylor & Francis: London, Vol. 4.

[49] Nassar, E. J., Ciuffi, K. J., Calefi, P. S. (2010). Europium III: different emission spectra in different matrices the same element, ISBN 978-1-61728-306-2. Chemistry Research and Applications, Editors: Harry K. Wright and Grace V. Edwards, Nova Science Publishers, Inc, New York, United State of America.

[50] Bandeira, L. C., Ciuffi, K. J., Calefi, P. S., Nassar, E. J. (2010). Silica matrix doped with calcium and phosphate by sol-gel. *Advances Bioscience and Biotechnology*, Vol. 1, No. 3, pp. 200-207.

[51] de Campos, B. M., Bandeira, L. C., Calefi, P. S., Ciuffi, K. J., Nassar, E. J., Silva, J. V. L., Oliveira, M., Maia, I. A. (2010). Protective coating materials on rapid prototyping by sol-gel, *Virtual and Physical Prototyping*, DOI: 10.1080/17452759.2010.491938. in press.

[52] Pereira, P. F. S., Matos, M. G., Ávila, L. R., Nassor, E. C. O., Cestari, A., Ciuffi, K. J., Calefi, P. S., Nassar, E. J. (2010). Red, Green and Blue (RGD) Emission doped $Y_3Al_5O_{12}$ (YAG) phosphors prepared by non-hydrolytic sol-gel route. *Journal of Luminescence*, Vol. 130, pp. 488-493.

[53] Cestari, A., Avila, L. R., Nassor, E. C. O., Pereira, P. F. S., Calefi, P. S., Ciuffi, K. J., Nakagaki, S., Gomes, A. C. P., Nassar, E. J. (2009). Characterization Of Glass Ionomer Dental Cements Prepared By A Non-Hydrolytic Sol-Gel Route. *Materials Research*, Vol 12, no 2, pp. 139-143.

[54] Nassar, E. J., Pereira, P. F. S., Ciuffi, K. J., Calefi, P. S. (2008). Photoluminescence Research Progress Chapter 10: Recent Development of Luminescent Materials Prepared by the Sol-Gel Process, 978-1-60456-538-6 Editors: Harry K. Wright and Grace V. Edwards, pp 265-285, Nova Science Publishers, Inc. New York, United State of America.

[55] Bandeira, L. C., Calefi, P. S., Ciuffi, K. J., Nassar, E. J., Salvado, I. M. M., Fernandes, M. H. F. V. (2011) Low Temperature Synthesis of Bioactive Materials, *Cerâmica*, in press.

[56] Kokubo, T. (1991). Bioactive glass ceramics: properties and applications, *Biomaterials*, Vol. 12, No 1, pp. 55-63.,

[57] Villacampa A.I., Ruiz J.M.G. (2000). Synthesis of a new hydroxyapatite-silica composite material. *Journal of Crystal Growth*, Vol. 11, pp. 111-115.

[58] Park, E., Condrate, R. A., Lee, D., Kociba, J., Gallagher, P. K. (2002). Characterization of hydroxyapatite: Before and after plasma spraying. *J. Mater. Sci.: Materials in Medicine*, Vol. 13, No. 2, 211-218.

[59] Petil, O., Zanotto, E. D., Hench, L. L. (2001). Highly bioactive P_2O_5-Na_2O-CaO-SiO_2 glass-ceramics. *J. Non-Cryst. Solids*, Vol. 292, No. 1-3, pp. 115-126.

[60] Li, J., Chem, Y., Yin, Y., Yao, F., Yao, K. (2007). Modulation of nano-hydroxyapatite size via formation on chitosan–gelatin network film in situ. *Biomaterials*, Vol. 28, No. 5, pp. 781-790.

[61] Sachlos, E., Wahl, D. A., Triffitt, J. T., Czernuszka, J. T. (2008). The impact of critical point drying with liquid carbon dioxide on collagen–hydroxyapatite composite scaffolds, *Acta Biomaterialia*, Vol. 4, No. 5, pp. 1322-1331.

[62] Liulan, L., Qingxi, H., Xianxu, H., Gaochun, H. (2007). Magnetic Properties and Intergranular Action in Bonded Hybrid Magnets, *J. Rare Earths*, Vol. 25, No. 3, pp. 336-340.

[63] Sachlos, E., Reis, N., Ainsley, C., Derby, B., Czernuszka, J. T. (2003). Novel collagen scaffolds with predefined internal morphology made by solid freeform fabrication, *Biomaterials*, Vol. 24, No. 8, pp. 1487-1497.

[64] Li, J., Habibovic, P., Yuan, H., van den Doel, M., Wilson, C. E., de Wijn, J. R., van Blitterswijk, C. A., de Groot, K. (2007). Biological performance in goats of a porous titanium alloy–biphasic calcium phosphate composite, *Biomaterials*, Vol. 28, No. 29, pp. 4209-4218.

[65] Hollander, D. A., von Walter, M., Wirtz, T., Sellei, R., Schmidt-Rohlfing, B., Paar, O., Erli, H. (2006). Structural, mechanical and in vitro characterization of individually structured Ti–6Al–4V produced by direct laser forming, *Biomaterials*, Vol. 27, No. 7, pp. 955-963.

[66] Lee, B. H., Abdullah, J., Khan, Z. A. (2005). Optimization of rapid prototyping parameters for production of flexible ABS object, *J. Mater. Processing Tehcn.*, Vol. 169, No. 1, pp. 54-61.

[67] Mostafa, N., Syed, H. M., Igor, S., Andrew, G. (2009). Performance Comparison of IP-Networked Storage, *Tsinghua Scie. & Techn.*, Vol. 14, No. 1, pp. 29-40.

[68] Galantucci, L. M., Lavecchia, F., Percoco, G. (2008). Study of compression properties of topologically optimized FDM made structured parts, *CIRP Annals – Manufacturing Techn.*, Vol. 57, No. 1, pp. 243-246.

[69] Bandeira, L. C., Calefi, P. S., Ciuffi, K. J., Nassar, E. J., Salvado, I. M., Fernandes, M. H. F. V., Silva, J. V. L., Oliveira, M., Maia, I. A. (2011). Calcium Phosphate Coatings By Sol-Gel On Acrylonitrile-Butadiene-Styrene Substrate. *Eclética Química*, submitted.

[70] Nassar E. J., Ciuffi K. J., Gonçalves R. R., Messaddeq Y., Ribeiro S. J. L. (2003). Filmes de titânio-silício preparados por "spin" e "dip-coating", Quim. Nova, Vol. 26, pp. 674-678.

[71] Bandeira L. C., de Campos B. M., de Faria E. H., Ciuffi K. J., Calefi P. S., Nassar E. J., Silva J. V. L., Oliveira M. F., Maia I. A. (2009). TG/DTG/DTA/DSC as a tool for studying deposition by the sol-gel process on materials obtained by rapid prototyping, *J. Thermal Analysis and Calorimetry*, Vol. 97, No. 1, pp. 67-70.

[72] Peter M., Binudal N. S., Soumya S., Nair S.V., Furuike T., Tamura H., Kumar R. J. (2010). Nanocomposite scaffolds of bioactive glass ceramic nanoparticles disseminated chitosan matrix for tissue engineering applications, *Carbohydrate Polymers*, Vol. 79, pp. 284-289.

[73] Bandeira, L. C., De Campos, B. M., Calefi, P. S., Ciuffi, K. J., Nassar, E. J., Silva, J. V. L., Oliveira, M. Maia, I. A. (2011). Coating On Organic Polymer With Macroporous Structure Prepared By Rapid Prototyping, *Journal of Nanostructured Polymers and Nanocomposites*, in press.

[74] Rocha, L. A., Molina, E. F., Ciuffi, K. J., Calefi, P. S., Nassar, E. J. (2007). Eu (III) as a probe in titânia thin films: the effect of temperature, *Materials Chemistry and Physics*, Vol. 101, No. 1, pp. 238-241.

[75] Rocha, L. A., Ciuffi, K. J., Sacco, H. C., Nassar, E. J. (2004). Influence on deposition speed and stirring type in the obtantion of titânia films, *Materials Chemistry and Physics*, Vol. 85, No. 2-3, pp. 245-250.

[76] Culbertson, B. M. (2001). Glass-ionomer dental restoratives. *Progress in Polymer Science*, Vol. 26, No. 4, pp. 577-604.

[77] Nicholson, J. W. (1998). Chemistry of glass-ionomer cements: a review. *Biomaterials*, Vol. 19, No. 6, pp. 485-494.

[78] Pires R., Nunes T. G., Abrahams I., Hawkes G. E., Morais C. M., Fernandez C. (2004). Stray-Field Imaging and Multinuclear Magnetic Resonance Spectroscopy Studies on the Setting of a Commercial Glass-Ionomer Cement. *J. Mater. Sci.: Mater. in Medicine*, Vol 15, pp. 201-208.

[79] Wright J. D., Sommerdijk N. A. M. (2001). Sol-Gel Materials Chemistry and Applications. ISBN 90-5699-326-7 1a Edição, Gordon and Breach Science Publishers; Amsterdam, Holanda.

[80] Yang Z., Lin Y. S. (2000). Sol-gel synthesis of silicalite/g-alumina granules. *Ind. Eng. Chem. Res.*, Vol. 39, pp. 944-4948.

[81] Stamboulis A., Hill R. G., Law R. V. (2004). Characterization of the structure of calcium alumino-silicate and calcium fluoro-alumino-silicate glasses by magic angle nuclear magnetic resonance (MAS-NMR). *J. Non-Cryst. Solids*, Vol 333, pp. 101-107.

[82] Stamboulis A., Law R. V., Hill R. G. (2004). Characterisation of commercial ionomer glasses using magic angle nuclear magnetic resonance (MAS-NMR). *Biomaterials*, Vol. 25, pp. 3907-3913.

[83] Pan J., Zhang H., Pan M. (2008). Self-assembly of Nafion molecules onto silica nanoparticles formed in situ through sol–gel process. *Journal of Colloid and Interface Science*, Vol. 326, pp. 55-60.

[84] Cestari, A., Bandeira, L. C., Calefi, P. S., Nassar, E. J., Ciuffi, K. J. (2009). Preparation of calcium fluoroaluminosilicate glasses containing sodium and phosphorus by the nonhydrolytic sol-gel method *Journal of Alloys and Compounds*, Vol. 472, pp. 299-306.

[85] Gorman, C. M., Hill, R. G. (2003). Heat-pressed ionomer glass-ceramics. Part I: an investigation of flow and microstructure *Dental Materials*, Vol. 19, No. 4, pp. 320–326.

[86] Griffin, S. G., Hill, R. G. (1999), Influence of glass composition on the properties of glass polyalkenoate cements. Part I: influence of aluminium to silicon ratio. *Biomaterials*, Vol. 20, No. 17, pp. 1579-1586.

[87] Phillips, R. W. (1993). Skinner Materiais Dentários, ISBN 85-7404-091-6. Editora Guanabara Koogan, 9a Edição, Rio de Janeiro, Brasil.

[88] Brentegani L. G., Bombonato K. F., Carvalho T. L. (1997). Histological evaluation of the biocompatibility of a glass-ionomer cement in rat alveolus. *Biomaterials*, Vol 18, pp. 137-140.

[89] Campos-Pinto M. M., de Oliveira D. A., Versiani M. A., Silva-Sousa Y. T., de Sousa-Neto M. D., da Cruz Perez D.E. (2008). Assessment of the biocompatibility of Epiphany root canal sealer in rat subcutaneous tissues. *Oral Surg Oral Med Oral Pathol Oral Radiol Endod*, Vol 105, pp. 77-81.,

[90] Garcia, L. da F. R., de Souza, F. de C. P. P., Teófilo, J. M., Cestari, A., Calefi, P. S., Ciuffi, K. J., Nassar, E. J. (2010). Synthesis and biocompatibility of na experimental glass ionomer cement prepared by a non-hydrolytic sol-gel method, *Brazilian Dental Journal*, Vol. 21, No. 6, pp. 499-507.

Development of DNA Based Active Macro-Materials for Biology and Medicine: A Review

Frank Xue Jiang[1], Bernard Yurke[2], Devendra Verma[3], Michelle Previtera[3],
Rene Schloss[3] and Noshir A. Langrana[3]

[1]*Department of Chemical and Biological Engineering,
Northwestern University, Evanston, IL*
[2]*Departments of Materials Science and Engineering and Electrical and Computer
Engineering, Boise State University, Boise, Idaho*
[3]*Departments of Biomedical Engineering and Mechanical and Aerospace Engineering,
Rutgers, The State University of New Jersey, Piscataway, New Jersey
USA*

1. Introduction

DNA was first discovered as the carrier of genetic information for the majority of the known living organisms, encoding the secret of life. Its delicate design based upon double helical structure and base pairing offers a stable and reliable media for storing hereditary codes, laying the foundation for the central dogma (Watson et al. 2003). The impact of this molecule is far reaching into scientific community and our society, as manifested in many fields, for instance, forensics (Budowle et al. 2003), besides medicine.

To date, a great deal of research effort has been directed towards understanding DNA's role in maintenance and expression of genome, and in the application of this understanding to biology and medicine, which is partly fueled by the market needs (e.g., DNA sequencer equipment market alone is expected to reach $450 million by 2010 (Saeks 2007)). For reviews on the development in this area, especially using DNA or RNA per se as therapeutic reagents in applications such as gene therapies, one is referred to a large number of reports (Blagbrough & Zara 2009, Cao et al. 2010, Patil et al. 2005, Ritter 2009). While this remains the center of the attention with the emergence of new subjects of knowledge including genetics and genomics, recent decades have witnessed increased interest in using DNA as structural components or guiding tools (LaBean & Li 2007) in developing novel materials thanks to DNA's many unique features. Among these features are its molecular recognition with only four bases (specificity and simplicity), stable structure held by stacking H-bonds and other weak forces and interactions (stability), and the ease in breaking of base-pairs and thus separating strands allowing modification different than covalent-bond based structures (reversibility and flexibility).

These attributes of DNA give rise to many favorable properties of DNA based macro-materials that are having and will have a wide range of applications. In synthesizing and constructing these DNA based structures, DNA has been used to provide template (e.g., (Aldaye et al. 2008, Niemeyer 2000)), serve as building block (e.g., (Ball 2005)), function as versatile linkages in the network (e.g., (Lin et al. 2004b, Um et al. 2006b)), and aid in the fabrication of the nano-, micro-, and macro-materials (e.g., (Alemdaroglu et al. 2008)). This is also of interest to the community of synthetic chemistry (Alemdaroglu & Herrmann 2007). The scope of the current and potential applications of DNA based materials ranges from DNA based electronics (Berashevich & Chakraborty 2008) and computing (Deaton et al. 1998) to novel material design (Dong Liu et al. 2007, Um et al. 2006a). The similar interest in using other three major types of macromolecules, namely, protein, lipids, carbonhydrates, as structural component for synthetic materials is also increasing (Ball 2005). For reviews in this regard particularly those on DNA based nanomaterials, readers are referred to the latest and comprehensive reviews by Seeman (Seeman 2007), Lu (Lu & Liu 2006, Lu & Liu 2007) and others (Alemdaroglu et al. 2008, Ball 2005, Condon 2006, Mrksich 2005, Niemeyer 2000). The focus of this review is the macroscopic materials designed, synthesized, and applied based on or inspired by DNA and the application of these materials specifically for biology and medicine.

Changes in the nanoscale structures can trigger macroscopic changes in the materials (Schneider & Strongin 2009). For these macro-materials, incorporation of DNA into the structural design confers a number of possibilities that would otherwise not be feasible. For instance, DNA imparts temperature dependent mechanical properties to structures crosslinked by them (Lin et al. 2004b), and unique aptamer interactions make possible phase transition at room temperature (Yang et al. 2008). For these materials, variation at nano-scale DNA structures can lead to sometimes dramatic changes in the bulk material properties, exemplifying 'little trigger' for 'big changes'. Among these DNA based macro-materials, of particular interest are a class of polymeric hydrogel materials, with the ever-increasing significance and promises along with the rapid development in the area of tissue engineering and biomaterials (Jiang et al. 2008b, Jiang et al. 2010c, Lin et al. 2006, Lin 2005, Lin et al. 2004b, Luo 2003, Um et al. 2006b). Additionally, mimicking *in vivo* tissue remodeling and property dynamics is of great importance in the reconstruction of the physiological conditions for cell growth and tissue repair, and DNA based macro-materials help contribute to address the issue thanks to modifications and alterations of the DNA based structures (Jiang et al. 2010b, Jiang et al. 2010c). Therefore, the review first sought to identify the key properties that are directly related to the design and synthesis of DNA based macro-materials and further recognizes the unique properties that result from incorporation of DNA in the structures of macroscopic materials. We then classified these DNA based macro-materials based upon the structural designs (i.e., DNA only, DNA as backbone, and DNA as crosslinker), and surveyed the current studies and potential application for each category of the materials from the literature. To aid in the further development of DNA based macro-materials, we summarized the key design parameters, considerations and major challenges. Lastly, we presented a conjecture on the potential directions.

2. Properties of DNA and DNA based structures

2.1 Properties of DNA

The properties of DNA can be classified in mainly three different levels: the sequence, the structure, and the folding pathway (Condon 2006). The composition or the sequence of DNA based on only four nucleotides, namely adenine (A), thymine (T), guanine (G), and cytosine (C), lays the basis for the primary structure, which also largely determines the secondary and tertiary structures of DNA. The complementarity underlying Watson-Crick base-pairing of A:T and G:C and the stacking forces leads to the classical double helical structure as well as other forms of secondary DNA structures (e.g., A and Z form of DNA). Watson-Crick base-paring is intricately orchestrated by a number of week forces, including hydrogen bonding, π-stacking, electrostatic forces, and hydrophobic effect (Aldaye et al. 2008). It also offers foundation for molecular recognition of DNA. Double stranded DNA is capable of self-folding into complex structures, enabling it to locomote and respond to the environment (Condon 2006). These three levels of structures give rise to some interesting and useful features or attributes.

2.1.1 Chemical properties

DNA is a water-soluble macromolecule, and synthetic DNA displayed good biocompatibility (Um et al. 2006b). It is generally stable under physiological conditions, but can be hydrolyzed by acid and alkali when pH changes. At the same time, DNA is a highly charged polymer mostly due to the phosphate group in the nucleotides. The flexibility of single-stranded DNA (ssDNA) and the relatively weak bonding between base pairs of duplex DNA allows interacting DNA strands to seek thermodynamically favored configurations, making possible programmed self assembly of complex structures (SantaLucia & Hicks 2004). G:C base-pair is more stable than A:C one due to the stronger hydrogen bond present, thus GC content markedly influence DNA properties.

2.1.2 Cleavage

DNA can be cleaved primarily based on three ways: hydrolysis, photochemistry and oxidative reactions (Biggins et al. 2006). In the natural living systems, hydrolysis is the primary mechanism. These mechanisms allow for different approaches in degradation of the DNA based materials or ways of protecting these materials from attacks. For example, in designing DNA only or DNA crosslinked macro-materials, the sequence of the DNA can be chosen in a way that it would be protected under physiological conditions while degraded upon pathological cues (e.g., bio-metal concentration (Jain et al. 1996)) releasing encapsulated therapeutic agents.

2.1.3 Enzymatic modification

DNA is susceptible to modification including cleavage and chemical deletion or addition by a large number of enzymes. Nature has created a rather delicate and precise machinery to manipulate, such as cutting, ligating, unwinding, folding, synthesizing, initiating, modifying, and deleting, DNA in vivo(Braun & Keren 2004, Watson et al. 2003). Many of the enzymes involved in the process have been identified, namely, restriction enzymes, ligase, helicase, gyrase, polymerase, primase, proofreading exonuclease, and a host of other enzymes.

2.1.4 Physical properties

Double stranded DNA is semi-flexible and can possess high rigidity. Upon based pairing, and DNA strands can be straightened, which underlying the design of a nano-actuator (Simmel & Yurke 2001) (**Figure 1**). Previous studies on DNA mechanics suggest that DNA strands can be considered as rigid rods with tensile modulus of hundreds of MPa (Smith et al. 1996) when the force applied is below certain threshold, and this leads to a smaller possibility of stretching DNA longitudinally.. Meanwhile, the energy to bend a DNA strand is inversely correlated to its length (Bustamante et al. 2003). It is emphasized here that as the earlier work pointed out, the physical properties of DNA are closely tied to its biological functions (Vologodskii & Cozzarelli 1994).

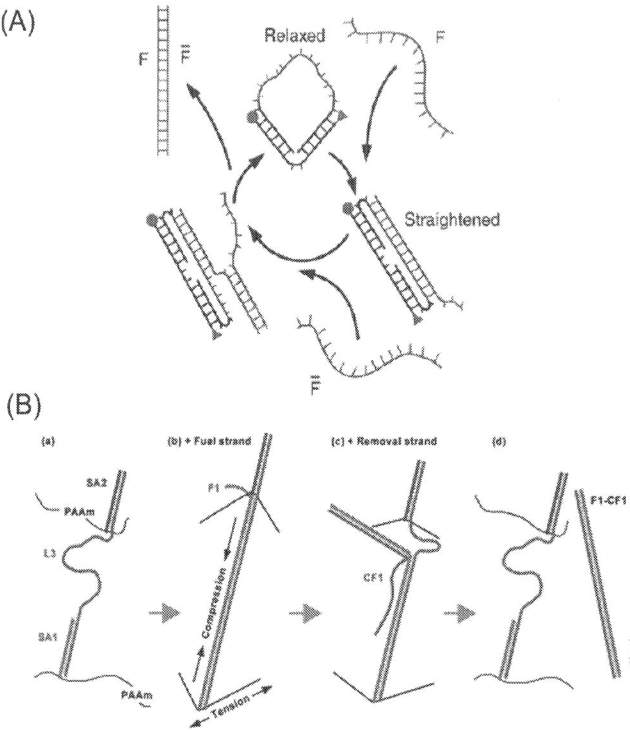

Fig. 1. Schematic of a DNA- based actuator and application to a DNA crosslinked macro-material. (**A**) In the nano-scale structure, at 'Relaxed' state, double-stranded (ds-) DNA is held by a single stranded (ss-) DNA, forming a loop structure. Upon the delivery of fuel strand (F) which base-pair with the ssDNA portion of the loop straightens the loop. With the introduction of the complementary of F, or $\bar{F}$,the fuel strand is displaced, resulting in restoration of the relaxed stated of the loop. Extracted from (Simmel & Yurke 2001), with publisher's permission. (**B**) Incorporation of this nano-structure to the formation of a polymer gel. (Ba) Initially, both the crosslinks and polymer chains are slack. (Bb) As the fuel strand hybridizes with the motor domain, the crosslink stiffens and compressive forces are generated, in turn tensing the polymer chains. (Bc) The removal strand hybridizes with the fuel strand at the toehold region. (Bd) Once the removal and fuel strands are fully hybridized, the gel reverts to its initial state. Extracted from (Lin et al. 2006), with publisher's permission.

2.1.5 Denaturing/Re-annealing

Upon being heated above its melting temperature (T_m), duplex DNA will separate into complimentary strands of ssDNA since the hydrogen bonds between the two strands and other stabilizing forces in the duplex can not withstand the separating forces. Once this occurs, DNA is called being degraded, denatured, or melted, and this process can be reversed if the melted DNA is cooled slowly, for which the term 're-association', 're-naturation', or 're-annealing' is used (Bart Haegeman 2008, Dhillon et al. 1980, Li et al. 2001, Smith et al. 1975). Four major factors dictate the rate of re-association: temperature, salt concentration, DNA concentration, and the length of DNA strand (Li et al. 2001). The optimal re-association temperature is approximately 20°C below melting temperature, and the presence of adequate amount of cations is necessary for re-annealing (Li et al. 2001). In this regard, the lack of proper re-associate conditions leads to the non-reversible change in the materials properties, such as that of EGDE crosslinked DNA gels (Topuz & Okay 2008).

Besides the aforementioned properties that are most relevant to the derivation of DNA based materials for biological application, DNA also possesses a multitude of other properties, including electronic and magnetic ones, that are attractive for applications including DNA based molecular electronics (Berashevich & Chakraborty 2008).

2.1.6 Self-assembly in DNA hydrogels

The remarkable molecular recognition capabilities of DNA make it a promising candidate for development of materials with highly complex structures (Chhabra et al. 2010, Um et al. 2006a). Xing at el. reported synthesis of pure DNA hydrogels, based on self-assembled DNA building blocks with more than two branches. These DNA hydrogels showed thermal and enzymatic responsive properties (Xing et al. 2011). DNA can also be covalently grafted onto synthetic polymers and serve as a cross-linker (Alemdaroglu & Herrmann 2007). The recognition of complementary DNA strands leads to cross-linking of polymer chains and causes hydrogel formation. Zhang et al. reported DNA hydrogels based on N-(fluorenyl-9-methoxycarbonyl)-D-Ala-D-Ala as the cross-linker, which exhibited gel-sol transition upon binding to its ligand (Zhang et al. 2003). Kang et al. developed a photo-responsive DNA-cross-linked hydrogel that exhibited sol-gel transition on exposure to different wavelengths of light. Specifically, photosensitive azobenzene moieties were incorporated into DNA strands, such that their hybridization to complementary DNAs responded differently to different wavelengths of light (Kang et al. 2011). They also showed the capability of such photo-responsive gels by controlling encapsulation and release of multiple drugs. Jiang et al. designed and developed DNA-polyacrylamide hydrogels based biomaterials, which exhibited the ability to increase and decrease its stiffness in-situ, depending on the DNA cross-linker (Jiang et al. 2008a, Jiang et al. 2010a, Jiang et al. 2010c).

2.2 Properties of DNA based macromaterials

DNA has also proven to be a useful material to give bulk materials added functionality. This is exemplified in the introduction of DNA nanostructures to the design of DNA based macromaterials (**Figure 1**) While we will discuss the classification of the DNA based macro-material in depth in the next section, here we survey the new and added functionality that are reported.

2.2.1 Adhesivity

The adhesive properties or the DNA based macromaterials are of significance when they are to be used for applications such as tissue repairs or wound healing where most cells in contact are anchorage-dependent. In a DNA crosslinked hydrogel material, it has been established that with the varying length of DNA crosslinker and different crosslink density, the surface ligand density is not noticeably modified (Jiang et al. 2008b).

2.2.2 Swelling

DNA based macromaterials particularly hydrogel, similar to other hydrogels, can swell in the aqueous conditions, which will be encountered in the in vivo applications. In a DNA-only gel system, it has been observed that in de-ionized water the gel can swell to over 6 times by volume (fiber length) (Lee et al. 2008). Up to one fold increase in weight has been observed for a DNA gel where dsDNA or ssDNA interacts with a cationic surfactant, CTAB (cetyltrimetrylammonium bromide), after swelling (Moran et al. 2007), which is in contrast to the case where proteins (e.g., lysozyme) replace CTAB. For a DNA crosslinked hydrogel, the observed swelling ratio reaches up to 4 times in volume (unpublished data).

2.2.3 Pore structure

In designing bio-scaffolds for tissue engineering applications, the size and range of the pores in the hydrogels is one of the most critical issues. Early investigation has primarily determined the range of pore size from ~20 um to 100 um suitable for cell growth and functioning (Chevalier et al. 2008), while in the drug delivery applications, pore size affects the size of the drug the delivery vehicle is capable of carrying and releasing (Lin & Metters 2006). Um and colleagues devised a hydrogel material based purely on DNA strands for cell encapsulation and reported survival and growth cell of CHO cells inside the hydrogel days after the culture. Aiming at the potential drug delivery applications, Liedl and coworkers examined a DNA crosslinked hydrogel and inferred the pore size from experimental investigation by using quantum dots (QDs) (Liedl et al. 2007). Interestingly, although the pore size of this hydrogel was found to be ~100 um, nanoparticles of 10 nm range can still be trapped. In this hydrogel, the pore structure depends on the length of the crosslinker, the nature of the polymer and interactions between the two. It is noted that in addition to the pore size/distribution and porosity that are generally of concern, pore interconnectivity, shape and uniformity are also of great significance in certain applications (Li et al. 2003).

2.2.4 Sol-gel transition

For DNA based macromaterials, particularly polymeric hydrogel material, gelation point exists between the solid and gel phases. For DNA crosslinked hydrogel, it is a function of crosslinking density, monomer concentration, and crosslinker length (Jiang et al. 2008a, Wei et al. 2008). At a pre-determined crosslinker length and monomer concentration, raising crosslinking density results in sol-gel transition, as reflected in high viscosity (Lin et al. 2004b) (**Figure 2**) also observed in other studies (Li et al. 2005). For DNA gels based on crosslinked DNA network (e.g., by EGDE) discontinuous phase transition has been reported (Amiya & Tanaka 1987, Topuz & Okay 2008).

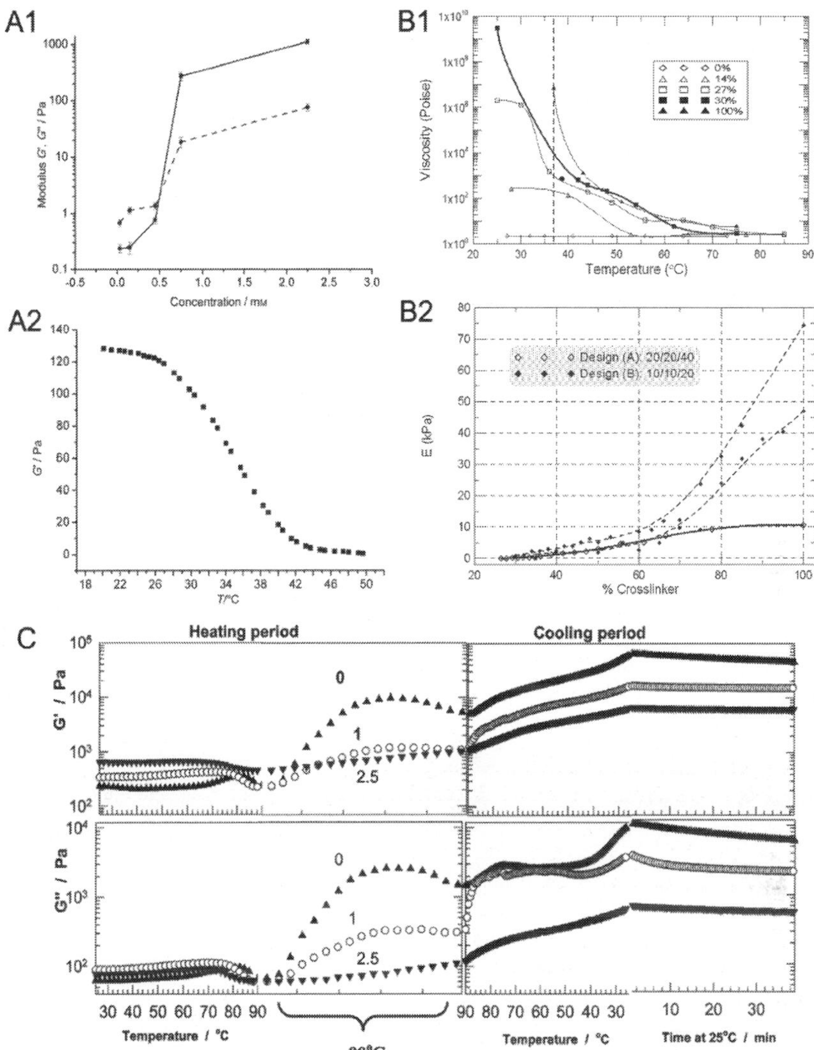

Fig. 2. Mechanical properties of DNA based macro-materials. (**A**) Rhelogy of DNA gels prepared from Y-shaped DNA at pH5.0. Dependence of storage (G′, solid line) and loss (G′′, dashed line) modulus on the concentration of the Y-shaped DNA unit (**A1**) at 25°C and on the temperature at the concentration of 0.60 mM of Y-shaped DNA unit (**A2**). Extracted from (Cheng et al. 2009). (**B**) Changes in viscosity of DNA gels of Design B with respect to the temperature at various levels of crosslinking (only show 0%, 14%, 27%, 30% and 100% for clarity) (**B1**) and changes in mechanical stiffness (modulus, E) with respect to the level of crosslinking for a DNA-crosslinked hydrogels of two designs (A) and (B) (Jiang et al. 2008b)

(B2). Open symbols and solid line represent stiffnesses of DNA gels of Design A (20/20/40 for length of SA1/SA2/L2) (extracted from (Lin et al. 2004b)) and solid symbols and dotted line represent stiffnesses of DNA gels of Design B (10/10/20). Note that at 37°C, the DNA gels are effectively solid. (C) Rhelogy of DNA solutions or gels with EGDE crosslinking (▲0%; O 1.0%; ▼2.5%) during heating-cooling cycles. ω= 1 Hz, strain is 0.01. Extracted from (Topuz & Okay 2008). Both images with publisher's permission.

2.2.5 Reversibility

As pointed in the previous section, DNA can be thermally degraded, and naturally bulk material based on DNA could experience property change along with DNA denaturing. Upon slow cooling and other proper conditions, DNA can re-anneal restoring the macro-material. For DNA crosslinked hydrogel, it is also possible to realize the reversible property change by introducing carefully design DNA strand bypassing the need of applying environmental stimuli such as light, pressure or temperature (Jiang et al. 2010b, Liedl et al. 2007, Lin et al. 2006). The key to this feature is branch migration based strand displacement (Lin et al. 2006) where a sticky end at the periphery of the DNA strands is necessary. Typically, the hybridization reaction occurs between two complementary DNA strands, and is affected by temperature and strand length. The process has a low rate constant several orders of magnitude less than the hybridization reaction (Reynaldo et al. 2000), and by designing a toehold, the process can increase dramatically (Yurke & Mills 2003). Yurke and Mills have determined that for a toehold length of eight bases, the exchange rate increases by six orders of magnitude (Yurke & Mills 2003). Branch migration takes place when a single-stranded DNA (ssDNA) competitively hybridizes with one strand of the DNA duplex starting at the sticky ends (or 'toehold'), and extends the hybridization until that strand is displaced entirely from the original DNA duplex (Watson et al. 2003). **(Figure 1B)** Essentially, since this strand has more complementary base pairs with the targeted ssDNA than do the side chains, generation of the doubled-stranded product is energetically favorable (Yurke & Mills 2003).

Consequently, the absence/presence of sticky ends offers off/on switch for the reversibility of gelation or possibility of structural modification with crosslinking density change. This special feature has fueled the interest in its drug delivery application (Liedl et al. 2007, Wei et al. 2008).

2.2.6 Mechanical properties

Mechanical properties including moduli have been investigated for various DNA based macromaterials **(Figure 2)**. DNA crosslinked hydrogels display temperature and crosslinking density dependent viscosity, mechanical property and gelation point **(Figure 2B)** (Jiang. et al. 2008a, Lin et al. 2004a). Chippada and colleagues developed formulation based on non-spherical inclusions, and made possible the probe of heterogeneity (variation with respect to. location) and anisotropy (difference with respect to direction) in the materials commonly seen in biological tissues (Chippada et al. 2009a, Chippada et al. 2009b). Other investigator used rheology and other techniques in mechanical characterization (Topuz & Okay 2008).

Increase in crosslinking density, microscopically straightens the single-stranded DNA side chain, and stiffens the micro-structure. Macroscopically, it is reflected in the increase in mechanical stiffness. The rigid dsDNA provides resistance also to compression, contributing to the creation of artificial tensegrity (Ghosh & Ingber 2007, Ingber 2006, Liu et al. 2004).

2.3 Approaches in characterization of DNA based macro-materials

Owning to the unique features from DNA, special considerations have to be taken in characterization of the DNA based macro-materials, which poses challenges and stimulated novel ways of probing.

2.3.1 DNA incorporation

Incorporation of the delivered DNA can be assessed indirectly by probing the residual DNA concentration where direct assessment is difficult, if not impossible (Jiang et al. 2010c). In this approach, a DNA strand with non-specific sequence was also included as a negative control to show that the only DNA strands with specific sequence can base-pair with the available DNA side chains on the polymer, and were truly incorporated into the network rather than pure diffusion. Additionally, in measuring DNA concentration, the differential in UV absorbance between ds- and ss-DNA can be used for the detection of crosslinking or de-crosslinking (see, for example, (Cheng et al. 2009, Topuz & Okay 2008)).

2.3.2 Mechanical properties

To investigate mechanical properties of the DNA based materials, a number of methods has been developed (Chippada et al. 2009a, Lin et al. 2004b, Topuz & Okay 2008). Lin and colleagues developed an inclusion based formulation to address the issue of limited availability of samples, sample preparation and intrusiveness associated with conventional testing apparatus (e.g. Instron, or dynamic mechanical analysis (Um et al. 2006b)) for these materials (Lin 2005, Lin et al. 2004b). Recently, along this line of work, nanoscale rods were deployed and new formation has been developed to assess the inhomogeneity and anisotropy of the hydrogel materials (Chippada et al. 2009a). Mechanical properties including stiffness can be used to infer the structure of the DNA based macro-structures. For instance, for a DNA crosslinked polyacrylamide hydrogel, the crosslinking density of DNA crosslinked hydrogel has been correlated to its mechanical stiffness for a specific crosslinker design, thus the choice of crosslinking density can be made aiming at specific mechanical stiffness range (Jiang et al. 2008b, Lin et al. 2004b).. Moreover, drastic change in the viscosity or rheology has been used as indicator as watershed between sol and gel-states

2.3.3 State of DNA strands

Fluorophore attached DNA strands have been previously deployed to examined the dynamics of DNA base-pairing. The mechanism behind this approach is that the distance change between two dyes, or fluorophore/quencher, can be probed by various techniques including FRET (Fluorescence resonance energy transfer), which indicates the state (e.g., bent or straightened) of the DNA strands (Simmel & Yurke 2001). Atomic force microscopy (AFM) is a powerful tool capable of resolving nano-scale features, and has been used to probe the DNA based structures (e.g., (Liu et al. 2004)) with limitations in resolution (a few nm) (Um et al. 2006b). Optical properties can also be used to monitor the state change (e.g., DNA binding to cations, DNA packing or denaturation) based on drug-hydrogel interactions by using circular dichroism (CD) along with other techniques such as polarized Raman spectroscopy(Lee et al. 2008, Tang et al. 2009).

3. Current DNA based macro-materials and applications in biology and medicine

Seeman and colleagues pioneered the work employing DNA as a structural material in creating nanodevices (Seeman 1981, Seeman 1982), and reported designs of nano-scale structures such as rings(Mao et al. 1997), cubes(Chen & Seeman 1991), and octahedral (Zhang & Seeman 1994). More investigators joined the effort stimulating the emergence of structural DNA nanotechnology (Douglas et al. 2009, Rothemund 2006, Seeman 2007, Yurke et al. 2000), particularly aptamers, DNAzymes, and molecular beacon (Condon 2006, Lu & Liu 2006, Wang et al. 2009).

Of particular interest is the fact that a number of the designed structures inspired by DNA offer a large variety of design parameters (e.g., sequence and DNA-protein interactions) to the nanotechnology engineers. When they are incorporated as part of the macrostructures such as a hydrogel network, by changing the design parameters at the nanoscale DNA structures, dramatic physical and chemical properties changes can be achieved at macro-level. Some of these properties and functionalities are highly desirable in biology and medicine. Moreover, due to the unique properties of DNA, *in situ* modifications of nano-level structures become possible, which often result in the dynamic properties of macro-level materials thus supply dynamic cues in biological applications. Furthermore, in realizing these changes, a great number of physical, chemical and biological stimuli can be employed together (Lu & Liu 2007), offering augmented flexibility in designs.

Owing to DNA's water solubility and the resemblance to the physiological environment, DNA based macro-materials particularly hydrogels is stimulating ever-increasing interest. Hydrogels are a class of hydrophilic polymers that possess both solid- and liquid-like properties, and they typically consist of an insoluble network of crosslinked polymer chains immersed in solvent. They have attracted great interest and have become ever-increasingly popular for many applications, including biomedical ones. Due to its hydrated nature, a hydrogel can better mimic the properties of the natural tissues and neural micro-environment that cells reside in. This has fueled the interest and development of hydrogels-based tissue engineering scaffolds. In addition, hydrogels generally respond to the environmental factors such as temperature and pH, and thus are among the candidates for the development of drug delivery systems. Based on the types of crosslinker, hydrogels can be categorized into two classes; gels with covalent junctions and gels with physical junctions (weak forces, physical entanglement or others), or more simply stated, chemical and physical gels. Natural polymers are synthesized by living organisms mostly through enzymatic processes, while synthetic polymers generally involve either condensation or addition approaches. Crosslinking yields a polymer network where polymer chains are inter-connected.

Along this line, a number of DNA based hydrogel materials have been devised and characterized, among which are those consist solely of DNA strands (Cheng et al. 2009, Mason et al. 1998, Um et al. 2006b), those with polymer backbone and DNA crosslinkers (Lin et al. 2004b, Nagahara & Matsuda 1996), and those with DNA as polymer backbone connected via physical or chemical bonds (Topuz & Okay 2008), where DNA 'nanoswitch' impart the hydrogel materials desired functionality and properties (**Figure 3**).

3.1 DNA based macro-materials

The majority of the DNA based macromaterials fall in the following three categories (**Table 1**) and examples of these materials are shown in **Figure 3**.

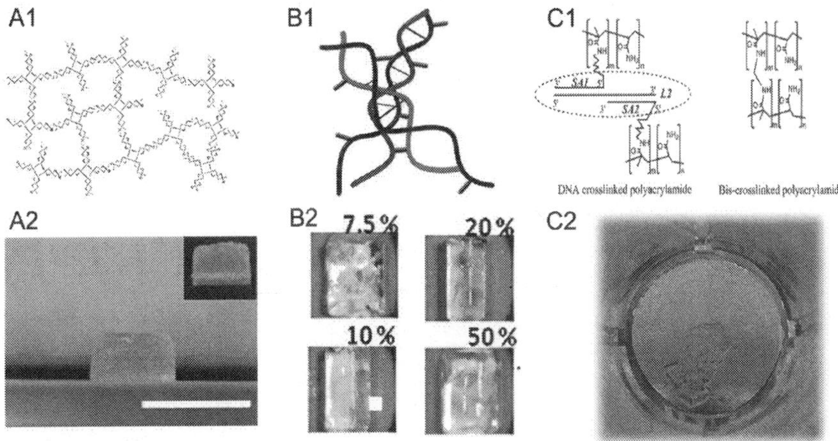

Fig. 3. Examples of DNA based macromaterials. (**A**) A DNA-only hydrogel: the schematic of the DNA structures (A1) and a prepared DNA gel (A2). Extracted from (Um et al. 2006b). (**B**) A EGDE crosslinked DNA gel (DNA as backbone): the schematic of the crosslinked structure (B1) with short thick black stick indicating EGDE bond and thin black stick indicating hydrogen bond (extracted from (Topuz & Okay 2008)) and resulting gels with varying EGDE content (B2) (extracted from (Topuz & Okay 2009)). (**C**) DNA crosslinked polyacrylamide hydrogel: the schematic showing the difference between DNA- and bis-crosslinked hydrogels (C1) (extracted from (Jiang et al. 2008b)) and a prepared gel (C2) in a well of a 24-well plate. All images with permission from publishers.

3.1.1 DNA-only network

The aqueous solution of DNA strands (~ 2,000 bps in length) can be viscous at high concentration before the critical overlap concentration is reached. Beyond this critical concentration, a weak gel can be formed due to the overlapping and entanglement of the DNA strands (Mason et al. 1998, Topuz & Okay 2008). Since the gelation point is reached based upon physical interactions rather than chemical bonds, this hydrogel is termed 'physical gel' (Mason et al. 1998, Topuz & Okay 2008). Though this approach has its advantages in the availability of the natural long double stranded DNA, the lack of controllability and stability limits their further application. Moreover, based on electrostatic interactions, by introducing hydrophilic ionic liquids, DNA hydrogel fibers have also been made (Lee et al. 2008). In this gel system, DNA strands compact into supercoils and bundle up forming aggregates, and give rise to new material properties such as stability and resistance to DNase digestion.

Luo group at Cornell University developed a hydrogel based entirely on DNA strand base pairing (**Figure 3A**). The synthesis involves two major steps: first, branched three- or four-armed 'X', 'Y', and 'T'-shaped DNA structures were synthesized from single-stranded DNA with partial complementarity based on DNA self-assembly; the sequence of the DNA strands were chosen and sticky ends were included such that it is available for enzymatic action; next, ligase, an enzyme capable of ligating DNA strands were deployed to connect the building blocks produced from the first step, thus forming a crosslinked DNA polymer network. The resulting hydrogel has been shown to have swelling and mechanical properties that are dependent on the initial concentration and the forms of DNA building blocks, and biodegradability determined also by the building blocks (Um et al. 2006b). The potential of applying this hydrogel for drug delivery application has also been demonstrated. Very recently, this group also reported that by incorporating linear plasmids into polymer network, this hydrogel is capable of generating natural proteins under cell-free conditions (Park et al. 2009).

By using the similar Y-shaped DNA building blocks, but a different mechanism to connect these building blocks, Cheng group also put forth a hydrogel design based entirely on DNA nanostructures (Cheng et al. 2009). Different than the approach by Luo and colleagues, enzymes are not needed in the synthesis. Rather, the sequence of DNA strands are designed that it contains C-rich domain to take advantage of the triple hydrogen bond formation which results in a crosslinker between two DNA building blocks. Because the formation of such crosslinkers is pH dependent, the resulting macroscopic hydrogel can be formed only at suitable pH and hence responsive to pH changes. This feature allows for a new scheme for drug delivery based on pH, which hold promises in cancer therapies particularly those associated with local pH changes. It is worthwhile noting that in these studies, relatively short synthetic single-stranded DNA is required and that the quantity of the samples is still limited (~20 µL) primarily due to the limited availability and cost in synthesis.

3.1.2 DNA as backbone

While DNA strands can be connected via enzymatic actions or base interactions where no other chemical entity is involved, they also can be crosslinked by other molecules via either physical or chemical interactions. In these materials, DNA constitutes the polymer backbone. As an example, physical DNA gels have been developed based on the interactions between DNA strands and sulfonium precursor of poly-phenylenevinylene (SP-PPV) (Tang et al. 2009). Positively charged SP-PPV resulting from polymerization at alkaline solution contributes to the hydrogel formation based on DNA/SP-PPV hybrids due to electrostatic interactions. This gel system has demonstrated interesting stability and resistance to heat or DNase attack, and the presence of DNA strands in gel network imparts the material unique biological properties. As a proof of concept, its optical properties have been shown to assist in monitoring drug delivery as illustrated in the recovery of fluorescence upon release of drugs (Tang et al. 2009). Further application of this system awaits the investigation on whether DNA will be shielded from enzymatic digestion under physiological conditions or diffuse out of gel network (Tang et al. 2009).

Besides physical crosslinking, DNA backbone can also be chemically crosslinked. Topuz and Okay (Topuz & Okay 2008) used ethylene glycol diglycidyl ether (EGDE) for this purpose (**Figure 3B**), since epoxide group of EGDE can react with amino group in the bases of two DNA strands, although the two bases can also be from the single strand. They discovered novel thermal properties. At low crosslinking density, dynamic moduli are altered in a non-reversible way when gels are subjected to heating and cooling, and this leads to a hydrogel with Young's modulus in the mega-Pascal (MPa) range. The increased physical entanglement upon heating and hydrogen bond formation at cooling were identified as the cause, although it is not clear whether controlling the kinetics of the DNA re-annealing could affect the process. Horkay and Basser examined the effect of ion strength and concentration on osmotic and mechanical properties of these DNA gels (Horkay & Basser 2004). DNA gel particles based on interactions between DNA and CTAB, a cationic surfactant, or lysozyme were developed by Moran and colleagues. In this physical gel, the electrostatic forces help stabilize the gel network (Moran et al. 2007).

3.1.3 DNA as crosslinker

DNA has long been used to provide bases for assembling microscopic structures into macroscopic objects by functioning as crosslinkers, and to give bulk materials added functionality (Lin et al. 2004b, Nagahara & Matsuda 1996, Neher & Gerland 2005). For instance, Mirkin and colleagues reported a method to organize colloidal gold nanoparticles and form aggregates (Mirkin et al. 1996) based on DNA crosslinking. The motivation in using DNA as linking reagents rather than the main building blocks or polymer backbone lies partly on the fact that in those cases large quantities of synthetic DNA are currently prohibitively expensive (Jiang et al. 2008a, Lin et al. 2004a, Mangalam et al. 2009), and that it is challenging to characterize these structures(Storhoff & Mirkin 1999).

By using DNA hybridization instead of covalent bonding to form crosslinks between polymer strands, hydrogel polymers have been given a temperature-dependent rigidity and thermal reversibility in crosslinking and gelation(Lin et al. 2004b, Nagahara & Matsuda 1996), and a number of new possibilities including *in situ* property change (Jiang et al. 2010c). In an early work (Nagahara & Matsuda 1996), poly(N,N-dimethylacrylamide-co-N-acryloxyloxysuccinimide) was reacted with 5′-amino-modified 10-mer oligonucleotides (oligoA or oligoT) to form polymer chains with short DNA side branches (**Figure 4**). Two different crosslinked structures (**Figure 4A**) were produced: in one of them oligoA branches from one solution of polymer chains hybridized with oligoT branches from the second solution, and in the other of them two oligoT branches hybridized with a third 20-mer OligoA strand. Gelation of the polymers as well as thermo-reversibility of crosslinking at elevated temperatures was demonstrated.

While the simple sequences used in (Nagahara & Matsuda 1996) preclude the formation of secondary structures (e.g., the hairpin structure), the possibility of off-alignment binding between two complementary sequences is high. Although perfect alignment of oligoA and oligoT strands is energetically favorable, misalignment by only a few bases may occur with little penalty. Such misalignments may result in mechanically weakened, kinked crosslinks. The probability of off-alignment binding between complementary DNA strands can be reduced by designing base sequences. Towards this end, by incorporating Acrydite™ modified oligonucleotides in PAM gels, Lin and colleagues (Lin et al. 2004b) illustrated and characterized reversible gelation and achieved a range of stiffness from a few hundred Pa to 10 kPa by varying crosslinker DNA density. They showed that sequence optimization is an effective method of enhancing the stability of DNA crosslinks (**Figure 4B**). In these gels, Acrydite™ modified oligonucleotides co-polymerize with acrylamide monomers to form polymer long chains with DNA side chains of specific length and sequence designated as SA1 and SA2. 'Crosslinker' oligonucleotides (L2) with a 'toehold' assume the functions of a crosslinker by hybridizing with SA1 and SA2 at the same time. By carefully designing another single-stranded DNA (ssDNA), also called "removal" DNA that is complementary to L2, one is able to reverse crosslinking process (**Figure 5C**). With this gel system, the pore structure upon reversible crosslinking was explored, giving rise to the potential application for controlled drug delivery (Liedl et al. 2007).

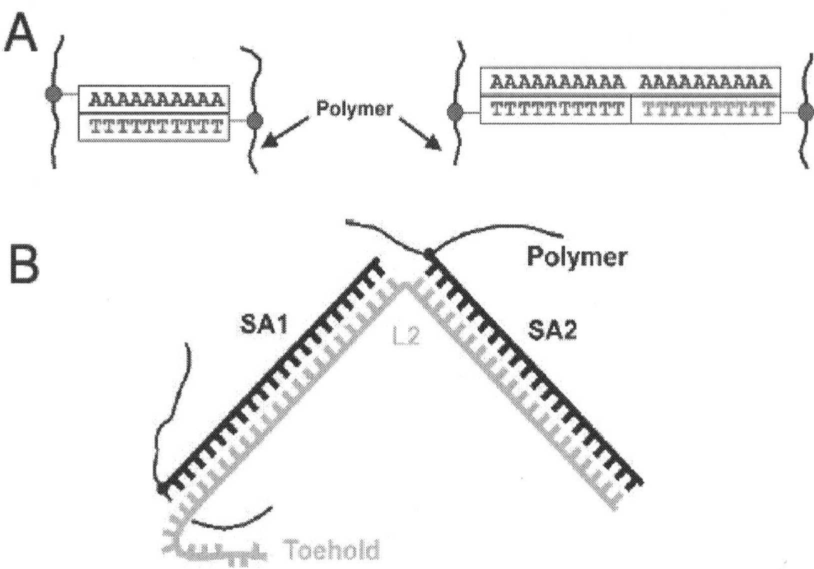

Fig. 4. Design of DNA crosslinked hydrogel. (**A**) The two DNA-crosslinked structures by Nagahara and Matsuda (Nagahara & Matsuda 1996). Oligonucleotides and their sequences are represented by the boxes and oligonucleotides-polymer connections by the dots. Extracted from (Nagahara & Matsuda 1996), with publisher's permission. (**B**) The design by Lin et al. (Lin et al. 2004b). SA1, SA2, and L2 represent DNA side chains and a crosslinker complementary to both DNA side chains. A toehold can be added for reversibility. Extracted from (Lin et al. 2006), with publisher's permission.

Replacing the covalently bound bis-crosslinks with paired DNA strands results in a gel possessing a number of potentially useful properties, such as thermal reversibility with a tunable melting temperature, reversibility of gelation without heating and without the need of initiator-catalyst system for re-gelation(Lin 2005, Lin et al. 2004b). More interestingly, by modifying the DNA crosslinking (i.e., oligonucleotide length or concentration), the mechanical properties of the gels can be engineered to take on particular values. Specifically, via delivery of more crosslinks, the DNA association/dissociation ratio could increase, resulting in a stiffened gel; in contrast, gels could be softened by lowering the crosslink density with removal DNA strands (designated as CL2, **Figure 5C**) that are complementary to L2. CL2 strands competitively base-pair with L2 strands and remove them from the gel network. The ease with which the mechanical properties of DNA crosslinked gels can be changed suggests that they would be useful in tissue engineering applications. This has generated interests in using DNA as crosslinking agent for various applications (Alemdaroglu & Herrmann 2007, Liedl et al. 2007, Murakami & Maeda 2005, Roberts et al. 2007, Wei et al. 2008, Yang et al. 2008). In addition, DNA has also been reported to crosslink organic network such as cellulose (Mangalam et al. 2009).

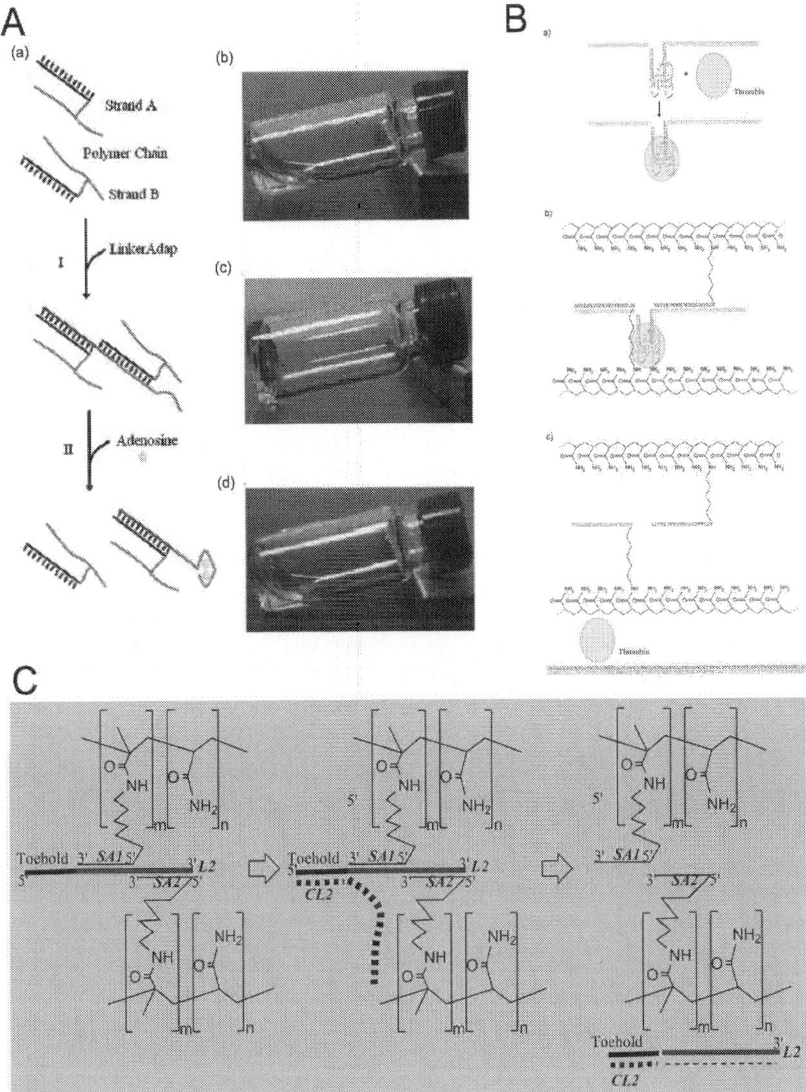

Fig. 5. Induce dynamic changes in DNA crosslinked hydrogels. **(A)** Addition of DNA crosslinks the hydrogel and upon delivery of adenosine which competitively binds to the crosslinker DNA, resulting in the reverse of gelation (a) as demonstrated in the transition from solution (b) to gel (c) and then back to solution (d). Extracted from (Yang et al. 2008) with publisher's permission. **(B)** DNA hydrogel capable of capturing and releasing thrombin based on thrombin-aptamer interactions. The end structure of DNA side chain A has high affinity to thrombin to form thrombin-aptamer complex (Ba), which can be capture

via DNA hybridization (Bb). Upon delivery of the ssDNA complementary to strand A, thrombin is released. Extracted from (Wei et al. 2008) with publisher's permission. (C) For a DNA crosslinked hydrogel (Jiang et al. 2010b), delivery of the 'removal' DNA strand complementary to the crosslinker DNA leads to de-crosslinking.

Dynamic materials can be used to manipulate cell behavior. In studies performed by Langrana and colleagues, DNA-crosslinked hydrogels (DNA hydrogels) were used as the underlying substrate to study the effects of dynamic mechanical cues on fibroblast behavior (Jiang et al. 2010c, Previtera et al. 2011). The DNA hydrogels have the ability to temporally change stiffness (Jiang et al. 2008a, Jiang et al. 2010c, Lin 2005, Lin et al. 2004a, Lin et al. 2005, Previtera et al. 2011). Upon a decrease or increase in DNA hydrogel stiffness, expansion or contraction forces are generated, respectively. The two properties cannot be decoupled (data unpublished). When grown on these dynamic hydrogels, fibroblast morphology is noticeably different compared to static hydrogels, which do not change in stiffness and thus do not generate forces (Jiang et al. 2010c, Previtera et al. 2011). GFP fibroblast became larger and more circular, compared to static conditions, when grown on DNA hydrogels that became softer and expanded (Previtera et al. 2011). Therefore, as the underlying substrate expands and softens, the GFP fibroblasts expand and become rounder morphology. This is in contrast to GFP fibroblast grown on dynamic hydrogels with increasing stiffness and contraction forces (Jiang et al. 2010c). These GFP fibroblasts became smaller and/or longer when compared to static hydrogels. However, these results depended on magnitude of hydrogel stiffness change (Jiang et al. 2010c).

3.2 Potential application of DNA based macro-materials
Three main areas of application are being explored by using these DNA based macromaterials (**Table 1**).

3.2.1 Biosensor| Actuator| Bioelectronics
Hydrogels synthesized from DNA nanostructures hold promises as biosensor (Cheng et al. 2009, Lin et al. 2004b), Simmel and Yurke designed a DNA-based actuator capable of switching between two physical states, which can potentially be used as motor to drive the nano-robot (**Figure 1**) (Simmel & Yurke 2001). This approach, together with others (Knoblauch & Peters 2004), can be adopted in hydrogel formation, giving rise to novel materials with changing properties upon 'fuel strand' delivery. Besides the potential uses of DNA based macromaterials in sensors and actuators, DNA's electronic properties and molecular recognition, feasibility of DNA manipulation at nano-scale, and the trend of miniaturization are driving the synergy between DNA and electronics. Braun and Keren (Braun & Keren 2004) put forth a scheme of constructing DNA based transistors, in which DNA is metallized and serves as a template for electronic circuit, which exemplifies DNA's impressive capability of information storage and molecular recognition mechanism. Incorporation of grafted oligonucleotides also leads to novel materials with high optical resolution, and can be potentially used in biosensing (Tierney & Stokke 2009) (**Figure 6A**).

3.2.2 Drug delivery vehicle
In response to various environmental factors, DNA may alter its secondary and tertiary structures, resulting in alterations in the bulk materials that are built upon them. Aiming at drug delivery application for cancer therapy, a great deal of effort has been made in

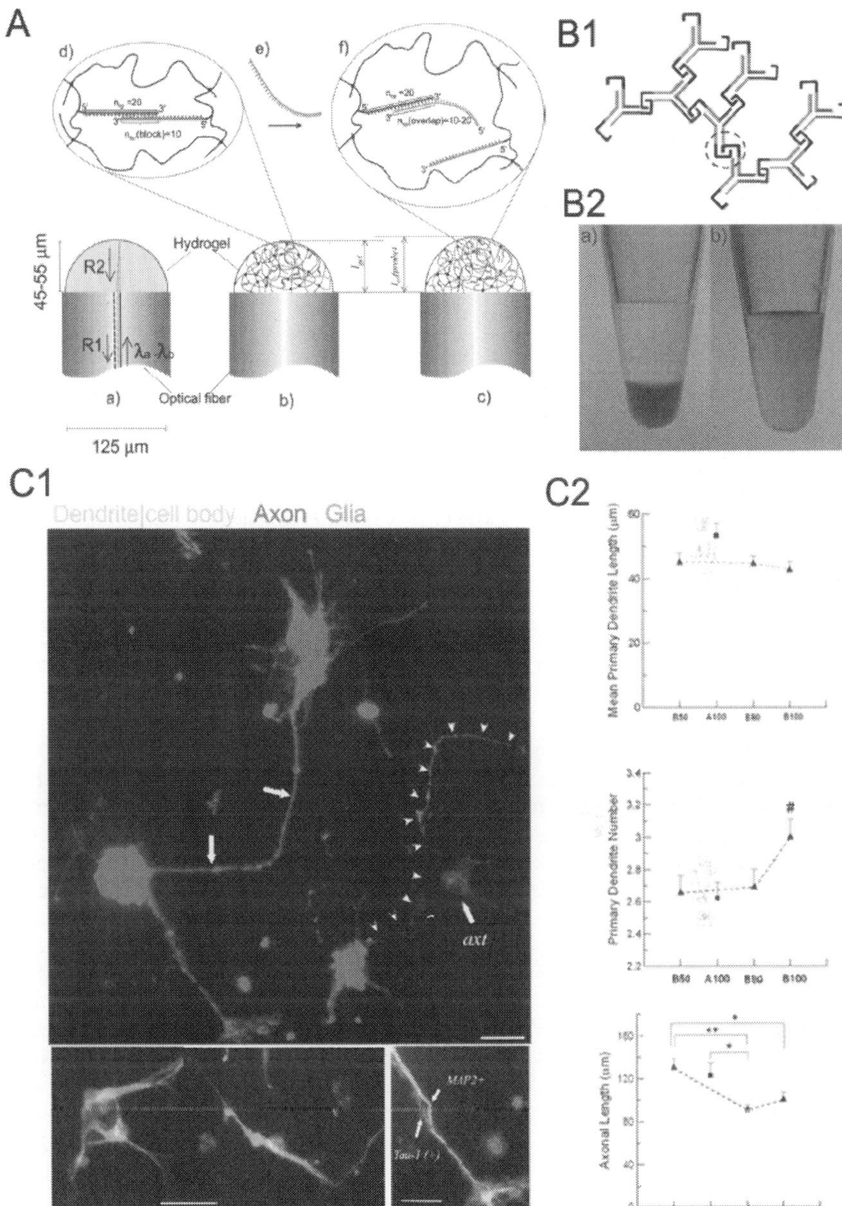

Fig. 6. Examples of application of DNA based macromaterials. (**A**) Schematic of hemispherical bio-sensitive hydrogel attached to the end of an optical fiber to determine

changes in the optical length for biosensing applications. Extracted from (Tierney & Stokke 2009). **(B)** DNA-only hydrogels based on branched Y-shaped DNA unit. Black i motif with cytosine-rich regions crosslinks adjacent Y units (B1). The DNA gel prepared from this design exhibited responsiveness to pH, in low pH where gold naoparticle (AuNP) was trapped in the gel (a) and at high pH gel dissociation leading to AuNP release (b) Extracted from (Cheng et al. 2009). **(C)** Neurite outgrowth on a DNA crosslinked hydrogel. Overlay of higher power images of MAP2 and Tau-1 stain reveals that axons and dendrites could reside closely in parallel with each other (C1). Red: Tau-1 immunostaining; Green: MAP2 immunostaining; Blue: GFAP immunostaining; Purple: DAPI staining. Scale bar is 50 □m. Comparison of neurite outgrowth, including mean primary dendrite length, primary dendrite number, and axonal length per neuron, on DNA gels of two designs. Extracted from (Jiang et al. 2008b). All images with publisher's permission.

designing responsive DNA gels. Among all the cues is pH due to the fact that certain cancer types are associated with local acidity (Gerweck & Seetharaman 1996). DNA motifs sensitive to changes in H+ concentration has been incorporated in the DNA based hydrogel to realize pH responsiveness. A DNA hydrogel in which gel-'drug' interactions are pH dependent was also proposed (Tang et al. 2009) **(Figure 6B)** along with others gels (Roberts et al. 2007) **(Table 2)**. In this design, the electrostatic interactions that retain drugs in the gel network can be reduced resulting in subsequent drug release (Tang et al. 2009). Besides pH, temperature may be another environmental trigger for drug release, particularly for those diseases with local temperature change (e.g., (Hildebrandt-Eriksen et al. 2002, Letchworth & Carmichael 1984)). Thermal responsiveness of the DNA hydrogel has been designed based on the temperature-dependent hybridization, sol-gel transition or physical properties (Costa et al. 2007, Lin et al. 2004b, Topuz & Okay 2008). Ion strength or concentration has also been explored to initiate drug release using DNA based macromaterials (Costa et al. 2006., Horkay & Basser 2004).

These hydrogels responsive to environmental factors hold promises in facilitating targeted delivery of therapeutic reagents, while their application has inherent limitation. First, their application is limited to where such environmental alterations exist; and second, their controllability is limited due to undesired environmental changes that may occur; third, their applicability is limited when temporal control in delivery is desired. Looking to expand the scope of application, some investigators attempted to develop dynamic DNA gel system without the need of environmental factors. DNA strand per se is naturally an ideal candidate. Lin and colleagues demonstrated possibility of triggering de-gelation by delivering ssDNA (Lin et al. 2006), and a similar scheme was adopted by Wei et al.. in designing a DNA gel capable of releasing proteins based on aptamer-thrombin interactions (Wei et al. 2008). Aiming at the same application relying on DNA aptamer-protein interactions, a latest study explored a hydrogel system capable of sustained protein release (Soontornworajit et al.). Diffusion profile and relationship between cargo size and pore size of this system were studied, and it was found that the nano-scale particles can be trapped even their size is smaller than the average pore size of the hydrogel network (Liedl et al. 2007). In addition to DNA strands, by using the similar system, adenosine has also been shown as the trigger for changes based on its interactions with aptamers (Yang et al. 2008). A recent work reported the enzyme triggered release of DNA in a polymer network with grafted DNA duplex (Venkatesh et al. 2009). This system is based on the conventional crosslinking but contains Acrydite modified DNA recognized by specific enzymes.

Additionally, DNA gels have been shown to be an ideal candidate for cell capsulation (Um et al. 2006b), potentially, serving as *in vivo* protein factory for protein synthesis and delivery (Park et al. 2009). Examples of the studies using DNA-only, DNA-as-backbone, and DNA crosslinked macromaterials on potential drug or gene delivery applications and the kinetics of release are in **Figure 7**.

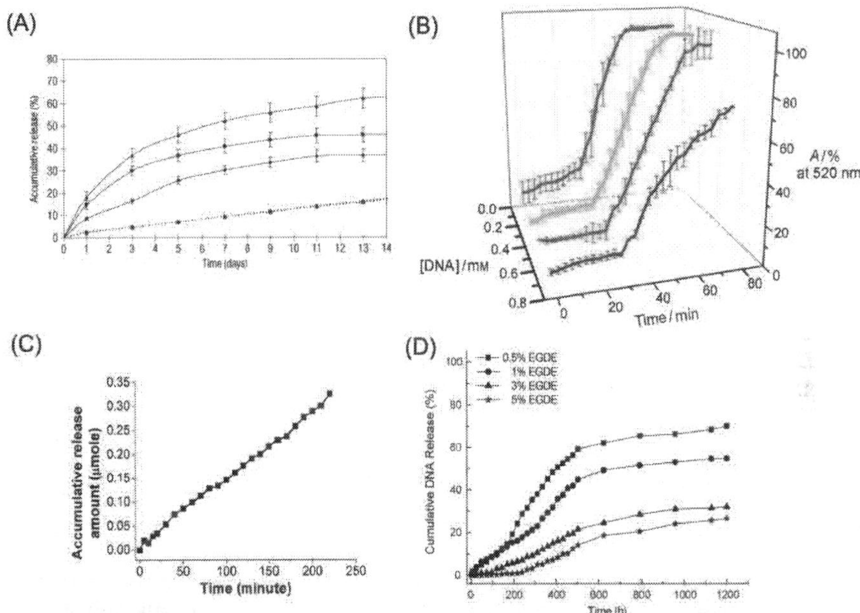

Fig. 7. Kinetics in the release of therapeutic agent using DNA based macromaterials. (**A**) Release of insulin (solid line) and CPT (camptothecin, dotted line) from a DNA-only gel. From top down, the lines indicate Y-, T-, X-, and T- DNA gels. Extracted from (Um et al. 2006b). (**B**) Release of gold nanoparticles from an aptamer-crosslinked hydrogel at interaction with cocaine. Extracted from (Zhu et al. 2010). (**C**) Release of antihypertensive nicardipine hydrochloride containing one nitro group from a physical DNA gel based on DNA and SP-PPV interactions. Extracted from (Tang et al. 2009) with publisher's permission (**D**) Cumulative release of DNA from a EGDE crosslinked DNA (as backbone) gel with various crosslinking density under sunlight. Extracted from (Costa et al. 2010). All images with publisher's permission.

3.2.3 Biomaterials/Tissue engineering

As mentioned in the previous discussion, hydrogel materials has been gaining increasing popularity due to its hydrated state mimicking natural tissues (Janmey et al. 2009, Nemir & West 2009, Uibo et al. 2009). Following this direction, one line of interest in applying DNA based macro-materials is to study cell-ECM interactions, an analog of tissue-biomaterials interplay. A DNA only gel system has been proved to possess cyto-biocompatibility by encapsulating CHO cells (Um et al. 2006b). Replacing the traditional bis-acrylamide crosslinker in a popular bis-gel system (Wang & Pelham 1998), DNA crosslinker of 20-50 nt long was used for the study of the effect of substrate stiffness on neurite outgrowth (Jiang et al. 2008b) (**Figure 6C**). In this system, difference in rigidity was created by varying length of the crosslinker, crosslinking density, or monomer concentration, among which crosslinking density can be modified via DNA strand delivery in situ. The potential of using these DNA crosslinked gels in tissue engineering application is promising (Chan & Mooney 2008, Ghosh & Ingber 2007).

The added advantages by using DNA based macromaterials were further demonstrated recently in subjecting cells to dynamic stiffness of the substrates (Jiang et al. 2010b, Jiang et al. 2010c). These studies were motivated by the fact that the micro-environment that cells reside in within natural tissues is dynamic and undergoes constant synthesis and degradation in both normal and pathological conditions (Lahann & Langer 2005, Mrksich 2005). Moreover, aging, development, external assault, and pathological processes can also lead to the alternations in the extracellular matrix (ECM) (Georges et al. 2007, Ingber 2002, Silver et al. 2003). In addition, at the tissue-implant interface, cells can actively modify surface of the implants, altering the stiffness of microenvironment of their own or other cells (Marquez et al. 2006). The changing stiffness could potentially make it possible to achieve optimal growth of a specific cell property (Jiang et al. 2008b) or direct stem cell differentiation (Engler et al. 2006) at different time points. These facts make it very desirable for the biomemetic materials to have the capability of undergoing controlled remodeling with respect to time. Previously, a limited number of attempts have yielded exciting findings (Chen et al. 2005, Lahann & Langer 2005, Mrksich 2005), in which dynamic changes were induced largely through application of environmental factors (e.g., temperature, pH, and electric field). However, the utility of these approaches in clinical setting could be problematic. With the unique hydrogen bond based crosslinking, DNA based and crosslinked materials, therefore, demonstrate time-dependent properties as reflected in swelling and mechanical modulus, and offer a feasible way of dynamically altering the macro-scale structure mimicking the *in vivo* conditions (**Figure 8**). Indeed, the initial results have indicated that encapsulated cells are viable in a DNA-only hydrogel, and in a DNA crosslinked hydrogel both mechano-sensitive cell types (e.g., fibroblast) (**Figure 9**) and neuron whose mechano-responses are being appreciated just recently respond to the changing stiffnesses, and the responses are specific to range and rate of changes and cell type (Jiang et al. 2010a, Jiang et al. 2010c). A summary of DNA based macromaterials with dynamic and responsive properties is presented in **Table 2**.

4. Design considerations in DNA based macro-materials

Different than other materials, DNA based macro-materials necessitate some unique considerations due to involvement of DNA nano-materials.

4.1 Stability
As pointed out in the last section, DNA strand can respond to a variety of environmental factors such as temperature, pH, and ion concentration and non-environmental factors such as exogenous DNA or enzyme. While it allows design of smart responsive materials, it also poses difficulties in maintaining the integrity of structures. Divalent or multi-valent cations

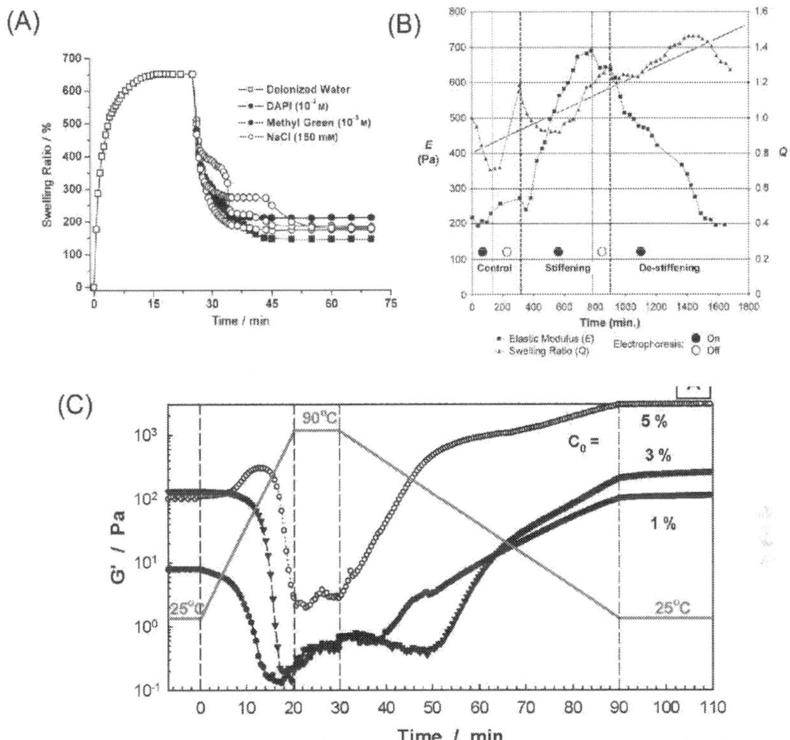

Fig. 8. Time history studies of DNA based macromaterials. (**A**) Swelling/Deswelling profile of a DNA-only hydrogel fiber. Swelling last till 25 min time point when different groove-binding molecules were introduced. Extracted from (Lee et al. 2008). (**B**) Time history plots of the elastic modulus and swelling ratio of a DNA crosslinked hydrogel. Both quantities were first measured throughout the control phase during which no DNA strands were added, first under the influence of electrophoretic power and then without it. Fuel strands were migrated into the gel during the stiffening phase and removal strands were introduced during the de-stiffening phase. Extracted from (Lin et al. 2006). (**C**) Changes in elastic modulus G′ and temperature of DNA (as backbone) gel with time at different DNA content. Extracted from (Orakdogen et al. 2010). All images with publisher's permission.

such as magnesium have been shown critical in both dsDNA stability and re-annealing. Thus using ion-containing buffer would be a better choice than deionized water in maintaining gel structure and integrity. Interestingly, the gel collapse has been observed for a EDGA crosslinked DNA gel system, where the form of dsDNA or ssDNA, DNA content, and co-solutes in the medium contribute to the kinetics (Costa et al. 2007). The thermal stability has been investigated in a number of studies. For a DNA only gel system, gels based on ssDNA were less stable than those made from dsDNA perhaps due to the synergistic effect of multiple strands, possibly due to distinct linear charge density, strand flexibility and hydrophobicity (Costa et al. 2007). DNA crosslinked polymeric hydrogel exhibited thermal reversibility and sol-gel transition, which is correlated to the thermal stability of the DNA base-pairing. As a result, in these gels DNA sequence has to be designed for desired melting temperature (Tm) by adjusting length of the strand, GC content, and/or thermal dynamics(Cheng et al. 2009, Lin et al. 2004a). It is noted that the critical temperature for the DNA based bulk material may be different from that of the involved DNA strands (Sun et al. 2005, Topuz & Okay 2008).

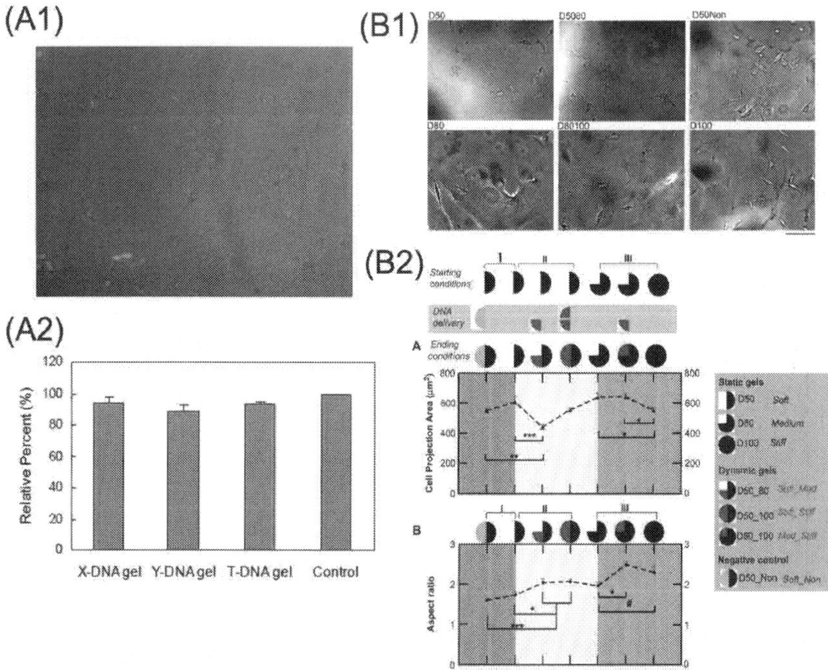

Fig. 9. Study of cellular behavior using DNA based macromaterials. (**A**) CHO cells in a DNA-only gel. Stained CHO cell encapsulated inside the gel (A1) and the majority of the cells were viable (A2). Extracted from (Um et al. 2006b) with publisher's permission. (**B**) L929 fibroblasts growth on dynamic substrates based on a DNA crosslinked hydrogel. (Upper panel) Typical morphology of L929 fibroblasts grown on DNA crosslinked hydrogels at Day 4, two days following DNA delivery. Scale bar is 100 mm. (Lower panel) Projection area and aspect ratio of L929 fibroblasts on DNA crosslinked hydrogels on Day 4. Extracted from (Jiang et al. 2010c) with publisher's permission.

Gels consisting of physically entangled DNA strands display resistance to DNase digestion (Lee et al. 2008). The hybrid between DNA strands and other polymer (such as SP-PPV (Tang et al. 2009)) also possess resistance to enzyme or heat. Thus, physical interactions between DNA strands and composite between DNA and other polymer may provide shield against enzymatic action.

4.2 Sequence design

Different to the majority of the materials based on DNA as backbones and some DNA-only gels where natural DNA (e.g., from salmon) was used, DNA crosslinked materials and DNA-only gels with designed DNA building blocks carry synthetic DNA. The sequence can be designed allowing added features. As we have discussed in the properties of DNA, the primary structure, i.e., the sequence or the order of nucleotides, of DNA primarily determines its secondary and tertiary structures, thus it is of significance to design sequence which gives the desired bulk material properties. Meanwhile, although two complementary DNA strands achieve their minimum energy state when they hybridize in the perfectly aligned configuration, DNA hybridization does not occur without error (Deaton et al. 1998). In addition, undesired interactions can occur between two strands as well as within a single strand. Such interactions typically involve the binding of complementary regions comprising only a small number of base pairs and include the formation of secondary structures such as the hairpin loop. In designing DNA sequences, it is desirable to decrease the number of possible mismatched hybridizations in order to maximize the efficiency of hybridization.

Design of a pair of equal-length sequences is essentially an optimization problem with the minimization of undesirable (e.g., off-alignment) interactions as the objective function. For instance, in the work by Lin et al. (Lin et al. 2004b), DNA sequence was generated by incorporating into the algorithm the following considerations: minimization of undesired interactions among strands and potential secondary structures, thermodynamic stability of the hybridized sequence pairs (e.g., GC content, terminal sequences, and hairpin structures (Lin et al. 2004b, SantaLucia & Hicks 2004)), and initiation of branch migration (e.g., length of sticky ends) (Deaton et al. 1998, Felsenfeld & Miles 2003). C-rich domain can be incorporated where it is desirable to have pH responsiveness (Cheng et al. 2009). It is noted, however, that due to the limitation on the current technology, and synthetic single-stranded DNA can have length up to 100 nt. The biological applications of these DNA crosslinked structures require additional caution. Examples include the ending sequence of the DNA strands, and the melting temperature needed to maintain the integrity of the DNA base structures.

4.3 Interaction between DNA and other entities

In the application of DNA based macro-materials for biology and medicine, DNA may potentially interact with an array of biologic entities such as protein, small molecules, other biopolymers, and endogenous DNA. The potential immunogenicity is also of concern. For dynamic DNA based macro-materials, the interactions between stimuli and DNA are also of interest. As an example, under physiological ion concentration, it was found that exchange between mono-(e.g., Na+) and bi-(e.g., Ca2+)valent cations affects volume, osmotic, and mechanical properties of a DNA gel consisting of DNA strands of ~2,000 bp. To avoid the unwanted biological effect, such as delivery DNA serving as anti-sense DNA, in the design of crosslinker DNA sequence, candidate sequences were screened by using a basic local alignment search tool (BLAST) algorithm which checks against the sequence in the genome of a specific specie and tissue type. To the same gel system, interactions between DNA aptamer and adenosine was explored as a way to initiate de-gelation (Yang et al. 2008), thus care needs to be exerted where such interactions are to be minimized in the presence of natural adenosine. Additionally, DNA strands can react with proteins and lipids (Liu et al. 2007). For example, DNA strands were reported to affect fibril formation of collagen matrix, and cation lipids (Liu et al. 2007). DNA-antibody interactions is another potential consideration in the design of DNA based macromaterials (Di Pietro et al. 2003). Of particular concern in the drug delivery applications are the drug-DNA interactions (Chaires & Waring 2001, Lu & Liu 2007).

4.4 Application of stimuli

Introduction of stimuli such as pH, ion concentration, or temperature may appear straight-forward, while it is potentially a concern for the delivery of large molecule such as ssDNA strands as cues. The kinetics and efficiency of delivery may be determined by the pore size of the structures, biochemical conditions, and interactions between DNA and other entities (e.g., soluble factors, inorganic compounds,) in the local microenvironment. To this end, more effective and delivery of ssDNA may be required in the clinical application. The thermal responses of the certain DNA gels merit attention due to the complexity in the changes of the material properties observed. For example, the alterations in materials properties induced by DNA denaturation and physical entanglement of resulting ssDNA may not be apparent (Topuz & Okay 2008).

4.5 Crosslinker parameters

In DNA as backbone gel system with EGDE as crosslinker, better stability but low dynamic moduli have been correlated to higher crosslinker content (Topuz & Okay 2008). Common design parameter for DNA based macro-material using synthetic DNA as crosslinker include DNA length and concentration and relative ratio of DNA and other components in the composite. Increased DNA concentration, or crosslinking density, causes materials to reach sol-gel transition and elevated mechanical stiffness beyond critical crosslinking density (Lin et al. 2004b). DNA length may be another design parameter, although its effect on bulk material properties was not noticeable when the length is in the 10 to 20 nt range (Jiang et al. 2008b). Lastly but not the least, one of the major hurdles of research and development of DNA based active materials using synthetic DNA is the relative high cost and limited availability of the synthetic forms of DNA (Jiang et al. 2008b, Lin et al. 2004b, Mangalam et al. 2009). Thus, this field of research awaits the development from other areas including synthetic chemistry and molecular biology to address this issue, and the trend has been towards the positive direction (Carlson 2009).

5. Outlook and potential directions

The progress outlined above has laid a solid foundation for the further development of the DNA based macro-material and for the further application of these materials in biology and medicine.

5.1 Inspiration from DNA nano-materials

The rapid development of DNA based nano-materials offers vast pool of ideas and hints, based on which novel macro-materials can be designed. For instance, from an ion-concentration based DNA-actuator (Fahlman et al. 2003), one could device a macromaterial based on formation of intermolecular guanine quartets. Another example is that DNA sliding, if tailored through the choice of base sequence in a periodic manner, may be useful in imparting unique properties to the resulting materials (Neher & Gerland 2005). Moreover, new stimuli for DNA nanostructures can be used for macromaterials. For instance, some DNA strands have been shown to interact with biometals (Goritz & Kramer 2005), thus the macromaterials constructed based on these DNA strands may have novel properties at the presence of physiological conditions. Other stimuli including light (Ogura et al. 2009), antibodies (Wiegel et al. 1987), proteins (Xie et al. 2007) and micelle (Ding et al. 2007) used in nanotechonlogy could be explored as the trigger for dynamic DNA based macro-materials. Along this line, DNA interstrand crosslinking from radical precursor independent of O2 (Greenberg 2005) may be of interest.

5.2 Refinement of the current designs

For DNA only system, multiple designs of the DNA building blocks can be incorporated for graded (with respect to time or location) control of the material properties. By combining physical and chemical crosslinking, gels with DNA backbones may achieve properties not seen in either system. Refinement of the DNA crosslinked hydrogel includes multi-step control by introducing multiple DNA crosslinker in a single system allowing multiple-step in increasing or decreasing crosslinking density. It also includes adding responsiveness to multiple cues by inclusion of DNA crosslinkers that are sensitive to stimuli including pH, temperature, and exogenous DNA strands. Responsiveness of these materials to different stimuli may be combined for the benefits of versatility and wider range of applications and control.

5.3 Force generation

It is possible to induce volume change of DNA based macromaterials as a way of generating forces in all three categories of DNA gel system (i.e., DNA-only, DNA as backbone, and DNA as crosslinker) (e.g., (Amiya & Tanaka 1987, Horkay & Basser 2004, Jiang et al. 2010c, Um et al. 2006b)) if the materials are implanted at injury site (e.g., spinal cord injury). This has been implicated to be useful in a myriad of applications including 'towed' (stretched) axonal regeneration (Bray 1984) in neural tissue engineering.

5.4 Dynamic porous scaffold

The porosity of the DNA-only gel system may be adjusted with DNA content and design for specific applications such as drug delivery or tissue engineering. For the DNA crosslinked macromaterials, the porosity and pore structure can be altered with the choice of crosslinking density, monomer concentration and monomer nature. For example, in constructing Acrydite-DNA crosslinked polymers (Jiang et al. 2008b, Lin et al. 2004b), the reactive end of the Acrydite-modified oligonucleotides contains vinyl group, thus besides polyacrylamide, poly-hydroxyethyl methacrylate (pHEMA), poly-hydroxy-propyl-methacrylamide(pHPMA), polymethyl methacrylate (pMMA), and copolymers (e.g. pHEMA-co-MMA and pHEMA-co-AEMA) are also candidates for DNA crosslinking (**Table 3**). These polymers are among the most studied non-biodegradable polymers for tissue engineering applications, including spinal cord injury research (Duconseille et al. 1998, Flynn et al. 2003, Lesny et al. 2002, Novikova et al. 2003) owing in part to their inhere biocompatibility (Ratner & Bryant 2004) and suitable pore size and porosity They have been engineered to carry neuro-trophic factors and present communicating porous structures (e.g., (Bakshi et al. 2004)), and to facilitate necrosis reduction, vasculature formation and axonal outgrowth across the graft-tissue interface (Dalton et al. 2002, Lesny et al. 2002, Yu & Shoichet 2005).

5.5 Controlled delivery

Previous work indicates that biomaterials based scaffold can provide enhanced gene delivery efficiency (De Laporte & Shea 2007). In the DNA crosslinked gel network, possibilities exist that by designing DNA sequence specific for an enzymatic action, the gel work can facilitate controlled release of the therapeutic reagent that is trapped in the gels (**Figure 9**). Venkatesh et al. illustrated that such enzymatic mechanism can be used for the delivery of DNA, though the release is not based on the change in the macro-material, but rather the by-product of restriction enzyme action (Venkatesh et al. 2009). Pore size and porosity of the gel network ought to design to facilitate such aim (Liedl et al. 2007).

5.6 DNA base-pairing

Although DNA's capability of binding complementary strands with high affinity is remarkable, it is not with limitations, as manifested in the errors in base-pairing and the hybridization kinetics (Condon 2006). While Nature has come up with elegant and complex machinery for error checking and correction in organisms, it remains a challenge in the synthetic DNA and it is much desirable to have such capability as well in the synthesis of DNA based active materials (Aldaye et al. 2008).

5.7 DNA modifications

It is promising to use DNA based materials as carrier for various protein-based therapeutics. For instance, biotin-labeled DNA (Kuzuya et al. 2009) can be incorporated in the gel network to attach streptavidin offering a means for protein separation, purification and potentially delivery. In this design, the 5′ end of the strands of DNA is biotinylated with a biotin-triethyleneglycol (TEG) residual. The effect of the environmental conditions and other factors on the biotin-streptavidin interactions could potentially be implemented as releasing mechanism. Realization of these promises hinges on the deep understanding of the DNA structure and properties and the interactions between DNA and others entities.

6. Concluding remark

Using DNA as a structural component has extended its functionality and significance from its critical biological roles, and has yielded DNA based macromaterials with DNA only, using DNA as backbone, and crosslinked by DNA. These DNA based macromaterials have benefited a great deal from the unique properties possessed by this molecule, and gained added functionalities and features such as thermal reversibility, sol-gel transition dependent on crosslinking density, and tunable mechanical stiffness. Currently, the application of these materials to the areas of bio-sensor/actuator, bioelectronics, drug delivery, and bioscaffold and tissue engineering is under investigation. There are a number of design considerations and parameters that are important to the success of applying these materials, which include stability, DNA sequence design, interactions between DNA and other molecules, and application of the stimuli in the materials. A wide range of applications await further development of these materials, particularly in the area of biology and medicine.

7. Acknowledgements

We want to express our appreciation to our collaborators in the previous projects, and apologize to those investigators whose work we were not able to cite appropriately due to space limitations.

Gel design		Properties/Features of the structures	Potential application	Ref
Backbone	**Crosslinker**			
DNA, branched structure (X, Y, T-shaped DNA)	DNA (Structure formation based on ligase-mediated reactions)	Tunable stiffness based on degree of branching Controllable swelling Biodegradable	Cell encapsulation Drug delivery Cell-free protein production	(Park et al 2009, Um et al 2006b)
Y shaped DNA strands	Interlocking motif based on protonation of C-rich domain	pH dependent stability; Fast responses to stimuli Sol-gel transition based upon DNA concentration	pH-sensitive drug-delivery system Tissue Engineering	(Cheng et al 2009)
Purified sodium DNA from calf thymus (~ 13,000 bp)	Physical entanglement between DNA strands	Entanglement occurs when reaching critical overlap concentration; Viscoelastic moduli dependence on DNA concentration	Understanding of biological processes such as mitosis	(Mason et al 1998)
Salmon DNA (20,000 bp)	Random physical entanglement between DNA strands	Stability for at least 3 months Resistance to DNase digestion	Drug delivery Tissue Engineering	(Lee et al 2008)
Salmon DNA (2000 bp) 9.3 % (w/w) in 4.0 mM sodium bromide solution at pH) 10.	ethylene glycol diglycidyl ether (EGDE). (epoxide groups on both ends react with amino group of base in nucleic acid)	At high cross-linker contents, no significant changes in the dynamic moduli were observed Physical gels exhibiting an elastic modulus in the order of MPa Strain hardening	Drug delivery Biosensor	(Orakdogen et al 2010, Topuz & Okay 2008, Topuz & Okay 2009)
Salmon DNA (2000 bp) 3 % (w/w)	EGDE	At low CaCl2 concentration, the gel volume gradually decreases as the CaCl2 concentration increases Exchange between mono- and di-valent cations causes volume change	Understanding of DNA-ion interactions under physiological conditions	(Horkay & Basser 2004)
Salmon DNA	electrostatic forces between DNA and poly(phenylenevinylene) (PPV) (Physical gel)	Swelling up to a hundred times, and Swelling degree decreases with increasing feed molar ratios of monomer to DNA and reaches a plateau Gel remains stable to heat, ultrasound, or DNase digestion	Drug delivery; Monitoring drug release	(Tang et al 2009)
Salmon DNA: 2000 bp 9 % (w/w)	EGDE (at different crosslinking density)	Shear-thinning behavior	Drug delivery; To study DNA–co-solute interactions;	(Costa et al 2006, Costa et al 2007, Costa et al 2006., Costa et al 2010)
Salmon DNA and cationic hydro-xyethyl celluloses based polymers	electrostatic forces between DNA and cations	Stochiometry, rheology shows a non-monotonic behavior with respect to charge ratio	Drug delivery	(Costa et al 2006, Costa et al 2006., Costa et al 2010)
Salmon DNA: 1000 bp	electrostatic forces between DNA and cationic surfactant or protein (e.g., lysozyme)	Different mechanisms for the interactions between ssDNA or dsDNA and cationic surfactant or protein Sustained DNA release	Gene delivery	(Moran et al 2007)

Row group labels: **DNA only** (rows 1–4), **DNA as backbone** (rows 5–10).

Table 1. Continued

Gel design		Properties/Features of the structures	Potential application	Ref
Backbone	**Crosslinker**			
Cellulose nanocrystals (CNXLs)	20 nt and 78 nt DNA	DNA grafting for ordered assembly of CNXLs Biocompatibility	Scaffold design for tissue engineering	(Mangalam et al 2009)
polyacrylamide	Thrombin associated ssDNA aptamer	Capture and release of thrombin based on its interaction with aptamer	Drug delivery vehicle Protein detection biosensor	(Wei et al 2008)
polyacrylamide	Bis and complementary DNA side chains	Optical property change due to alterations in DNA crosslinking	Biosensor	(Tierney & Stokke 2009)
polyacrylamide	The third DNA strand complementary to DNA side chain	Controlled drug release Quantum dots being trapped even when the average pore size is larger.	Drug delivery of nanoscale agents	(Liedl et al 2007)
Polyacrylamide	Aptamer	Adenosine induced gel dissociation Fast response	Drug delivery	(Yang et al 2008)
Polyacrylamide	Aptamer	Color change upon aptamer-target interactions	Visual detection platform Drug delivery	(Zhu et al 2010)
polyacrylamide	The third DNA strand complementary to DNA side chain	Temperature dependent gelation point; Tunable mechanical stiffness dependent on crosslinking density Reversible gelation via delivery of 'removal' DNA	Drug delivery Scaffold design for tissue engineering Substrate for study of cell-ECM interactions	(Jiang et al 2008b, Jiang et al 2010b, Jiang et al 2010c, Lin et al 2006, Lin et al 2004b)

(Row group label: **DNA as crosslinker**)

*Note: bp: base-pair for double-stranded DNA; nt: nucleotide for single-stranded DNA

Table 1. Summary of DNA based hydrogel materials.

Gel structure	Stimuli	Reversible?	Characterization	Potential Application	Ref
Y shaped building block and interlocking forces	pH	Y	Rheological properties AuNP release DNA structures using PAGE Gel structures using SEM	Drug delivery Bio-sensing	(Cheng et al 2009)
DNA strands and SP-PPV, and electrostatic force between the two	pH	TBD	Swelling degree Resistance to DNase	Drug delivery Monitoring drug release	(Tang et al 2009)
EGDE crosslinked long DNA strand	Temperature	N	Dynamic rheological measurements	Drug delivery Biosensor	(Topuz & Okay 2008)
DNA gel formation based on branched DNA nano-structure	nuclease	TBD (e.g., with nuclease)	AFM examination of pore structure DNA gel mass	Cell encapsulation and delivery	(Um et al 2006b)
DNA (20 bp) crosslinked polyacrylamide	ssDNA	Y	Indirect measurement Diffusion (Quantum dots tracing)	Drug delivery for nanoscale agents	(Liedl et al 2007)
DNA (10 and 20 bp) crosslinked polyacrylamide	ssDNA	Y	Gel mechanical properties (stiffness/ viscosity) Concentration of residual DNA outside gel network	Tissue engineering substrate and scaffold stem cell expansion platform Delivery vehicle	(Jiang et al 2010c, Lin et al 2006)
Thrombin associated aptamer as crosslinker	ssDNA	Y	DNA gel electrophoresis Viscosity	Drug delivery Controlled release	(Wei et al 2008)
DNA (10 and 20 bp) crosslinked polyacrylamide	Adenosine	Y	Absorbance of gold nanoparticles in gels	Drug delivery (target specific)	(Yang et al 2008)
DNA crosslinked polyacrylamide	Target interacting with aptamer	Y	Gel color change	Visual detection Drug delivery	(Zhu et al 2010)

Note: AuNP: Gold nano-particle; ssDNA: single-stranded DNA. PAGE: polyacrylamide gel electrophoresis. SEM: scanning electron microscopy. TBD: To be determined

Table 2. Summary of DNA base materials with dynamic and responsive properties.

DNA X-linked polymer	Gel preparation	Chemical structure	Notes
pHEMA, poly-(hydroxyethyl methacrylate) (Bakshi et al. 2004, Carone & Hasenwinkel 2006, Flynn et al. 2003)	Crosslink: L2 (replacing EGDMA) HEMA (Aldrich, St. Louis, MO) (20%-60%) Initiator-catalyst (APS-TEMED)		Pore size: 10 μm to 20 μm
pHPMA, poly-(hydroxypropyl-methacrylamide) (Duconseille et al. 1998, St'astny et al. 2002)	Crosslink: L2 (replacing DMHA) HPMA monomer Initiator (AIBN)		Pore size: 10 μm to 50 μm

Table 3. DNA crosslinked vinyl polymers such as pHEMA and pHPMA.

8. Abbreviation

dsDNA: double-stranded DNA; ssDNA: single-stranded DNA; ECM: extracellular matrix; bp: basepair; nt: nucleotide.

9. Reference

Aldaye FA, Palmer AL, Sleiman HF. 2008. Assembling materials with DNA as the guide. *Science* 321: 1795-9

Alemdaroglu FE, Alemdaroglu NC, Langguth P, Herrmann A. 2008. DNA Block Copolymer Micelles - A Combinatorial Tool for Cancer Nanotechnology. *Advanced Materials* 20: 899-902

Alemdaroglu FE, Herrmann A. 2007. DNA meets synthetic polymers--highly versatile hybrid materials. *Org Biomol Chem* 5: 1311-20

Amiya T, Tanaka T. 1987. Phase Transitions in Cross-Linked Gels of Natural Polymers. *Macromolecules* 20: 1162-64

Bakshi A, Fisher O, Dagci T, Himes BT, Fischer I, Lowman A. 2004. Mechanically engineered hydrogel scaffolds for axonal growth and angiogenesis after transplantation in spinal cord injury. *J Neurosurg Spine* 1: 322-9

Ball P. 2005. Synthetic biology for nanotechnology. *Nanotechnology* 16: R1-R8

Bart Haegeman DV, Jean-Jacques Godon and Jérôme Hamelin. 2008. DNA reassociation kinetics and diversity indices: richness is not rich enough. *Oikos*

Berashevich J, Chakraborty T. 2008. How the surrounding water changes the electronic and magnetic properties of DNA. *J Phys Chem B* 112: 14083-9

Biggins JB, Prudent JR, Marshall DJ, Thorson JS. 2006. A continuous assay for DNA cleavage using molecular break lights. *Methods Mol Biol* 335: 83-92

Blagbrough IS, Zara C. 2009. Animal models for target diseases in gene therapy—Using DNA and siRNA delivery strategies. *Pharmaceutical research* 26: 1-18

Braun E, Keren K. 2004. From DNA to transistors. *Advances in Physics* 53: 441 — 96

Bray D. 1984. Axonal growth in response to experimentally applied mechanical tension. *Dev Biol* 102: 379-89

Budowle B, Allard MW, Wilson MR, Chakraborty R. 2003. Forensics and mitochondrial DNA: applications, debates, and foundations. *Annu Rev Genomics Hum Genet* 4: 119-41

Bustamante C, Bryant Z, Smith SB. 2003. Ten years of tension: single-molecule DNA mechanics. *Nature* 421: 423-7

Cao S, Cripps A, Wei MQ. 2010. New strategies for cancer gene therapy: Progress and opportunities. *Clinical and Experimental Pharmacology and Physiology* 37: 108-14

Carlson R. 2009. The changing economics of DNA synthesis. *Nat Biotechnol* 27: 1091-4

Carone TW, Hasenwinkel JM. 2006. Mechanical and morphological characterization of homogeneous and bilayered poly(2-hydroxyethyl methacrylate) scaffolds for use in CNS nerve regeneration. *J Biomed Mater Res B Appl Biomater*

Chaires J, Waring M. 2001. *Drug-nucleic acid interactions.* Academic press.

Chan G, Mooney DJ. 2008. New materials for tissue engineering: towards greater control over the biological response. *Trends in biotechnology* 26: 382-92

Chen CS, Jiang X, Whitesides GM. 2005. Microengineering the Environment of Mammalian Cells in Culture. *MRS Bulletin* 30: 194-201

Chen JH, Seeman NC. 1991. Synthesis from DNA of a molecule with the connectivity of a cube. *Nature* 350: 631-3

Cheng E, Xing Y, Chen P, Yang Y, Sun Y, et al. 2009. A pH-triggered, fast-responding DNA hydrogel. *Angew Chem Int Ed Engl* 48: 7660-3

Chevalier E, Chulia D, Pouget C, Viana M. 2008. Fabrication of porous substrates: a review of processes using pore forming agents in the biomaterial field. *J Pharm Sci* 97: 1135-54

Chhabra R, Sharma J, Liu Y, Rinker S, Yan H. 2010. DNA Self-assembly for Nanomedicine. *Advanced Drug Delivery Reviews* 62: 617-25

Chippada U, Langrana N, Yurke B. 2009a. Complete mechanical characterization of soft media using nonspherical rods. *J Appl Phys* 106: 63528

Chippada U, Yurke B, Georges PC, Langrana NA. 2009b. A nonintrusive method of measuring the local mechanical properties of soft hydrogels using magnetic microneedles. *J Biomech Eng* 131: 021014

Condon A. 2006. Designed DNA molecules: principles and applications of molecular nanotechnology. *Nat Rev Genet* 7: 565-75

Costa D, Hansson P, Schneider S, GraÃ§a Miguel M, Lindman Br. 2006. Interaction between Covalent DNA Gels and a Cationic Surfactant. *Biomacromolecules* 7: 1090-95

Costa D, Miguel MG, Lindman B. 2007. Responsive polymer gels: double-stranded versus single-stranded DNA. *J Phys Chem B* 111: 10886-96

Costa D, Santos Sd, Antunes FE, Miguel MG, Lindman B. 2006. Some novel aspects of DNA physical and chemical gels. *ARKIVOC* 4: 161-72

Costa D, Valente AJM, Pais AACC, Miguel MG, Lindman B. 2010. Cross-linked DNA gels: Disruption and release properties. *Colloids and Surfaces A: Physicochemical and Engineering Aspects* 354: 28-33

Dalton PD, Flynn L, Shoichet MS. 2002. Manufacture of poly(2-hydroxyethyl methacrylate-co-methyl methacrylate) hydrogel tubes for use as nerve guidance channels. *Biomaterials* 23: 3843-51

De Laporte L, Shea LD. 2007. Matrices and scaffolds for DNA delivery in tissue engineering. *Adv Drug Deliv Rev* 59: 292-307

Deaton R, Garzon M, Murphy RC, Rose JA, Franceschetti DR, Stevens SE. 1998. Reliability and Efficiency of a DNA-Based Computation. *Physical Review Letters* 80: 417

Dhillon SS, Rake AV, Miksche JP. 1980. Reassociation Kinetics and Cytophotometric Characterization of Peanut (Arachis hypogaea L.) DNA. *Plant Physiol* 65: 1121-27

Di Pietro SM, Centeno JM, Cerutti ML, Lodeiro MF, Ferreiro DU, et al. 2003. Specific antibody-DNA interaction: a novel strategy for tight DNA recognition. *Biochemistry* 42: 6218-27

Ding K, Alemdaroglu FE, Borsch M, Berger R, Herrmann A. 2007. Engineering the structural properties of DNA block copolymer micelles by molecular recognition. *Angew Chem Int Ed Engl* 46: 1172-5

Dong Liu X, Yamada M, Matsunaga M, Nishi N. 2007. Functional Materials Derived from DNA. *Functional Materials and Biomaterials*: 149-78

Douglas SM, Dietz H, Liedl T, Hogberg B, Graf F, Shih WM. 2009. Self-assembly of DNA into nanoscale three-dimensional shapes. *Nature* 459: 414-8

Duconseille E, Woerly S, Kelche C, Will B, Cassel JC. 1998. Polymeric hydrogels placed into a fimbria-fornix lesion cavity promote fiber (re)growth: a morphological study in the rat. *Restor Neurol Neurosci* 13: 193-203

Engler AJ, Sen S, Sweeney HL, Discher DE. 2006. Matrix elasticity directs stem cell lineage specification. *Cell* 126: 677-89

Fahlman RP, Hsing M, Sporer-Tuhten CS, Sen D. 2003. Duplex Pinching: A Structural Switch Suitable for Contractile DNA Nanoconstructions. *Nano Letters* 3: 1073-78

Felsenfeld G, Miles HT. 2003. The Physical and Chemical Properties of Nucleic Acids. *Annual Review of Biochemistry* 36: 407-48

Flynn L, Dalton PD, Shoichet MS. 2003. Fiber templating of poly(2-hydroxyethyl methacrylate) for neural tissue engineering. *Biomaterials* 24: 4265-72

Georges PC, Hui JJ, Gombos Z, McCormick ME, Wang AY, et al. 2007. Increased stiffness of the rat liver precedes matrix deposition: implications for fibrosis. *Am J Physiol Gastrointest Liver Physiol* 293: G1147-54

Gerweck LE, Seetharaman K. 1996. Cellular pH gradient in tumor versus normal tissue: potential exploitation for the treatment of cancer. *Cancer Res* 56: 1194-8

Ghosh K, Ingber DE. 2007. Micromechanical control of cell and tissue development: implications for tissue engineering. *Advanced Drug Delivery Reviews* 59: 1306-18

Goritz M, Kramer R. 2005. Allosteric control of oligonucleotide hybridization by metal-induced cyclization. *J Am Chem Soc* 127: 18016-7

Greenberg MM. 2005. DNA interstrand cross-links from modified nucleotides: mechanism and application. *Nucleic Acids Symp Ser (Oxf)*: 57-8

Hildebrandt-Eriksen ES, Christensen T, Diemer NH. 2002. Mild focal cerebral ischemia in the rat. The effect of local temperature on infarct size. *Neurol Res* 24: 781-8

Horkay F, Basser PJ. 2004. Osmotic observations on chemically cross-linked DNA gels in physiological salt solutions. *Biomacromolecules* 5: 232-7

Ingber DE. 2002. Mechanical signaling and the cellular response to extracellular matrix in angiogenesis and cardiovascular physiology. *Circ Res* 91: 877-87

Ingber DE. 2006. Cellular mechanotransduction: putting all the pieces together again. *The FASEB journal* 20: 811

Jain A, Alvi NK, Parish JH, Hadi SM. 1996. Oxygen is not required for degradation of DNA by glutathione and Cu(II). *Mutation Research/Fundamental and Molecular Mechanisms of Mutagenesis* 357: 83-88

Janmey PA, Winer JP, Murray ME, Wen Q. 2009. The hard life of soft cells. *Cell Motil Cytoskeleton* 66: 597-605

Jiang FX, Yurke B, Firestein BL, Langrana NA. 2008a. Neurite outgrowth on a DNA crosslinked hydrogel with tunable stiffnesses. *Annals of Biomedical Engineering* 36: 1565-79

Jiang FX, Yurke B, Firestein BL, Langrana NA. 2008b. Neurite outgrowth on a DNA crosslinked hydrogel with tunable stiffnesses. *Ann Biomed Eng* 36: 1565-79

Jiang FX, Yurke B, Schloss RS, Firestein BL, Langrana NA. 2010a. Effect of Dynamic Stiffness of the Substrates on Neurite Outgrowth by Using a DNA-Crosslinked Hydrogel. *Tissue Engineering Part A* 16: 1873-89

Jiang FX, Yurke B, Schloss RS, Firestein BL, Langrana NA. 2010b. Effect of dynamic stiffness of the substrates on neurite outgrowth by using a DNA-crosslinked hydrogel. *Tissue Eng Part A* (In press)

Jiang FX, Yurke B, Schloss RS, Firestein BL, Langrana NA. 2010c. The relationship between fibroblast growth and the dynamic stiffnesses of a DNA crosslinked hydrogel. *Biomaterials* 31: 1199-212

Kang H, Liu H, Zhang X, Yan J, Zhu Z, et al. 2011. Photoresponsive DNA-Cross-Linked Hydrogels for Controllable Release and Cancer Therapy. *Langmuir*

Knoblauch M, Peters W. 2004. Biomimetic actuators: where technology and cell biology merge. *Cellular and molecular life sciences* 61: 2497-509

Kuzuya A, Kimura M, Numajiri K, Koshi N, Ohnishi T, et al. 2009. Precisely programmed and robust 2D streptavidin nanoarrays by using periodical nanometer-scale wells embedded in DNA origami assembly. *Chembiochem* 10: 1811-5

LaBean TH, Li H. 2007. Constructing novel materials with DNA. *Nano Today* 2: 26-35

Lahann J, Langer R. 2005. Smart Materials with Dynamically Controllable Surfaces. *MRS Bulletin* 30: 185-88

Lee CK, Shin SR, Lee SH, Jeon JH, So I, et al. 2008. DNA hydrogel fiber with self-entanglement prepared by using an ionic liquid. *Angew Chem Int Ed Engl* 47: 2470-4

Lesny P, De Croos J, Pradny M, Vacik J, Michalek J, et al. 2002. Polymer hydrogels usable for nervous tissue repair. *J Chem Neuroanat* 23: 243-7

Letchworth GJ, Carmichael LE. 1984. Local tissue temperature: a critical factor in the pathogenesis of bovid herpesvirus 2. *Infect Immun* 43: 1072-9

Li R, McCoy BJ, Diemer RB. 2005. Cluster aggregation and fragmentation kinetics model for gelation. *J Colloid Interface Sci* 291: 375-87

Li S, De Wijn JR, Li J, Layrolle P, De Groot K. 2003. Macroporous biphasic calcium phosphate scaffold with high permeability/porosity ratio. *Tissue Eng* 9: 535-48

Li Y, White J, Stokes D, Sayler G, Sepaniak M. 2001. Capillary electrophoresis as a method to study DNA reassociation. *Biotechnol Prog* 17: 348-54

Liedl T, Dietz H, Yurke B, Simmel F. 2007. Controlled trapping and release of quantum dots in a DNA-switchable hydrogel. *Small* 3: 1688-93

Lin CC, Metters AT. 2006. Hydrogels in controlled release formulations: network design and mathematical modeling. *Adv Drug Deliv Rev* 58: 1379-408

Lin D, Langrana N, Yurke B. 2005. Inducing reversible stiffness changes in DNA-crosslinked gels. *Journal of Materials Research* 20: 1456-64

Lin DC. 2005. *Design and properties of a new dna-crosslinked polymer hydrogel*. Rutgers University, Piscataway, NJ

Lin DC, Yurke B, Langrana NA. 2004a. Mechanical properties of a reversible, DNA-crosslinked polyacrylamide hydrogel. *Journal of Biomechanical Engineering* 126: 104

Lin DC, Yurke B, Langrana NA. 2004b. Mechanical properties of a reversible, DNA-crosslinked polyacrylamide hydrogel. *J Biomech Eng* 126: 104-10

Liu D, Wang M, Deng Z, Walulu R, Mao C. 2004. Tensegrity: construction of rigid DNA triangles with flexible four-arm DNA junctions. *J Am Chem Soc* 126: 2324-5

Liu XD, Yamada M, Matsunaga M, Nishi N. 2007. Functional Materials Derived from DNA In *Functional Materials and Biomaterials*, pp. 149-78

Lu Y, Liu J. 2006. Functional DNA nanotechnology: emerging applications of DNAzymes and aptamers. *Curr Opin Biotechnol* 17: 580-8

Lu Y, Liu J. 2007. Smart nanomaterials inspired by biology: dynamic assembly of error-free nanomaterials in response to multiple chemical and biological stimuli. *Acc Chem Res* 40: 315-23

Luo D. 2003. The road from biology to materials. *Materials Today* 6: 38-43

Mangalam AP, Simonsen J, Benight AS. 2009. Cellulose/DNA Hybrid Nanomaterials. *Biomacromolecules*

Mao C, Sun W, Seeman NC. 1997. Assembly of Borromean rings from DNA. *Nature* 386: 137-8

Marquez JP, Genin GM, Pryse KM, Elson EL. 2006. Cellular and matrix contributions to tissue construct stiffness increase with cellular concentration. *Ann Biomed Eng* 34: 1475-82

Mason TG, Dhople A, Wirtz D. 1998. Linear Viscoelastic Moduli of Concentrated DNA Solutions. *Macromolecules* 31: 3600-03

Mirkin CA, Letsinger RL, Mucic RC, Storhoff JJ. 1996. A DNA-based method for rationally assembling nanoparticles into macroscopic materials. *Nature* 382: 607-9

Moran MC, Miguel MG, Lindman B. 2007. DNA gel particles: particle preparation and release characteristics. *Langmuir* 23: 6478-81

Mrksich M. 2005. Dynamic Substrates for Cell Biology. *MRS Bulletin* 30: 180-84

Murakami Y, Maeda M. 2005. DNA-responsive hydrogels that can shrink or swell. *Biomacromolecules* 6: 2927-9

Nagahara S, Matsuda T. 1996. Hydrogel formation via hybridization of oligonucleotides derivatized in water-soluble vinyl polymers. *Polym Gels Networks* 4: 111-27

Neher RA, Gerland U. 2005. DNA as a programmable viscoelastic nanoelement. *Biophys J* 89: 3846-55

Nemir S, West JL. 2009. Synthetic Materials in the Study of Cell Response to Substrate Rigidity. *Ann Biomed Eng*

Niemeyer CM. 2000. Self-assembled nanostructures based on DNA: towards the development of nanobiotechnology. *Curr Opin Chem Biol* 4: 609-18

Novikova LN, Novikov LN, Kellerth JO. 2003. Biopolymers and biodegradable smart implants for tissue regeneration after spinal cord injury. *Curr Opin Neurol* 16: 711-5

Ogura Y, Nishimura T, Tanida J. 2009. Self-Contained Photonically-Controlled DNA Tweezers. *Applied Physics Express* 2: 025004

Orakdogen N, Erman B, Okay O. 2010. Evidence of Strain Hardening in DNA Gels. *Macromolecules* 43: 1530–38

Park N, Um SH, Funabashi H, Xu J, Luo D. 2009. A cell-free protein-producing gel. *Nat Mater* 8: 432-7

Patil SD, Rhodes DG, Burgess DJ. 2005. DNA-based therapeutics and DNA delivery systems: a comprehensive review. *The AAPS Journal* 7: 61-77

Previtera ML, Trout K, Chippada U, Schloss R, Langrana NA. 2011. *Fibroblast behavior on tunable gels with decreasing elasticity.* Presented at ASME 2011 Summer Bioengineering Conference.

Ratner BD, Bryant SJ. 2004. BIOMATERIALS: Where We Have Been and Where We are Going. *Annual Review of Biomedical Engineering* 6: 41-75

Reynaldo LP, Vologodskii AV, Neri BP, Lyamichev VI. 2000. The kinetics of oligonucleotide replacements. *J Mol Biol* 297: 511-20

Ritter T. 2009. Gene therapy in transplantation: Toward clinical trials. *Current Opinion in Molecular Therapeutics* 11: 504-12

Roberts MC, Hanson MC, Massey AP, Karren EA, Kiser PF. 2007. Dynamically Restructuring Hydrogel Networks Formed with Reversible Covalent Crosslinks. *Advanced Materials* 19: 2503-07

Rothemund PW. 2006. Folding DNA to create nanoscale shapes and patterns. *Nature* 440: 297-302

Saeks J. 2007. DNA Sequencing Equipment and Services Markets, Kalorama Information, New York, NY

SantaLucia J, Jr., Hicks D. 2004. The thermodynamics of DNA structural motifs. *Annu Rev Biophys Biomol Struct* 33: 415-40

Schneider HJ, Strongin RM. 2009. Supramolecular interactions in chemomechanical polymers. *Acc Chem Res* 42: 1489-500

Seeman NC. 1981. *Nucleic Acid Junctions: Building Blocks for Genetic Engineering in Three Dimensions.* pp. 269-277. New York: Adenine Press.

Seeman NC. 1982. Nucleic acid junctions and lattices. *J Theor Biol* 99: 237-47

Seeman NC. 2007. An overview of structural DNA nanotechnology. *Mol Biotechnol* 37: 246-57

Silver FH, DeVore D, Siperko LM. 2003. Invited Review: Role of mechanophysiology in aging of ECM: effects of changes in mechanochemical transduction. *J Appl Physiol* 95: 2134-41

Simmel FC, Yurke B. 2001. Using DNA to construct and power a nanoactuator. *Physical Review E* 63: 041913

Smith MJ, Britten RJ, Davidson EH. 1975. Studies on nucleic acid reassociation kinetics: reactivity of single-stranded tails in DNA-DNA renaturation. *Proc Natl Acad Sci U S A* 72: 4805-9

Smith SB, Cui Y, Bustamante C. 1996. Overstretching B-DNA: the elastic response of individual double-stranded and single-stranded DNA molecules. *Science* 271: 795-9

Soontornworajit B, Zhou J, Shaw MT, Fan TH, Wang Y. 2010. Hydrogel functionalization with DNA aptamers for sustained PDGF-BB release. *Chem Commun (Camb)* 46: 1857-9

St'astny M, Plocova D, Etrych T, Kovar M, Ulbrich K, Rihova B. 2002. HPMA-hydrogels containing cytostatic drugs. Kinetics of the drug release and in vivo efficacy. *J Control Release* 81: 101-11

Storhoff JJ, Mirkin CA. 1999. Programmed Materials Synthesis with DNA. *Chemical Reviews* 99: 1849-62

Sun M, Pejanovic S, Mijovic J. 2005. Dynamics of Deoxyribonucleic Acid Solutions As Studied by Dielectric Relaxation Spectroscopy and Dynamic Mechanical Spectroscopy. *Macromolecules* 38: 9854-64

Tang H, Duan X, Feng X, Liu L, Wang S, et al. 2009. Fluorescent DNA-poly(phenylenevinylene) hybrid hydrogels for monitoring drug release. *Chem Commun (Camb)*: 641-3

Tierney S, Stokke BT. 2009. Development of an oligonucleotide functionalized hydrogel integrated on a high resolution interferometric readout platform as a label-free macromolecule sensing device. *Biomacromolecules* 10: 1619-26

Topuz F, Okay O. 2008. Rheological Behavior of Responsive DNA Hydrogels. *Macromolecules* 41: 8847-54

Topuz F, Okay O. 2009. Formation of Hydrogels by Simultaneous Denaturation and Cross-Linking of DNA. *Biomacromolecules* 10: 2652-61

Uibo R, Laidmae I, Sawyer ES, Flanagan LA, Georges PC, et al. 2009. Soft materials to treat central nervous system injuries: evaluation of the suitability of non-mammalian fibrin gels. *Biochim Biophys Acta* 1793: 924-30

Um SH, Lee JB, Park N, Kwon SY, Umbach CC, Luo D. 2006a. Enzyme-catalysed assembly of DNA hydrogel. *Nature Materials* 5: 797-801

Um SH, Lee JB, Park N, Kwon SY, Umbach CC, Luo D. 2006b. Enzyme-catalysed assembly of DNA hydrogel. *Nat Mater* 5: 797-801

Venkatesh S, Wower J, Byrne ME. 2009. Nucleic acid therapeutic carriers with on-demand triggered release. *Bioconjug Chem* 20: 1773-82

Vologodskii AV, Cozzarelli NR. 1994. Conformational and thermodynamic properties of supercoiled DNA. *Annu Rev Biophys Biomol Struct* 23: 609-43

Wang K, Tang Z, Yang CJ, Kim Y, Fang X, et al. 2009. Molecular engineering of DNA: molecular beacons. *Angew Chem Int Ed Engl* 48: 856-70

Wang YL, Pelham RJ, Jr. 1998. Preparation of a flexible, porous polyacrylamide substrate for mechanical studies of cultured cells. *Methods Enzymol* 298: 489-96

Watson JD, Baker TA, Bell SP, Gann A, Levine MA, Losick R. 2003. *Molecular Biology of the Gene*. Benjamin Cummings.

Wei B, Cheng I, Luo KQ, Mi Y. 2008. Capture and release of protein by a reversible DNA-induced sol-gel transition system. *Angew Chem Int Ed Engl* 47: 331-3

Wiegel FW, Geurts BJ, Goldstein B. 1987. Crosslinking and gelation between linear polymers: DNA-antibody complexes in systemic lupus erythematosus. *Journal of Physics A: Mathematical and General* 20: 5205-18

Xie J, Fan R, Meng Z. 2007. Protein oxidation and DNA-protein crosslink induced by sulfur dioxide in lungs, livers, and hearts from mice. *Inhal Toxicol* 19: 759-65

Xing Y, Cheng E, Yang Y, Chen P, Zhang T, et al. 2011. DNA HYDROGELS: Self-Assembled DNA Hydrogels with Designable Thermal and Enzymatic Responsiveness. *Advanced Materials* 23: 1116-16

Yang H, Liu H, Kang H, Tan W. 2008. Engineering target-responsive hydrogels based on aptamer-target interactions. *J Am Chem Soc* 130: 6320-1

Yu TT, Shoichet MS. 2005. Guided cell adhesion and outgrowth in peptide-modified channels for neural tissue engineering. *Biomaterials* 26: 1507-14

Yurke B, Mills A. 2003. Using DNA to Power Nanostructures. *Genetic Programming and Evolvable Machines* 4: 111-22

Yurke B, Turberfield AJ, Mills AP, Jr., Simmel FC, Neumann JL. 2000. A DNA-fuelled molecular machine made of DNA. *Nature* 406: 605-8

Zhang Y, Gu H, Yang Z, Xu B. 2003. Supramolecular hydrogels respond to ligand-receptor interaction. *Journal of the American Chemical Society* 125: 13680-81

Zhang Y, Seeman NC. 1994. Construction of a DNA-truncated octahedron. *J Am Chem Soc,* 116: 1661-69

Zhu Z, Wu C, Liu H, Zou Y, Zhang X, et al. 2010. An aptamer cross-linked hydrogel as a colorimetric platform for visual detection. *Angew Chem Int Ed Engl* 49: 1052-6

3

Porous Apatite Coating on Various Titanium Metallic Materials via Low Temperature Processing

Takamasa Onoki

Department of Materials Science, Graduate School of Engineering,
Osaka Prefecture University
Japan

1. Introduction

Nowadays, hydorxyapatite (HA) is widely used as bioceramics in reconstructive surgery, in dentistry and as drug delivery materials due to the good biocompatibility and osteoconductivity [Hench, 1998]. One of the limitations for the usage of these materials is their low mechanical strength. Thus, many researchers focuses on the development of new biomaterials, that combine the osteoconductive characteristics of bioactive ceramics with sufficient strength and toughness for load-bearing applications. The combination of high strength of the metals with osteoconductive properties of bioactive ceramics makes HA coated metallic implants, which titanium (Ti) or its alloys was mainly used, very attractive for the loaded-bearing applications in orthopedic and dental surgery[Long & Rack, 1998]. A plasma spraying method has been conventionally employed for the HA coating. However, this method has some problems (e.g. a poor coating-substrate adherence, lack of HA crystallinity) for the long-term performance and lifetime of the implants [Aoki, 1994]. Therefore, new HA coating methods hava attracted great interests in recent years for replacing the high temperature techniques like plasma spraying.

Hydrothermal hot-pressing (HHP) method is a possible processing route for producing a ceramic body at relatively low temperatures (under 300°C) [Yamasaki et al., 1986]. The compression of samples under hydrothermal conditions accelerates densification of inorganic materials. It is known that the water of crystallization in calcium hydrogen phosphate dihydrate (CaHPO$_4$ · 2H$_2$O; DCPD) is slowly lost below 100°C [Peelen et al., 1991]. If the released water can be utilized as a reaction solvent during the HHP treatment, it is to be expected that the joining HA to metal can be achieved simultaneously under the hydrothermal condition, in addition to the synthesis and solidification of HA through the chemical reaction as follows [Hosoi et al., 1996]:

$$6CaHPO_4 \cdot 2H_2O + 4Ca(OH)_2 \rightarrow Ca_{10}(PO_4)_6(OH)_2 + 18H_2O \qquad (1)$$

In our previous reports, we have proposed a HHP method for bonding HA ceramics and pure Ti, Ti alloys, Magnesium alloy, and Ti based bulk metallic glass by hydrothermal hot-pressing techniques [Onoki et al., 2003a, 2003b, 2005, 2006, 2008a, 2008b 2009a, 2009b, 2010a, 2010b, 2010c, 2011].

Ti-based alloys are beneficial for biomedical applications due to their low density, excellent biocompatibility, and corrosion resistance. Combining the advantages of both bulk metallic glass and Ti-based alloy, Ti-based bulk metallic glasses are expected to be applied as a new type of biomaterial. However, it is well-known that the surface of bulk metallic glasses, which are bioinert, must be bioactive to use as bone replacing medical/dental materials as well as Ti and its alloys. Recently, it was reported a concept called "Growing Integrated Layer" [GIL] that improves adhesion performance without cracking and peeling the ceramic coatings [6]. In particular, if the metallic glass or alloy contains a very reactive component like Ti, it can grow on the bulk metallic materials with its "root" in the bulk. This was named the "Growing Integrated Layer" or "Graded Intermediate Layer [GIL]" and the "Growing Integration Process [GIP]" for its formation process. Multiple layered, laminated, integrated, graded, and diffused coatings have been investigated to decrease the stress accumulation, which however, it is not easy, particularly when the interface is sharp. Even widely diffused interface(s) of larger micron sizes are preferable for joining and coating bulk ceramics on metallic materials. Such a GIL of oxide films grown from the "seed," i.e., the most reactive component in the bulk metallic materials is interesting as a novel process of oxide film formations, especially because the oxide film can be fabricated in a solution at such low temperatures as RT-200°C when chemical and/or electrochemical potentials are added. Thermal stress accumulation can be avoided in those low temperature formations of the ceramic film on the metal.

In this chapter, some hydrothermal technologies as low temperarure process were described as HA ceramics bonding/coating methods and surface modification of various Ti metallic materials, especially pure Ti and Ti-based bulk metallic glass.

2. Bonding HA ceramics and pure Ti

2.1 Hyrothermal hot-pressing

In this study, DCPD used as a starting powder was prepared by mixing 1.0M calcium nitrate solution (99.0%; Ca(NO)$_3$ · 4H$_2$O, KANTO CHEMICAL CO., INC., Japan) and 1.0M diammonium hydrogen phosphate solution (98.5%; (NH$_4$)$_2$HPO$_4$; KANTO CHEMICAL CO., INC., Japan). The mixing was carried out at a room temperature (approximately 20°C). In order to control the pH value of the mixing solution, acetic acid (99.5%; KANTO CHEMICAL CO., INC., Japan) and ammonia solution (28.0-30.0%; KANTO CHEMICAL CO., INC., Japan) were added. The value of pH was kept around 8.5 initially, and then changed to 6.0 using the acetic acid and ammonia solution after the mixing in order to prevent the formation of impurities and to produce pure DCPD powders. It has been shown from preliminary tests that when no control of the pH was conducted the synthesized DCPD contained the impurities such as CaHPO$_4$ (DCPA), and amorphous calcium phosphate (ACP). The precipitate from the mixture was filtered and washed with deionized water and acetone. The washed filter cake was oven-dried at 50°C for 24 hours, and then the dried cake was ground to a powder. No impurity in the synthesized DCPD used was detected by powder X-ray diffraction. The synthetic DCPD and calcium hydroxide (95.0%; Ca(OH)$_2$; KANTO CHEMICAL CO., INC., Japan) were mixed in a mortar for 60min with a Ca/P ratio of 1.67 which was stoichiometric ratio of HA.

A commercially available pure Ti rod (Nilaco, Japan, diameter: 20mm, thickness: 10mm, purity: 99.5% JIS Grade1), 20mm in diameter, was used in this experiment. The Ti rod was cut into disks with a thickness of 10mm. The disks were cleaned in deionized water and acetone by using an ultrasonic cleaner. The Ti surfaces were finished using 1500# emery paper. After the surface finish with emery paper, the titanium disks were washed again by deionized water, and then dried in air. The powder mixture and Ti disks were placed into the middle of the autoclave simultaneously, as shown in Fig.1.

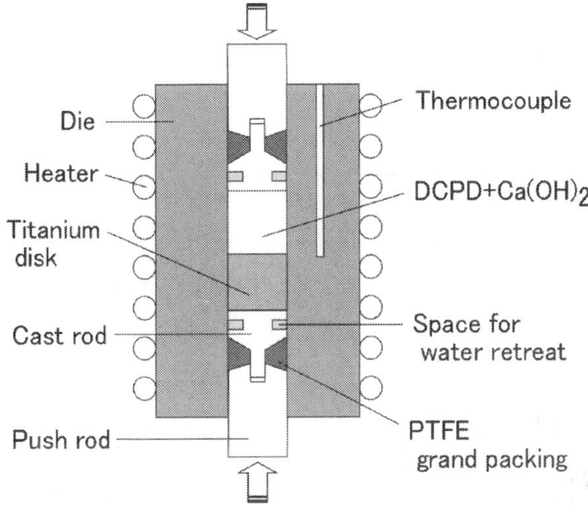

Fig. 1. Schematic illustration of the autoclave for Hydrothermal Hot-Pressing(HHP).

The autoclave made of steel has a pistons-cylinder structure with an inside diameter of 20 mm. The pistons possess escape space for hydrothermal solution squeezed from the sample, and this space regulates the appropriate hydrothermal conditions in the sample. A grand packing of polytetrafluoroethylene (PTFE) is placed between a cast rod and push rod. The PTFE was used to prevent leakage of the hydrothermal solutions. The stainless steel (SUS304) autoclave has pistons within a cylindrical structure with an inside diameter of 20 mm. The pistons enable the hydrothermal solution squeezed from the sample to escape, and this regulates the appropriate hydrothermal conditions in the sample. Polytetrafluoroethylene (PTFE) is packed between a cast rod and a push rod. The PTFE was used to prevent leakage of the hydrothermal solutions. Pressure of 40 MPa was initially applied to the sample through the push rods from the top and bottom at room temperature. After initial loading the autoclave was heated to 150°C at 10°C/min with a sheath-type heater, and then the temperature was kept constant for two hours. The axial pressure was kept at 40 MPa during the hydrothermal hot-pressing treatment. After the HHP treatment, the autoclave was naturally cooled to room temperature, and the sample was removed from the autoclave.

2.2 Adhesion properties evaluation

3-point bending tests were conducted to obtain an estimate of the fracture toughness for the HA/Ti interface as well as for the HA ceramics made by the HHP method. Core-based specimens were used for the fracture toughness tests following the ISRM suggested method[Hashida, 1993]. The configuration of the core-based specimen is schematically shown in Fig.2. In order to measure the interface toughness, stainless steel-rods were glued to the HA/Ti body and solidified HA body using epoxy resin in order to prepare standard core specimens specified in the ISRM suggested method. A pre-crack was introduced in the HA/Ti interface, as shown in Fig.2. The depth and width of the pre-crack were 5mm and 50μm, respectively. In order to determine the fracture toughness of the HA only, HA specimens were sandwiched and glued with stainless steel rods. In this type specimens, a pre-crack was introduced in the center of HA ceramics. The specimens were loaded at a cross-head speed of 1mm/min until a fast fracture took place. The stress intensity factor K was employed to obtain an estimate of the fracture property of the HA/Ti interface and the HA ceramics, using the following equation:

$$K = 0.25(S/D)\cdot Y_c'\cdot(F/D^{1.5}) \tag{2}$$

where D is diameter of the specimen(20mm), $S(=3.33D)$ is supporting span, F is load, Y_c' is the dimensionless stress intensity factor. The value of Y_c' can be found in the literature reference[Hashida, 1993]. Y_c' was fixed 7.0 due to the initial crack depth, as shown in Fig.2. The critical stress intensity K_c was computed from peak load at the onset of fast fracture. It should be mentioned here that the formula given in Eq.(2) is derived under the assumption of isotropic and homogeneous materials. The HA/Ti bonded specimen used in this study consists of 2 or 3 kinds of the materials. While exact anisotropic solution is needed for the quantitative evaluation of the stress intensity factor, the isotropic solution in Eq.(2) is used to obtain an estimate of the fracture toughness for the HA and HA/Ti specimens.

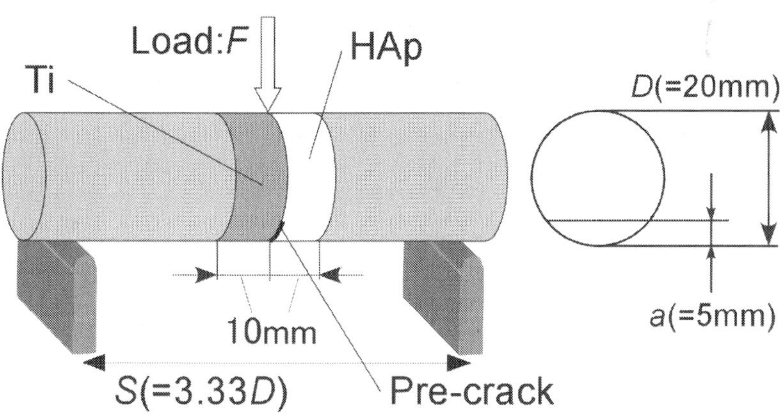

Fig. 2. Schematic illustration of 3-point bending test.

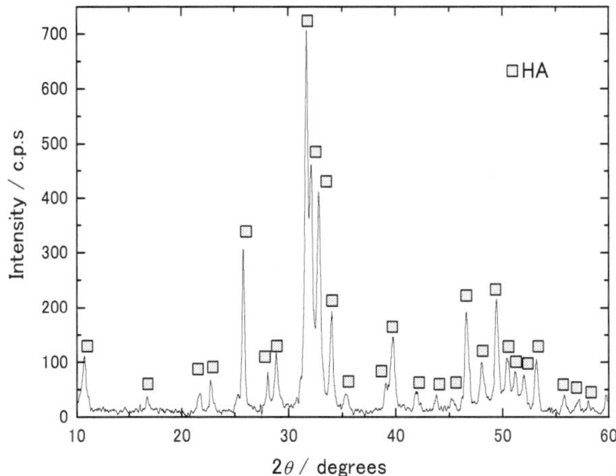

Fig. 3. X-ray diffraction pattern of the HA ceramics.

2.3 HA/Ti Bonding and its adhisive properties

It is seen that the shrinkage started at approximately 90°C. This temperature is close to the dehydration temperature of DCPD. Thus, it is thought that the shrinkage is initiated by the dehydration of DCPD. The shrinkage rate became larger with the increasing temperature, and then the shrinkage rate became smaller. The shrinkage continued during the HHP treatment. The pressure was held at 40MPa constant for the whole period of HHP treatment. As given in Fig.3, X-ray diffraction analysis showed that the DCPD and $Ca(OH)_2$ powder materials were completely transformed into HA by the HHP treatment. As demonstrated in Fig.4, the HA ceramics could be bonded to the Ti disks at the low temperature of 150°C using the above-mentioned HHP treatment. The density of the HA ceramics prepared by the HHP in this study was 1.9 g/cm^3. In addition to the DCPD powder, three different types of powders were used as a starting material: HA and β-tri calcium phosphate (β-TCP). No bonding with a Ti disk was observed, when the above starting powders were used and treated by the HHP under the conditions of 150°C and 40MPa. Thus, DCPD was the only starting material that produced HA/Ti bonded bodies among the precursors for HA used in this study.

Fig.5 shows a photograph of the fracture surface in the bonded HA/Ti body after 3-point bending test. It can be noted that the crack initiated from the pre-crack tip and propagated not along the HA/Ti interface, but into the HA. This observation suggests that the fracture toughness of the HA/Ti interface is close to or higher than that of the HA ceramics only. The critical stress intensity factor K_c was 0.30 $MPam^{1/2}$ for the HA ceramics, and 0.25 $MPam^{1/2}$ for the bonded HA/Ti. The toughness data are the average value obtained from at least five specimens. The K_c value for the bonded HA/Ti body gives a slightly lower value than that of the HA ceramics only. The difference in K_c data is potentially due to the residual stress induced by the thermal expansion mismatch between HA and Ti. The K_c value achieved for the pure Ti was close to the highest value obtained for the Ti alloys in our research [Onoki et al. 2003]. The fracture appearance in Fig.5 may suggest that the interface toughness should be equal or higher than that of the HA ceramics only. While further development is needed to improve the fracture property of the solidified HA, the HHP treatment may have the advantage over the plasma-spraying technique in the preparation of thermodynamically stable HA without decomposition. The above results demonstrate the usefulness of the HHP method for bonding HA and Ti in order to produce a bioactive layer in biomaterials.

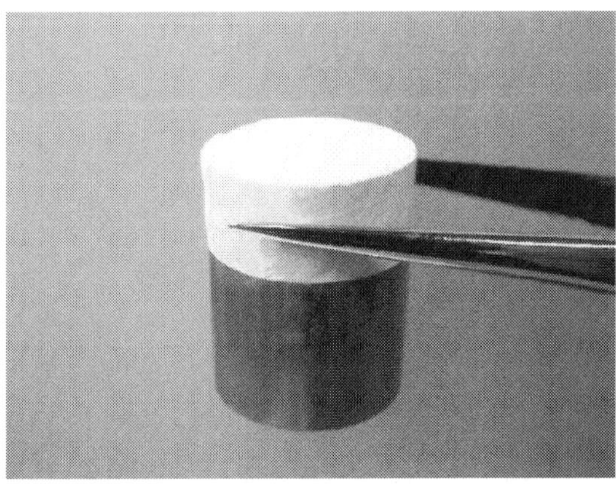

Fig. 4. Photograph of the bonded body of HA ceramics and Ti metal (20mm in diameter).

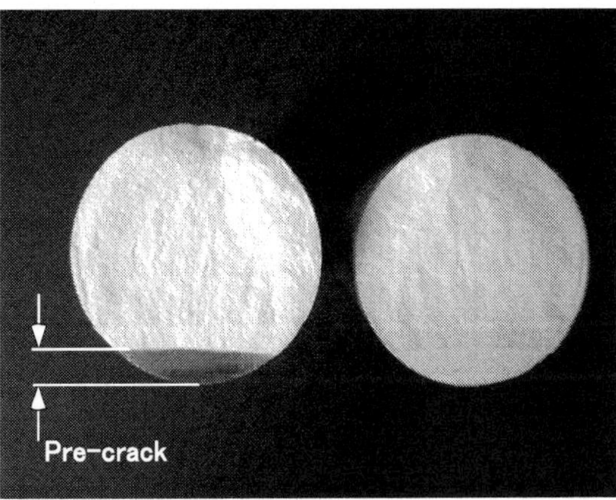

Fig. 5. Photograph of the fracture surface in a HA/Ti specimen after 3point bending test. Note that the crack propagated into HA.

2.4 Ti surface analysis by X-ray photoelectron spectroscopy (XPS)

Traditionally, Ti and its alloys have been reported to be bioinert. When embedded in the human body, a fibrous tissue encapsulates the implant isolating it from the surrounding bone forms. Since conventional metals as biomaterials are usually covered with metal oxides, surface oxide films on metals play an important role not only against corrosion but also regarding tissue compatibility. The composition of surface oxide film varies according to environmental changes, although the film is macroscopically stable. Passive surfaces co-exist in close contact with electrolytes, undergoing a continuous process of partial dissolution and re-precipitation from the microscopic viewpoint [Kelly, 1982]. In this sense, the surface composition constantly changes according to the environment. The film on titanium consists of amorphous or low-crystalline and non-stoichiometric TiO_2 [Kelly, 1982]. The surface oxide film of titanium just after polishing in water contains not only Ti^{4+} but also Ti^{3+} and Ti^{2+} [Beck, 1973; Kelly, 1982]. Hydrated phosphate ions are adsorbed by a hydrated titanium oxide surface during the release of protons [Hanawa, 1992]. Calcium ions are adsorbed by phosphate ions adsorbing on a titanium surface, and, eventually, calcium phosphate is formed. The above mentioned phenomena are characteristic in titanium and titanium alloys [Hanawa, 1992]. In this regard, an anatase-like structure is effective for apatite nucleation [Wei et al., 2002ab], whereas the naturally formed oxide film on titanium surface is mainly amorphous. The ability of titanium in order to form calcium phosphate on itself is one of the reasons for its better hard-tissue compatibility than those of other metals. This property is applied to the surface modification of titanium and its alloys to improve hard-tissue compatibility. In the case of alkali-heat-treated titanium, calcium and phosphate are orderly deposited, and calcium deposition is the pre-requisite for phosphate deposition [Yang et al., 1999].

It is easily expected that the bonding properties of the HA/Ti interface prepared by the HHP method can be depended on the Ti surface conditions. As preliminary experimental results, Ti surface finished in wet environment can be achieved bondong to HA ceramics through the above mentioned HHP method. However, Ti surfaces before the HHP processing have not been investigated precisely. The present study aims to investigate Ti surface properties through X-ray photoelectron spectroscopy (XPS) analysis. Particular attention was been paid to chemical composition and oxidation states of Ti in surface films in order to explain the bonding mechanism of Ti and HA ceramics by the HHP.

A commercially available pure Ti rod was used. The Ti surfaces were finished using 1500# emery paper in *air* and *water* conditions. After the emery paper finish, the Ti disks were cleaned in deionized water and ethanol by using an ultrasonic cleaner, and then dried in air. XPS was performed with an electron spectrometer (ULVAC-PHI, ESCA1600). All binding energies given in this paper are relative to the Fermi level, and all spectra were excited with the monochromatized Al Kα line (1486.61 eV). The spectrometer was calibrated against Au $4f_{7/2}$ (binding energy, 84.07 eV) and Au $4f_{5/2}$ (87.74 eV) of pure gold and Cu $2p_{3/2}$ (932.53 eV), Cu $2p_{1/2}$ (952.35 eV), and Cu Auger $L_3M_{4,5}M_{4,5}$ line (kinetic energy, 918.65 eV) of pure copper. The energy values were based on published data [Asami & Hashimoto, 1977]. The reproducibility of the results was confirmed several times under the same conditions.

In order to clarify the surface related chemical characteristics of the Ti, XPS analysis was performed for the specimens as-polished mechanically in *air* or *water* environments. The XPS spectra of the specimens over a wide binding energy region exhibited peaks of Ti 2p, O

1s and C 1s. The C1s spectrum showing a peak at around 285.0 eV arose from a contaminant hydrocarbon layer covering the topmost surface of the specimens. The spectra of the O 1s electron energy regions about *air* and *water* finishing are obtained. In particular, the O 1s spectrum was composed of at least two overlapping peaks which were so-called OM and OH oxygen in water finishing Ti specimens. The OM oxygen corresponds to O^{2-} ion in oxyhydroxide and/or oxide, and the OH oxygen is oxygen linked to protons in the surface film, being composed of OH- ions and bound water in the surface films. As shown in Fig.6(a) and (b), The decomvoluted XPS spectra of the O 1s region contained three peaks originating from oxide (O^{2-}), hydroxide or hydroxyl groups (OH-), and hydrate and/or adsorbed water (H_2O) [Beck,1973; Kelly, 1982]. Calculated fitting curves area of the samples finished in *air* and *water* environment was summarized in table.1. From these results, hydroxide or hydroxyl groups (OH-) of the water finishing sample increased compared with the air finishing sample. There is no distinguished difference between *air* and *water* finishing specimens within the Ti 2p spectra.

XPS characterization revealed differences in the Ti surfaces properties between water and air in finishing circumstances before the HHP treatment, as shown in Fig. 6(a) and (b). Compared with the air finishing Ti samples, the O 1s region XPS spectra of the water finishing Ti samples was significantly assigned hydroxide or hydroxyl groups (OH-) and hydrate and/or adsorbed water (H_2O). The HA/Ti bonding via the HHP processing could be achieved in only water finished Ti samples, as shown in Fig.4. On the other hand, the Ti samples finishing in air circumstance could not achieved the bonding to the HA ceramics. From these results, the bonding model between Ti and HA ceramics through the HHP method is suggested and summarized in below.

Calcium, phosphate and hydrate ions within the HHP autoclave solution are adsorbed on a titanium surface, and eventually calcium phosphate is formed, and then solidified as hydroxyapatite ceramics, as shown in Fig. 4. Similar model has been reported by Kokubo *et al* during apatite derived from Simulated Body Fluid (SBF) [Kokubo & Takadama, 2006]. It was guessed that the HA/Ti bonding behavior depended on amount of hydroxide or hydroxyl groups (OH-) and hydrate and/or adsorbed water (H_2O). In order to bond Ti and HA ceramics through the HHP techniques, it is important factors that hydroxide or hydroxyl groups (OH-) and hydrate and/or adsorbed water (H_2O) are remained on Ti surfaces. There is a threshold of amount of hydroxide or hydroxyl groups (OH-) between air and water finished surfaces.

	O^{2-}	OH^-	H_2O
(a) Air	62.50%	18.20%	19.30%
(b) Water	58.20%	24.60%	17.20%

Table 1. Calculated fitting curve area portion of O1s XPS spectra shown in Fig.6(a) and (b).

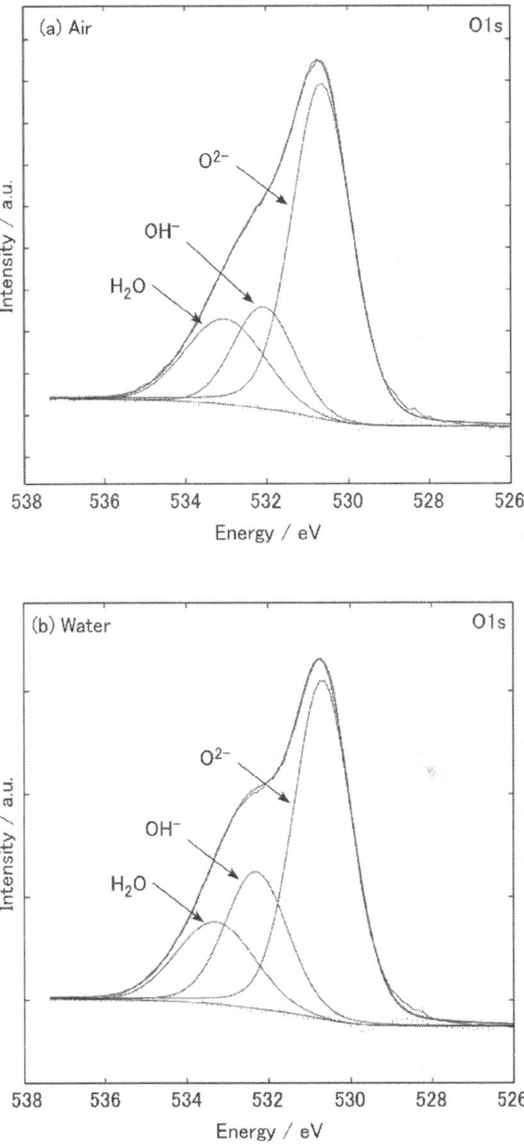

Fig. 6. O 1s XPS spectra of Ti surfaces finished in air (a) and water (b) environments (after curve fitting analysis).

3. Surface modification of metallic materials

3.1 Pure Ti

Recently, it was reported a concept called "Growing Integrated Layer" [GIL] that improves adhesion performance without cracking and peeling the ceramic coatings. Formation of a bioactive titanium dioxide (TiO_2) hydro-gel layer has been shown to improve the nucleation of calcium phosphate during chemical deposition. The TiO_2 layer could be prepared by alkaline [Wei et al., 2002ab; Kim et al., 2000; Wen et al., 1998ab], H_2O_2 [Ostuki et al., 1997; Kaneko et al., 2001; Rohanizadeh et al., 2004], sol-gel [Oswald et al., 1999] or heat treatment methods [Rohanizadeh et al., 2004]. It is demonstrated that the treatment of Ti with a NaOH solution followed by heat treatment at 873 K forms a crystalline phase of sodium titanate layer on the Ti surface resulting in improved adhesion of apatite coating prepared by incubation in simulated body fluid (SBF) [Wei et al., 2002ab; Kim et al., 2000; Takadama et al., 2001]. The authors concluded that release of the sodium ions from the sodium titanate layer causes formation of Ti-OH groups that react with the calcium ions from the SBF and form calcium titanate, which then could act as nucleation sites for apatite crystal formation [Takadama et al., 2001]. Alkali treatment results in the formation of a TiO_2 layer leading to a negatively charged surface, which in turn attracts cations such as calcium ions [Wen et al., 1998a]. Etching with acid followed by alkali treatment was also investigated to combine the surface roughness increase due to acid treatment and formation of a TiO_2 bioactive layer [Wei et al., 1998b]. TiO_2 could be also prepared using H_2O_2 alone or a mixture of acid/H_2O_2 or metal chlorides/H_2O_2 solutions [[Ostuki et al., 1997; Kaneko et al., 2001; Rohanizadeh et al., 2004]. Thus, it is expected that the interface strength of HA/Ti bonded bodies prepared by the HHP method can be improved, if the above-mentioned bioactive layer can be formed on the Ti surface. The present study examines the effects of various surface modification of Ti with $5M$ NaOH solution on the HA/Ti bonding behavior via the HHP method.

A commercially available pure Ti rod (99.5%; Nilaco, Japan), 20mm in diameter, was used. The Ti rod was cut into disks with a thickness of 10mm. And the disks were cleaned in deionized water and acetone by using an ultrasonic cleaner. The Ti surfaces were finished using 1500# emery paper. The Ti surface without a special surface modification is referred to as "NORMAL" surface. After the emery paper finish, the titanium disks were washed by deionized water, and then dried in air. In order to form bioactive TiO_2 and sodium titanate layer on the Ti surface, some of the "NORMAL" disks were treated with alkali solution (5M NaOH) under the same conditions used in the literatures [Wei et al., 2002ab; Kim et al., 2000; Takadama et al., 2001]. The Ti disks were immersed in the NaOH solution for 24 hours. Some of the Ti disks were further heat–treated at 873 K for 1 hour after the NaOH immersion for 24 hours. It has been reported that the reaction layer formed by the NaOH immersion at 333 K for 24 hours was easily detached by adhesive tape [Kim et al., 1997]. In order to avoid the detachment of the reaction layer, the Ti needed to be heated up to 873 K for 1 hour in an air. The Ti disk with the heat treatment at 873 K is labeled as "HEAT" surface, and without the heat treatment is labeled as "IMMER" surface. In our previous study [Onoki et al., 2003], the surface modification of Ti alloys (Ti-15Mo-5Zr-3Al and Ti-6Al-2Nb-1Ta) with NaOH solution (5M NaOH) at 323 K for 2 hours could significantly improve the bonding strength of the HA/Ti. Thus, it is expected that the same surface treatment on the Ti surface may enhance the bonding strength of the HA/Ti. Some of the "NORMAL" Ti disks were placed into a small vessel with the NaOH solution and heated up to 323 K for 2 hours. Ti disk treated with the hydrothermal NaOH solution is referred to as "HYDRO" surface. After the above-mentioned treatments with NaOH solution, the Ti disks were washed by deionized water, and then dried in air. The surface conditions of Ti disks are summarized in Table 2.

In order to characterize the Ti surfaces, the surfaces were observed using scanning electron microscopy (SEM: HITACHI FE-SEM S-4300, Japan) and were examined using FT-Raman spectroscopy. The microprobe instrument used for the FT-Raman spectroscopy consisted of a spectrometer (JOBIN YVON-HORIBA SPEX) fitted with a microscope (OLYMPUS-BX30, Japan) which had spatial resolution on the sample close to 1μm. The 632.8nm line of an He-Ne laser was used as excitation, focused in a spot of approximately 1μm diameter, with an incident power of 2mW.

Type	NaOH Treatment		873K
	Temp.(K)	Time(h)	Heat
NORMAL	-	-	-
IMMER	333	24	-
HEAT	333	24	1h
HYDRO	423	2	-

Table 2. Conditions of the surface modifications of the Ti disks.

In order to characterize the reaction products formed on the Ti surface products, observation by SEM of the Ti surfaces treated with the NaOH solution were conducted for the "NORMAL", "IMMER", "HEAT" and "HYDRO" specimens, respectively. Furthermore, analytical results of Raman spectroscopy analysis were obtained of the "NORMAL", "IMMER", "HEAT" and "HYDRO" specimens, respectively.

The surface morphology of the "NORMAL" specimen exhibits finishing lines in the nearly horizontal direction, which were formed at the stage of surface polishing with the emery paper. Raman spectra for the "NORMAL" specimen shows no distinct peak. The surface morphology of the "IMMER" specimen has a very fine network structure (0.1μm scale) with finishing lines in the perpendicular direction. In the Raman spectra of the "IMMER" specimen, the peaks of TiO_2 (both of anatase and rutile phase) and sodium titanate ($Na_2Ti_5O_{11}$) could be detected. The surface morphology of the "HEAT" specimen has a sponge-like structure and no finishing lines are visible on it. In the Raman spectra of the "HEAT" specimen, the peaks of TiO_2 (only rutile phase) and sodium titanate ($Na_2Ti_5O_{11}$) could be detected. Indeed, it has been reported in the literature [Kim et al., 1997] that in a Ti specimen treated with NaOH solution dehydration of the reaction layer took place by the post heating at 873 K and that the anatase phase as observed in the "IMMER" specimen was changed into rutile phase entirely.

The surface morphology of the "HYDRO" specimen has a needle like structure, as shown in Fig.7. In the Raman spectra of the "HYDRO" specimen, the peaks of TiO_2 (both of anatase and rutile phase) and sodium titanate ($Na_2Ti_5O_{11}$) could be detected as shown in Fig.7. As reported in the literature [Kim et al., 1997], we have also observed that the reaction layer of the "IMMER" specimen was easily peeled off by adhesive tape, and that the post heating at 873 K was necessary in order to prevent the delamination of the reaction layer from the Ti specimen. However, it was shown that in the "HYDRO" specimen no delamination of the reaction layer took place in peeling tests with adhesive tape, without post heating process. It is shown from the Raman spectra that the reaction layers formed on the Ti surface are essentially composed of the identical chemical compounds both for the "IMMER" and "HYDRO" specimens.

3.2 HA bonding behavior

For all the Ti surface modifications, the HA ceramics was successfully bonded to the Ti using the HHP method. The NaOH treatments on Ti surface were useful in obtaining HA/Ti bonding body regardless of the any NaOH treatment conditions (see Table3). The fracture toughness for the HA/Ti specimens was obtained from 3-point bending tests. The fracture toughness data are given in Table 3 along with the reaction products characterized by the Raman spectra. The fracture toughness data are the average value obtained from at least five specimens. The facture toughness, K_c for the "IMMER" and "HEAT" specimens shows a slightly lower value than that of the "NORMAL" specimens. The fracture toughness data of the "HYDRO" specimens is higher than that of "NORMAL" specimens. In the literature [Kim et al., 1997], SBF soaking tests were conducted on "IMMER" Ti specimen and "HEAT" Ti specimen at 310 K. The growth rate of bone-like apatite on the Ti surface was shown to be lager for the "IMMER" specimen than that in the "HEAT" specimen. Based on the above observation, it has been concluded that the formation of anatase in the reaction layer produced by the treatment with 5M NaOH solution accelerates the growth rate of bone-like apatite in the Ti surface. On the other hand, it was shown that the reaction layer of "IMMER" specimen had poor adhesion with Ti substrates in contrast with "HEAT" specimen. The lower fracture toughness in the "IMMER" specimen may be due to the poor adhesion property, in spite of the formation of anatase in the reaction layer. The plausible cause for the lower fracture toughness in the "HEAT" specimen may be attributed to the lack of anatase, even though the adhesion property was improved significantly through the post heat treatment. As described above, no delamination of the reaction layer occurred for the "HYDRO" specimen in the tests conducted using adhesive tape. The higher adhesion property in conjunction with the formation of anatase may provide the explanation for the larger fracture toughness value in the "HYDRO" specimen.

It was revealed that the hydrothermal surface modification with the NaOH solution was the most effective technique among the surface treatment methods used in this study in the improvement of the bonding strength of the HA/Ti interface produced by the HHP method. The fracture toughness for the "HYDRO" specimen showed 40% increase with respect to that of the "NORMAL" specimen.

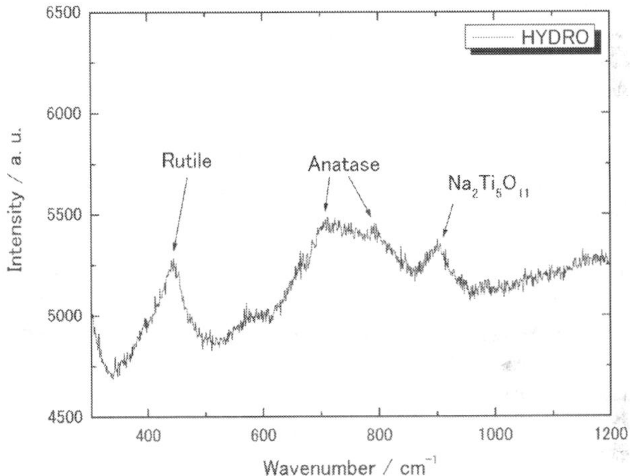

Fig. 7. SEM Micrograph and FT-Raman Spectra of the "HYDRO" Ti surface.

Type	Titanium Oxide		Sodium Titanate	K_c (MPam$^{1/2}$)
	Anatase	Rutile		
NORMAL	–	–	–	0.25
IMMER	O	O	O	0.22
HEAT	–	O	O	0.21
HYDRO	O	O	O	0.35

Table 3. Characterization of the Ti surface products and the interface fracture toughness.

3.3 Ti based Bulk metallic glass

Bulk glassy alloys are promising materials for structural and functional uses due to their superior properties compared to their crystalline counterparts [Greer, 1995; Inoue, 2000; Ashby & Greer, 2006]. These alloys are known to exhibit high hardness, high tensile strength, and good fracture toughness. The unique properties of bulk glassy alloys make them extremely attractive for biomedical applications. The mechanical deformation behavior of biological materials, which is characterized by a high recovery of strain (above 2%) after deformation, is very different from that of common metallic materials [Li et al., 2000]. Another problem concerning metallic implants in orthopedic surgery is the mismatch of Young's modulus between a human bone and metallic implants. The bone is insufficiently loaded due to a mismatch called stress-shielding. From the viewpoint of the requirements toward implant materials for hard tissue replacement, a biomaterial with low elastic modulus is required. Glassy alloys have lower Young's modulus and an extremely high elastic limit of 2%. Since bulk glassy alloys have a unique ability to flex elastically with the natural bending of bone, they distribute stresses more uniformly. Faster healing rates will result from reduced stress shielding effects while minimizing stress concentrators.

Ingots of Ti-based bulk metallic glass ($Ti_{40}Zr_{10}Cu_{36}Pd_{14}$: BMG) were prepared by arc-melting the pure elements with purities above 99.9% in an argon atmosphere [Zhu et al., 2007]. Cylindrical rods (5mm diameter) were prepared by copper mold casting method. And cut into disks with a thickness in 2 mm. The BMG surfaces were finished using emery paper. After the surface finish with the emery paper, the BMG disks were degreased prior to hydrothermal-electrochemical experiment. After sonicated in acetone and rinsed with distilled water, disks were dried at ambient temperature. Glassy structure of the BMG was examined by X-ray diffraction patterns (XRD).

Recently, it was reported a concept called "Growing Integrated Layer" [GIL] that improves adhesion performance without cracking and peeling the ceramic coatings. In order to produce GIL on surface of the BMG, hydrothermal-electrochemical treatment was conducted, as shown in Fig.8 [Yoshimura et al., 2008]. The BMG substrate was used as the working anode and platinum substrate was used as the cathode. The distance between electrodes was kept at 4 cm. The active anodic surface area immersed in electrolyte was 0.707 cm². GIL was fabricated by the hydrothermal-electrochemical treatment at 90 °C for 120 minutes in aqueous solutions of 5M NaOH as an electrolyte. A constant electric current of 0.5 mA/cm² was applied between electrodes. After hydrothermal-electrochemical treatments, the specimens were removed from the electrolyte, washed with distilled water and then dried at 80 °C for 2 hours in air. The GIL produced by the hydrothermal-Electrochemical method was observed by an scanning electron microscopy (SEM: Hitachi S-4500, Japan) in surface morphology and a cross section view. Surface products were removed from the BMG and observed by a transmission electron microscopy (TEM: Hitachi H-9000, Japan).

The BMG disks were examined by powder X-ray diffractions patterns. Broadened XRD patterns denote a glassy nature. For $Ti_{40}Zr_{10}Cu_{36}Pd_{14}$, a glassy state was confirmed as well as the literature [*Zhu et al.*, 2007]. Surface morphology and cross section view of the GIL by the SEM observation were displayed in Fig.9 (a) and (b), respectively. The surfaces after the hydrothermal-electrochemical treatments had nanometer-scale meshed structure. The mesh products might be consisted of amorphous nano-rods from the results of SEM observations as shown in Fig.9. The nano-rods had amorphous structure and no crystalline phases like as

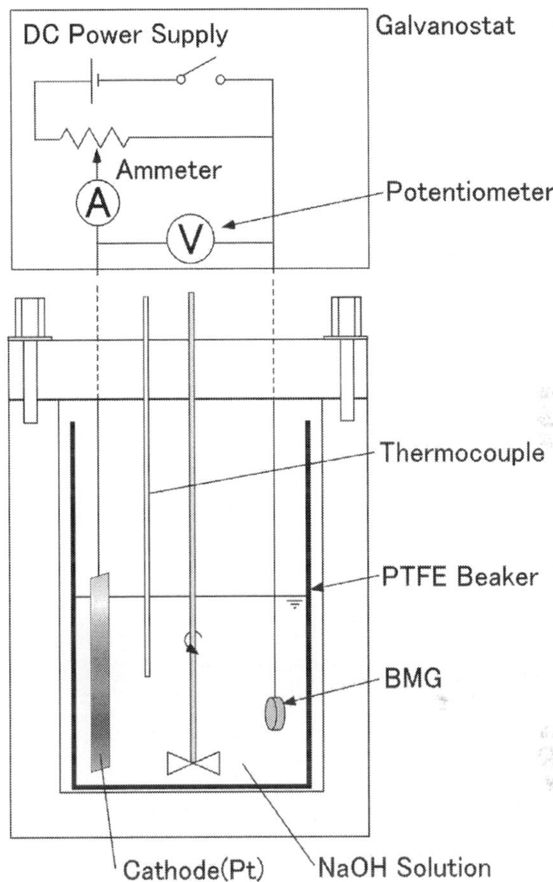

Fig. 8. Schematic illustration of experimental equipment for hydrothermal-electrochemical treatment.

titanate nanotubes [Onoki et al., 2009a]. Thin Film mode XRD analysis supported that the surface products formed on the BMG had the amorphous phase. Depth of the nano-meshed structure was approximately 1μm. It was supposed that the GIL was constructed between the BMG substrate and the nano-mesh materials.

Bonding hydroxyapatite ceramics and the BMGs could be achieved through the HHP treatment, like as shown in Fig.10. Because the BMG is essentially bioinert, the BMG without surface treatments has no HA bulk ceramics adhesion. It is easily guessed that the HA adhesive ability of the BMG surfaces is derived from the GIL made by the hydrothermal-electrochemical treatment. The bonding body was shaped in 5mm diameter structure like as Fig.4 by using grinder. In case of the BMG without the GIL, the bonding sample was easily separated into the HA ceramics and the BMG. However, the HA/BMG bonding body with the GIL was keep adhesive during the 5mm grinding process. The interface between the BMG and HA ceramics might have sufficient mechanical strength.

Since glass transition and crystallization onset temperatures of about $Ti_{40}Zr_{10}Cu_{36}Pd_{14}$ have been reported at 396 and 445°C, respectively [Zhu et al., 2007], the uniquely mechanical properties of the bulk metallic glasses must be lost in case of adding over 400°C heat treatment. Lower temperature techniques (under 400°C) are required for surface bioactivity treatments for retaining the uniquely mechanical properties of the bulk metallic glasses. The operating temperature (maximum 150°C) of the above hydrothermal treatments are low enough. A series of hydrothermal techniques is expected to be one of the most useful methods for creating bioactivity to Ti-based bulk metallic glasses surfaces. Moreover, hydrothermal techniques are appropriate for compositing bulk metallic glassy materials and ceramics to provide complementary functions with other materials due to the low temperature of hydrothermal processing. By using other hydrothermal techniques [Onoki et al., 2006], HA ceramics coating on Ti-based BMG could be achieved.

As shown in Figs. 10(a) and (b), we observed near the interface between the HA and the BMGs with the hydrothermal-electrochemical treatments for 40 and 120 minutes, respectively. Amorphous nano-meshed structure on the BMG surface was disappeared during the hydrothermal hot-pressing (HHP) process for bonding HA ceramics and the BMGs. If HA precipitates into the nano-meshed structure, the nano structure can be watched in these SEM micrographs. However, there was no fragmented or remained nano-mesh. It is known that the surface amorphous products processed by low temperature solution are meta-stable phase and easily dehydrated and hydrolyzed [Kim et al., 1997; Onoki et al., 2008]. Consequently, it is guessed the amorphous products were consumed or decomposed during the HHP treatments. An intermediate layer between HA ceramics and the BMG was revealed in the SEM micrographs as shown in Figs. 10(a) and (b). It is observed that the greater hydrothermal-electrochemical treatment time, the thicker the intermediate layer became. Compared with Fig.9 and 10, the thickness of the GIL and the intermediate layer were estimated in almost same value, respectively. It was concluded that the intermediate layer was the GIL remained on the BMG. The intermediate layer would play a role of relaxation of thermal expansion misfit and bonding layer between the BMG and the HA ceramics. The BMG with the hydrothermal-electrochemical treatments for 10 minutes could not achieved the bonding the BMG and HA ceramics through the HHP. The 10minutes treated BMG had insufficient thickness of the intermediate layer. It is speculated that sufficient thickness of GIL is the most important factor for the bonding BMG and HA ceramics by the HHP method. From above shown Results and discussion, it is demonstrated that the hydrothermal techniques (hydrothermal-electrochemical treatment and hydrothermal hot-pressing) are necessarey and useful for bonding between BMG and HA ceramics.

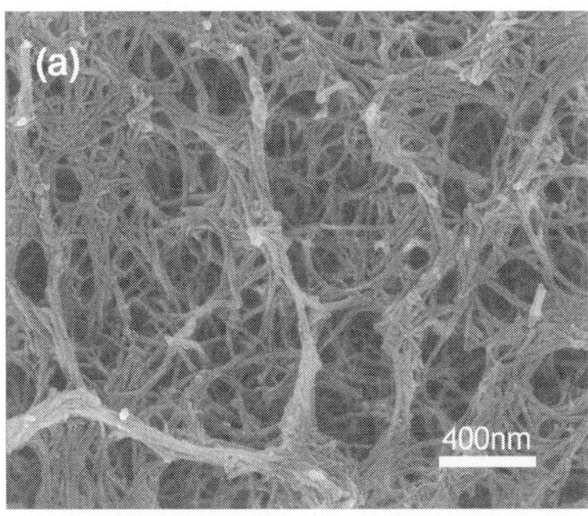

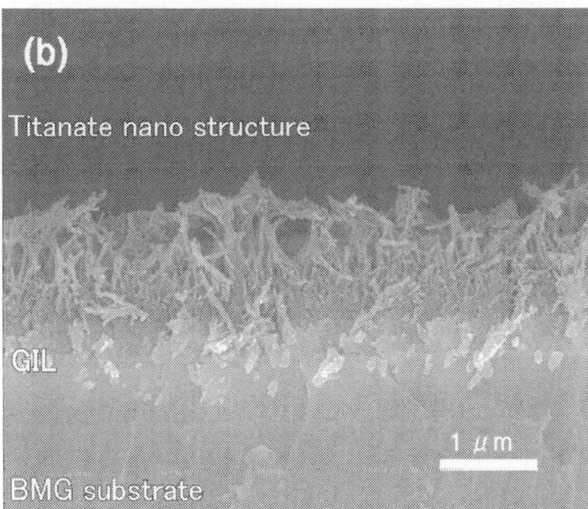

Fig. 9. SEM micrograph of BMG surface through hydrothermal-electrochemical treatment for 120 minutes (a) and in a cross sectional view(b), respectively.

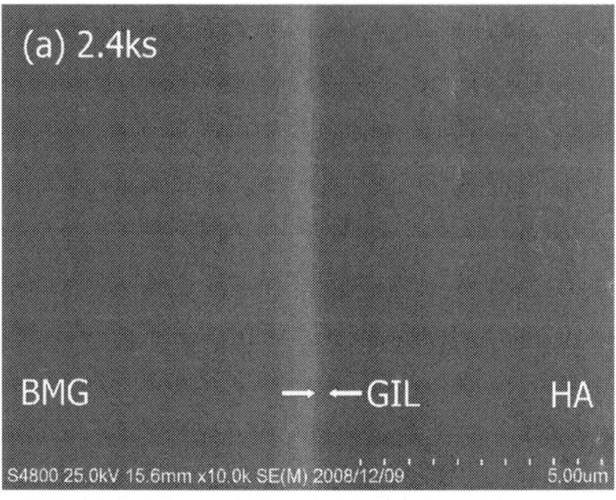

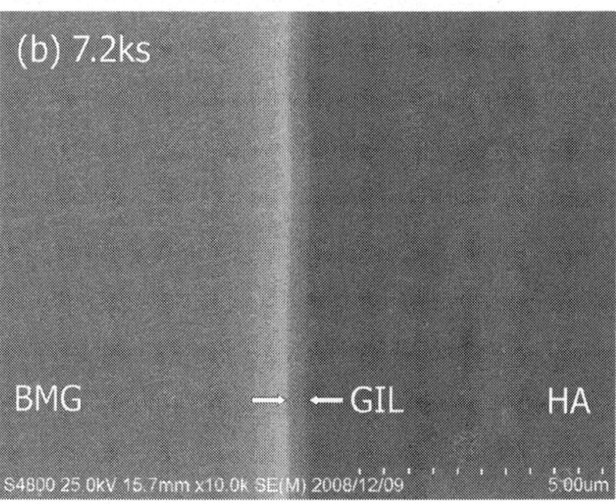

Fig. 10. SEM micrograph of the interface between HA ceramics and BMG disk with hydrothermal-electrochemical treatments for 40 (a) and 120 minutes(b), respectively.

4. HA coating by Double layered capsule hydrothermal hot-pressing

4.1 Introduction and using materials

In previous sections, it has been proposed a HHP method for bonding HA ceramics and pure Ti, Ti alloys and Ti based BMG by hydrothermal hot-pressing. HA ceramics was bonded to the *flat* surface of a Ti disk using the HHP method at 150°C and 40MPa. The thickness of the HA ceramics bonded was about 10mm. It has been shown from three point bending tests that the fracture strength of the HA/Ti interface was equal to or higher than that of the HA ceramics. Additionally, X-ray diffraction analyses revealed that the HA ceramics bonded had high crystallinity without any decomposition and impurity. However, the HA bonding was achieved only on *flat* surfaces of Ti because of uniaxial pressing. In order for the hydrothermal method to be applicable to orthopedic and dental implant materials, we should develop a method for coating thin HA layer onto curved surface. This section describes a new methods for coating HA ceramics layer on Ti rod at the low temperature as low as 135°C by using the newly developed double layered capsule hydrothermal hot-pressing (DC-HHP) method, which utilizes isostatic pressing under hydrothermal conditions.

The synthetic DCPD and calcium hydroxide (95.0%; $Ca(OH)_2$; KANTO CHEMICAL CO., INC., Japan) were mixed in a mortar for 60min with a Ca/P ratio of 1.67 which was stoichiometric ratio of HA. A commercially available pure Ti rod (99.5%; Nilaco, Japan), 1.5mm in diameter, was used in this study. The Ti rod was cut into a length of approximately 20mm. Ti surfaces were finished using #1500 emery paper. The rods were cleaned in deionized water and acetone by using an ultrasonic cleaner.

Recently, it has been reported that surface modifications for forming bonelike apatite can induce the high bioactivity of bioinert materials in simulated body fluid (SBF) [Kokubo et al., 2004]. In our previous research [Onoki et al., 2003], bonding HA ceramics and Ti alloys (Ti-15Mo-5Zr-3Al and Ti-6Al-2Nb-1Ta) was achieved by the HHP method through the surface modification of Ti alloys with alkali solution (5*M* NaOH). It is reported that the surface treatment of the Ti alloys with the alkali solution was very effective to improve the fracture strength of the interface between HA and Ti alloys produced by the HHP method. Based on the above results, the Ti rods used in this study were treated with 5*M* NaOH solution after the emery paper finish. The hydrothermal treatment with the NaOH solution was conducted at 150°C for 2 hours using a small vessel (volume: 7.5ml). After the surface treatment, the Ti rods were washed by deionized water, and then dried in air.

4.2 Double layered capsule hydrothermal hot-pressing (DC-HHP) method

A new technique was developed in this study in order to prepare HA coating layers on a cylindrical rod with the objective of applying the hydrothermal hot-pressing method to a substrates with more complicated configurations, as shown in Fig.11. The newly developed method uses a cylindrical capsule having the double layered structure, which is subjected to isostatic pressing under hydrothermal conditions. A schematic illustration of the capsule in a cross section view is shown in Fig.12.

Firstly, the Ti rod and the powder mixture of DCPD and $Ca(OH)_2$ were placed into a tube made of polyfluoroethylene (FEP). The weight of the powder mixture put into was approximate 0.1g. The initial diameter and thickness of the FEP tube was 1.8mm and 100μm. The FEP shrinks thermally by approximately 25% at around 130°C. The powder mixture was loaded into the FEP tube such that the Ti rod was concentrically positioned with respect to the tube axis. Both the ends of the FEP tube were fastened with paper staples. The sample assemblage encapsulated using the FEP tube is called "capsule I" in this study. Secondly, the capsule I was further encapsulated using a poly-vinylidene-chloride (PVC; 11μm thickness, Asahi-KASEI, Japan) film. Between the capsule I and PVC film, alumina powder

(3.0 μm diameter; Buehler Ltd., USA) was placed. The thickness of the alumina powder layer was approximately 3mm. The capsule prepared using the PVC film is called "capsule II". Then, the PVC film was sealed off using a thermo-compression method. As expected, the sealing of the capsule II was found to be crucially important for the subsequent hydrothermal treatment. No solidification and bonding of the HA ceramic layer was observed when there was a pre-existing defect in the thermally bonded PVC film and the water used for pressure application seeped into the capsule. Thus, the double layered capsule was vacuumed prior to sealing the capsule II, and left for 1 hour at the laboratory in order to check the seal tightness. Initially, a semi-permeable membrane for water was used instead of the FEP tube. However, it was observed that the fastened ends of the FEP tube could act as a narrow water flow-path. Thus, the thermally shrinkable FEP tube was used for capsule I in this study. The excessive water released from the reaction of DCPD and $Ca(OH)_2$ penetrates through the FEP tube ends into the pore space of alumina powder layer, maintaining the appropriate hydrothermal condition inside the capsule I. Thus, the alumina layer serves as an escape space for water as well as a medium for pressure application.

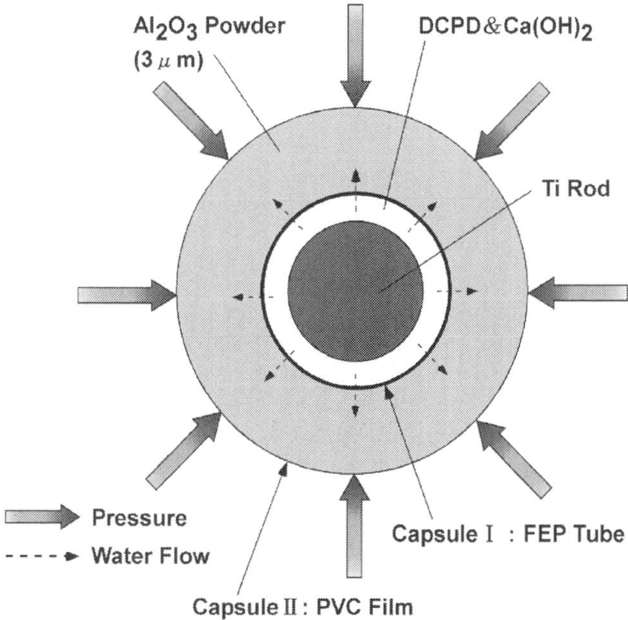

Fig. 11. Schematic illustration of the concept of the double layered capsule hydrothermal hot-pressing (DC-HHP) method.

The capsules prepared were put into a batch type high temperature and pressure vessel (volume: 300ml, SUS316L, AKICO, Japan) for hydrothermal treatment. A schematic illustration of the hydrothermal treatment is shown in Fig.13. The deionized water in the vessel serves as a medium for isostatic pressing (mechanical compaction). The vessel was heated up to 135°C at a heating rate of 5°C/min, and then the temperature was kept constant. The maximum allowable temperature of the PVC film used for capsule II is 140°C, and the treatment temperature was set to be 135°C due to the temperature limitation. The pressure was kept at 40MPa using pressure regulator. After the treatment, the vessel was naturally cooled down to a room temperature, and the capsules removed from the vessel. In order to investigate the effects of treatment time on the HA coating properties, the treatment time was varied within the range of 3-24 hours.

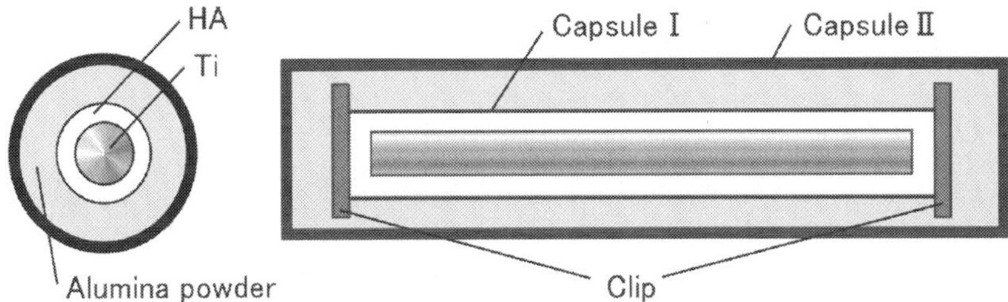

Fig. 12. Schematic illustrations of a double layered capsule in a cross section view.

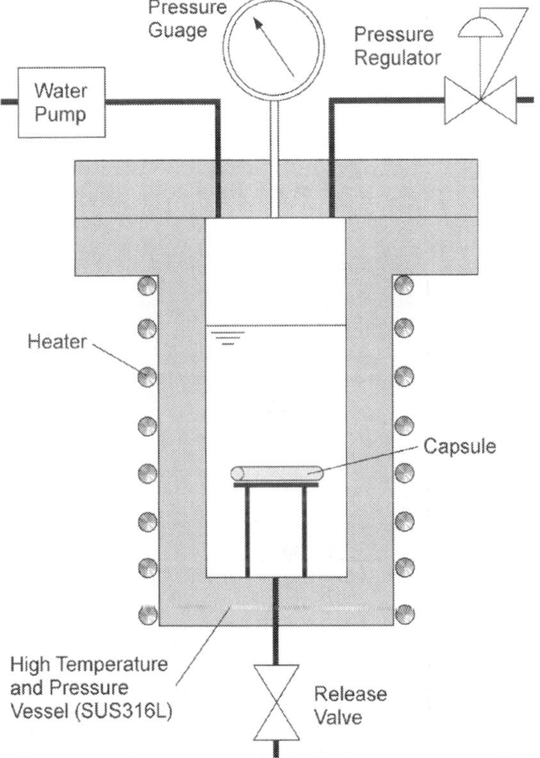

Fig. 13. Schematic illustration of the hydrothermal method and the capsule used in this study.

4.3 Adhesion properties evaluation

Pull-out tests were conducted in order to evaluate the adhesion strength of the HA ceramics coating. A schematic illustration of the pull-out testing method used is drawn in Fig.14. The HA coated samples were embedded into an epoxy resin (Sumitomo 3M, Japan) placed in a sample attachment. According to the supplier, the tensile and shear strength of the epoxy resin used were measured to be approximately 29.4 and 11.8 MPa, respectively. And then the protruding part of the HA coating was removed with a grinder and a knife. In order to identify the crystals in the treated specimens, the removed HA coating material was applied to power X-ray diffraction analysis (XRD; MX21, Mac Science, Japan) with $CuK\alpha$ radiation 40kV 40mA at a scanning speed of 3.00°/min with a scanning range (2θ) from 10° to 40°. The microstructure of the coating surface was observed by scanning electron microscopy (SEM; S4300, Hitachi, Japan).

The specimens were loaded with an Instron-type testing machine at a cross-head speed of 0.5mm/min until the Ti rods were pulled out entirely. The sample attachment was connected to a universal joint. The upper part of the Ti rod was gripped with a manual wedge grip and then loaded with the testing machine through a load-cell. The load P and cross-head displacement δ were recorded during the tests. In order to evaluate the adhesion properties of the HA coating on Ti rod quantitatively, the shear strength and fracture energy were calculated from the results of the pull-out testing. The shear fracture strength, τ is computed using the following equation:

$$\tau = \frac{P}{\pi dL} \qquad (3)$$

where d is the diameter of Ti rods(1.5mm), L is the embedment length of Ti rods, P is the load. The fracture energy, G is calculated using the following equation:

$$G = \frac{A}{\pi dL} \quad \left(A = \int_0^{\delta_c} P d\delta \right) \qquad (4)$$

where δ is the cross-head displacement and A is the area under the load versus cross-head displacement up to the complete pull out displacement, δ_c.

4.4 Results and discussion

As demonstrated in Fig.15, HA ceramic layers could be coated to the all surface of Ti rods at the low temperature of 135°C using the above-mentioned DC-HHP method. The thickness of the HA coating layers could be controlled by the volume of the HA starting powder placed in capsule I, and the thickness range achieved in this study was 10 μm–1 mm. The experimental results for the thickness of 50 μm will be presented below. When the seal tightness of capsules II was imperfect, the water placed in the vessel was observed to penetrate into the inner space of the capsules and the HA coating could not be achieved. It was critically important that the inner space of the capsules was vacuumed prior to the hydrothermal treatment in order to check and ensure the seal tightness of the capsules.

XRD profiles of the HA coating layers are shown in Fig.16 for the different treatment times. Only the peaks for HA are observed for the treatment time of 12 and 24 hours, whereas the HA layer prepared with the treatment time of 3 and 6 hours includes the phases of the starting materials or the precursors. Thus, it is understood that the treatment time longer

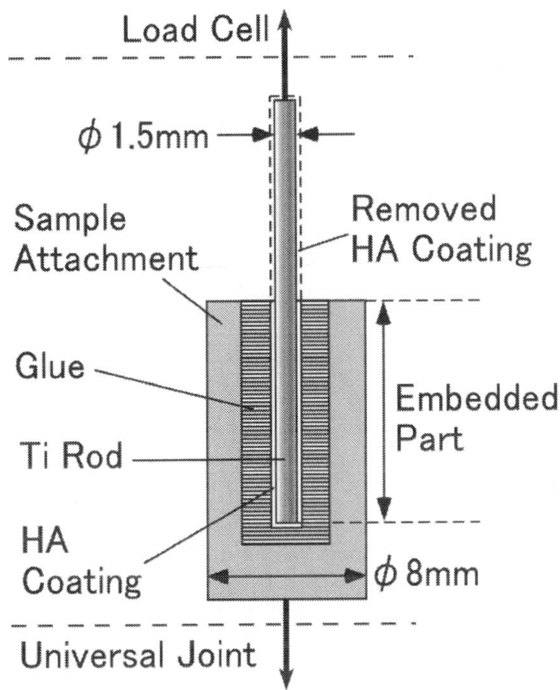

Fig. 14. Schematic illustration of the pull-out testing method in a cross section view.

than 12 hours was required to convert the starting powders to HA entirely. The crystallinity of the HA coating is observed to increase with the increasing treatment time, as indicated by the intensity of the peaks. Furthermore, the XRD analysis shows that the low temperature hydrothermal method induced no chemical decomposition unlike high temperature methods such as a plasma spraying method.

Fig. 15. Photograph of the sample of HA ceramic coating on Ti rod.

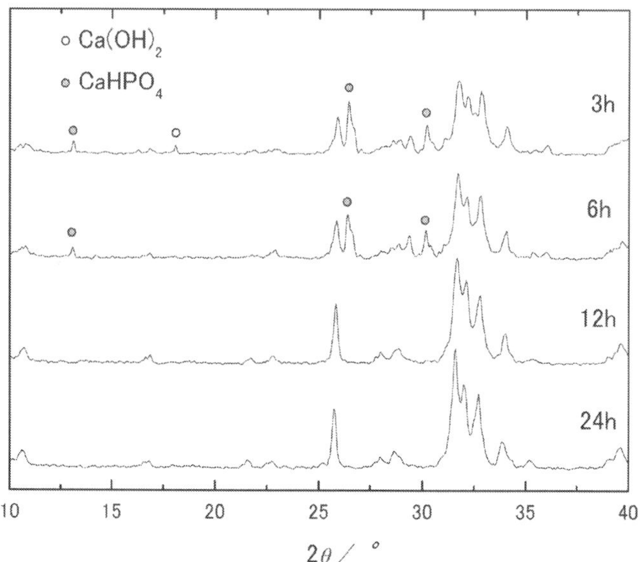

Fig. 16. XRD profiles of the HA coatings treated for 3,6,12 and 24hours, respectively.

The load-displacement curves obtained from the pull-out tests are shown for the treatment time of 24 hours in Fig.17. Fig.18 shows the appearance of the specimen after the pull-out testing for the treatment time of 24 hours. The right-hand side of the specimen show the region where the HA coating was completely stripped off from the Ti rod before the pull-out test. The white part of the specimen (the left-hand side) was initially embedded in the epoxy resin and corresponds to the region that was pulled out after the test. It is clearly seen that the surface of the embedded Ti rod (the left-hand side) is completely covered with the remaining HA layer. An SEM micrograph of the pull-out part in Fig.18(a) is given in Fig.18(b). The appearance of the scratched surface indicates that significant abrasion of the HA layer took place in the pull-out process. It is demonstrated from Fig.18(a) that the crack propagation took place not along the HA/Ti interface, but in the HA coating layer. Tthe crack path was always located within the HA coating irrespective of the different treatment times. This observation suggests that the fracture toughness of the HA/Ti interface is close to or higher than that of the HA ceramics only. If the interface between the HA and Ti rod was held just by interfacial friction with no significant chemical bond, no HA ceramics would remain on the Ti rods after the pull-out testing.

In all the load-displacement curves, there is observed so-called pop-in behavior where the load sharply decreases at the load level indicated by arrow and then increases again until the peak load is reached. The pop-in behavior is considered to correspond to the onset of the crack propagation. The load ascending part after the pop-in may suggest that the initiated crack deviated from its original crack orientation and propagated into the HA coating away from the HA/Ti interface region. The deviation of the crack path from the HA/Ti interface region may induce an additional frictional resistance to the crack propagation in the pull-out process. Indeed, significant abrasion of the HA coating layers was observed on the fracture surface after the pull-out tests, as shown in Fig.18(a).

The post-peak of the load-displacement curves is characterized by the load descending part due to a stable pull-out process of the Ti rod. It is seen that the critical displacement at which the applied load become zero is close to the embedded length of the Ti rod (approximately 10 mm).

In order to evaluate quantitatively the bonding characteristics of the HA coatings, the shear strength and fracture energy were calculated from the load-displacement curves of the pull-out tests. Two shear strength parameters were calculated in this study. Hereafter, the shear strength parameters computed from the pop-in load, and the peak load are designated by τ_i and τ_{max}, respectively. The calculated shear strength of each surface modifying conditions in Table.3 are plotted in Fig.19 as a function of the treatment time. The data of τ_{max} shown in Fig.19 are the averaged results obtained from at least 4 specimens.

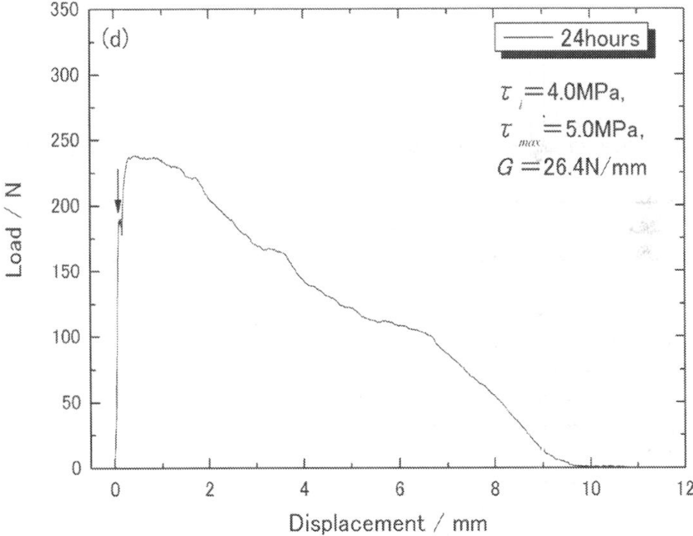

Fig. 17. Load-Displacement curve of the pull-out testing for the HA coatings treated for 24 hours.

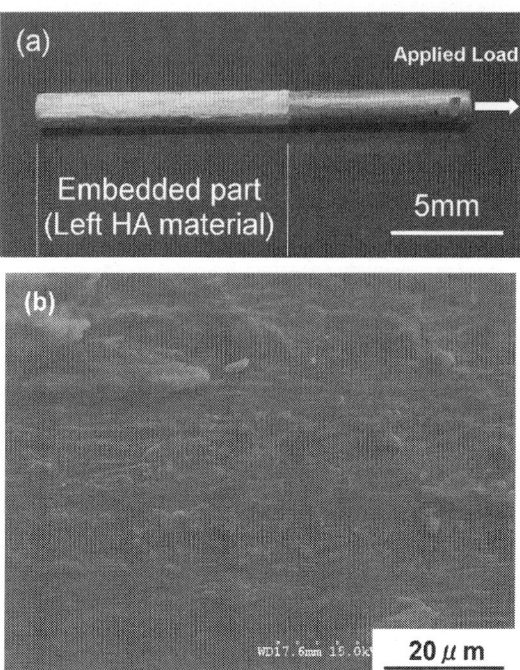

Fig. 18. Photograph of the specimen after pull-out testing (treated for 24hours) and SEM micrograph.

It is seen that the τ_i gives an almost constant value (4.1 to 4.6 MPa) irrespective of the treatment time, even though there is a slight increase in the τ_i for the treatment time of 6 and 12 hours. The τ_{max} initially increased and then gradually decreased when the treatment time was longer than 6 hoursThe "HYDRO" specimens showed the highest strength in the DC-HHP as well as the HHP processing. It has been shown in our recent experiment that the adhesion of the HA ceramics bonded on Ti using the conventional HHP method decreased also for longer treatment times [Onoki et al, 2003b]. It should be remembered that treatment time longer than 12 hours was needed for the complete conversion of the starting materials into pure HA, and some starting materials remained in the solidified layers for the shorter treatment times. The above-mentioned experimental results suggest that there may be an optimal hydrothermal treatment time for the production of pure and stronger HA coatings. The adhesion properties measurements and XRD analyses indicates that the treatment time of 12 hours is the suitable condition for the HA coating by DC-HHP method under 135 °C and 40 MPa. Our approximate estimates using the Archimedes' method have shown that the density of the HA coating initially increased until the treatment time of 6 hours and then gave a constant value for the longer treatment time (the measured density was 1.6 g/cm³ for 3hours, and 1.9g/cm³ for the longer treatment time). Thus, the initial increase in the adhesion properties may be due to the formation of HA and its densification. The XRD profiles in Fig.16 suggests that the crystal size of the HA for the treatment time of 12 hours may be larger than that for 24 hours. This observation may provide a possible explanation for the decreased adhesion properties for the longer treatment time. However, more detailed examination is needed in order to discuss the reason for the presence of such an optimal treatment time. Now investigation of the bonding mechanism in the HHP process is now in progress.

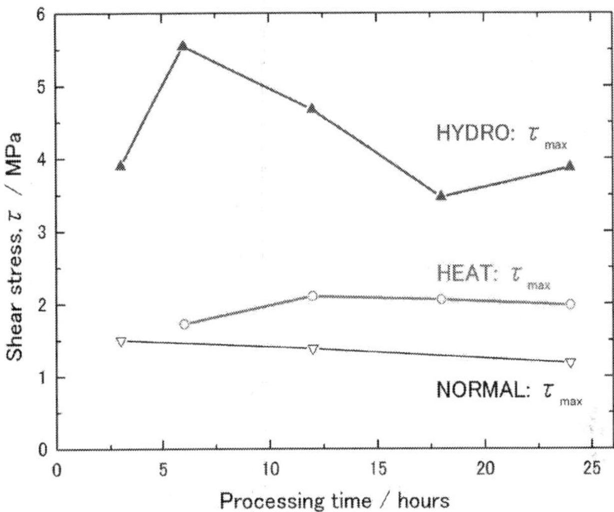

Fig. 19. Relationship between HA coating adherence and hydrothermal treatment time.

5. Micro structure of HA ceramics by HHP method

SEM photograph of the detached HA layer surface is shown in Fig.20. As exemplified in the figure, the microstructure of the HA layers prepared in this study is composed of the number of pores. The maximum dimension of the pores was observed to be approximately 20 μm. The microstructure of the HA layers prepared with the different treatment times was similar to the one shown in Fig.20. The density of the HA layers as determined by the Archimedes method was approximately 1.9 g/cm³, regardless of the different treatment times. The relative density was calculated to be approximately 60 %, assuming the theoretical density of HA (3.16 g/cm³). HA ceramics have been also synthesized using the conventional hydrothermal hot-pressing (HHP) method in our previous section, where a uniaxial load was applied to the starting powers using upper and lower loading rods in order to produce high pressure environments. The density of the HA ceramics prepared by the conventional HHP method has been measured to be approximately 1.9 g/cm³, when the treatment temperature and pressure were the same as the conditions used in this study. The agreement in the density suggests that it may be possible to produce suitable hydrothermal conditions employing the HHP and the DC-HHP method developed in this study. It is seen that the density of the HA layers prepared falls in the range of those for human bone (1.6-2.1 g/cm³). Therefore, it may be possible to adjust and tailor the elastic modulus of HA coatings using the DC-HHP method and to mitigate the stress shielding due to the misfit in the elastic modulus [Hench, 1998]. The above-mentioned physical properties and the porous microstructure of the HA ceramic coatings prepared by the DC-HHP method may be beneficial to enhance the osteoconductivity and osteointegrativity of orthopedic and dental implant materials in comparison with dense HA ceramic coatings. It was confirmed that pigmented ink fully penetrated the HA ceramics made by the HHP, and that the HA ceramics had open pore structure.

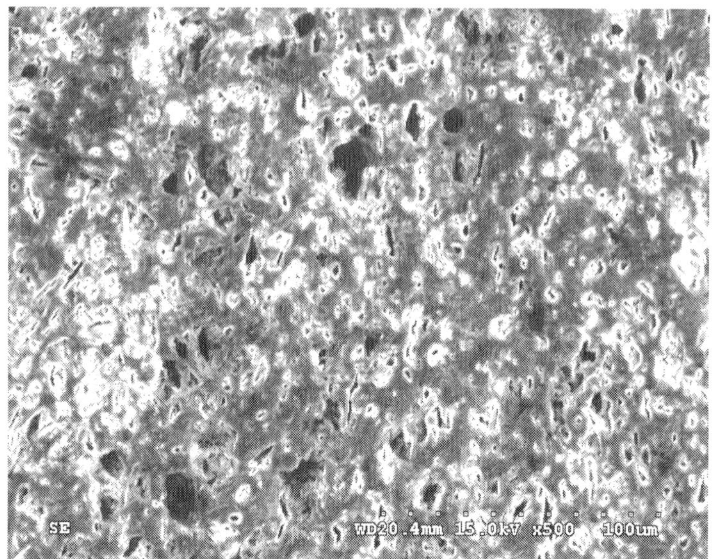

Fig. 20. SEM Photograph of the HA surface produced by the HHP method.

6. Conclusions

In this chapter, Ti surfaces were characterized with X-ray photoelectron spectroscopy (XPS) in order to explain the mechanism of bonding Ti and hydroxyapatite (HA) ceramics by hydrothermal hot-pressing (HHP). XPS characterization revealed differences in the Ti surfaces properties between water and air in finishing circumstances. Compared with the air finishing Ti samples, the O 1s region XPS spectra of the water finishing Ti samples was significantly assigned hydroxide or hydroxyl groups (OH-) and hydrate and/or adsorbed water (H_2O). It was clarified that Ti surfaces finished in only water environment could achieve the bonding to HA ceramics.

Secondly,it was investigated how Ti surface modifications affect the interface fracture toughness in HA/Ti bodies prepared by the hydrothermal hot-pressing method. It was revealed that hydrothermal treatment technique, in which the Ti specimen was exposed to 5M NaOH at 423 K for 2hours, was most effective. The hydrothermal method provided the highest interface fracture toughness (0.35 MPam$^{1/2}$) among the Ti surface modification methods used in this study, while the interface fracture toughness for the other two methods was slightly lower than that obtained for the specimen with no Ti surface modification (0.25 MPam$^{1/2}$). The enhancement of the interface fracture toughness for the hydrothermal surface modification was probably due to the presence of anatase formed on the Ti surface and the good adhesion in the reaction layer. It was shown that the first successful attempt to form direct bonding between the Ti-based bulk metallic glass: $Ti_{40}Zr_{10}Cu_{36}Pd_{14}$ and HA bulk ceramics. The bonding could be obtained in only cases of the BMG with the GIL. It was demonstrated that a series of hydrothermal techniques could be very useful for bonding bulk ceramics and bulk metallic materials. The surface of the Ti-based BMG can be made bioactive by coating bioactive ceramics like as hydroxyapatite (HA) through the low temperature techniques in the range of RT and 150°C. In order to form a growing integrated layer (GIL) on the BMG surface for improving adhesive properties to HA ceramics, the BMG substrates needs to be treated in 5mol/L NaOH solution at 90°C for 120 minutes by hydrothermal-electrochemical techniques. Hydrothermal hot-pressing (HHP) treatment (150°C, 40MPa, 2hours) of the BMG and powder mixture of $CaHPO_4 \cdot 2H_2O$ and $Ca(OH)_2$ is appropriate way for bonding the BMG and HA ceramics because of low operating temperature.

Additionally, it was demonstrated that HA ceramics could be coated to Ti rods at the temperature as low as 135°C by using the newly developed double capsule HHP method (DC-HHP method). Thickness of the HA coating prepared was approximately 50μm. No chemical decomposition and no impurity was observed in the XRD analyses of the HA coating prepared by DC-HHP method. The HA coating layer was shown to have a porous microstructure with the density of 1.9 Mgm-3 and the relative density of approximately 60 %, which is relatively close to those of human bone. In order to evaluate the adhesion properties of the HA coatings on the Ti rods, pull-out tests were conducted. It was revealed that the crack propagated not along the HA/Ti interface but within the HA ceramic layer in the pull-out tests. The fracture property of the HA/Ti interface was suggested to be close to or higher than that of the HA ceramics. The shear strength obtained from the pull-out tests was in the range of 4.0-5.5 MPa. Finally, it was demonstrated that various hydrothermal techniques were very useful and effective to hydroxyapatite ceramics coating on various Ti metallic materials.

7. Acknowledgment

These works were partly supported by "Grant-in-Aid for Cooperative Research Project of Nationwide Joint-Use Research Institutes on Development Base of Joining Technology for New Metallic Glasses and Inorganic Materials" and "Grant-in-Aid for Young Scientists (B), 18760516" from the Ministry of Education, Science, Sports, and Culture of Japan.

8. References

Aoki, H. (1994). *Medical Applications of Hydroxyapatite*, Ishiyaku Euro-America, ISBN1-56386-023-6, Tokyo.

Asami, K. & Hashimoto, K. (1977). The X-ray photo-electron spectra of several oxides of iron and chromium. *Corrosion Sci.*, 17, 559-570.

Ashby, M. F. & Greer, A. I. (2006). Metallic glasses as structural materials. *Scripta. Mater.*, 53, 321-326.

Bavykin, D. V., Friedrich, J. M. & Walsh, F. C. (2006). Protonated titanates and TiO_2 nanostructured materials: synthesis, properties, and applications. *Adv. Mater.*, 18, 2807-2824.

Beck, T. R. (1973). Electrochemistry of freshly-generated titanium surfaces I. Scraped-rotating-disk experiments. *Electrochem. Acta.*, 18, 807-814.

Greer, A. L. (1995). Metallic glasses. *Science*, 267, 1947-1953.

Hench, L. L. (1998). Bioceramics. *J. Am. Ceram. Soc.*, 81, 1705-1733.

Hanawa, T. & Ota, M. (1992). Characterization of surface film formed on titanium in electrolyte using XPS. *Appl. Surf. Sci.*, 55, 269-276.

Hashida,T. (1993). Fracture toughness testing of core-based specimens by acoustic emission. *Int. J. Rock Mech. Min. Sci. Geomech. Abstr.*, 30, 61-69.

Hosoi, K., Hashida,T., Takahashi, H., Yamasaki, N., & Korenaga, T. (1996). New Processing Technique for Hydroxyapatite Ceramics by the Hydrothermal Hot-Pressing Method. *J. Am. Ceram. Soc.*, 79, 2771-2774.

Inoue, A. (2000). Stabilization of metallic supercooled liquid and bulk amorphous alloys. *Acta Mater.*, 48, 279-306.

Kaneko, S., Tsuru, K., Hayakawa, S., Takemoto, S., Ohtsuki, C., Ozaki, T., Inoue, H. & Osaka, A. (2001). In vivo evaluation of bone-bonding of titanium metal chemically treated with a hydrogen peroxide solution containing tantalum chloride. *Biomaterials*, 22, 875-881.

Kelly, E. J. (1982). Electrochemical behavior of titanium. *Mod. Aspect. Electrochem.*, 14, 319-424.

Kim, H. M., Miyaji, F., Kokubo, T. & Nakamura, T. (1997). Effect of heat treatment on apatite-forming ability of Ti metal induced by alkali treatment. *J. Mater. Sci.: Mater. Med.*, 8, 341-347.

Kim, H. M., Kokubo, T., Fujibayashi, S., Nishiguchi, S. & Nakamura, T. (2000). Bioactive macroporous titanium surface layer on titanium substrate. *J. Biomed. Mater. Res.*, 52, 553-557.

Kokubo, T. & Takadama, H. (2006). How useful is SBF in predicting in vivo bone bioactivity? *Biomaterials*, 27, 2907-2915.

Kokubo, T., Kim, H. M., Kawashita, M. & Nakamura, T. (2004). Bioactive metal: preparation and properties. *J. Mater. Sci. Mater. Med.*, 15, 99-107.

Li, B. Y., Rong, L. J. & Li, Y. Y. (2000). Stress-strain behavior of porous Ni-Ti shape memory intermetallics synthesized from powder sintering. *Intermetallics*, 8, 643-646.

Long, M. & Rack, H. (1998). Titanium alloys in total joint replacement--a materials science perspective. *Biomaterials*, 19, 1621-1639.

Ohtsuki, C., Iida, H., Hayakawa, S. & Osaka, A. (1997). Bioactivity of titanium treated with hydrogenperoxide solution containing metal chlorides. *J. Biomed. Mater. Res.*, 35, 39-47.

Onoki, T., Hosoi, K. & Hashida,T. (2003a). JOINING HYDROXYAPATITE CERAMICS AND TITANIUM ALLOYS BY HYDROTHERMAL METHOD. *Key Eng. Mater.*, 240-242, 571-574

Onoki, T., Tanaka, M., Hosoi, K. & Hashida, T. (2003b) Proceedings of the 15th Symposium on Functionally Graded Materials, ISBN 4-9901902-0-3, Sapporo, Japan, Nov. 20-21 2003, 1-4.

Onoki, T., Hosoi, K. & Hashida, T., (2005). New Technique for Bonding Hydroxyapatite Ceramics and Titanium by Hydrothermal Hot-pressing Method. *Scr. Mater.*, 52, 767-770.

Onoki, T. & Hashida, T. (2006). New method for hydroxyapatite coating of titanium by the hydrothermal hot isostatic pressing technique. *Surf. Coat. Tech.*, 200, 6801-6807.

Onoki, T., Hosoi, K., Hashida, T., Tanabe, Y., Watanabe, T., Yasuda, E. & Yoshimura, M. (2008a). Effects of titanium surface modifications on bonding behavior of hydroxyapatite ceramics and titanium via hydrothermal hot- pressing. *Mater. Sci. Eng. C*, 28, 207-212.

Onoki, T., Wang, X., Zhu, S., Hoshikawa, Y., Sugiyama, N., Akao, M., Yasuda, E., Yoshimura, M. & Inoue, A. (2008b). Bioactivity of Titanium-based Bulk Metallic Glass Surfaces via Hydrothermal Hot-pressing Treatment. *J. Ceram. Soc. Japan*, 116, 115-117.

Onoki, T., Wang, X., Zhu, S., Sugiyama, N., Hoshikawa, Y., Akao, M., Matsushita, N., Nakahira, A., Yasuda, E., Yoshimura, M. & Inoue, A. (2009a). Effects of growing integrated layer [GIL] formation on bonding behavior between hydroxyapatite ceramics and Ti-based bulk metallic glasses via hydrothermal techniques. *Mater. Sci. Eng. B*, 161, 27-30.

Onoki, T., Higashi, T., Wang, X., Zhu, S., Sugiyama, N. Hoshikawa, Y., Akao, M. Matsushita, N., Nakahira, A. Yasuda, E., Yoshimura, M.& Inoue, A. (2009b). Interface structure between Ti-based bulk metallic glasses and hydroxyapatite ceramics jointed by hydrothermal techniques. *Mater. Trans.*, 50, 1308-1312.

Onoki, T. & Nakahira, A. (2010a). Effests of titanium polishing environments on bonding behavior of hydroxyapatite ceramics and titanium by hydrothermal hot-pressing. *Mater. Sci. Eng. B*, 173, 72-75.

Onoki, T., Kuno, T, Nakahira, A. & Hashida, T., (2010b). Effects of titanium surface treatment on adhesive properties of hydroxyapatite ceramics coating by double layered capsule hydrothermal hot-pressing. *J. Ceram. Soc. Japan*, 118, 530-534.

Onoki, T. & Yamamoto, S. (2010c). Hydroxyapatite ceramics coating on magnesium alloy via a double layered capsule hydrothermal hot-pressing. *J. Ceram. Soc. Japan*, 118, 749-752.

Onoki,T., Yamamoto, S., Onodera, H. & Nakahira, A. (2011). New technique for bonding hydroxyapatite ceramics and magnesium alloy by hydrothermal hot-pressing method. *Mater. Sci. Eng. C*, 31, 499-502.

Oswald, M., Hessel, V. & Riedel, R. (1999). Formation of ultra-thin ceramic TiO_2 films by the Langmuir–Blodgett technique – a two-dimensional sol-gel process at the air–water interface. *Thin Solid Films*, 339, 283-289.

Rohanizadeh, R., Al-Sadeq, M. & LeGeros, R.Z. (2004). Preparation of different forms of titanium oxide on titanium surface: Effects on apatite deposition. *J. Biomed. Mater. Res.*, 71A, 343-352.

Takadama, H., Kim, H. M., Kokubo, T. & Nakamura, T. (2001). An X-ray photoelectron spectroscopy study of the process of apatite formation on bioactive titanium metal. *J. Biomed. Mater. Res.*, 55, 185-193.

Wen, H. B., De Wijn, J. R., Cui, F.Z. & De Groot, K. (1998a). Preparation of bioactive Ti6Al4V surfaces by a simple method. *Biomaterials*, 19, 215-221.

Wen, H.B., Van den Brink, J., De Wijn, J.R., Cui, F.Z. & De Groot, K. (1998b). Crystal growth of calcium phosphate on chemically treated titanium. *J. Cryst. Growth*, 186, 616-623.

Wei, M., Kim, H.M. Kokubo, T. & Evans, J.H. (2002a). Optimising the bioactivity of alkaline-treated titanium alloy. *Mater. Sci. Eng. C*, 20, 125-134.

Wei, M., Uchida, M., Kim, H. M., Kokubo,T. & Nakamura, T. (2002b). Apatite-forming ability of CaO-containing titania. *Biomaterials*, 23, 167-172.

Yamasaki, N., Yanagisawa, K., Nishioka, M. & Kanahara, S. (1986). hydrothermal hot-pressing method: apparatus and application. *J. Mater. Sci. Lett.*, 5, 355-356

Yang, B. C., Weng, J., Li, X. D. & Zhang, X. D. (1999). The order of calcium and phosphate ion deposition on chemically treated titanium surfaces soaked in aqueous solution. *J. Biomed. Mater. Res.* 47, 213-219.

Yoshimura, M., Onoki, T., Fukuhara, M., Wang, X., Nakata, K. & Kuroda, T. (2008). Formation of Growning Integrated Layer [GIL] between Ceramics and Metallic Materials for Improved Adhesion Performance. *Mater. Sci. Eng. B*, 148, 2-6.

Zhu, S. L., Wang, X. M., Qin, F. X. & Inoue, A. (2007). A new Ti-based bulk glassy alloy with potential for biomedical application. *Mater. Sci. Eng. A*, 459, 233-237.

4

Fabrication and Antibacterial Performance of Graded Nano–Composite Biomaterials

Anping Xu and Dongbin Zhu
*School of Mechanical Engineering, Hebei University of Technology, Tianjin,
China*

1. Introduction

Solid freeform fabrication (SFF), also termed as layered manufacturing, produces parts directly from a computer model without part-specific tooling and human intervention (Sachlos et al., 2003; Leong et al., 2003; Kim et al., 2009; Dwivedi and Kovacevic, 2004; Cai et al., 2003; Tian et al., 2002; Guo et al., 2002; Cawley, 1999; Calvert et al., 1998; Alimardani and Toyserkani, 2008; Alemohammad et al., 2007; Li, 2005; Bryant et al., 2003). It has great potential to fabricate functionally graded materials (FGMs).

Through selective slurry extrusion (SSE) based technique of SFF, Xu et al. successfully fabricated multi-material dental crown (Xu et al., 2005) and further proposed a novel approach (as shown in Figure 1), termed as equal distance offset (EDO), to representing and process planning for SFF of functionally graded materials so as to meet the requirement of modelling and fabricating 3D complex shaped FGM objects (Xu and Shaw, 2005).

In EDO (Xu and Shaw, 2005), a neutral arbitrary 3D CAD model is adaptively sliced into a series of 2D layers. Within each layer, 2D material gradients are designed and represented via dividing the 2D shape into several sub-regions enclosed by iso-composition contours. If needed, the material composition gradient within each of the sub-regions can be further specified by applying the equal distance offset algorithm to each sub-region. Using this approach, an arbitrary-shaped 3D FGM object with linear or non-linear composition gradients can be represented and fabricated via suitable SFF machines. The process planning for SFF of FGM objects is shown in Figure 2.

In recent years, the inkjet colour printing based SFF technology has been of great interests in tissue engineering (Calvert et al., 2007; Xu et al., 2006; Saunders et al., 2008; Sanchez et al., 2008; Roth, 2004; Hasenbank et al., 2008; Cui and Boland, 2009), as it can fabricate 3D complex shaped graded material with smooth gradients (Wang and Shaw, 2006).

Human teeth have some very good properties such as high hardness and wearability, good heat insulation, high strength, etc. These properties are related to its graded structure. Figure 3 (Tooth anatomy, (2011). http://www.mydr.com.au/first-aid-self-care/tooth-anatomy) shows a drawing of a healthy tooth cut in half lengthways, which shows the layers of the tooth and its internal structure, as well as how the tooth relates to the gum and surrounding jaw bone.

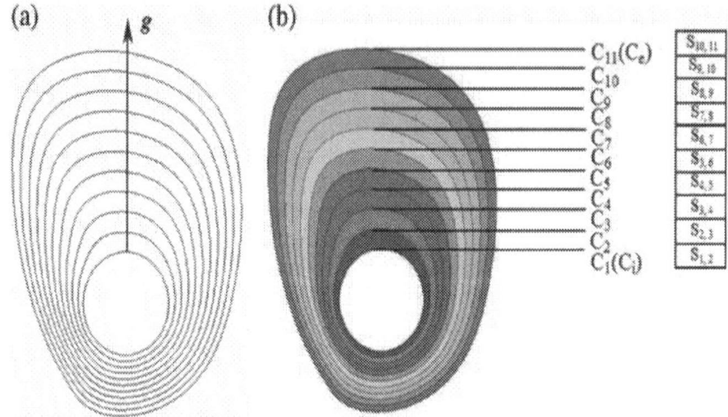

Fig. 1. EDO approach to representing functionally graded material objects.

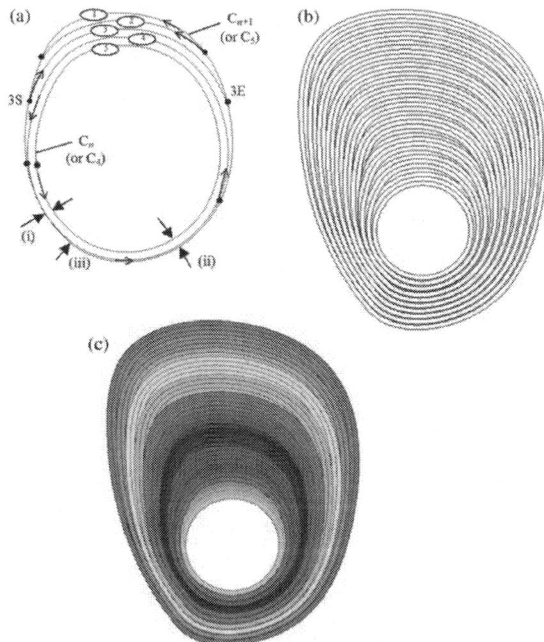

Fig. 2. Process planning for SFF of FGM objects.

A normal tooth can be divided into three parts: the crown, the neck and the root. The crown is the part of the tooth that is visible above the gum (gingiva). The neck is the region of the tooth that is at the gum line, between the root and the crown. The root is the region of the tooth that is below the gum. Some teeth have only one root, for example, incisors and canine (eye) teeth, whereas molars and premolars have 4 roots per tooth.

The crown of each tooth has a coating of enamel, which protects the underlying dentine. Enamel is the hardest substance in the human body, harder even than bone. It gains its hardness from tightly packed rows of calcium and phosphorus crystals within a protein matrix structure. Once the enamel has been formed during tooth development, there is little turnover of its minerals during life. Mature enamel is not considered to be a 'living' tissue.

The major component of the inside of the tooth is dentine. This substance is slightly softer than enamel, with a structure more like bone. It is elastic and compressible in contrast to the brittle nature of enamel. Dentine is sensitive. It contains tiny tubules throughout its structure that connect with the central nerve of the tooth within the pulp. Dentine is a 'live' tissue.

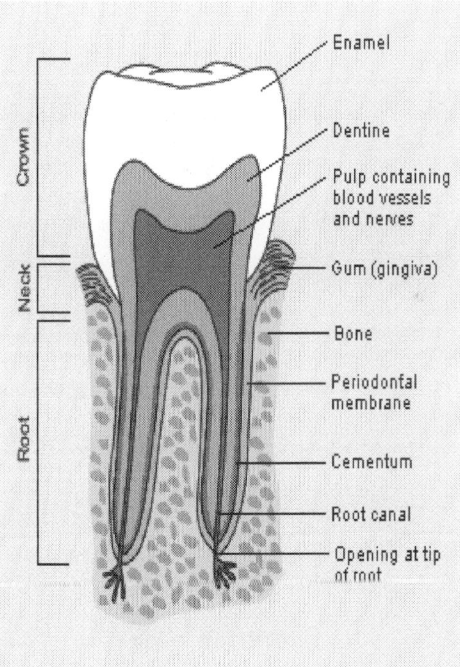

Fig. 3. Internal structure of a normal tooth.

Below the gum, the dentine of the root is covered with a thin layer of cementum, rather than enamel. Cementum is a hard bone-like substance onto which the periodontal membrane attaches. This membrane bonds the root of the tooth to the bone of the jaw. It contains elastic fibres to allow some movement of the tooth within its bony socket.

The pulp forms the central chamber of the tooth. The pulp is made of soft tissue and contains blood vessels to supply nutrients to the tooth, and nerves to enable the tooth to sense heat and cold. It also contains small lymph vessels which carry white blood cells to the tooth to help fight bacteria.

The extension of the pulp within the root of the tooth is called the root canal. The root canal connects with the surrounding tissue via the opening at the tip of the root. This is an opening in the cementum through which the tooth's nerve supply and blood supply enter the pulp from the surrounding tissue.

Unfortunately, dental caries is one of the most important problems in human oral diseases (Namba et al., 1982). It is known that *streptococcus mutans* play one of most important roles in dental caries (Jessica et al., 2007). During metabolizing carbohydrates, *streptococcus mutans* produces organic acid, which can induce the demineralization of tooth surface and results in dental caries (Ooshima et al., 2000).

Some literatures disclosed that the dental plaque pH after a sucrose rinse can decrease to 4.5 or even 4 (Hefferren et al., 1981; Thylstrup et al., 1986). A pH of higher than 6 is considered to be the safe area, a plaque pH of 6.0~5.5 is the potentially cariogenic area, and pH of 5.5~4 is the cariogenic or dangerous area for cavity formation. Therefore, the maintenance of a higher pH value in the plaque is very important for the anticaries.

The fluoride has been used for about five decades in caries prevention. However, it is difficult to control its quantity to a proper level (Nakajo et al., 2008); excessive fluoride is harmful to human body and insufficient fluoride will not take effect for anticaries action. In recent years much attention has been paid to developing fluoride-free techniques that can prevent human teeth from caries(Scherp et al., 1971; Allakera and Ian Douglas, 2009).

Some natural products were reported to be candidates of new anticariogenic substances (Shouji et al., 2000; Matsumoto et al., 1999). However, there is not scientific evidence about the effect of natural tourmaline on the *streptococcus mutans*.

Tourmaline is a kind of electropolar mineral belonging to the trigonal space group of R3m, whose general chemical formula can be written as $XY_3Z_6Si_6O_{18}(BO_3)_3W_4$, where X is Na^+, Ca^{2+}, K^+, or vacancy; Y is Mg^{2+}, Fe^{2+}, Mn^{2+}, Al^{3+}, Fe^{3+}, Mn^{3+}, Cr^{3+}, Ti^{4+} or Li^+; Z is Al^{3+}, V^{3+}, Cr^{3+}, or Mg^{2+}; and W is OH^-, F^-, or O^{2-} (Castañeda et al., 2006). Its Crystal structure of tourmaline is shown in Figure 4 (Fuchs et al., 1998).

The most important feature among the electric properties of the tourmaline is the possession of spontaneous and permanent poles, which would produce an electric dipole, especially in a small powder with a diameter of several microns or less (Jin et al., 2003). Therefore, a strong electric field exists on the surface of a tourmaline powder (Nakamura et al., 1994; Zhu et al., 2008). The electric field effect of tourmaline powders can influence the redox potential of water and regulated the pH value of solution (Xia et al., 2006). The maintenance of pH value is very important for the growth and metabolism of bacteria (Esgalhado et al., 1995).

Therefore, the graded composite materials were fabricated with nano-tourmaline and nano-hydroxyapatite powders by direct inkjet colour printing for the study of inhibiting function on *streptococcus mutans*. Moreover, the mechanism by which tourmaline inhibits the growth and acid production of *streptococcus mutans* was also explained from the view of pH value.

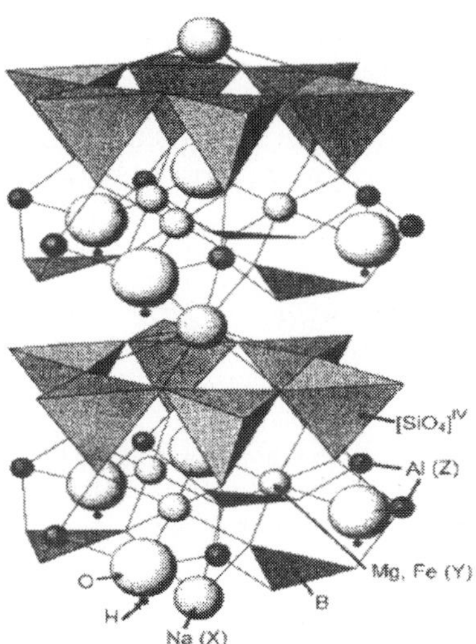

Fig. 4. Crystal structures of tourmaline.

2. Printing setup

The printer consists of a printing table and two piezoelectric XJ126 monochromatic print head (Xaarjet Ltd., Cambridge, UK). There are a total of 134 nozzles viewable on the nozzle plate. The active nozzles are numbered 1 to 126, with nozzle 1 referenced to the datum features on the printhead (as shown in Figure 5), linking a ME1+ continuous ink supply system (Epson, Japan) for the different inks.

Fig. 5. XJ126 printhead (Courtesy of Xaar. See http://www.xaar.com/xaar126.aspx).

A PC104 Programmable Multi-Axis Controller (Delta Tau Ltd. California, USA) for the sliding table and print head movement, and a XaarJet HPC for print head operation were incorporated. The control program was XUSB application in windows XP. The XJ126 print head is a 126 nozzle, piezoelectric drop-on-demand array of print width 17.14 mm and nozzle diameter 50 µm. The sliding table moved at 50 mm/s for printing. The gap between the nozzle plate and the printing surface was maintained at 1 mm by Z displacement. The graded colour pattern was written using Adobe Photoshop software and converted to monochrome binary image files, in device-independent bitmap format, for printing pattern on the substrate. The processing flow of an image is shown in Figure 6.

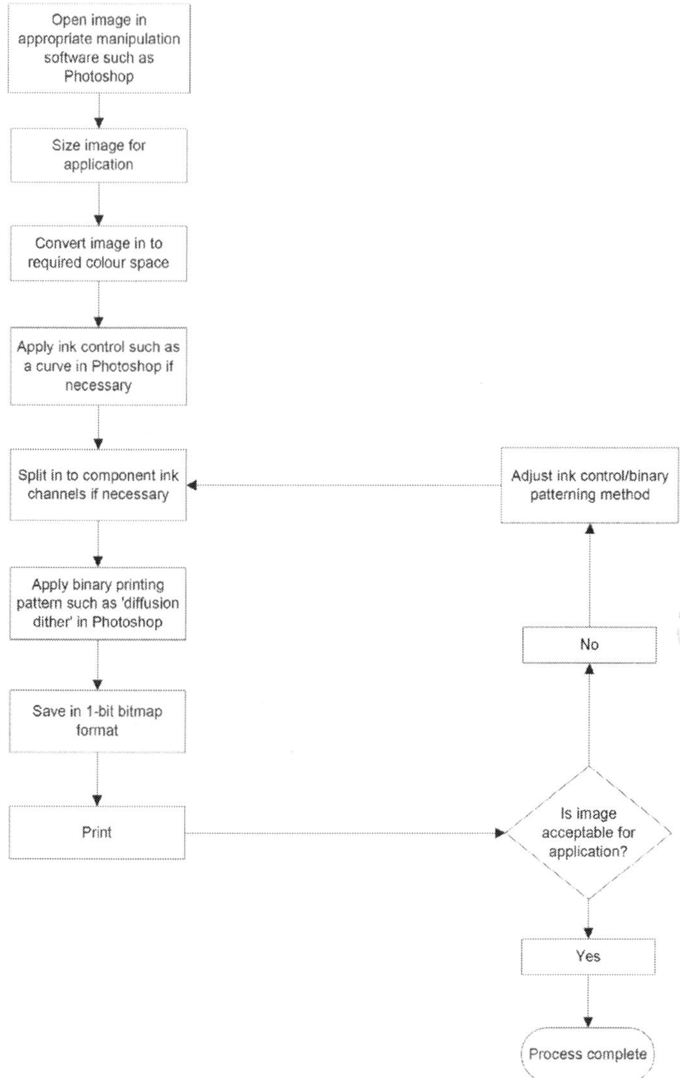

Fig. 6. Processing flow of an image (Courtesy of XAAR, Guide to Operation).

3. Materials and methods

Tourmaline powders were from the Inner Mongolia Autonomous Region of China, whose chemical compositions were analyzed by EDAX Phoenix energy dispersion spectroscope (EDS) and given in the mass ratio as follows: Al_2O_3, 34.98%; B_2O_3, 10.94%; K_2O, 0.036%; Na_2O, 0.91%; CaO, 2.534%; MgO, 0.2%; SiO_2, 34.6%; Fe_2O_3, 15.8%. The preparation of nano-hydroxyapatite powders was reported in Ref. (Shih et al., 2004). The streptococcus mutans was purchased from Tianjin Medical University, China. Brain heart infusion (BHI) broth, mitis salivarius agar (MSA) and phenol red broth were purchased from Hangzhou Tianhe Microorganism Reagent Co., Ltd., China. Other reagents were analytical.

These tourmaline and hydroxyapatite inks with ethyl alcohol as the carrier and oleic acid as the dispersant had a solid loading of 15 and 20 wt%, respectively. After mixing the powders in ethyl alcohol, the suspensions were kept in the ultrasonicator for 1 h. Then, sedimentation experiments for more than 24 h confirmed that this was indeed the case. To provide the visual appreciation of the graded materials fabricated, the tourmaline suspension was mixed with 20 wt% of the commercial XaarJet magenta ink, whereas the hydroxyapatite suspension was mixed with 20 wt% of the XaarJet cyan ink. The *streptococcus mutans* can be cultured on the graded materials.

The compositional profiles were determined using a scanning electron microscope (Philips XL30, Amsterdam, Holland) equipped with energy dispersive spectrometry (EDS). The X-ray diffraction (XRD) analysis was performed on a Philips-X'Pert TMD diffractometer with Cu Ka radiation (λ= 0.154056 nm). The patterns were scanned over the 2θ angular range 10~60° at a scan rate of 0.03° 2θ s^{-1}.

The growth of *streptococcus mutans* was examined at 37 °C in 0.95 ml of BHI broth containing 1% glucose and various concentrations of tourmaline. These tubes were inoculated with 0.1 ml of an overnight culture grown in the BHI broth, and incubated at 37 °C for 1 day. The optical density (OD) of cells was measured spectrophotometrically at 600 nm.

The filter-sterilized tourmaline was added to 0.95 ml of the phenol red broth containing 1% glucose, which was then inoculated with 0.05 ml of the seed culture of *streptococcus mutans*. The cultures were incubated at 37 °C for 1 day, and the pH of the cultures was determined using a precise pH meter (PHS-3B, Shanghai Precision & Scientific Instrument Co., Ltd., China).

Fifteen grams of tourmaline were added with 100 ml of bacterial suspensions. The initial pH of the medium was changed between 5.0 and 11.0, whereas the temperature was constant at 37 °C. The pH of the bacterial suspension at different time was measured with precise pH meter mentioned above.

Ten replicates were made for each test. Data were analyzed using the statistical package for social sciences (SPSS). The data were expressed by the mean ± S.D. Differences between means of the experimental and control groups were evaluated by the Student's t-test. The level of significance for statistical analyses was 0.05.

4. Results and discussion

XRD patterns of the tourmaline powder is shown in Figure 7. From Figure 7, we can see that the main crystal structure of the tourmaline powder does not change apparently and no other phase of tourmaline can be found before and after printing, indicating that the crystal structures of tourmaline powders were not changed. Through calculation, the crystal volume had no change either.

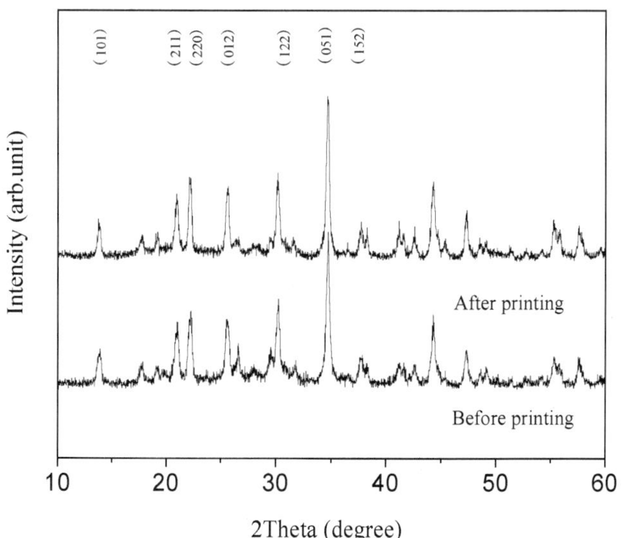

Fig. 7. XRD pattern of tourmaline powders before and after printing.

Graded materials were printed using 10 overprinted layers at a printer defined resolution of 185 dot-per-inch (dpi). As shown in Figure 8(a), the tourmaline in the two graded materials of dimensions 10 mm×10 mm increases with a linear gradient from the left to the right. And in Figure 8(b), the tourmaline in the elliptical two graded object increases with a linear gradient from the center to the edges. These compositional profiles of graded materials match that of the design very well.

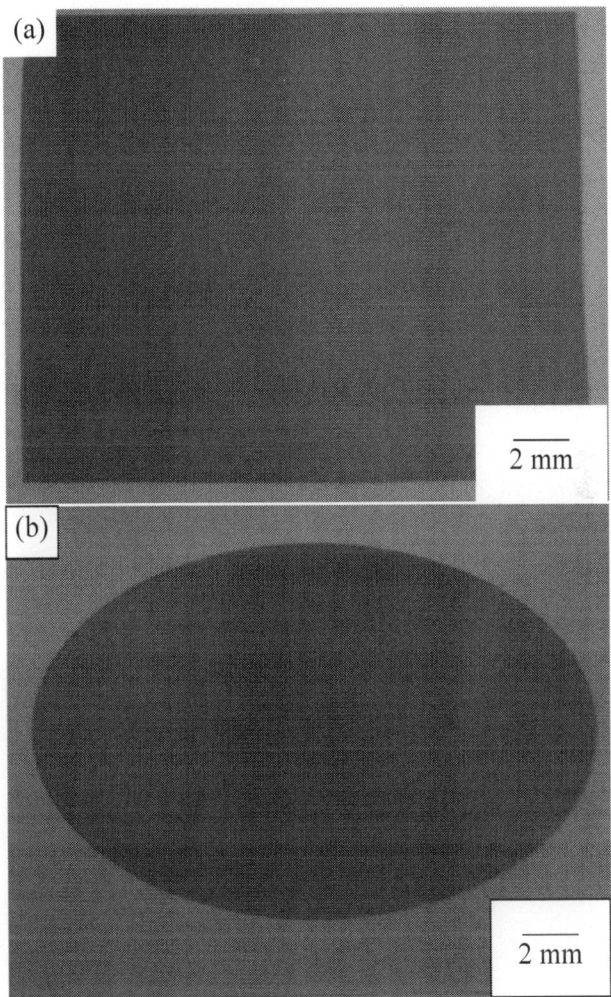

Fig. 8. Graded materials with a linear gradient of increasing tourmaline (a) from the left to the right and (b) from the centre to the edges.

The *streptococcus mutans* was cultured on the graded materials in Figure 8(a). The compositional profiles of the graded materials before and after culturing *streptococcus mutans* determined by using an energy-dispersive spectrometer are shown in Figure 9.

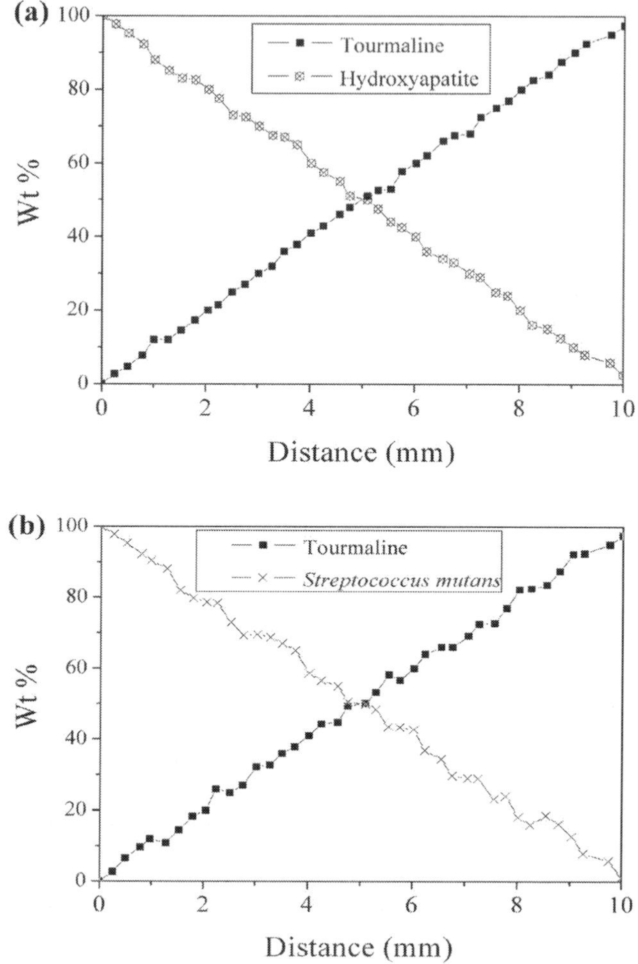

Fig. 9. Graded materials with a linear graded distribution (a) before and (b) after culturing *streptococcus mutans*.

As shown in Figure 9(a), before culturing *streptococcus mutans*, the graded compositional profiles vary in a linear distribution and match the designed composition. After cultured on the graded materials, the *streptococcus mutans* shows a reverse distribution to that of tourmaline (Figure 9(b)). To further investigate the effect of tourmaline on *streptococcus mutans*, the bacteria were exposed to 10, 30, 70, 100, 150, and 200 mg/ml of tourmaline culture. Figure 10 shows the effect of tourmaline on the growth of *streptococcus mutans*.

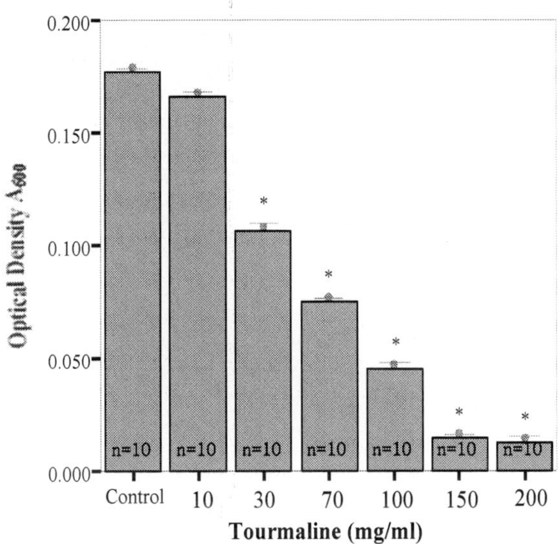

Fig. 10. Effect of tourmaline on the growth of *streptococcus mutans*. *p<0.05 is statistically significant as determined by the Student's *t*-test for the mean values different from the control group.

The bacteria were inoculated to BHI broth with each concentration of tourmaline and anaerobically incubated for 1 day at 37°C. The optical density of A600 was read by a spectrophotometer. From Figure 10, it can be seen that the tourmaline has an antibacterial activity against *streptococcus mutans* in a dose dependent manner, and exhibits significant inhibition at concentrations higher than 30 mg/ml compared to the control group ($p < 0.05$). *Streptococcus mutans* grow in plaque, and releases various organic acids during metabolizing carbohydrates. The organic acids demineralise tooth surfaces and initiate the dental caries. The effect of tourmaline on acid production of *streptococcus mutans* is shown in Table 1.

Concentration (mg/ml)	n	pH
Control	10	5.14 ± 0.023
10	10	5.15 ± 0.028
30	10	5.90 ± 0.072*
70	10	6.87 ± 0.018*
100	10	7.97 ± 0.017*
150	10	8.50 ± 0.202*
200	10	8.52 ± 0.130*

Note: *p<0.05 was statistically significant as determined by the Student's *t*-test for the mean values different from the control group.

Table 1. Effect of tourmaline on acid production of *streptococcus mutans*.

The decrease of pH was significantly inhibited in the presence of tourmaline (30–200 mg/ml) compared to the control group. Except antibacterial activity of tourmaline, the reason may be that the strong electric field of tourmaline would affect the water molecule in the culture and produce plenty of negative ions, which can neutralize the organic acid in the culture.

The comparison of Figure 8 and Table 1 clearly shows that the antibacterial performance of tourmaline has the similar tendency of increasing pH value of the bacterial suspensions. This result gives the support to the suggestion given by Dashper and Reynolds (Dashper and Reynolds, 2000) that *streptococcus mutans* exhibit significantly lower maximum culture OD at pH 7.1 compared with at pH 6.0 and 6.3.

The pH value is very important for the growth and metabolism of bacteria (Xia et al., 2006; Matsumoto et al., 1999). In what follows, we will give the mechanism by which tourmaline acts on the inhibition of *streptococcus mutans* from the view of pH value.

Figure 11 shows the effect of tourmaline on the different pH values of the bacterial suspensions.

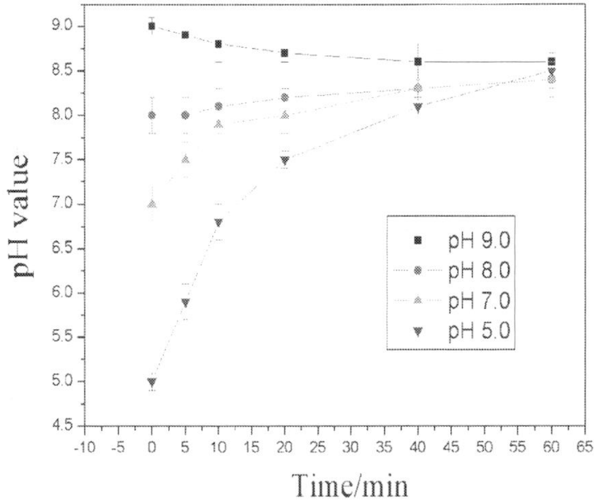

Fig. 11. Effect of tourmaline on the different pH values of the bacterial suspensions.

In Figure 11, 15 grams of tourmaline were added with 100 ml of bacterial suspensions. The initial pH of the medium was changed between 5.0 and 11.0, whereas the temperature was constant at 37°C.

It can be seen that the pH value was shifted toward 8.5 irrespective of whether it was acidic or alkaline. The increase rate of pH value in acidic solution was than the decrease rate in basic solution. Surface adsorption and ion exchange adsorption of H^+ helped the increase rate of pH value in acidic solution more than the decrease rate in basic solution.

The electrode characteristics of tourmaline particles influenced the redox potential of water and regulated the pH value of solution to about 8.5, which was consistent with the results of

Yoshitake et al. (Yoshitake et al., 1996), Zhan et al. (Zhan et al., 2006) and Xia et al. (Xia et al., 2006). The mechanism of which is that the *streptococcus mutans* was remained in the culture of pH at about 8.5 self-adjustment induced by tourmaline. The maintenance of pH value was very important for the growth and metabolism of bacteria. This culture inhibits growth and acid production of *streptococcus mutans*.

5. Conclusions

In this chapter, the graded biomaterial can be fabricated by inkjet printing. Moreover, to the best of our knowledge, this study is the first to use self-adjustment pH value property of tourmaline for inhibiting the growth and acid production of *streptococcus mutans*. As the results show, tourmaline can inhibit the growth and acid production of *streptococcus mutans*, and exhibits significant inhibition at concentrations 30~200 mg/ml, resulting from pH at about 8.5 self-adjustment induced by tourmaline. Therefore, this material has an inhibitory effect on *streptococcus mutans*, and can be used as a promising biomaterial for preventing human teeth from caries.

6. Acknowledgements

The authors would like to thank the "211" Project of Hebei University of Technology, and the Natural Science Foundation of Hebei Province under Grant Numbers: E2004000052 and E2006000039, and the National Natural Science Foundation of China under Grant Number: 60973079, for the financial support.

7. References

Sachlos, E., Reis, N., Ainsley, C., Derby, B., and Czernuszka, J.T., (2003). Novel Collagen Scaffolds with Predefined Internal Morphology Made by Solid Freeform Fabrication, Biomaterials, Vol. 24, No. 8, 1487-1497

Leong, K.F., Cheah, C.M., and Chua, C.K., (2003). Solid Freeform Fabrication of Three-dimensional Scaffolds for Engineering Replacement Tissues and Organs. Biomaterials, Vol. 24, No. 13, 2363-2378

Kim, H.C., Choi, K.H., Doh, Y.H., and Kim, D.S., (2009). Fabrication of Parts and Their Evaluation Using a Dual Laser in the Solid Freeform Fabrication System. Journal of Materials Processing Technology, Vol. 209, No. 10, 4857-4866

Dwivedi, R. and Kovacevic, R., (2004). Automated Torch Path Planning Using Polygon Subdivision for Solid Freeform Fabrication Based on Welding. Journal of Manufacturing Systems, Vol. 23, No. 4, 278-291

Cai, K., Guo, D., Huang, Y., and Yang, J., (2003). Solid Freeform Fabrication of Alumina Ceramic Parts through a Lost Mould Method. Journal of the European Ceramic Society, Vol. 23, No. 6, 921-925

Tian, J., Zhang, Y., Guo, X., and Dong, L., (2002). Preparation and Characterization of Hydroxyapatite Suspensions for Solid Freeform Fabrication. Ceramics International, Vol. 28, No. 3, 299-302

Guo, D., Cai, K., Nan, C., Li, L.T., and Gui, Z.L., (2002). Gelcasting based solid freeform fabrication of piezoelectric ceramic objects. Scripta Materialia, Vol. 47, No. 6, 383-387

Cawley, J.D., (1999). Solid Freeform Fabrication of Ceramics. Current Opinion in Solid State and Materials Science, Vol. 4, No. 5, 483-489

Calvert, P., O'Kelly, J., and Souvignier, C., (1998). Solid Freeform Fabrication of Organic-inorganic Hybrid Materials. Materials Science and Engineering: C, Vol. 6, No. 2-3, 167-174

Alimardani M. and Toyserkani, E., (2008). Prediction of Laser Solid Freeform Fabrication Using Neuro-fuzzy Method. Applied Soft Computing, Vol. 8, No. 1, 316-323

Alemohammad, H., Toyserkani, E., and Paul, C.P., (2007). Fabrication of Smart Cutting Tools with Embedded Optical Fiber Sensors Using Combined Laser Solid Freeform Fabrication and Moulding Techniques. Optics and Lasers in Engineering, Vol. 45, No. 10, 1010-1017

Li, X.X., Wang, J.W., Shaw, L.L., and Cameron, T.B., (2005). Laser Densification of Extruded Dental Porcelain Bodies in Multi-material Laser Densification Process," Rapid Prototyping Journal, Vol. 11, No. 1, 52-58

Bryant, Frances D., Sui, G., Leu, Ming C., (2003). A Study on Effects of Process Parameters in Rapid Freeze Prototyping, Rapid Prototyping Journal, Vol. 9, No. 1, 19-23

Xu, A.P., Qu, Y.X., Wang, J.W., and Shaw, L.L., (2005). Design for Solid Freeform Fabrication of Dental Restoration. Proceedings of the International Conference on Mechanical Engineering and Mechanics in Nanjing, China, October 26-28, 1216

Xu, A.P. and Shaw, L.L., (2005). Equal Distance Offset Approach to Representing and Process Planning for Solid Freeform Fabrication of Functionally Graded Materials. Computer- Aided Design, Vol. 37, No. 12, 1308-1318

Calvert, P., (2007). Inkjet Printing Technology Offers a Way to Create Three-dimensional Biological Structures for Studying Cell Interactions and Artificial Organs. Science, Vol. 318, 208-209

Xu, T., Gregory, C.A., Molnar, P., Cui, X.F., Jalota, S., Bhaduri, S.B., and Boland, T., (2006). Viability and Electrophysiology of Neural Cell Structures Generated by the Inkjet Printing Method. Biomaterials, Vol. 27, No. 19, 3580-3588

Saunders, R.E., Gough, J.E., and Derby, B., (2008). Delivery of Human Fibroblast Cells by Piezoelectric Drop-on-demand Inkjet Printing. Biomaterials, Vol. 29, No. 2, 193-203

Sanchez, V., Madec, M.B., and Yeates, S.G., (2008). Inkjet Printing of 3D Metal-insulator-metal Crossovers. Reactive and Functional Polymers, Vol. 68, No. 6, 1052-1058

Roth, E.A., Xu, T., Das, M., Gregory, C., Hickman, J.J., and Boland, T., (2004). Inkjet Printing for High-throughput Cell Patterning. Biomaterials, Vol. 25, No. 17, 3707-3715

Hasenbank, M.S., Edwards, T., Fu, E., Garzon, R., Kosar, T.F., Look, M., Mashadi-Hossein, A., and Yager, P., (2008). Demonstration of Multi-analyte Patterning Using Piezoelectric Inkjet Printing of Multiple Layers. Analytica Chimica Acta, Vol. 611, No.1, 80-88

Cui X. and Boland T., (2009). Human Microvasculature Fabrication Using Thermal Inkjet Printing Technology. Biomaterials, Vol. 30, No. 31, 6221-6227

Wang, J. and Shaw, L.L., (2006). Fabrication of Functionally Graded Materials via Inkjet Color Printing. Journal of the American Ceramic Society, Vol. 89, No. 10, 3285-3289

Tooth anatomy, http://www.mydr.com.au/first-aid-self-care/tooth-anatomy, being valid as of April 2011.

Namba, T., Tsunezuka, M., and Hattori, M., (1982). Dental Caries Prevention by Traditional Chinese Medicines. Part II. Potent Antibacterial Action of Magnoliae Cortex Extracts against Streptococcus Mutans. Planta Medica, Vol. 44, No. 2, 100-106

Jessica Welin-Neilands and Gunnel Svensäter, (2007). Acid Tolerance of Biofilm Cells of Streptococcus Mutans. Applied and Environmental Microbiology, Vol. 73, No. 17, 5633-5638

Ooshima, T., Osaka, Y., Sasaki, H., Osawa, K., Yasuda, H., Matsumura, M., Sobue, S., and Matsumoto, M., (2000). Caries Inhibitory Activity of Cacao Bean Husk Extract in In-vitro and Animal Experiments. Archives of Oral Biology, Vol. 45, No. 8, 639-645

Hefferren, J.J., Koehler, H.M., Foods, Nutrition and Dental Health. Park Forest South, IL: Pathotox Publishers; 1981, p.138.

Thylstrup, A., Fejerskov, O., Textbook of Cariology. Denmark, Munksgaard: Copenhagen; 1986. 145-146.

Nakajo K., Imazato S., Takahashi Y., Kiba W., (2009). Fluoride Released from Glass-ionomer Cement is Responsible to Inhibit the Acid Production of Caries-related Oral Streptococci. Dent. Mater., Vol. 25, No. 6, 703-708

Scherp, H.W., (1971). Dental Caries: Prospects for Prevention. Science, Vol. 173, No. 4003, 1199-1205

Allaker, R.P., Douglas, C.W.I., (2009). Novel Anti-microbial Therapies for Dental Plaque-related Diseases. International Journal of Antimicrobial Agents, Vol. 33, 8-13

Shouji, N., Takada, K., Fukushima, K., and Hirasawa, M., (2000). Anticaries Effect of a Component from Shiitake. Caries Res., Vol.34, No.1, 94-98

Matsumoto, M., Minami, T., Sasaki, H., Sobue, S., Hamada, S., and Ooshima, T., (1999). Inhibitory Effects of Oolong Tea Extract on Caries-Inducing Properties of Mutans Streptococci. Caries Res., Vol. 33, No. 6, 441-445

Castañeda, C., Eeckhout, S.G., Costa, G.M., Botelho, N.F., Grave, E.D., (2006). Effect of Heat Treatment on Tourmaline from Brazil. Phys. Chem. Miner.,Vol. 33, No. 3, 207-216

Fuchs, Y., Lagache, M., Linares, J., (1998). Fe-tourmaline Synthesis under Different T and f_{o2} Conditions. American Mineralogist, Vol. 83, 525-534

Jin, Z.Z., Ji, Z.J., Liang, J.S., Wang, J., Sui, T.B., (2003). Observation of Spontaneous Polarization of Tourmaline. Chinese Physics, Vol. 12, No. 2, 222-225

Nakamura, Terutaro, Fujishiro, Koji, Kubo, Tetujiro, Iida, Masamori, (1994). Tourmaline and lithium niobate reaction with water. Ferroelectrics, Vol. 155, No. 1-4, 207-212

Zhu, D.B., Liang, J., Ding, Y., Xue, G., Liu, L., (2008). Effect of Heat Treatment on Far Infrared Emission Properties of Tourmaline Powders Modified with a Rare Earth. Journal of the American Ceramic Society, Vol. 91, No. 8, 2588-2592

Xia, M.S., Hu, C.H., Zhang, H.M., (2006). Effects of Tourmaline Addition on the Dehydrogenase Activity of Rhodopseudomonas Palustris. Process Biochemistry, Vol. 41, No. 1, 221-225

Esgalhado, M.E., Roseiro, J.C., Collaco, M.T.A., (1995). Interactive Effects of pH and Temperature on Cell Growth and Polymer Production by Xanthomonas Campestris. Process Biochemistry, Vol. 30, No. 7, 667-671

XAAR, (2006). Guide to Operation of XJ126 High Performance Ink Jet Printhead. Cambridge, United Kingdom

Xaar 126 Product datasheet, http://www.xaar.com/xaar126.aspx, being valid as of April 2011.

Shih, W.J., Chen, Y.F., Wang, M.C., Hon, M.H., (2004). Crystal Growth and Morphology of the Nano-sized Hydroxyapatite Powders Synthesized from CaHPO4·2H2O and CaCO3 by Hydrolysis Method. Journal of Crystal Growth, Vol. 270, No. 1-2, 211-218

Dashper, S.G., and Reynolds, E.C., (2000). Effects of Organic Acid Anions on Growth, Glycolysis, and Intracellular pH of Oral Streptococci. J. Dent. Res., Vol. 79, No. 1, 90-96

Yoshitake, N., Ayumu, Y., Kazuya, O., (1996). pH Self-Controlling Induced by Tourmaline. Journal of Intelligent Material Systems and Structures, Vol. 7, No. 3, 260-263

Zhan, J., Ge, B., Wang, P., Wang, Z., Zheng, C., Huang, B., and Jiang, M., (2009). Study on the Microstructure of Natural Tourmaline and Mechanism of Its Influence on pH Value of Water. Functional Materials, Vol. 40, No. 4, 556-559

5

Self–Assembled Peptide Nanostructures for Biomedical Applications: Advantages and Challenges

Jaime Castillo-León, Karsten B. Andersen and Winnie E. Svendsen
DTU Nanotech, Technical University of Denmark,
Denmark

1. Introduction

Over the last 20 years, self-assembled nanostructures based on peptides have been investigated and presented as biomaterials with an impressive potential to be used in different bionanotechnological applications such as sensors, drug delivery systems, bioelectronics, tissue reparation, among others. Several advantages (mild synthesis conditions, relatively simple functionalization, low-cost and fast synthesis) confirm the promise of these biological nanostructures as excellent candidates for such uses.

Through self-assembly, peptides can give rise to a range of well-defined nanostructures such as nanotubes, nanofibers, nanoparticles, nanotapes, gels and nanorods. However, there are several challenges that have yet to be extensively approached and solved. Issues like controlling the size during synthesis, the stability in liquid environments and manipulation have to be confronted when trying to integrate these nanostructures in the development of sensing devices or drug-delivery systems. The fact that these issues present difficulties is reflected in the low number of devices or systems using this material in real applications.

The present chapter discusses these challenges and presents possible solutions. A review of the state-of-the-art work concerning the use of peptide self-assembled structures in biomedical applications is given. Additionally, our findings regarding the on-chip synthesis of peptide self-assembled nanotubes and nanoparticles, their controlled manipulation, as well as electrical and structural characterizations are introduced. Our latest results showing the interaction of peptide self-assembled structures with cells for the development of a combined sensing/cell culture platform and the use of these material in clean-room processes together with the stability of the biological structures in liquid are also presented.

2. Peptide nanostructures formed by self-assembly

The field of biological self-assembly is very diverse and the structures formed can vary tremendously in both shape and size. For this reason, a full description of all possible self-assembled structures and the monomers forming them is beyond the scope of this chapter. Rather, the focus will be on the applications and challenges that one needs to be aware of when working with such structures. For this, it is important to have a certain understanding of the process behind the formation and this section therefore provides a brief introduction to the concepts behind self-assembly along with a short description of the most important structures that can be formed though self-assembly. Reviews have been written about each of the different structures and we therefore by no means claim to provide an exhaustive account of these configurations.

The structures formed by hydrophobic dipeptides and those formed by diphenylalanine will be given special attention since they are able to give rise to nanotubes, nanofibers and nanoparticles depending on the formation conditions. Furthermore, these structures will serve as model materials throughout the chapter to illustrate the challenges faced when working with the self-assembly of structures. Table 1 gives an overview of some self-assembly structures. We recommend that the interested reader consult the references in the table for further information.

Structure	Example of Monomer	Relevant references
Nanotubes	Hydrophobic dipeptides, cyclic peptide, linear peptides	(Bong et al., 2001, Gorbitz, 2001, Nelson et al., 1997)
Nanofibers	Hydrophobic peptides, beta sheets	(Kim et al., 2010, Scanlon & Aggeli, 2008, Wiltzius et al., 2009)
Nanoparticles	Boc – Diphenylalanine	(Adler-Abramovich & Gazit, 2008, Bohr et al., 1997, Krebs et al., 2007, Nigen et al., 2010, Slotta et al., 2008)
Nanotapes	P_{11}-II	(Fairman & Akerfeldt, 2005, Fishwick et al., 2003)
Gels	K24	(Aggeli et al., 1997, Fairman & Akerfeldt, 2005)

Table 1. Overview of several structures that can be formed through self-assembly. For each category, examples of monomers capable of forming the particular structure are given. Further information regarding the different structures can be found in the articles mentioned in the last column.

Even though the specific configuration of the different monomers, as well as the structures formed from them, vary a lot both in shape and size, the overall driving force behind the formation procedure is the same: entropy. Self-assembly of the monomers occurs when it is thermodynamically favorable for them to be incorporated in the particular self-assembled structures rather than be exposed to the solvent from all sides. The structures listed in Table 1 are described in more detail below.

2.1 Nanofibers

Self-assembling nanofibers have long been a focal point for researches all around the world and for good reason. These structures have received much attention as potential building blocks for the next generation of biosensors. They have been considered both as a fabrication material and as important components in the final device (as electrode modification or as the central part of a biological field effect transistor (BioFET)).

Many different structures have been shown to self-assembly into nanofiber-like conformations. The most well-known of these are undoubtedly the amyloid fibrils. These beta sheets of aminoacids stack together in aggregates to form long insoluble fibrils. The insolubility of the structures can be harmful in the body and, for instance, Alzheimer's disease is caused by such an aggregation of the amyloid beta 42 protein fragments. However, in biosensor applications the insolubility of the self-assembled nanofibers is highly desirable since it will insure the long-term stability of the sensor.

2.2 Nanotubes

Overall, the shape of nanotubes is very similar to that of the nanofibers discussed above; the difference being that the nanotubes are hollow. Nanotubes can therefore be used in much of the same applications as nanofibers and can furthermore be employed in implementations where the cavity inside the structures is loaded with drugs or for reducing metal ions to form nanowires of metal covered with a peptide shell for easy functionalization purposes (Reches & Gazit, 2003).

Nanotubes can be obtained from a large variety of monomers such as cyclic peptides, as demonstrated in (Ghadiri et al., 1993) and (Tarek et al., 2003), linear peptide fragments such as phenylene ethynylene oligomers (Kim et al., 2010, Slotta et al., 2008) and disc-shaped motifs. Another structure that shares the tubular configuration is that of the rosette nanotubes formed from a heteroaromatic bicyclic base (Fenniri et al., 2002). Figure 1 offers illustrations of these different formation processes.

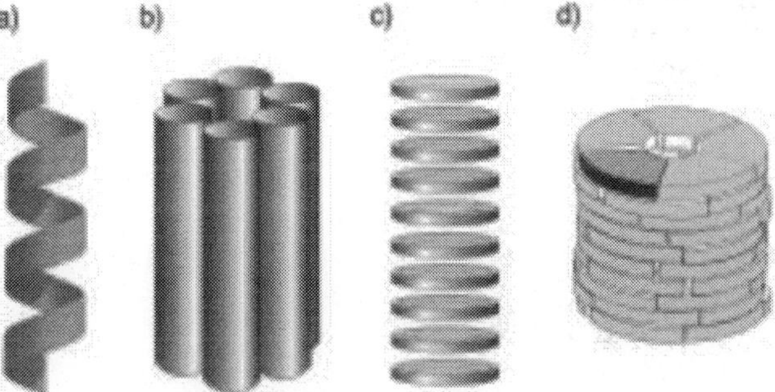

Fig. 1. Examples of self-assembly processes that can give nanotube structures: coiled peptide fragment similar to the alpha helices in proteins, supramolecular assembly of tubular structures to form larger tubes, stacking of cyclic peptide motifs, and an arrangement of disc-shaped motifs in larger nanotubes. Reprinted from (Bong et al., 2001).

2.3 Nanoparticles

The field of nanoparticles is somewhat diverse and covers the small well-defined structures formed by different monomers. Such structures range from nanospheres with a hollow core to various solid structures. The hollow particles have received much attention as possible drug delivery candidates and their non-hollow counterparts as biological variants of nanobeads.

2.4 Nanotapes

Peptide nanotapes are formed from the stacking of peptide beta sheets as described in (Fishwick et al., 2003). The formed tape structure often interacts to form double layers with similar nanotape structures. When the concentration of these nanotapes exceeds a certain threshold, they tend to form hydrogels as described below (Aggeli et al., 1997).

2.5 Hydrogels

Hydrogels based on peptide structures constitute interesting "smart" materials in which the properties of the gel change depending on various parameters such as pH, ionic strength, temperature and salinity (Aggeli et al., 2003). These hydrogels can be formed by a number of different self-assembled structures such as for instance the nanotapes mentioned above (Fairman & Akerfeldt, 2005).

2.6 Hydrophobic dipeptides

Several structures formed through a self-assembly process have been described very briefly above. To provide the reader with an idea of the structure of such self-assembled materials, a more thorough description of the conformations formed by the hydrophobic dipeptide diphenylalanine is presented below.

Hydrophobic dipeptides, and in particular diphenylalanine, are rather unique monomers since – depending on the formation condition – they can self-assemble into nanotubes or nanofibers. Moreover, with addition of a protective group at the amino group of the peptide, they can also form spherical nanoparticles. The structures formed from this material will be used to illustrate the various topics covered in the present chapter and are therefore described in more detail.

The diphenylalanine monomer has named a full class of self-assembling molecules known as the FF class in which the monomers have in common that they are hydrophobic dipeptides with somewhat bulky side-chains. Some of the monomers in this structure class are illustrated in Figure 2. This section offers only a brief description of the assembled structure, whereas a more detailed one of the self-organizing process from hydrophobic dipeptides can be found in (Aggeli et al., 2003, Gorbitz, 2007, 2001, 2006).

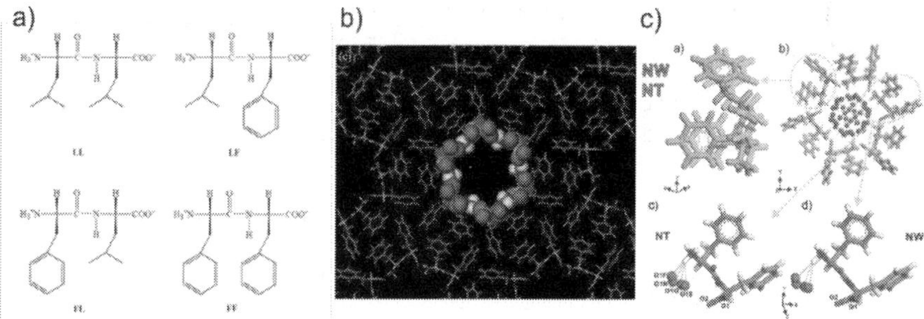

Fig. 2. Structures of different hydrophobic dipeptides that form nanotubes according to the diphenylalanine peptide shown in the lower right corner. Reprinted with permission from (Aggeli et al., 2003, Gorbitz, 2007, 2001).

Figure 2a) illustrates different examples of hydrophobic dipeptides, ranging from the bulky diphenylalenine to the somewhat smaller dileucine. All of the structures shown in the figure have in common that the side-chains of both the involved amino acids are somewhat hydrophobic - hence the name and the fact that they as any peptides have a hydrophilic backbone along with free amino- and carboxylic groups in either end of the molecule. The most favorable configuration of the peptide is with two side-chains sticking out on opposite sides. However, when both side-chains possess hydrophobic properties, it is thermodynamically favorable to have them both on the same side of the peptide backbone, thereby facilitating the shielding of these hydrophobic regions. Figure 2b provides the structure of the nanotube formed by this monomer and Figure 2c compares the position of the monomers in the nanotubes and nanofibers.

The nanotubes are formed in solution by the addition of the peptide monomers to water, whereas the nanofibers are grown from an amorphous peptide film on a substrate in an aniline vapor. To form nanoparticles with this particular monomer, a tertbutoxycarbonyl group is attached to the amino group of the dipeptide (Adler-Abramovich & Gazit, 2008). When these modified dipeptides are emerged into a mixture of ethanol and water the nanoparticles are formed within minutes.

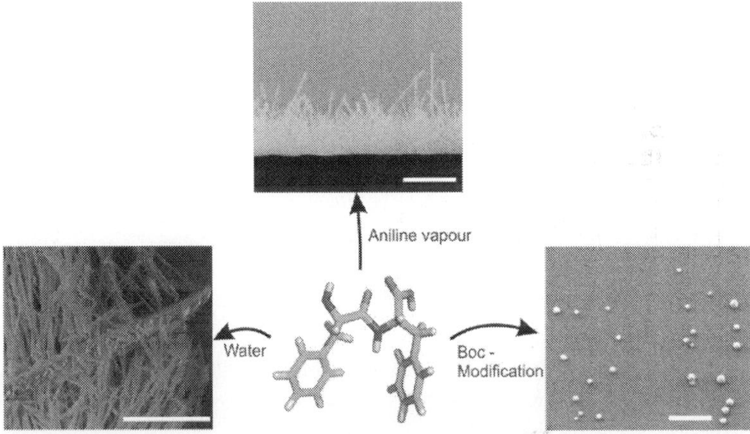

Fig. 3. SEM images of the three different structures that can be obtained from dipeptide diphenylalanine. Right) The nanotubes are formed in aqueous solution. Top) The peptide nanofibers that are grown from an amorphous film in an aniline vapor. Left) Nanoparticles formed from a modified version of the peptide monomer where the amino group is protected by a Boc group. The scale bar in the images corresponds to 1 μm.

Figure 3 illustrates the structure of the nanotubes formed by the diphenylalanine dipeptides. On the left one can see an image of how these smaller nanotubes assemble into larger microcrystals with radii up to a couple of micrometers. Both scale bars correspond to 1 μm and hence the single nanotube formed from diphenylalanine has a diameter around 200 nm. The top image presents a SEM image of the peptide fibers grown from a substrate. Here, it can be seen that the nanofibers are somewhat similar in width and length (both depend on the condition during formation). Finally, the right image shows the nanoparticles formed from the Boc modified diphenylalanine. The size of these spherical nanoparticles ranges from 50 nm to hundreds of nanometers.

All of the organization processes mentioned above are driven by entropy through hydrophobic - hydrophilic interactions and are further stabilized by hydrogen bonding between the individual peptide fragments of which the structures consists. This fundamental knowledge will prove very important when the stability of the structures is assessed later in this chapter. The next section is devoted to different approaches for characterizing such structures. In order to understand the results, a general idea behind the formation process is necessary.

3. Characterization of self-assembled peptide nanostructures

Before one starts working with new materials it is important to characterize the materials in question to determine in which application the structures can be used. If for instance the structure turns out to have isolating properties, it may prove challenging to employ it in biological field effect transistors and one should either modify the materials or find another application. This section presents different means of characterizing the self-organizing structures with respect to their electrical, physical and chemical properties. Of course, these characteristics can be determined in a multitude of ways and the optimal approach would be to verify the results using several techniques investigating the same properties. A few examples on how such characterizations can be conducted are described below. The structures formed by diphenylalanine monomers presented in the previous section will be used as an illustrative case.

3.1 Atomic Force Microscopy (AFM)

An atomic force microscope (AFM) can be used in many different modes for the characterization of a number of parameters ranging from the configuration of the formed structures over the conductivity of these to the determination of the Young modulus. This section describes a few of these characterization techniques. The simplest task involves investigating the geometry of the formed structures directly through any of the standard AFM imagining techniques. When the outer geometry of the formed structures has been determined this information can be used later in the investigation. One can for instance employ AFM as an electrical force microscope (EFM) where a potential difference is applied between the tip and the sample. This would determine whether or not the structure is hollow. In such an approach the structure itself is first imaged using standard AFM techniques, after which the tip is raised some nanometers above the surface and kept at this distance using the data from the AFM measurement. The cantilever is then vibrated and the phase shift of the cantilever, which is strongly dependent on the material between the AFM tip and the substrate, is monitored. As a result, the appearance of the signal will differ depending on whether the sample is hollow or solid. A more detailed description of this technique can be found in (Clausen et al., 2011, Clausen et al., 2008).

AFM can also be utilized to determine the conductance of the formed structures using the AFM tip and gold electrodes placed on the substrate in a standard two-point electrical measurement. For this type of investigation, the structures need to be placed in a way where one end of them is on top of a metal electrode on the substrate and the other end extends over the isolating material. The AFM tip can then be positioned on top of the part of the structure that extends outside the metal surface. This way, the distance over which the I-V curve is measured can be varied and, by following the evolution of the signal, the contact resistance can be isolated from the resistance of the structure itself. If the contact resistance dominates the measurement, the resistance of the structure should not change with the distance. However, if the contact resistance can be neglected, the resistance should drop with the distance.

When the geometrical shape of the structures is known, AFM can be used in combination with a modeling approach to determine the Young modulus of the structure in question. The AFM tip is positioned on top of the structure and pressed into it. From the force distance curves, the Young modulus can be derived directly by a theoretical model as described in (Niu et al., 2007) or by a comparison to a simulation as presented in (Kol et al., 2005). One of the main criteria for a successful characterization is knowledge of the overall structure of the material. Since the models that the data are compared to are strongly dependent on the employed structure, the results would vary greatly for different assumptions. If for instance a structure is assumed hollow in the model when it is in fact solid, the interpretation of the force distance curves would lead to a much larger Young modulus than the real value.

The last example involves the direct observation of the thermal stability of the structures under dry conditions as described in (Sedman et al., 2006). In this approach, the AFM tip is placed on top of the structure to be investigated and the position of the tip is monitored as its temperature is increased. The structure thereby degenerates, which causes the tip to move downward as an indication that the temperature limit for this particular structure has been reached.

3.2 Electron microscopy and related techniques

Another widely used characterization tool is the electron microscope, including techniques ranging from scanning electron microscopy (SEM) to transmission electron microscopy (TEM) and related approaches such as Focused Ion Beam (FIB) milling. This section gives a few examples on how such methods can be used to characterize self-organizing structures. As for the AFM analysis, the simplest starting point for any investigation is to determine the geometry of the structure directly. Again, the surface shape of the formed structures can be readily visualized. However, more information can be gained from a thorough SEM analysis when combined with other methods.

By using TEM, it can be easily verified whether the formed structures contain cavities or not. In TEM, high-energy electrons are sent through the material so that the intensity of each point in the image corresponds to the number of electrons passing through that specific point in the sample. In such images – if the structure is indeed hollow – its central part will appear lighter than its walls (Clausen et al., 2011).

In combination with a simple hotplate, the thermal stability of the structures can be investigated as described in (Adler-Abramovich et al., 2006). In this approach, the formed structures are heated to a specific temperature and then investigated with SEM to determine whether they seem unharmed or not. This kind of investigation takes advantage of the good resolution of SEM combined with the fast imaging possibilities.

Finally, the strength of the structure can be investigated using SEM with a built-in FIB source. The structure to be analyzed is simply placed inside a FIB SEM and one measures the amount of time it takes to mill through the sample. By comparing the time to values known for other materials, it is possible to get an idea of the stability of the structure. If a FIB SEM is not available, similar experiments can be conducted in a reactive ion etching (RIE) machine as described in (Larsen et al., 2011).

3.3 Other characterization techniques

Not every parameter of the structures can be determined using these two classes of instruments and many other techniques are available for the characterization of the parameters mentioned above with others. This section presents a few of these tools and focuses on the characterization of parameters that have not been brought up above.

Even though many structures are very stable under dry conditions, the results may be quite different when the same structures are submerged in a liquid. To test the stability of the formed configurations when wet, one needs to find a way of monitoring them when submerged. When this is not possible or when clearer results are demanded, the solution in which the structures are submerged can be monitored for the concentration of the monomers. For this purpose, a sample containing the structures is placed in a fresh solution with no structures or monomers present. Samples are then taken from the solution at different times so that the concentration of the monomers in the formerly fresh solution can be monitored as a function of time. (The concentration of the monomers can be measured using for instance HPLC). If the concentration of the monomers increases continuously over time until equilibrium is reached, this is an indication of the structures being dissolved in that solution. However, if the concentration of the monomers remains constant throughout the experiment then either the structures do not dissolve or dissolve very quickly. The process is described in more detail in (Andersen et al., 2011).

This section has focused on how standard lab tools and traditional imaging equipment can be used in the characterization of various parameters for self-assembled structures. Other standard characterization techniques such as Fourier Transformed Infrared spectroscopy (FTIR), x-ray diffraction (XRD), Differential scanning calorimetry (DSC), Thermogravimetric analysis (TGA) etc. can be employed on self-assembled structures as well as on traditional materials (Ryu & Park, 2008).

4. Advantages of self-assembled peptide nanostructures for biomedical purposes

There are many applications in the biomedical field where self-assembled peptide nanostructures could play an important role as part of biosensing platforms, as efficient drug-delivery systems, as contrast image agents or as a hydrogels for tissue reparation. This section presents the advantages that make this biomaterial such a promising candidate for such applications.

4.1 Synthesis

One of the more interesting properties of self-assembled peptide nanostructures is the fact that their synthesis takes place under non-harsh conditions. These types of biological supramolecular structures are normally fabricated at room temperature, aqueous environments and without using specialized equipment. These parameters mark a huge difference between these biological nanomaterials and nanomaterials traditionally used in nanotechnology such as carbon nanotubes or silicon nanowires of which fabrication implies high temperatures, specialized equipment and in some cases clean-room facilities increasing their production cost.

Additionally, the synthesis of these self-assembled peptide nanostructures varies from a few seconds to several days of incubation. Depending of the building block used, different shapes can be obtained. A special case is the short aromatic dipeptide, diphenylalanine. By varying the synthesis conditions, nanotubes, nanofibers or nanoparticles can be generated (Reches & Gazit, 2006). All these mild fabrication conditions are reflected in the low cost of the fabrication process.

Recently, our group reported on the on-chip fabrication of self-assembled peptide nanotubes and nanoparticles (Castillo-León et al., 2011). These on-chip fabricated structures displayed a more uniform size and shape as compared to counterparts prepared according to the traditional method. Figure 4 shows the microfluidic chip used for the synthesis of self-assembled peptide nanotubes and nanoparticles. In less than a minute, hundreds of these on-chip fabricated self-assembled structures were obtained.

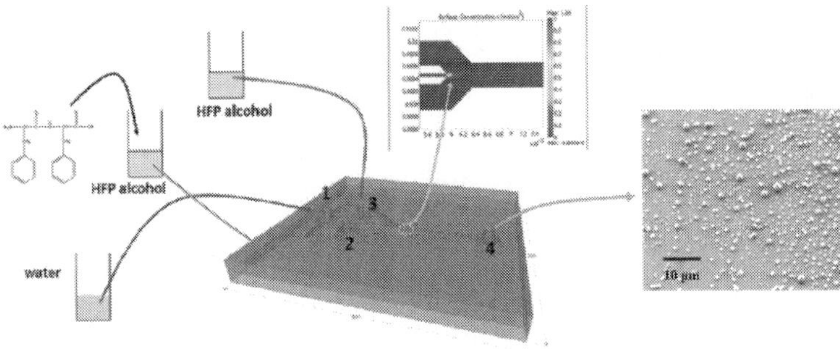

Fig. 4. An illustration of the designed microfluidic chip. The inset shows a simulation of the mixing of different solutions inside the merging channel, where the nanostructures are expected to be formed. SEM image of the on-chip formed self-assembled nanoparticles. Reprinted from (Castillo-León et al., 2011) Copyright (2011), with permission from Elsevier.

As mentioned before, the fabrication of self-assembled peptide nanostructures can, in almost all the cases, be done at room temperature under aqueous conditions. However, two interesting approaches that involve the use of temperatures between 100 and 220 °C were presented by Ryu and Park for the fabrication of vertically well-aligned peptide nanofibers and Adler-Abramovich et al. for the formation of peptide nanotube arrays by chemical vapor deposition (Adler-Abramovich et al., 2009, Ryu & Park, 2008). In both cases the synthesis was performed under non-aqueous conditions, giving nicely organized nanostructure arrays.

As presented in Table 1, there is a long list of peptide monomers that can be used to synthesize self-assembled nanostructures. The final shape and size of the obtained nanostructure will depend on the choice of peptide monomers as well as on synthesis parameters such as pH, solvent polarity, temperature, etc.

4.2 Functionalization

In order to use these biological nanostructures as contrast imaging agents or as part of a biosensing platform they need to be decorated with appropriate functional molecules providing them with specific properties. Functional compounds such as antibodies, magnetic or metallic particles, enzymes, quantum dots or fluorescent compounds have been incorporated into the structure of self-assembled peptide nanostructures (Reches & Gazit, 2007). Ryu and co-workers developed photoluminescent peptide nanotubes by the in-situ incorporation of luminescent complexes composed of photosensitizers such as salicylic acid, cf. Figure 5 (Ryu et al., 2009). Based on this idea, the same group later developed an optical biosensor for the detection of neurotoxins and compounds such as glucose and hydrogen peroxide (Kim et al., 2011, Kim et al., 2011).

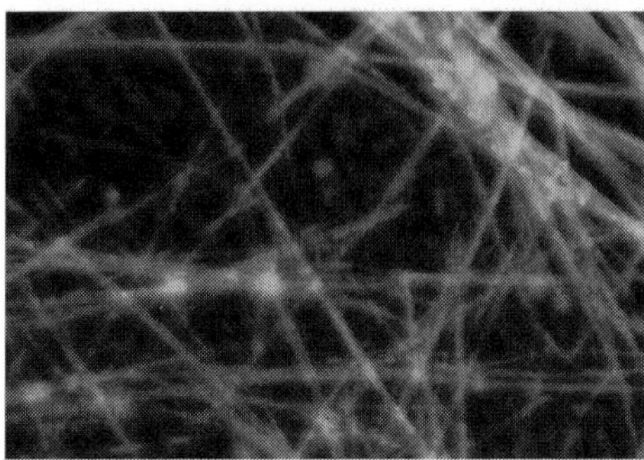

Fig. 5. Optical image of fluorescent self-assembled peptide nanostructures. The nanotubes were functionalized with salicylic acid in order to enhance its fluorescence following the method developed by (Ryu et al., 2009).

An interesting approach for the functionalization of biological self-assembled nanostructures was developed by Kasotakis et al. It involved using a self-organized peptide building block as a scaffold for the systematic introduction of metal-binding residues at specific locations within the structure. By employing an octapeptide from the fiber protein of adenovirus, three new cysteine-containing octapeptides were designed. These synthesized fibrils were able to efficiently bind silver, gold, and platinum nanoparticles (Kasotakis et al., 2009). The metallic-decorated fibers are considered being used in photothermal therapy and in the development of surface-enhanced Raman scattering (SERS) biosensors for detecting DNA by taking advantage of the Raman signal enhancement due to the presence of silver and gold particles. Figure 6 displays transmission electron microscopy (TEM) images of metallic decorated nanofibrils.

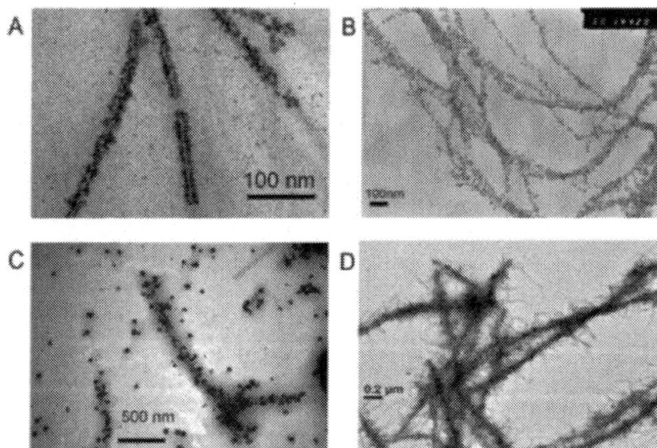

Fig. 6. TEM images of peptide fibrils after incubating with a platinum solution. (A) Fibrils formed from NSGAITIG peptide, scale bar = 100 nm. (B) the NCGAITIG peptide, scale bar = 100 nm. (C) the CNGAITIG peptide, scale bar = 500 nm. (D) CSGAITIG peptide, scale bar = 0.2 µm. Reprinted with permission from (Kasotakis et al., 2009) Wiley 2009.

Another example of the advantages of functionalization of self-assembled peptide nanostructures was presented by Jayawarna and colleagues; they introduced chemical functionality into Fmoc-peptide hydrogels by adding Fmoc-protected amino acids with varying side groups. In this way, an improved 3D in-vitro cell culture system was developed. The functionalized gel showed an enhanced compatibility with the cell culture. Additionally, the properties of the gel could be tuned depending of the type of cell used in order to obtain even better results in terms of compatibility (Jayawarna et al., 2009).

4.3 Biocompatibility

Despite the high levels of attention given to this type of biomaterial, an advanced study to evaluate the biocompatibility and immunogenicity of these nanostructures is still lacking. Such an investigation will bring important information that will define the possibility to use this biomaterial in applications such as drug-delivery systems or tissue reparation in humans. The available studies are limited to growth cells and tissues in the presence of self-assembled peptide nanofibers or hydrogels and assessed whether the exposed cell or tissue growth was affected by the presence of the biomaterial. Next, a couple of examples will be presented and more studies involving the interaction between self-assembled nanostructures and cells or tissues will be listed in Table 2.

Mahler et al. fabricated a self-assembled hydrogel using the Fmoc-diphenylalanine peptide. This hydrogel was used as a cellular support to grow Chinese hamster ovary (CHO) cells. The cellular viability was analyzed using a 3-(4,5-dimethylthiazolyl-2)-2,5-diphenyltetrazolium bromide (MTT) assay. The obtained results demonstrated that cells grown on the hydrogel scaffold showed a viability of more than 90%(Mahler et al., 2006).

In another study, Jayawarna and co-workers investigated the use of Fmoc-dipeptides with different aminoacids as three-dimensional cell culture platforms. Their work confirmed the use of this peptide building block as a good option for cell culture. The investigation involved growing chondrocyte, human dermal fibroblasts and 3T3 fibroblast cells in two and three dimensions (Jayawarna et al., 2006, Jayawarna et al., 2008, Jayawarna et al., 2009, Jayawarna et al., 2007).

Liebmann and collaborators synthesized self-assembling peptide hydrogels to generate patterned 3D cell cultures within patterned silicon microstructures (Liebmann et al., 2007). In their study, astrocyte, MDCK and COS7 cells were used. The results proved how the peptide-derivative hydrogel simplifies both handling and loading of the gel to the microstructures and how the composition and matrix density of the gel could be fully controlled by chemical tailoring processes.

Our group utilized a nanoforrest of self-assembled peptide nanofibers previously used for electrochemical detection of dopamine as a cellular support for the growth of HeLa and PC12 cells (Sasso et al., 2011). As shown in Figure 7, the cells grew and divided without difficulty, proving that the presence of peptide nanofibers did not hinder cell growth. The next step in this study will be to combine the sensing properties of the functionalized nanofibers with the cell culture characteristics in order to develop a combined biosensing/cell culture platform where the cells releasing dopamine will be in intimate contact with the biorecognition element.

A more detailed and complete biocompatibility and immunogenicity assessment will clarify the risks of biomedical use of these materials in humans. More examples of employing self-assembled peptide hydrogels in applications involving interaction with cells or tissues are presented at the end of the chapter in Table 2.

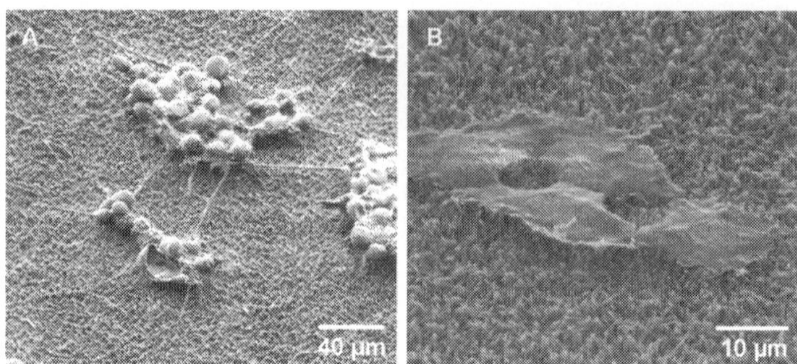

Fig. 7. SEM image of the growth of A) PC12 and B) HeLA cells on a self-assembled peptide nanofiber forest.

5. Challenges when using self-assembled peptide nanostructures for biomedical purposes

Despite the many advantages making self-assembled peptide nanostructures a great candidate for various biomedical applications, challenges regarding size control during synthesis, manipulation and immobilization as well as stability in liquid environments need to be overcome. In certain applications such as the development of drug delivering systems, the size of the nanostructures as well as their stability in different solutions (buffers, acid or basic solutions) is of vital relevance in order to obtain a correct behaviour. If the nanostructure selected for drug-cargo is bigger than the target cell, the delivery process will be difficult and will most probably fail. The same will occur if the biological nanostructure is unstable in liquid solutions causing its structure to disappear after a few seconds in contact with the solution (MaHam et al., 2009). This example highlights the importance of controlling some of these parameters during the synthesis and application of self-assembled structures for biomedical purposes.

This section discusses challenges regarding size control during synthesis, stability in liquid environments, manipulation and conductivity. Moreover, some alternatives to overcome them are presented.

5.1 Size control

As previously mentioned, the control of the size of self-assembled peptide nanostructures is very important in applications such as biosensors or drug-delivery systems. The presence of nanostructures of equivalent dimensions will be reflected in the fabrication of biosensing platforms with similar and reproducible characteristics.

Due to the nature of the self-assembled process, the control of their size during the synthesis is extremely difficult to obtain. However, changes in the fabrication parameters or the use of templates could help to obtain self-assembled nanostructures of similar dimensions.

With polycarbonate membranes as templates, Porrata and co-workers were able to synthesize peptide nanotubes with controlled diameters. The diameters of the peptide nanotubes were as small as 50 nm Porrata (Porrata et al., 2002). A different approach was later applied by Han et al. in order to fabricate self-assembled peptide nanostructures with defined structures and sizes by changing the polarity of the solvents used during the fabrication process (Han et al., 2008). Their results demonstrated that by altering the polarity of the solvent used during the synthesis, diverse morphologies ranging from nanotubes to nanoribbons and nanofibers could be achieved. Despite the importance of their work with regard to understanding natural self-assembly into complicated architectures, their approach allows only a structural control and not one of the sizes of the obtained structures. An elegant study to obtain size-tuneable assemblies of peptide nanofibers was presented by Park et al. One-dimensional self-assembled peptide nanofibers with tuneable sizes were obtained by changing the initial concentration of the monomer used to fabricate them and by controlling the pH environment. Peptide nanofibers with diameters from 110 to 600 nm and lengths from 1.95 to 18 nm were obtained following this method (Park et al., 2009).

5.2 Stability in liquid environments

The use of self-assembled peptide nanostructures in biomedical applications were supramolecular structures will be in direct contact with liquid solutions, e.g., their use as drug delivery systems, imaging contrast agents or as biosensors, requires a good stability in the solution in question. In order to illustrate the importance of this point, we will take as an example the self-assembled nanostructures based on the aromatic dipeptide diphenylalanine. They have been presented as versatile materials with potential uses in different fields like biosensing, microelectronics, drug-delivery, and tissue reparation among others (Scanlon & Aggeli, 2008). However, it has been noted that the nanotubes formed by the diphenylalanine peptide were not very stable once they were in contact with liquid solutions such as water or phosphate buffers, i.e., liquids that are commonly use in biomedical studies (Andersen et al., 2011, Ryu & Park, 2008).

A detailed study was done to evaluate the behavior of these structures under liquid conditions using solvents of common use for biomedical purposes. The investigation made it clear that the structures were completely unstable when dissolved in water, phosphate buffer or methanol, cf. Figure 8. These results will certainly limit their use in many applications. Surprisingly, the stability of nanofibers fabricated using the same dipeptide showed completely different results; they were stable in the presence of these liquids (Andersen et al., 2011).

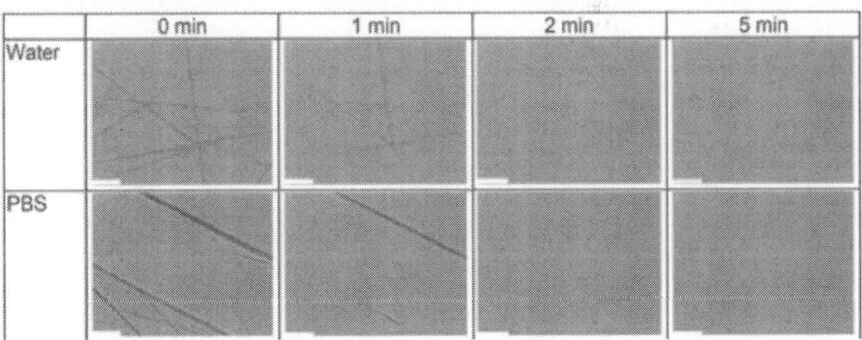

Fig. 8. Microscopy images of the peptide nanotubes dissolved in distilled water and PBS (pH 7,4). From these images, the nanotubes were found to lose their structure in some of the solutions in which it has been claimed that they were stable. All scale bars in the images correspond to 20 μm. Reprinted from (Andersen et al., 2011)- Reproduced by permission of The Royal Society of Chemistry.

As in the case of the size control, the focus of this study was to change the parameters during the synthesis in order to obtain more stable nanotube structures. When changing the pH value of the solution used for the self-assembly of the peptide nanotubes, more stable structures were obtained. The nanotubes fabricated using a buffer of pH 3 instead one of pH 7 displayed a better stability in water and phosphate buffer. This is just a first step towards the fabrication of stable diphenylalanine nanotubes. More studies are required in order to evaluate the implications of the pH change on the properties of this biological nanostructure.

It is also worth mentioning that the discovery of the instability of the diphenylalanine nanotubes in water gave rise to the idea to use these structures as an etching mask material for the rapid fabrication of silicon wires (Larsen et al., 2011). This application is beyond the scope of the present chapter but we mention it to encourage the reader to see the positive side of the results that could initially appear as negative.

5.3 Manipulation

The development of biosensing platforms using self-assembled peptide nanostructures requires the connection between these structures and the appropriate transducers (electrodes). Due to their dimensions, the manipulation of micro and nano-scale biomaterials is a critical issue. Linking of our macroscopic world to the nanoscopic one of nanotubes, nanofibers and nanoparticles is a technological challenge. Fortunately, thanks to advances in micro- and nano-fabrication, several techniques and instruments have emerged to help scientists overcome the obstacles of size when interacting with tiny biological entities (Castillo et al., 2009, Castillo et al., 2011).

Apart from being able to find, grasp, push or move these biological objects to a desired location it is important to avoid any damage to the nanostructure as a result of the interaction between the manipulation instrument and the object to be moved. Altering or changing the structure of the self-assembled nanostructure during the manipulation process could affect the behavior of these structures used in a specific biomedical application.

The manipulation of self-assembled peptide nanostructures could involve the direct contact of the instrument as well as the biological structure, like when using atomic force microscopy (AFM) tips. Another option is to perform the manipulation through non-contact methods as in the case of dielectrophoresis (DEP) or optical tweezers. Next, examples of available manipulation methods are presented.

The thermomechanical manipulation of aromatic peptide nanotubes was demonstrated by Sedman et al. at the University of Notthingham (UK). Using an atomic force microscope as a thermomechanical lithographic tool, Sedman and co-workers were able to manipulate self-assembled nanotubes formed by two aromatic peptides; diphenylalanine and dinapthylalanine. Indents and trenches were thermally etched into the nanostructures, suggesting their possible use as nano-barcodes (Sedman et al., 2009).

Magnetic alignment is a versatile contact-free manipulation method that is effective over the whole sample. With this technique, the manipulation is achieved by magnetic beads that are attached or engulfed by the biological entity to be manipulated. Reches and Gazit aligned nanotubes assembled from diphenylalanine monomers using a external magnetic field as shown in figure 9 (Reches & Gazit, 2006).

However, a principal drawback of magnetic manipulation is the fact that the aligning forces are very small. Consequently, the alignment will occur only when the molecules present in the sample are very large and contain moieties with high susceptibility as in the case of aromatic units.

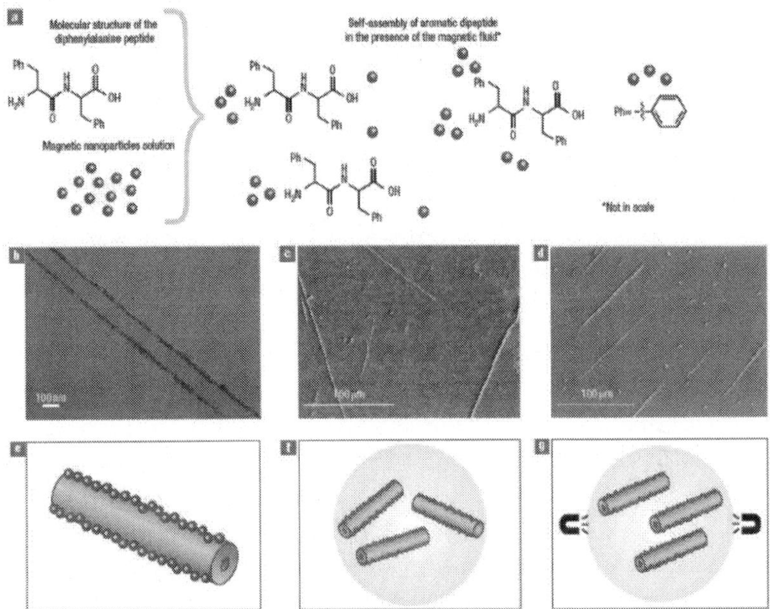

Fig. 9. The self-assembly of the diphenylalanine-based peptide nanotubes in the presence of a ferrofluid and their exposure to an external magnetic field resulting in the control over their horizontal alignment. (a) Schematic representation of the dipeptide monomers self-assembled in the presence of a ferrofluid solution containing magnetite nanoparticles approx. 5 nm in diameter. (b) TEM image of a self-assembled peptide tube coated with magnetic particles. (c) Low-magnification SEM micrograph of the self-assembled magnetic tubes. (d) Horizontal arrangement of the self-assembled magnetic tubes after their exposure to a magnetic field, observed by low-magnification SEM. (e) Schematic representation of the self-assembled magnetic tubes. (f,g) Schematic representations of the magnetic tubes randomly oriented before exposure to the magnetic field (f) and horizontally aligned upon exposure to the magnetic field (g). Reprinted by permission from Macmillan Publishers Ltd: Nature Nanotechnology, (Reches & Gazit, 2006), copyright (2006).

Another contact-free method used for the manipulation of self-assembled nanostructures is dielectrophoresis. Dielectrophoresis occurs when a polarizable object is exposed to an inhomogeneous electric field, so that the Coulomb forces induced on the charges on each half of the dipole differ, causing a net force on the object. The manipulation of self-assembled peptide nanotubes was reported by our group in 2008. Using micro-patterned gold electrodes, peptide nanotubes were manipulated by adjusting the amplitude and frequency of the applied voltage, cf. Figure 10. The electrical characterization of the immobilized peptide nanotubes was studied, both for single peptide nanotubes and bundles of them (Castillo et al., 2008).

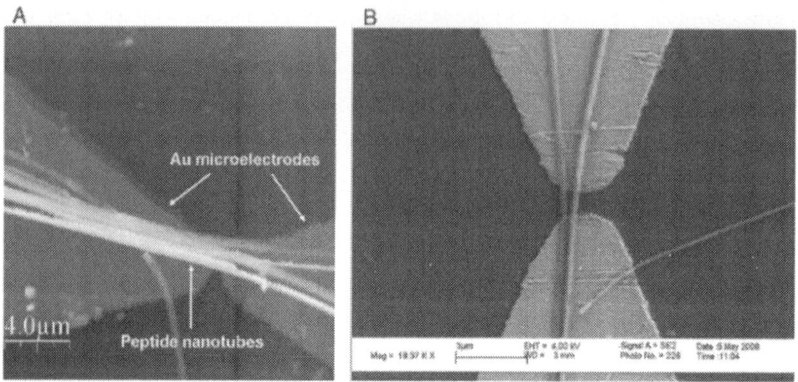

Fig. 10. (A) AFM image of the SAPNT bundles immobilized on Au electrodes using DEP. (B) SEM image of two amyloid peptide nanotubes immobilized on top of gold microelectrodes. Reprinted with permission from (Castillo et al., 2008) Wiley 2008.

At the same time, de la Rica and co-workers used DEP for the incorporation of antibody functionalized peptide nanotubes on top of gold electrodes for the development of a label-free pathogen detection chip. The peptide nanotubes were modified with antibodies against the herpes simplex virus type 2 (HSV-2), and thus the binding of the virus to its antibody was monitored by a capacitance change between the electrodes (de la Rica et al., 2008).

Zhao and Matsui developed a method for the accurate immobilization of antibody-functionalized peptide nanotubes on protein-patterned arrays by optimizing their ligand-receptor interactions (Zhao & Matsui, 2007). Peptide nanotubes self-assembled from bolaamphiphile peptide monomers and bis(N-α-amido-glycylglycine)-1,7-heptane dicarboxylate were coated with antibodies anchored on the amide groups of the nanostructure surface via hydrogen bonding. The antibody-modified nanotubes where then attached to complementary antigen-patterned surfaces as a function of antigen concentration.

Finally, Kumara et al. demonstrated the use of laser tweezers for the trapping and manipulation of self-assembled flagella protein nanotubes. Nanotubes with diameters below 50 nm were optically trapped using a biologically infrared wavelength (1064 nm) laser tweezer without affecting the nanostructure (Kumara et al., 2006).

In summary, manipulation of self-assembled peptide nanostructure was possible without altering the properties or the structure of these biological entities. Contact and contact-free techniques are available for the precise positioning of these samples on desired locations for their integration into more complicated structures such as biosensing platforms.

5.4 Conductivity

The low conductivity of self-assembled peptide nanostructures limits their use in the development of sensing or diagnosis platforms without involving a functionalization step that introduces compounds to help increase their conductivity. Self-assembled peptide nanotubes and nanofibers synthesized from various sources displayed low conductivity values and high resistance ($R>10^{14}\ \Omega$) (Castillo et al., 2008, Scheibel et al., 2003). Conductive polymers, enzymes, or metallic particles are some of the materials used to functionalize the self-assembled nanostructures thereby giving rise to an increase in electrical current conductivity. After the functionalization step, the functionalized nanostructures can be integrated with transducers and used for the detection of compounds of biomedical relevance. Following changes in current, potential or capacitance as a result of the interaction between the analyte and the biorecognition element the presence and quantity of these samples can be confirmed.

Scheibel et al. fabricated conducting nanowires thanks to a controlled self-assembly of amyloid fibers and selective metal deposition (Scheibel et al., 2003). Nanowires containing cysteine residues were used to covalently link colloidal gold nanoparticles on the surface of these nanostructures. And additional metal was then deposited by chemical enhancement of the colloidal gold by reductive deposition of metallic silver and gold from salts. The final biotemplated metal wires showed an increase in conductivity from the pA- to the mA-range, figure 11.

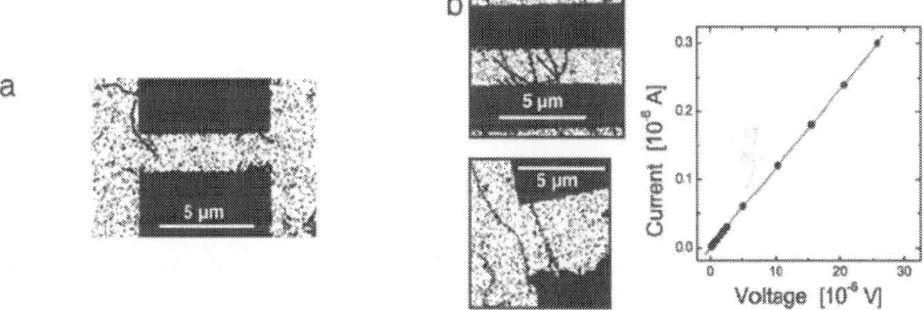

Fig. 11. Electrical behavior of NM-templated metallic fibers. (a) Gold nanowires not bridging the gap between two electrodes did not conduct. (b) Gold nanowires bridging the gap between two electrodes (*Left*) exhibited linear I–V curves (*Right*), demonstrating ohmic conductivity with a low resistance of R =86 Ω (the same for each). Such an ohmic response is indicative of continuous, metallic connections across the sample. Reprinted from Scheibel, T. et al. PNAS, 100, 2003, 4527-4532. Copyright (2003) National Academy of Sciences, U.S.A.

In a different approach, Sasso et al. used a conductive polymer, polypyrrole, in order to increase the conductivity of self-assembled peptide nanofibers. The presence of the conductive polymer on the surface of the nanofibers increased their conductivity and rendered possible their use as the biorecognition element in a dopamine biosensor. By employing this sensor, dopamine detection was possible with a detection limit of 3.1 µM; a value close to the dopamine concentration in in-vivo systems (Sasso et al., 2011).

Enzyme-modified peptide nanotubes have been utilized for the detection of compounds of biomedical relevance such as glucose, ethanol, or hydrogen peroxide as demonstrated by several groups. The enzymes were connected to the self-assembled peptide nanostructures through different matrices such as poly(allylamine hydrochloride) (Cipriano et al., 2010) or glutaraldehyde (Yemini et al., 2005, Yemini et al., 2005). Finally the modified peptide nanotubes were immobilized on the surface of metallic electrodes using polymer matrices like polyethyleneimine. Amperometric detection was used to detect and quantify the presence of the previously mentioned compounds. Based on a similar approach, Yang and co-workers developed a glucose sensor using glucose oxidase functionalized peptide nanofibers. The developed biosensor was able to electrochemically detect glucose in a concentration range required for clinical applications (Yang et al., 2009).

In a different approach, de la Rica et al. employed anti-E. coli and IgG antibodies to functionalize peptide nanotubes against *E. coli* and *S. typhi* cells, which are involved in several diseases. Using these functionalized nanostructures, they fabricated a reusable pathogen biochip able to detect analyte cells in the range from 10^2 to 10^4 cells within one hour. The capturing of the cells by the antibodies was monitored by following changes in the capacitance (de la Rica et al., 2010).

As presented in this section, several alternatives are available in order to improve the conductivity of self-assembled peptide nanostructures and impulse their use as biorecognition elements in biosensors or diagnosis platforms. The permanent advances in surface chemistry and synthesis of new conductive polymers will accelerate the integration of biological self-assembled nanostructures in biosensing applications.

6. Applications

All the previously mentioned advantages make self-assembled peptide nanostructures a very promising biomaterial with huge potential in several biomedical applications. These applications have been reviewed in numerous articles (Aggeli et al., 2001, de la Rica & Matsui, 2010, Gao & Matsui, 2005, Gazit, 2007, Hauser & Zhang, 2010, Kyle et al., 2010, Kyle et al., 2008, Rajagopal & Schneider, 2004, Scanlon & Aggeli, 2008, Woolfson & Ryadnov, 2006, Yan et al., 2010), and a detailed description of some of these applications has already been given in this chapter. In order to provide a broader vision of the possibilities of using these biological nanostructures with biomedical purposes, a list of different self-assembled peptide nanostructures is presented in Table 2. The table lists the type of structure and their use.

Self-assembled nanostructure	Application	Reference
Hydrogel	3D scaffold to stimulate pre-osteoblast (MC3T3-E1) cell attachment and growth	(Zhang et al., 2009)
Hydrogel	Extracellular matrix to grow primary human dermal fibroplasts	(Kyle et al., 2010)
Hydrogel	Injectable delivery	(Branco et al., 2009)
Hydrogel	Therapeutic effects as a wound dressing for the treatment of deep second degree burns in rats	(Meng et al., 2009)
Hydrogel	Carrier for therapeutic proteins	(Koutsopoulos et al., 2009)
Hydrogel	Drug delivery of pyrene	(Li et al., 2009)
Hydrogel	Gene delivery	(Rea et al., 2009)
Hydrogel	Promoting axonal growth in the injured spinal cord	(Ueda et al., 2008)
Hydrogel	Cartilage tissue engineering	(Kisiday et al., 2002)
Hydrogel	Implantation of cardiac progenitor cells	(Tokunaga et al., 2010)
Nanotubes	Detection of neurotoxins	(Kim et al., 2011)

Self-assembled nanostructure	Application	Reference
Nanotubes	Pathogens detection	(de la Rica et al., 2008, de la Rica et al., 2010)
Nanotubes	Immunosensor	(Cho et al., 2008)
Nanotubes	Glucose, ethanol and hydrogen peroxide detection	(Yemini et al., 2005, Yemini et al., 2005)
Nanotubes	Antiviral agent	(Horne et al., 2005)
Nanotubes	Controlled drug release	(Chen et al., 2011)
Nanotubes	Hydrogen peroxide detection	(Cipriano et al., 2010)
Nanotubes	Drug delivery	(von Maltzahn et al., 2003)
Nanofibers	Copper detection	(Viguier et al., 2011)
Nanofibers	Dopamine detection	(Sasso et al., 2011)
Nanofibers	*Yersinia pestis* detection	(Men et al., 2010)
Nanofibers	Glucose detection	(Yang et al., 2009)

Table 2. Biomedical applications of self-assembled peptide nanostructures.

7. Conclusions

As presented throughout this chapter, self-assembled peptide nanostructures are biomaterials with numerous advantages making them excellent candidates for applications with biomedical purposes, including biosensing, diagnosis, drug delivery, and tissue reparation, among others. The biological nanostructures are synthesized under mild conditions, at low cost and rapidly, and can be easily functionalized with different functional compounds (antibodies, enzymes, quantum dots, fluorescent molecules, metallic and magnetic particles). However, it is important to be aware of the challenges involved when working with these self-assembled nanostructures. Several options to solve issues related to size control, manipulation, low conductivity and stability under liquid conditions have been presented. Despite these disadvantages, self-assembled nanostructures are already playing an important role in applications such as tissue reparation, drug delivery and biosensing. Future work should be directed towards a more complete study of the biocompatibility and immunogenicity of this biomaterial in order to clarify the consequences of an exposure to it.

8. Acknowledgements

The European Community (BeNatural/NMP4-CT-2006-033256) and the Danish Agency for Science Technology and Innovation (FTP 271-08-0968) are gratefully acknowledged for financial support.

9. References

Adler-Abramovich, L., Aronov, D., Beker, P., *et al.* (2009). Self-assembled arrays of peptide nanotubes by vapour deposition. *Nature Nanotechnology*, 4: 849-54, 1748-3387

Adler-Abramovich, L. and Gazit, E. (2008). Controlled patterning of peptide nanotubes and nanospheres using inkjet printing technology. *Journal of Peptide Science*, 14, 2: 217-23, 1075-2617

Adler-Abramovich, L., Reches, M., Sedman, V. L., Allen, S., Tendler, S. J. B. and Gazit, E. (2006). Thermal and chemical stability of diphenylalanine peptide nanotubes: Implications for nanotechnological applications. *Langmuir*, 22, 3: 1313-20, 0743-7463

Aggeli, A., Bell, M., Boden, N., *et al.* (1997). Responsive gels formed by the spontaneous self-assembly of peptides into polymeric beta-sheet tapes. *Nature*, 386, 6622: 259-62, 0028-0836

Aggeli, A., Bell, M., Carrick, L. M., *et al.* (2003). pH as a trigger of peptide beta-sheet self-assembly and reversible switching between nematic and isotropic phases. *Journal of the American Chemical Society*, 125, 32: 9619-28, 0002-7863

Aggeli, A., Nyrkova, I. A., Bell, M., *et al.* Exploiting peptide self-assembly to engineer novel biopolymers: Tapes, ribbons, fibrils and fibres. 1-17. In: Aggeli A, Boden N, Zhang S, eds. 2001. Self-Assembling Peptide Systems in Biology, Medicine and Engineering. Springer, 0-7923-7090-2, Dordrecht.

Andersen, K. B., Castillo-Leon, J., Hedstrom, M. and Svendsen, W. E. (2011). Stability of diphenylalanine peptide nanotubes in solution. *Nanoscale*, 3, 3: 994-98, 2040-3364

Bohr, H., Kuhle, A., Sorensen, A. H. and Bohr, J. (1997). Hierarchical organization in aggregates of protein molecules. *Zeitschrift Fur Physik D-Atoms Molecules and Clusters*, 40, 1-4: 513-15, 0178-7683

Bong, D. T., Clark, T. D., Granja, J. R. and Ghadiri, M. R. (2001). Self-assembling organic nanotubes. *Angewandte Chemie-International Edition*, 40, 6: 988-1011, 1433-7851

Branco, M., Wagner, N., Pochan, D. and Schneider, J. (2009). Release of model macromolecules from self-assembling peptide hydrogels for injectable delivery. *Biopolymers*, 92, 4: 318-18, 0006-3525

Castillo-León, J., Rodriguez-Trujillo, R., Gauthier, S., Jensen, A. C. Ø. and Svendsen, W. E. (2011). Micro-"factory" for self-assembled peptide nanostructures. *Microelectronic Engineering*, 88, 1685-1688, 01679317

Castillo, J., Dimaki, M. and Svendsen, W. E. (2009). Manipulation of biological samples using micro and nano techniques. *Integrative Biology*, 1, 1: 30-42, 1757-9694

Castillo, J., Svendsen, W. E. and Dimaki, M. (2011). *Micro and Nano Techniques for the Handling of Biological Samples*, CRC Press, 9781439827437, New York

Castillo, J., Tanzi, S., Dimaki, M. and Svendsen, W. (2008). Manipulation of self-assembly amyloid peptide nanotubes by dielectrophoresis. *Electrophoresis*, 29, 24: 5026-32, 0173-0835

Chen, Y., Song, S., Yan, Z., Fenniri, H. and Webster, T. J. (2011). Self assembled rosette nanotubes encapsulate and slowly release dexamethasone. *International Journal of Nanomedicine*, 6: 1035-44, 1178-2013

Cho, E. C., Choi, J. W., Lee, M. Y. and Koo, K. K. (2008). Fabrication of an electrochemical immunosensor with self-assembled peptide nanotubes. *Colloids and Surfaces a-Physicochemical and Engineering Aspects*, 313: 95-99, 0927-7757

Cipriano, T. C., Takahashi, P. M., de Lima, D., *et al.* (2010). Spatial organization of peptide nanotubes for electrochemical devices. *Journal of Materials Science*, 45, 18: 5101-08, 0022-2461

Clausen, C. H., Dimaki, M., Panagos, S. P., *et al.* (2011). Electrostatic force microscopy of self-assembled peptide structures. *Scanning*, DOI: 10.1002/sca.20231, 1932-8745

Clausen, C. H., Jensen, J., Castillo, J., Dimaki, M. and Svendsen, W. E. (2008). Qualitative Mapping of Structurally Different Dipeptide Nanotubes. *Nano Letters*, 8, 11: 4066-69, 1530-6984

de la Rica, R. and Matsui, H. (2010). Applications of peptide and protein-based materials in bionanotechnology. *Chemical Society Reviews*, 39, 9: 3499-509, 0306-0012

de la Rica, R., Mendoza, E., Lechuga, L. M. and Matsui, H. (2008). Label-Free Pathogen Detection with Sensor Chips Assembled from Peptide Nanotubes. *Angewandte Chemie-International Edition*, 47, 50: 9752-55, 1433-7851

de la Rica, R., Pejoux, C., Fernandez-Sanchez, C., Baldi, A. and Matsui, H. (2010). Peptide-Nanotube Biochips for Label-Free Detection of Multiple Pathogens. *Small*, 6, 10: 1092-95, 1613-6810

Fairman, R. and Akerfeldt, K. S. (2005). Peptides as novel smart materials. *Current Opinion in Structural Biology*, 15, 4: 453-63, 0959-440X

Fenniri, H., Deng, B. L., Ribbe, A. E., Hallenga, K., Jacob, J. and Thiyagarajan, P. (2002). Entropically driven self-assembly of multichannel rosette nanotubes. *Proceedings of the National Academy of Sciences of the United States of America*, 99: 6487-92, 0027-8424

Fishwick, C. W. G., Beevers, A. J., Carrick, L. M., Whitehouse, C. D., Aggeli, A. and Boden, N. (2003). Structures of helical beta-tapes and twisted ribbons: The role of side-chain interactions on twist and bend behavior. *Nano Letters*, 3, 11: 1475-79, 1530-6984

Gao, X. Y. and Matsui, H. (2005). Peptide-based nanotubes and their applications in bionanotechnology. *Advanced Materials*, 17, 17: 2037-50, 0935-9648

Gazit, E. (2007). Self-assembled peptide nanostructures: the design of molecular building blocks and their technological utilization. *Chemical Society Reviews*, 36, 8: 1263-69, 0306-0012

Ghadiri, M. R., Granja, J. R., Milligan, R. A., McRee, D. E. and Khazanovich, N. (1993). Self-assembling organic nanotubes based on a cylci peptide architecture. *Nature*, 366, 6453: 324-27, 0028-0836

Gorbitz, C. H. (2007). Microporous organic materials from hydrophobic dipeptides. *Chemistry-a European Journal*, 13, 4: 1022-31, 0947-6539

Gorbitz, C. H. (2001). Nanotube formation by hydrophobic dipeptides. *Chemistry-a European Journal*, 7, 23: 5153-59, 0947-6539

Gorbitz, C. H. (2006). The structure of nanotubes formed by diphenylalanine, the core recognition motif of Alzheimer's beta-amyloid polypeptide. *Chemical Communications*, 22: 2332-34, 1359-7345

Han, T. H., Park, J. S., Oh, J. K. and Kim, S. O. (2008). Morphology Control of One-Dimensional Peptide Nanostructures. *Journal of Nanoscience and Nanotechnology*, 8, 10: 5547-50, 1533-4880

Hauser, C. A. E. and Zhang, S. G. (2010). Designer self-assembling peptide nanofiber biological materials. *Chemical Society Reviews*, 39, 8: 2780-90, 0306-0012

Horne, W. S., Wiethoff, C. M., Cui, C. L., *et al.* (2005). Antiviral cyclic D,L-alpha-peptides: Targeting a general biochemical pathway in virus infections. *Bioorganic & Medicinal Chemistry*, 13, 17: 5145 53, 0968-0896

Jayawarna, V., Ali, M., Jowitt, T. A., *et al.* (2006). Nanostructured hydrogels for three-dimensional cell culture through self-assembly of fluorenylmethoxycarbonyl-dipeptides. *Advanced Materials*, 18, 5: 611-14, 0935-9648

Jayawarna, V., Richardson, S. M., Gough, J. and Ulijn, R. (2008). Self-assembling peptide hydrogels: Directing cell behaviour by chemical composition. *Tissue Engineering Part A*, 14, 5: P347, 1937-3341

Jayawarna, V., Richardson, S. M., Hirst, A. R., *et al.* (2009). Introducing chemical functionality in Fmoc-peptide gels for cell culture. *Acta Biomaterialia*, 5, 3: 934-43, 1742-7061

Jayawarna, V., Smith, A., Gough, J. E. and Ulijn, R. V. (2007). Three-dimensional cell culture of chondrocytes on modified di-phenylaianine scaffolds. *Biochemical Society Transactions*, 35: 535-37, 0300-5127

Kasotakis, E., Mossou, E., Adler-Abramovich, L., *et al.* (2009). Design of Metal-Binding Sites Onto Self-Assembled Peptide Fibrils. *Biopolymers*, 92, 3: 164-72, 0006-3525

Kim, J., Han, T. H., Kim, Y. I., *et al.* (2010). Role of Water in Directing Diphenylalanine Assembly into Nanotubes and Nanowires. *Advanced Materials*, 22, 5: 583-87, 0935-9648

Kim, J. H., Lim, S. Y., Nam, D. H., Ryu, J., Ku, S. H. and Park, C. B. (2011). Self-assembled, photoluminescent peptide hydrogel as a versatile platform for enzyme-based optical biosensors. *Biosensors & Bioelectronics*, 26, 5: 1860-65, 0956-5663

Kim, J. H., Ryu, J. and Park, C. B. (2011). Selective Detection of Neurotoxin by Photoluminescent Peptide Nanotubes. *Small*, 7, 6: 718-22, 1613-6810

Kisiday, J., Jin, M., Kurz, B., *et al.* (2002). Self-assembling peptide hydrogel fosters chondrocyte extracellular matrix production and cell division: Implications for cartilage tissue repair. *Proceedings of the National Academy of Sciences of the United States of America*, 99, 15: 9996-10001, 0027-8424

Kol, N., Adler-Abramovich, L., Barlam, D., Shneck, R. Z., Gazit, E. and Rousso, I. (2005). Self-assembled peptide nanotubes are uniquely rigid bioinspired supramolecular structures. *Nano Letters*, 5, 7: 1343-46, 1530-6984

Koutsopoulos, S., Unsworth, L. D., Nagaia, Y. and Zhang, S. G. (2009). Controlled release of functional proteins through designer self-assembling peptide nanofiber hydrogel scaffold. *Proceedings of the National Academy of Sciences of the United States of America*, 106, 12: 4623-28, 0027-8424

Krebs, M. R. H., Devlin, G. L. and Donald, A. M. (2007). Protein particulates: Another generic form of protein aggregation? *Biophysical Journal*, 92, 4: 1336-42, 0006-3495

Kumara, M. T., Srividya, N., Muralidharan, S. and Tripp, B. C. (2006). Bioengineered flagella protein nanotubes with cysteine loops: Self-assembly and manipulation in an optical trap. *Nano Letters*, 6, 9: 2121-29, 1530-6984

Kyle, S., Aggeli, A., Ingham, E. and McPherson, M. J. (2010). Recombinant self-assembling peptides as biomaterials for tissue engineering. *Biomaterials*, 31, 36: 9395-405, 0142-9612

Kyle, S., Riley, J. M., Aggeli, A., Ingham, E. and McPherson, M. J. (2008). Self-assembling peptides for scaffolds in regenerative medicine: Production using recombinant DNA technology. *Tissue Engineering Part A*, 14, 5: OP247, 1937-3341

Larsen, M., Andersen, K., Svendsen, W. and Castillo-León, J. (2011). Self-Assembled Peptide Nanotubes as an Etching Material for the Rapid Fabrication of Silicon Wires. *BioNanoScience*, 1, 1: 31-37, 2191-1649

Li, F., Wang, J., Tang, F. S., *et al.* (2009). Fluorescence Studies on a Designed Self-Assembling Peptide of RAD16-II as a Potential Carrier for Hydrophobic Drug. *Journal of Nanoscience and Nanotechnology*, 9, 2: 1611-14, 1533-4880

Liebmann, T., Rydholm, S., Akpe, V. and Brismar, H. (2007). Self-assembling Fmoc dipeptide hydrogel for in situ 3D cell culturing. *Bmc Biotechnology*, 7: 1-11, 1472-6750

MaHam, A., Tang, Z. W., Wu, H., Wang, J. and Lin, Y. H. (2009). Protein-Based Nanomedicine Platforms for Drug Delivery. *Small*, 5, 15: 1706-21, 1613-6810

Mahler, A., Reches, M., Rechter, M., Cohen, S. and Gazit, E. (2006). Rigid, self-assembled hydrogel composed of a modified aromatic dipeptide. *Advanced Materials*, 18, 11: 1365-70, 0935-9648

Men, D., Zhang, Z. P., Guo, Y. C., *et al.* (2010). An auto-biotinylated bifunctional protein nanowire for ultra-sensitive molecular biosensing. *Biosensors & Bioelectronics*, 26, 4: 1137-41, 0956-5663

Meng, H., Chen, L. Y., Ye, Z. Y., Wang, S. T. and Zhao, X. J. (2009). The Effect of a Self-Assembling Peptide Nanofiber Scaffold (Peptide) When Used as a Wound Dressing for the Treatment of Deep Second Degree Burns in Rats. *Journal of Biomedical Materials Research Part B-Applied Biomaterials*, 89B, 2: 379-91, 1552-4973

Nelson, J. C., Saven, J. G., Moore, J. S. and Wolynes, P. G. (1997). Solvophobically driven folding of nonbiological oligomers. *Science*, 277, 5333: 1793-96, 0036-8075

Nigen, M., Gaillard, C., Croguennec, T., Madec, M. N. and Bouhallab, S. (2010). Dynamic and supramolecular organisation of alpha-lactalbumin/lysozyme microspheres: A microscopic study. *Biophysical Chemistry*, 146, 1: 30-35, 0301-4622

Niu, L. J., Chen, X. Y., Allen, S. and Tendler, S. J. B. (2007). Using the bending beam model to estimate the elasticity of diphenylalanine nanotubes. *Langmuir*, 23, 14: 7443-46, 0743-7463

Park, J. S., Han, T. H., Oh, J. K. and Kim, S. O. (2009). Size-Dependent Isotropic/Nematic Phase Transition Behavior of Liquid Crystalline Peptide Nanowires. *Macromolecular Chemistry and Physics*, 210, 16: 1283-90, 1022-1352

Porrata, P., Goun, E. and Matsui, H. (2002). Size-controlled self-assembly of peptide nanotubes using polycarbonate membranes as templates. *Chemistry of Materials*, 14, 10: 4378-81, 0897-4756

Rajagopal, K. and Schneider, J. P. (2004). Self-assembling peptides and proteins for nanotechnological applications. *Current Opinion in Structural Biology*, 14, 4: 480-86, 0959-440X

Rea, J. C., Gibly, R. F., Barron, A. E. and Shea, L. D. (2009). Self-assembling peptide-lipoplexes for substrate-mediated gene delivery. *Acta Biomaterialia*, 5, 3: 903-12, 1742-7061

Reches, M. and Gazit, E. (2007). Biological and chemical decoration of peptide nanostructures via biotin-avidin interactions. *Journal of Nanoscience and Nanotechnology*, 7, 7: 2239-45, 1533-4880

Reches, M. and Gazit, E. (2003). Casting metal nanowires within discrete self-assembled peptide nanotubes. *Science*, 300, 5619: 625-27, 0036-8075

Reches, M. and Gazit, E. (2006). Controlled patterning of aligned self-assembled peptide nanotubes. *Nature Nanotechnology*, 1, 3: 195-200, 1748-3387

Reches, M. and Gazit, E. (2006). Designed aromatic homo-dipeptides: formation of ordered nanostructures and potential nanotechnological applications. *Physical Biology*, 3, 1: S10-S19, 1478-3967

Ryu, J., Lim, S. Y. and Park, C. B. (2009). Photoluminescent Peptide Nanotubes. *Advanced Materials*, 21, 16: 1577-81, 0935-9648

Ryu, J. and Park, C. B. (2008). High-Temperature Self-Assembly of Peptides into Vertically Well-Aligned Nanowires by Aniline Vapor. *Advanced Materials*, 20, 19: 3754-578, 0935-9648

Sasso, L., Vedarethinam, I., Emnéus, J., Svendsen, W. E. and Castillo-León, J. (2011). Self-Assembled Diphenylalanine Nanowires for Cellular Studies and Sensor Applications *J Nanosci Nanotechnol: Accepted*, DOI: 10.1166/jnn.2011.4534, 1533-4880

Scanlon, S. and Aggeli, A. (2008). Self-assembling peptide nanotubes. *Nano Today*, 3, 3-4: 22-30, 1748-0132

Scheibel, T., Parthasarathy, R., Sawicki, G., Lin, X. M., Jaeger, H. and Lindquist, S. L. (2003). Conducting nanowires built by controlled self-assembly of amyloid fibers and selective metal deposition. *Proceedings of the National Academy of Sciences of the United States of America*, 100, 8: 4527-32, 0027-8424

Sedman, V. L., Adler-Abramovich, L., Allen, S., Gazit, E. and Tendler, S. J. B. (2006). Direct observation of the release of phenylalanine from diphenylalanine nanotubes. *Journal of the American Chemical Society*, 128, 21: 6903-08, 0002-7863

Sedman, V. L., Allen, S., Chen, X. Y., Roberts, C. J. and Tendler, S. J. B. (2009). Thermomechanical Manipulation of Aromatic Peptide Nanotubes. *Langmuir*, 25, 13: 7256-59, 0743-7463

Slotta, U. K., Rammensee, S., Gorb, S. and Scheibel, T. (2008). An engineered spider silk protein forms microspheres. *Angewandte Chemie-International Edition*, 47, 24: 4592-94, 1433-7851

Tarek, M., Maigret, B. and Chipot, C. (2003). Molecular dynamics investigation of an oriented cyclic peptide nanotube in DMPC bilayers. *Biophysical Journal*, 85, 4: 2287-98, 0006-3495

Tokunaga, M., Liu, M. L., Nagai, T., *et al.* (2010). Implantation of cardiac progenitor cells using self-assembling peptide improves cardiac function after myocardial infarction. *Journal of Molecular and Cellular Cardiology*, 49, 6: 972-83, 0022-2828

Ueda, Y., Ishii, K., Toyama, Y., Nakamura, M. and Okano, H. (2008). Self-assembling peptide scaffold promotes the axonal growth in the injured spinal cord. *Neuroscience Research*, 61: S93-S93, 0168-0102

Viguier, B., Zór, K., Kasotakis, E., *et al.* (2011). Development of an Electrochemical Metal-Ion Biosensor Using Self-Assembled Peptide Nanofibrils. *ACS Applied Materials & Interfaces*, 3: 1594-1600, 1944-8244

von Maltzahn, G., Vauthey, S., Santoso, S. and Zhang, S. U. (2003). Positively charged surfactant-like peptides self-assemble into nanostructures. *Langmuir*, 19, 10: 4332-37, 0743-7463

Wiltzius, J. J. W., Landau, M., Nelson, R., *et al.* (2009). Molecular mechanisms for protein-encoded inheritance. *Nature Structural & Molecular Biology*, 16, 9: 973-U98, 1545-9985

Woolfson, D. N. and Ryadnov, M. G. (2006). Peptide-based fibrous biomaterials: some things old, new and borrowed. *Current Opinion in Chemical Biology*, 10, 6: 559-67,

Yan, X. H., Zhu, P. L. and Li, J. B. (2010). Self-assembly and application of diphenylalanine-based nanostructures. *Chemical Society Reviews*, 39, 6: 1877-90, 0306-0012

Yang, H., Fung, S. Y., Pritzker, M. and Chen, P. (2009). Ionic-Complementary Peptide Matrix for Enzyme Immobilization and Biomolecular Sensing. *Langmuir*, 25, 14: 7773-77, 0743-7463

Yemini, M., Reches, M., Gazit, E. and Rishpon, J. (2005). Peptide nanotube-modified electrodes for enzyme-biosensor applications. *Analytical Chemistry*, 77, 16: 5155-59, 0003-2700

Yemini, M., Reches, M., Rishpon, J. and Gazit, E. (2005). Novel electrochemical biosensing platform using self-assembled peptide nanotubes. *Nano Letters*, 5, 1: 183-86, 1530-6984

Zhang, F., Shi, G. S., Ren, L. F., Hu, F. Q., Li, S. L. and Xie, Z. J. (2009). Designer self-assembling peptide scaffold stimulates pre-osteoblast attachment, spreading and proliferation. *J Mater Sci-Mater Med*, 20, 7: 1475-81, 0957-4530

Zhao, Z. and Matsui, H. (2007). Accurate immobilization of antibody-functionalized peptide nanotubes on protein-patterned Arrays by optimizing their ligand-receptor interactions. *Small*, 3, 8: 1390-93, 1613-6810

6

Preparation of Nanocellulose with Cation–Exchange Resin Catalysed Hydrolysis

Huang Biao, Tang Li-rong, Dai Da-song,
Ou Wen, Li Tao and Chen Xue-rong
Fujian Agriculture and Forestry University,
China

1. Introduction

Cellulose is the most abundant organic compound on earth and is present in a wide variety of living species, such as animals, plants and bacterial (Tingaut et al., 2010; Elazzouzi-Hafraoui et al., 2008). This linear polymer is constituted of repeating β-D-glucopyranosyl units joined by 1→4-glycosidic linkages. The molecules of cellulose are stabilized laterally by hydrogen bonds between hydroxyl groups and oxygens of adjacent molecules. However, they can be broken chemically under strong aqueous acid or high temperature. Manipulating cellulose molecules on the nanometer scale to create the nanocellulose of excellent properties has become a hotspot of cellulose science. As for nanocellulose, it is currently believed that at least one of its dimension is lower than 100nm. Moreover, nanocellulose exhibits the property of certain gels or fluids under normal conditions. Compared with microcrystalline cellulose, nanocellulose presents very attractive properties such as low density, high chemical reactivity, high strength and modulus, and high transparency (Nogi et al., 2009; Lee et al., 2008; Pääkko et al., 2007; Siró & Plackett, 2001). Therefore, nanocellulose has a great potential for use as filler in nanocomposites and have attracted a great deal of interest recently. Nanocellulose has been reported to improve the mechanical properties by incorporating into a wide range of polymer matrices, including poly(3-hydroxybutyrate), hydroxypropyl cellulose, poly(L-lactide), waterborne polyurethane, poly(3,4-ethylenedioxythiophene), polyvinyl acetate, poly(o-ethoxyaniline). The composite applications may be for use as coatings and films, paints, foams, packaging. (Cai et al., 2011; Zimmermann et al., 2010; Pei et al., 2010; Wang et al., 2010; Mendez & Weder, 2010; de Rodriguez et al., 2006; Medeiros et al., 2008). Moreover, the potential of nanocellulose applications in the area of paper and paperboard manufacture is obvious. Nanocellulose are expected to enhance the fiber-fiber bond strength and, hence, have a strong reinforcement effect on paper materials. Nanocellulose may be useful as a barrier in grease-proof type of papers and as a wet-end additive to enhance retention, dry and wet strength in commodity type of paper and board products. Nanocellulose also can be used as a low calorie replacement for today's carbohydrate additives used as thickeners, flavour carriers and suspension stabilizers in a wide variety of food products and is useful for producing fillings, crushes, chips, wafers, soups, gravies, puddings etc. On the other hand, the food applications were early recognised as a highly interesting application field for nanocellulose due to the rheological behaviour of the nanocellulose gel (Wikipedia, 2011).

The preparation of nanocellulose derived from wood was introduced more than two decades ago (Aulin et al., 2009). Although wood is one of the main resources for the cellulose, competition from different sectors such as the building products and furniture industries and the pulp and paper industry, as well as the combustion of wood for energy, makes it challenging to supply all users with the quantities of wood needed at reasonable cost (Siró & Plackett, 2001). Besides wood, nanocellulose also could be prepared from many agricultural residue and corps, such as cotton, hemp, sisal, bagasse, wheat straw. Therefore, nanocellulose will be key to the development of higher-value agricultural residue products and could find economic interest (de Mesquita et al., 2010). In literature, there are many reports on nanocellulose prepared from diverse non-wood sources including wheat straw (Panthapulakkal et al., 2006; Kaushik et al., 2010; Alemdar & Sain, 2008), potato tuber cells (Dufresne et al., 2000), sisal (Morán et al., 2008; de Rodriguez et al., 2006) and banana rachis (Zuluaga et al., 2009).

Sulfuric acid hydrolysis of cellulose is a well-known process to remove amorphous regions, leaving the crystalline segments intact and leading to the formation of high purity single crystals (de Mesquita et al., 2010). Ion-exchange resins have been used commercially as solid acid catalysts in many areas, such as alkylation with olefins, alkyl halides, alkyl esters, isomerization, transalklation and nitration. Compared with liquid acid, the main advantages of cation-exchange resin include reduced equipment corrosion, ease of product separation, less potential contamination in waste streams and recycle of the catalyst (Harmer & Sun, 2001).

Morever, the energy of ultrasound is transferred to the polymer chains through a process called cavitation, which is the formation, growth, and violent collapse of cavities in the water. The energy provided by cavitation in this so-called sonochemistry is approximately 10-100kJ/mol, which is within the hydrogen bond energy scale (Tischer et al., 2010). It can accelerate hydrogen ions to penetrate into the cellulose amorphous chains, promoting the hydrolytic cleavage of the glycosidic bonds (Zhao & Feng, 2007; Filson & Dawson-Andoh, 2009; Wang & Cheng, 2009).

In this study, we aim to present an environmentally and economically novel way to prepare nanocellulose from microcrystalline cellulose by using cation-exchange resin as catalyst with ultrasonic-assisted hydrolysis. Response surface methodology and Box-behnken statistical experiment design method were employed for modeling and optimization of the influence of operating variables on the yield of nanocellulose. In addition, the characterization for morphologies, structure, spectrum properties and rehological behaviors of nanocellulose were also investigated.

2. Experimental section

2.1 Materials

Microcrystalline Cellulose (MCC) used in the experiment was purchased from Shandong Ruitai Chemicals Co., Ltd. The catalyst, NKC-9 cation-exchange resin (NKC-9), was provided by the Chemical Plant of Nankai University of China. It is a macroreticular copolymer styrene-divinyl benzene in H+ form and has the following properties: exchange capacity (mmol/g [H+])≥4.7, pearl size of 0.45-1.25 mm, true wet density of 1.20-1.30 g/mL. The resin was firstly pretreated by distilled water to eliminate some impurities and then dried in an oven at 50 °C for 24 h. Dried resin was used for further experimental studies.

2.2 Sample preparation

Nanocellulose was isolated from MCC by means of cation-exchange resin hydrolysis. 3 g dried MCC and 30 g ion exchange resin were put into 250 ml distilled deionized water. The suspension was stirred and sonicated at 40-60°C for 150-210min. Then the ion exchange resin was separated from cellulose suspension. The resulting suspension was centrifuged several times at 12000 rpm and washed with distilled deionized water until the supernatant became turbid, and then the nanocellulose was collected.

2.3 Experimental design and statistical analysis

The Box-Behnken experimental design method was used to determine the effects of major operating variables on the yield of nanocellulose and to find the combination of variables in order to produce maximum nanocellulose yields. The advantage of Box-Behnken design is that it has only three levels, coded –1, 0, and +1 for low, middle and high concentrations, respectively. This experimental design reduced the number of experiments, so it is more efficient and easier to arrange and to interpret in comparison to others (Majumder et al., 2009). Therefore, this statistical technique was adopted in this study.

The experiments at ratio of NKC-9 to MCC (5: 1, 10: 1 and 15: 1), temperature (40, 50 and 60°C), time (150, 180 and 210min) were employed simultaneously covering the spectrum of variables for the percentage of nanocellulose yield in the Box-Behnken Design. As presented in Table 1, the experimental design involved three parameters (X_1, X_2 and X_3), each at three levels, coded –1, 0, and +1 for low, middle and high concentrations, respectively. A second-order polynomial equation was used to express the responses as a function of the independent variables as follows:

$$Y = B_0 + \sum_{i=1}^{n} B_i X_i + \sum_{i=j=1}^{n} B_{ij} X_i X_j \tag{1}$$

Where Y represents the measured response variables, three variables are involved and hence n takes the value 3. B_0 is the constant coefficient, B_is are the linear coefficients, B_{ij}s are the interaction coefficients.

Factor	Symbols & level[b]		
	-1	0	+1
ratio of resin X_1	5	10	15
temperature X_2 (°C)	40	50	60
time X_3 (min)	150	180	210

[b] $x_1 = (X_1 - 10)/5$; $x_2 = (X_2 - 50)/10$; $x_3 = (X_3 - 180)/30$.

Table 1. Code and level of factors chosen for the trials.

2.4 Characterization of nanocellulose

Electron microscopy was conducted to observe the surface of MCC and nanocellulose. Samples were mounted on metal stubs by double side adhesive tape and examined with FEI XL30 ESEM-FEG field emission scanning electron microscopy (FESEM). The nanostructure of nanocellulose was examined in a transmission electron microscope (TEM), Tecnai G2F20 FETEM (FEI Co. Ltd., USA) at an acceleration voltage of 200 kV. The X-ray diffraction

(XRD) patterns were recorded by a X'Pert Pro MPD X-ray diffractometer equipped with Cu Kα radiation (λ= 0.154 nm). Fourier transform infrared spectroscopy (FTIR) was used to examine any changes in the chemical structure of samples. A Nicolet 380 (Thermo electron Instruments Co., Ltd., USA) was used to obtain the spectra of each sample. The rheological behavior of sample was examined by DV-III+pro rheometer (Brookfield Engineering Laboratories, Inc., USA) with SC4-34 spindle.

3. Results and discussion

3.1 Optimization of hydrolysis conditions for nanocellulose

The effect of process variables like ratio of NKC-9 to MCC, temperature and time on the preparation of nanocellulose was investigated by means of response surface methodology, Box-Behnken Design (BBD). Table 2. shows the coded value of the variables and the yield of nanocellulose (response). The whole design consisted of 17 experimental points carried out in random order. Five replicates at the centre of the design were used to estimate a pure error sum of squares. The data obtained was analyzed by applying multiple regression analysis method based on Eq. (1). The predicted response Y for nanocellulose yield was obtained and shown as:

$$Y = 50.68 + 0.15X_1 - 1.11X_2 + 1.67 X_3 + 0.74X_1 X_2 - 0.35X_1 X_3 - 1.07X_2 X_3 - 3.85X_1^2 - 3.83X_2^2 - 5.17 X_3^2 \tag{2}$$

In this equation, Y is the predicted response variable, i.e., the yield of nanocellulose (%), X_1, X_2 and X_3 are the independent variables in coded units, i.e., ratio of NKC-9 to MCC, temperature and time, respectively.

The data obtained from Eq. (2) are significant. It is verified by F-value and the analysis of variance (ANOVA) by fitting the data of all independent observations in response surface quadratic model. The summary of the analysis of variance (ANOVA) of the results of the quadratic model fitting are shown in Table 3. ANOVA is indispensable to testing the significance and adequacy of the model. The corresponding variables would be more significant if the absolute F-value becomes greater and the p-value becomes smaller (Chen et al., 2010). The model F-value of 42.80 implies that the model is significant. There is only a 0.01% chance that a "model F-value" this large could occur due to noise. Value of "Prob > F" less than 0.05 indicates that the model terms are significant. In this case, X_2, X_3, X_2X_3, X_1^2, X_2^2 and X_3^2 are significant model terms. However, ratio of NKC-9 to MCC (X_1) and interaction terms (X_1X_2 and X_1X_3) had a negative effect on Y.

The "Lack of fit F-value" of 4.79 implies that the lack of fit is not significant relative to the pure error. There is a 8.23% chance that a "Lack of fit F-value" this large could occur due to noise. The determination coefficient(R^2), a measure of the goodness of fit of the model, was very significant at the level of 98.22%, the model was unable to explain only 1.78% of the total variations. The value of adjusted R^2, was also very high at the level of 0.9592, indicating high significance of the model.

The effect of hydrolysis conditions on the yield of nanocellulose is shown in Table 2 by the coefficient of the second-order polynomials. To visualize and identify the type of interactions between test variables, the two and three dimensional contour plots are shown in Fig. 1, including ratio of NKC-9 to MCC and temperature, ratio of NKC-9 to MCC and time, as well as temperature and time. The circular contour plots indicate that the interaction

Trial No.	X_1	X_2 (°C)	X_3 (min)	Yield of NCC, Y (%)	
				Experimental	Predicted
1	0(10)	0(50)	0(180)	50.38	50.68
2	0(10)	-1(40)	1(210)	45.86	45.54
3	-1(5)	-1(40)	0(180)	43.71	44.71
4	0(10)	0(50)	0(180)	50.63	50.68
5	0(10)	-1(40)	-1(150)	40.79	40.05
6	0(10)	0(50)	0(180)	51.64	50.68
7	0(10)	0(50)	0(180)	50.38	50.68
8	0(10)	0(50)	0(180)	50.38	50.68
9	1(15)	-1(40)	0(180)	43.46	43.52
10	-1(5)	0(50)	1(210)	44.22	43.54
11	1(15)	0(50)	-1(150)	39.82	40.49
12	1(15)	0(50)	1(210)	42.88	43.13
13	-1(5)	1(60)	0(180)	41.07	41.01
14	1(15)	1(60)	0(180)	43.77	42.78
15	0(10)	1(60)	-1(150)	39.66	39.98
16	-1(5)	0(50)	-1(150)	39.76	39.50
17	0(10)	1(60)	1(210)	40.43	41.17

Table 2. Experimental designs and results.

between the corresponding variables is negligible. An elliptical or saddle nature of the contour plots indicates significance of the interactions between the corresponding (Majumder et al., 2009; Fu et al., 2007).

As can be seen in the plots, the two and three dimensional contour plots with NKC-9 to MCC and temperature had a circular nature, indicating a negligible interactive effect on the yield between the two independent variables (Fig.1A). The yield of nanocellulose increased with rise in ratio of NKC-9 to MCC and temperature.

Factors	SS [a]	DF [b]	MS [c]	F-Value	Prob.(P)>F
Model	303.22	9	33.69	42.80	<0.0001
X_1	0.17	1	0.17	0.22	0.6559
X_2	9.88	1	9.88	12.55	0.0094
X_3	22.34	1	22.34	28.38	0.0011
X_1X_2	2.19	1	2.19	2.78	0.1395
X_1X_3	0.49	1	0.49	0.62	0.4559
X_2X_3	4.62	1	4.62	5.87	0.0459
X_1^2	62.36	1	62.36	79.21	<0.0001
X_2^2	61.81	1	61.81	78.52	<0.0001
X_3^2	112.36	1	112.36	142.73	<0.0001
Residual	5.51	7	0.79		
Lack of Fit	4.31	3	1.44	4.79	0.0823
Pure Error	1.20	4	0.30		
Total	308.73	16			

[a] SS, Sum of Squares. [b] DF, Degrees of Freedom. [c] MS, Mean Square.

Table 3. Analysis of variance (ANOVA) for the the yield of nanocellulose.

As shown in Fig.1, in the design boundary, each response surface plot had a clear peak and the corresponding contour plot had a clear maximum, which means that the maximum hydrogen yield could be achieved inside the design boundaries (Ghosh & Hallenbeck, 2010). The yield of nanocellulose increased with increasing ratio of NKC-9 to MCC, temperature and time to the optimal levels, and then decreased with a further increase in these parameters. The optimized conditions for maximum nanocellulose yield can be obtained by Design-expert software. The optimum values of the test variables in uncoded units obtained were ratio of NKC-9 to MCC 9.97: 1, temperature 48.30°C, time 189.00min. At these optimized conditions, the model predicted that the maximum yield of nanocellulose is 50.93%.

As to the actual experimental condition, some conditions were modified as follows: ratio of NKC-9 to MCC 10: 1, temperature 48°C and time 189min. To confirm the model adequacy for predicting maximum yield, three replicates experiments under modified conditions were conducted and the maximum yield of nanocellulose obtained was 50.04% which agreed well with the predicted value.

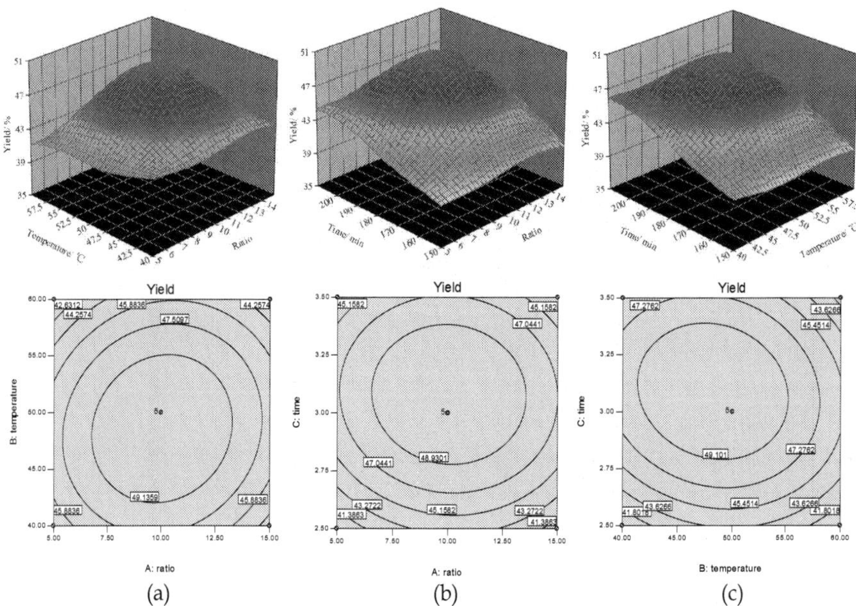

Fig. 1. Two and three dimensional contour plots for the maximum yield of nanocellulose. RSM plots were generated with the data shown in Table 2. (a) Yield of nanocellulose (%) as a function of ratio of NKC-9 to MCC and temperature. (b) Yield of nanocellulose (%) as a function of ratio of NKC-9 to MCC and time. (c) Yield of nanocellulose (%) as a function of temperature and time.

The interactions between ratio of NKC-9 to MCC and time are not perfectly elliptical (Fig.1(b)). The yield of nanocellulose increased with time rising up to certain level, beyond which yield declined slightly. This may be due to that over reaction time completely digests the cellulose so as to yield its component sugar molecules. Fig.1(c) shows the effect of temperature and time on the yield of nanocellulose. The result had an elliptical nature, indicating a significant interactive effect of the two independent variables on the yield.

3.2 Morphology of microcrystalline cellulose and nanocellulose

Fig.2 shows the morphologies of NCC at different concentrations. After hydrolysis and centrifugation process, the NCC with the concentration of 1% is obtained as shown in Fig.2(a). The micrograph of NCC in Fig.2(b) presents the glossy hydrogel at the concentration of 9%. The appearance of a stable gel is an obvious indication of the presence of NCC. The NCC powder is obtained by freeze drying (Fig.2(c)). It exhibits white metallic luster colour. As is shown in Fig.2(d), evaporation of aqueous suspensions of NCC at room temperature produces solid films with perfect optical transparency.

(a)   (b)

(c)   (d)

Fig. 2. Macrostructure of NCC.

Figure.3 compares the morphology of MCC (cf. Fig.3(a)) and NCC (cf. Fig.3(b) and Fig.3(c)). The diameter of MCC is around 15µm, the length is about 20-80µm. After treatment, a remarkable outcome of this novel nanocellulose extraction method can be seen in Fig.3(b) and Fig.3(c). This image also shows that nanocellulose with uniform diameter of approximately 2-24 nm can be obtained after hydrolysis, and these nanocellulose can form a very fine network. The diameter distribution of nanocellulose was determined with FEG-TEM (cf. Fig.3(c)), and the result is plotted in the following figures.

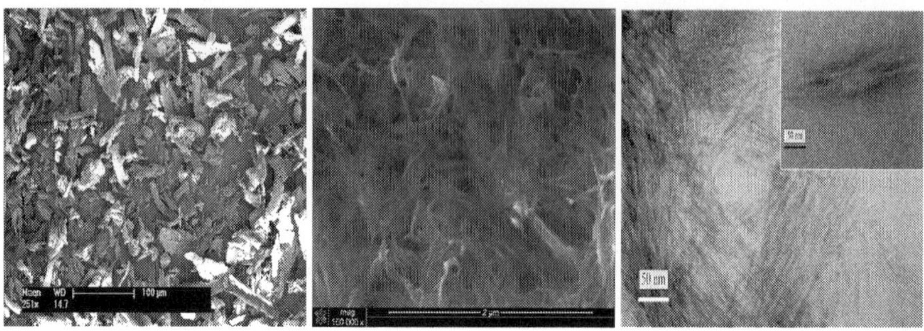

Fig. 3. Morphology of (a) MCC (by FEG-SEM, x 251); (b) NCC (by FEG-SEM, x 100 00) and (c) NCC (by FEG-TEM).

3.3 X-ray diffraction analysis

X-ray crystallography was used to compare the crystallinity of MCC and NCC. Results of X-ray powder diffraction photograph from MCC and NCC are shown in Fig.4(a) and 4(b), respectively. Fig.4 shows that the major crystalline peak which represents the cellulose crystallographic plane (002, Bragg reflection) occurs at 2θ = 22.667° and 2θ = 22.521° for MCC and NCC, respectively. Compared with MCC (crystallinity 75.2%), after cation-exchange resin catalytic hydrolysis, the crystallinity of nanocellulose is 84.26%. The increase in crystallinity may be due to the removal of amorphous regions in the cellulose. This leads to the realignment of cellulose molecules. This also indicates that the nanocellulose obtained by this novel isolation may be more effective in achieving higher reinforcement for composite materials.

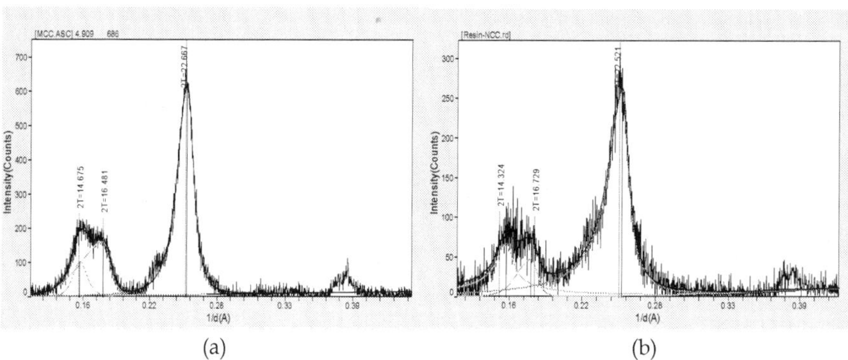

(a)   (b)

Fig. 4. XRD patterns of (a) NCC and (b) MCC.

3.4 FTIR spectroscopy analysis

Fig.5 shows the FTIR spectra of MCC and NCC obtained after the cation-exchange resin catalytic hydrolysis. In general, the FTIR spectrum of nanocellulose is very similar to that of MCC. The dominant peaks in the region between 3600 and 2800 cm^{-1} are due to the stretching vibrations of C--H and O$-$H (Alemdar et al., 2008). The peaks at 3347 cm^{-1} and

2900 cm^{-1} are attributed to the stretching vibrations of O--H and symmetric stretching vibrations of C--H, respectively. The bands in the 1430 cm^{-1} region are due to the C-H deformation vibrations of CH2 (Ibrahima et al., 2010). The absorbency at 1058 cm^{-1} in the spectrums of MCC and nanocellulose is associated with the C--O stretching vibration. The observation of the peaks at 1112 and 1165 cm^{-1} can be attributed to the C--O and C--C stretching vibration of cellulose ether. And the absorption band at 895 cm^{-1}, for MCC and NCC, respectively, may be regarded as anomeric carbon of cellulose, that is, β-D-glucopyranosyl, which indicates the typical structure of cellulose.

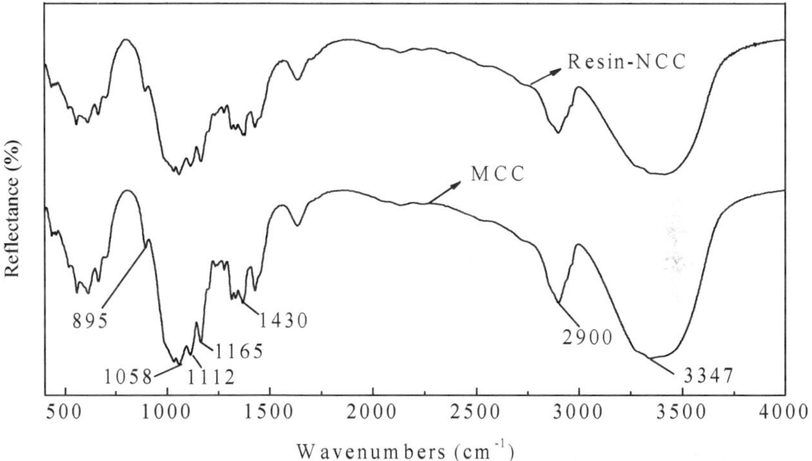

Fig. 5. FTIR spectrum of MCC and NCC.

3.5 Rheological properties of nanocellulose suspension

Fig.6 illustrates the variation of the shear stress of the suspensions with the function of concentration at various shear rates. It shows that the shear stress increases with the rising concentration of NCC. The equilibrium flow curves, τ and γ (cf. Fig. 7(a)), η and γ (cf. Fig. 7(b)), of NCC suspensions at various concentrations (2.5%, 3.5%, 4.5%) are shown in Fig.7. All the suspensions display non-Newtonian flow behavior. According to the Ostwald de Waele law, deviation of the flow curve from New-Tonian behavior can be quantified with the flow behavior index, the smaller the value, the greater the degree of shear thinning. The values consistency coefficient and flow behavior for NCC suspension at various concentration are summarized in Table 4. Given a negative value for n, the value for the suspension containing 4.5% NCC is very unusual. This also can be seen from fig. 9a, which shows that shear stress drops when the shear rate increase from 2.0 s^{-1} to 5.0 s^{-1}. It should be noted that this result is quite different from the presented reports. This may be due to the formation of NCC gel. These suspensions exhibit a shear-thinning behaviour, decrease of viscosity with increasing shear rate (cf. Fig. 7(b)).

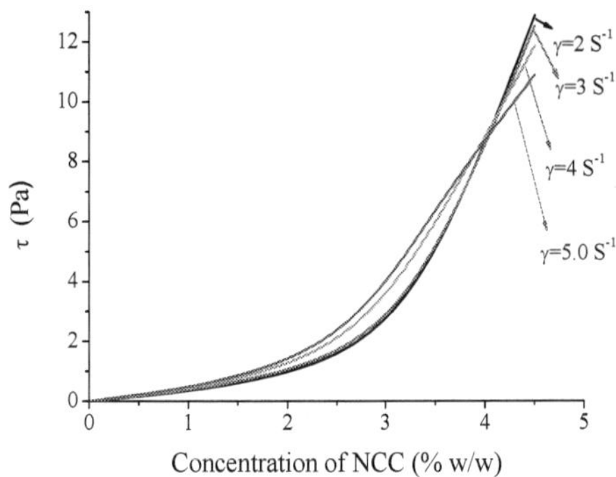

Fig. 6. Variation of the stress with the function of concentration of NCC at various shear rates.

Concentration of NCC Suspension (%)	Flow behavior index, n	Consistency coefficient (Pa s^n)
2.5	-0.199	15.1
3.5	0.478	2.98
4.5	0.273	0.699

Table 4. Power low constants for NCC suspension at various concentrations.

The effect of temperature on the rheological behaviour of the NCC suspensions is also studied. In general, the effect of temperature on rheological properties needs to be documented. Fig 8 shows the effect of temperature on the viscosity of NCC suspensions at various concentrations. The effect of temperature on viscosity of fluid foods at a specified shear rate can be described by the Arrhenius relationship:

$$\eta = Ae^{\frac{E_a}{RT}}$$

(3)

where η is the viscosity (Pa·s), A is a constant (Pa·s), T is the absolute temperature (K), R is the gas constant (8.315 J mol^{-1} K^{-1}), and Ea is the activation energy (J/mol). Generally, the magnitudes of Ea and A can be determined from regression analysis of 1/T versus ln η. It shows that the viscosity decreases when the temperature increases in the range of 33.5-68 °C for the suspensions at various concentrations. This indicates that the correlation between viscosity and temperature may agree with Arrhenius equation in this range of temperature. The results are summarized in Table 5. As we know, the low Ea values mean that the effect of temperature on the considered parameter is small. The activation energy value (Ea) of NCC suspension will amount to a maximum value when the concentration is 3.5%, which indicates that the decrease in viscosity with temperature rising was more pronounced in this concentration. But after 68 °C, the relationship between temperature and viscosity displays different curves at various concentrations. The viscosity also decreases with the increase of temperature for 3.5% NCC suspension, but for 2.5% and 4.5%, the opposite can be observed. This may be due to the swelling of NCC in the water when the temperature increases. It can be observed from Fig.9 that, with temperature rising, the degree of shear thinning is greater.

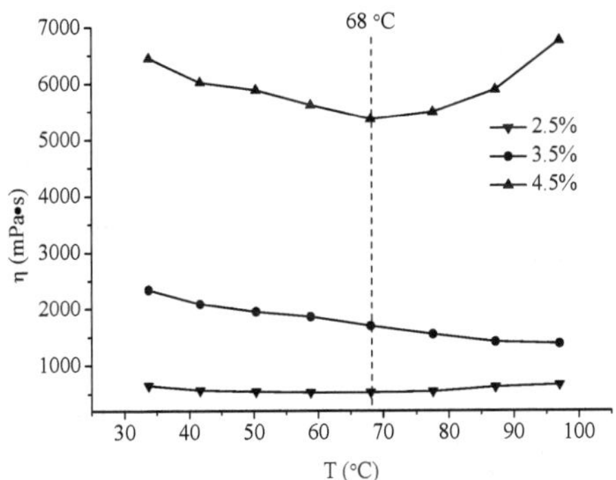

Fig. 8. Variation of the viscosity with a function of the temperature of NCC suspensions at various concentrations.

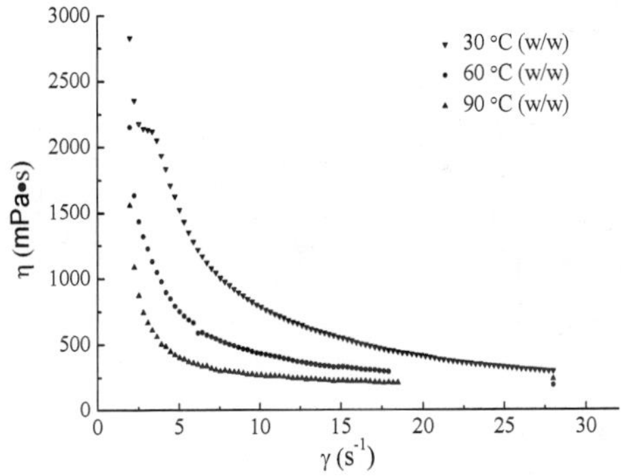

Fig. 9. Variation of the viscosity with a function of shear rate of NCC suspension at various temperatures.

Concentration of NCC (%)	A (Pa·s)	Ea (J · mol⁻¹)	R²
2.5	0.044	6750	0.88
3.5	0.112	7708	0.90
4.5	1.126	4433	0.98

Table 5. Activation energies (Ea) of NCC suspension at different concentrations (%).

4. Conclusions

NCC is prepared from MCC by cation-exchange resin hydrolysis with the aid of ultrasonification treatment. Compared with liquid acid and alkali chemicals, the main advantages of cation-exchange resin include reduced equipment corrosion, ease of product separation, less potential contamination in waste streams. With RSM and the corresponding Box-Behnken design, the optimal conditions for the production is determined as follows: ratio of NKC-9 to MCC 10: 1, temperature 48 °C, time 189 min. In the optimal conditions, the experiment yield of NCC was 50.04 %, which agrees with the predicted value. NCC prepared by cation-exchange resin hydrolysis presents the interconnected web-like structure, with its diameter mainly in the range of 2-24nm and length in hundreds. The spectrums of FTIR and XRD indicate that the obtained NCC still preserves the basic chemical structure of cellulose. Analyses of the rheological properties of NCC indicate that the suspension is the shear thinned pseudoplastic fluid with solid stability.

5. Acknowledgement

This work was supported by the National Natural Science Foundation of China (Grant NO. 30972312), and Natural Science Foundation of Fujian Province (Grant No. 2010J01270).

6. References

Alemdar, A. & Sain, M. (2008). Biocomposites from wheat straw nanofibers: Morphology, thermal and mechanical properties, *Composites Science and Technology*, Vol. 68, pp. 557-565.

Alemdar, A. & Sain M. (2008). Isolation and characterization of nanofibers from agricultural residues—Wheat straw and soy hulls, *Bioresource Technology*, Vol. 99, pp. 1664-1671.

Aulin, C., Johansson, E., Wågberg, L., et al. (2010). Self-organized films from cellulose I nanofibrils using the layer-by-layer technique, *Biomacromolecules*, Vol. 11, pp. 872-882.

Cai, Z. J., Y, G. & Kim, J. (2011). Biocompatible nanocomposites prepared by impregnating bacterial cellulose nanofibrils into poly(3-hydroxybutyrate), *Current Applied Physics*, Vol. 11, pp. 247-249.

Chen, X. P., Wang, W. X., Li, S. B., et al. (2010). Optimization of ultrasound-assisted extraction of Lingzhi polysaccharides using response surface methodology and its inhibitory effect on cervical cancer cells, *Carbohydrate Polymers*, Vol. 80, pp. 944-948.

de Mesquita, J. P., Donnici, C. L. & Pereira, F. V. (2010). Biobased nanocomposites from layer-by-layer assembly of cellulose nanowhiskers with chitosan, *Biomacromolecules*, Vol. 11, pp. 473-480.

de Rodriguez, N. L. G., Thielemans, W. & Dufresne, A. (2006). Sisal cellulose whiskers reinforced polyvinyl acetate nanocomposites, *Cellulose*, Vol. 13, pp. 261-270.

Dufresne, A., Dupeyre, D. & Vignon, M. R. (2000). Cellulose microfibrils from potato tuber cells: Processing and characterization of starch-cellulose microfibril composites, *J. Appl. Polymer Sci*, Vol. 76, pp. 2080-2092.

Elazzouzi-Hafraoui, S., Nishiyama, Y., Putaux, J. L., et al. (2008). The shape and size distribution of crystalline nanoparticles prepared by acid hydrolysis of native cellulose, *Biomacromolecules*, Vol. 9, No. 1, pp. 57-65.

Filson, P. B. & Dawson-Andoh, B. E. (2009). Sono-chemical preparation of cellulose nanocrystals from lignocellulose derived materials, *Bioresource Technology*, Vol. 100, No. 7, pp. 2259-2264.

Fu, J. F., Zhao, Y. Q. & Wu, Q. L. (2007). Optimising photoelectrocatalytic oxidation of fulvic acid using response surface methodology, *Journal of Hazardous Materials*, Vol. 144, pp. 499-505.

Ghosh, D. & Hallenbeck, P. C. (2010). Response surface methodology for process parameter optimization of hydrogen yield by the metabolically engineered strain *Escherichia coli* DJT135, *Bioresource Technology*, Vol. 101, pp. 1820-1825.

Harmer, M. A. & Sun, Q. (2001). Sold acid catalysis using ion-exchange resins, *Applied Catalysis A: General*, Vol. 221, pp. 45-62.

Ibrahima, M. M., Dufresneb A, El-Zawawya, W. K., et al. (2010). Banana fibers and microfibrils as lignocellulosic reinforcements in polymer composites, *Carbohydrate polymers*, Vol. 81, No. 4, pp. 811-819.

Kaushik, A., Singh, M. & Verma, G. (2010). Green nanocomposites based on thermoplastic starch and steam exploded cellulose nanofibrils from wheat straw, *Carbohydrate Polymers*, Vol. 82, No. 2, pp. 337-345.

Lee, S. Y., Mohan, D. J., Kang, I. A., et al. (2009). Nanocellulose Reinforced PVA Composite Films: Effects of Acid Treatment and Filler Loading, Fibers and Polymers, Vol. 10, No. 1, pp. 77-82.

Majumder, A., Singh, A. & Goyal, A. (2009). Application of response surface methodology for glucan production from *Leuconostoc dextranicum* and its structural characterization, *Carbohydrate Polymers*, Vol. 75, pp. 150-156.

Medeiros, E. S., Mattoso, L. H., Bernardes-Fiho, R., et al. (2008). Self-assembled films of cellulose nanofibrils and poly(o-ethoxyaniline), *Colloid Polymer Science*, Vol. 286, pp. 1265-1272.

Mendez, J. D. & Weder, C. (2010). Synthesis, electrical properties, and nanocomposites of poly(3,4-ethylenedioxythiophene) nanorods, *Polymer Chemistry*, Vol. 1, No. 8, pp. 1237-1244.

Morán, J. I., Alvarez, V. A., Cyras, V. P., et al. Extraction of cellulose and preparation of nanocellulose from sisal fibers, *Cellulose*, 2008, Vol. 15, No. 1, pp. 149-159.

Nogi, M., Iwamoto, S., Nakagaito, A. N., et al. (2009). Optically Transparent Nanofiber Paper, *Advanced materials*, Vol. 21, No. 16, pp. 1595-1598.

Pääkko, M., Ankerfors, M., Kosonen, H., et al. (2007). Enzymatic hydrolysis combined with mechanical shearing and high-pressure homogenization for nanoscale cellulose fibrils and strong gels, *Biomacromolecules*, Vol. 8, No. 6, pp. 1934-1941.

Panthapulakkal, S., Zereshkian, A. & Sain, M. (2006). Preparation and characterization of wheat straw fibers for reinforcing application in injection molded thermoplastic composites, *Bioresource Technology*, Vol. 97, pp. 265-272.

Pci, A., Zhou, Q. & Berglund, L. A. (2010) Functionalized cellulose nanocrystals as biobased nucleation agents in poly(L-lactide) (PLLA) – crystallization and mechanical property effects, *Composites Science and Technology*, Vol. 70, pp. 815-821.

Siró, I. & Plackett, D. (2001). Microfibrillated cellulose and new nanocomposite materials: a review, *Cellulose*, Vol. 17, No. 3, pp. 459-494.

Tingaut, P., Zimmermann, T. & Lopez-Suevos, F. (2010). Synthesis and characterization of bionanocomposites with tunable properties from poly(lactic acid) and acetylated microfibrillated cellulose, *Biomacromolecules,* Vol. 11, No. 2, pp. 454-464.

Tischer, P. C. S. F., Sierakowski, M. R., Westfahl, H., et al. (2010). Nanostructural reorganization of bacterial cellulose by ultrasonic treatment, *Biomacromolecules,* Vol. 11, No. 5, pp. 1217-1224.

Wang, S. & Cheng, Q. (2009). A Novel Process to isolate fibrils from cellulose fibers by high-intensity ultrasonication, *Journal of applied polymer science,* Vol. 113, No. 2, pp. 1270-1275.

Wang, Y. X., Tian, H. F. & Zhang, L. N. (2010). Role of starch nanocrystals and cellulose whiskers in synergistic reinforcement of waterborne polyurethane, *Carbohydrate Polymers,* Vol. 80, pp. 665-671.

Wikipedia. (2011). Nanocellulose, In: *Wikipedia,* accessed on 24 April 2011, Available from: http://en.wikipedia.org/wiki/Nanocellulose.

Zhao, H. P. & Feng, X. Q. (2007). Ultrasonic technique for extracting nanofibers from nature materials, *Applied physics letters,* Vol. 90, No. 7, 073112 (1-2).

Zimmermann, T., Bordeanu, N. & Strub E. (2010). Properties of nanofibrillated cellulose from different raw materials and its reinforcement potential, *Carbohydrate Polymers,* Vol. 79, pp. 1086-1093.

Zuluaga, R., Putaux, J. L., Cruz, J., et al. (2009). Cellulose microfibrils from banana rachis: Effect of alkaline treatments on structural and morphological features, Carbohydrate Polymers, Vol. 76, No. 1, pp. 51-59.

7

The Role of Biodegradable Engineered Scaffold in Tissue Engineering

Ghassem Amoabediny[1,2], Nasim Salehi-Nik[1,2] and Bentolhoda Heli[1,2]
*[1]Department of Biomedical Engineering, Research Centre for New Technologies in Life Science Engineering, University of Tehran,
[2]Department of Chemical Engineering, Faculty of Engineering, University of Tehran,
Iran*

1. Introduction

Tissue engineering is fundamentally described as the generation of three-dimensional (3D) artificial tissues. Its consequential task is to regenerate human tissue or to develop cell-based substitutes in order to restore, reconstruct or improve tissue functions (Pörtner et al., 2005; Ellis, 2005). Proper processing of biological and mechanical functionality is monumental for tissue engineered structures, the ones which are not mainly sufficient enough yet. Acquiring the solution for this problem demands intensive researches and studies in every aspects and steps of TE (Sengers et al., 2007). As a matter of fact, creating a functional tissue requires efficient growth of various types of cells on a 3D scaffolds and the bulk production of one cell seems not to be adequate (Ellis et al., 2005).

The principal function of a scaffold is to direct cell behavior such as migration, proliferation, differentiation, maintenance of phenotype, and apoptosis by facilitating sensing and responding to the environment via cell–matrix and cell–cell communications (Tabesh et al., 2009). Therefore, having such abilities provides scaffolds seeded with a special type of cell as an important part of tissue engineering and regenerative medicine. The scaffold design and fabrication are major areas of biomaterial research, since biomaterial scaffold can create substrate within which cells are instructed to form a tissue or an organ in a highly controlled way. In this chapter, it is tried to provide an inclusive survey of biopolymers to be used as scaffolds for tissue engineering, fabrication methods and engineering challenges such as mass transfer and mechanical strength. In the proceeding, these factors are reviewed in vascular and nerve systems.

2. Scaffold considerations

2.1 Requirements of appropriate scaffold materials for tissue engineering

Scaffold design and fabrication are major areas of biomaterial research and they are also important areas for tissue engineering and regenerative medicine research. Scaffold provides the necessary support for cells to proliferate and maintain their differentiated functions, and its architecture defines the ultimate shape of a new organ.

An ideal scaffold should possess the following characteristics to bring about the desired biological response (1) the scaffold should possess inter-connecting pores of appropriate scale to favor tissue integration and vascularization, (2) be made from material with controlled biodegradability or bio-resorbability, (3) appropriate surface chemistry to favor cellular attachment, differentiation and proliferation, (4) possess adequate mechanical properties to match the intended site of implantation and handling, (5) should not induce any adverse response and, (6) be easily fabricated into a variety of shapes and sizes (Liu et al., 2007; Sachlos et al., 2003).

Due to control scaffold degradation and mechanical integrity, cell-scaffold interaction as well as cell functon, one must have access to a range of materials. Therefore, an appropriate fabrication method is required with which it is possible to have a structure with different independent parameters and materials (Yarlagadda et al., 2005).

It is worth to mention that degradation of synthetic polymers, both *in vitro* and *in vivo* conditions, releases by-products. For example, for PLLA releasing Lactic acid during degradation, causes reducing the pH, which further accelerates the degradation rate due to autocatalysis which later affects cellular function. (Sachlos & Czernuszka, 2003, as cited Reed and Gilding, 1981)

In addition to degradation rate and by-products, certain physical characteristics of the scaffolds must be considered when designing a substrate to be used in tissue engineering applications. For instance, in order to allow proper cell attachment, the scaffold must have a large surface area which can be achieved by creating a highly porous polymeric foam. In these foams, the pore size should be large enough to allow cells to penetrate through the pores, to maximize nutrient and oxygen diffusion, interstitial fluid and blood flow into the interior of the scaffold, to manipulate tissue differentiation (Yarlagadda et al., 2005, as cited Le Huec et al. 1995; Tsuruga et al., 1997). These characteristics (porosity and pore size) often depend on the material and method of scaffold fabrication (Mikos&Temenoff, 2000, as cited Mooney et al., 1999; Nam et al. 2000)

2.2 Decent materials for scaffolds fabrication

In order to have an effective function, an ideal scaffold must possess the optimum structural parameters, conductivity to the cellular activities leading to neo-tissue formation; these include cell penetration and migration into the scaffold, cell attachment onto the scaffold substrate, cell spreading and proliferation and cell orientation. Such scaffold design parameters are now described with reference to these cellular activities. One of the first considerations when designing a scaffold for tissue engineering is the choice of material. The three main material types which have been successfully investigated to be applied in developing scaffolds include (i) natural polymers, (ii) synthetic polymers, and (iii) ceramics (Willerth & Sakayama-Elbert, 2007; Radulescu et al., 2007).

2.2.1 Natural materials

Natural polymers commonly derived from protein or carbohydrate polymers have been used as scaffolds for the growth of several tissue types. In the area of tissue engineering, for example, scientists and engineers look for scaffolds on which it may successfully grow cells to replace damaged tissue. Typically, it is desirable for these scaffolds to be: biodegradable, non-toxic/non-inflammatory, mechanically similar to the tissue to be replaced, highly porous, encouragement of cell attachment and growth, easy and cheap to manufacture, and capable of attaching with other molecules (Elmstedt, 2006; Cuy, 2004). Here some examples of natural polymers that have been previously studied for biomaterials application are reviewed.

- Collagen

Collagen is considered by many scientists as an ideal scaffold or matrix for tissue engineering as it is the major protein component of the extracellular matrix. It provides support to connective tissues such as skin, tendons, bones, cartilage, blood vessels, and ligaments in its native environment, and also interacts with cells in connective tissues and transduces essential signals for the regulation of cell anchorage, migration, proliferation, differentiation, and survival. Collagen is defined by high mechanical strength, good biocompatibility, low antigenicity and ability of being cross-linked, and tailored for its mechanical degradation and water uptake properties; Twenty-seven types of collagens have been identified so far, but collagen type I is the most abundant and the most investigated for biomedical applications (Chunlin et al., 2004).

Collagen may also be processed into a variety of formats including porous sponges, gels, and sheets. It can be cross-linked with chemicals to make it stronger or to alter its degradation rate (Cuy, 2004). However, for medical applications, the implantation of foreign cells causes immunological problems. Collagen has potential uses as follows (Matin, 2004):
- Collagen gel matrix maintains its shape following cell seeding and culture,
- Highly permeable bio-scaffold design,
- Production of tissue implants for reconstructive/cosmetic surgery applications, and
- Generation of spinal cord repair implants.

- Chitosan

Chitosan is a cationic polymer obtained from chitin comprising copolymers of β (1→4)-glucosamine and N-acetyl-D-glucosamine. Chitin is a natural polysaccharide found particularly in the shell of crustacean, cuticles of insects and cell walls of fungi and is the second most abundant polymerized carbon found in nature (Khor & Lim, 2003).

This polymer has many suitable properties. It can be used for wound dressing, drug delivery, and tissue engineering (cartilage, nerve and liver tissue) applications. These properties include: (Willerth et al., 2007):
- Minimal foreign body reaction,
- Mild processing conditions (synthetic polymers often need to be dissolved in harsh chemicals; chitosan will dissolve in water based on pH),
- Controllable mechanical/biodegradation properties (such as scaffold porosity or polymer length), and
- Availability of chemical side groups for attachment to other molecules.

Chitosan has already been investigated for adoption in the engineering of cartilage, nerve, and liver tissue. Current difficulties applying chitosan as a polymer scaffold in tissue engineering include low strength and inconsistent behavior with seeded cells (Madihally and Matthew, 1999). Fortunately, chitosan may be easily combined with other materials in order to increase its strength and cell-attachment potential. Mixtures with synthetic polymers such as poly (vinyl alcohol) and poly (ethylene glycol), or natural polymers such as collagen, have already been produced. These combinations promise improving the performance of the combined construct over the behavior of either component alone (Cuy, 2004).

- Agarose/alginate

Agarose and alginate are linear polysaccharides obtained from seaweed and algae, respectively. Both polygosaccharides must undergo extensive purification to prevent immune responses after implantation (Willerth et al., 2007). Moreover, encapsulation of certain cell types into alginate beads may actually enhance cell survival and growth. In addition, alginate has been explored to function in liver, nerve, heart, and cartilage tissue engineering.

Unfortunately some drawbacks to alginate include mechanical weakness and poor cell adhesion. In order to overcome these limitations, the strength and cell behavior of alginate have been enhanced by mixtures with other materials, including the natural polymers agarose and Chitosan (Cuy, 2004). Mohan et al. described the preparation and characterization of alginate sponges to be used as scaffolds in tissue engineering. They fabricated highly porous 3D scaffolds from cheaply available sodium alginate, which exhibits good biocompatibility. The scaffold fabricated by a combination of freeze drying and particulate leaching, showed increased porosity and pore size. Better pore characteristics and swelling properties may permit more cell invasion and nutrient supply. Moreover alginate is thermally stable, non-cytotoxic and biodegradable (Mohan and Nair, 2005).

- Fibronectin

Fibronectin (FN) is a glycoprotein which exists outside cells and on the cell surface. It also exists in blood, other body fluids and on the cell surfaces of connective tissue. This protein associates with the other proteins of the extra cellular matrix (ECM) like fibrinogen, collagen, glycosaminoglycans and with suitable receptors which are in the cell membrane (Ebner et al., 2006). Fibronectin is composed of tandem repeats of three distinct types (I, II and III) of individually folded modules.

2.2.2 Synthetic materials

Synthetic polymers have been widely used for over 20 years as surgical sutures, with long established clinical success and many are approved for human use by the Food and Drug Administration (FDA). The polymers which approved by FDA are as follows: PCL, PLLA, PLGA, PEG and PGA.

However current synthetic polymers do not possess a surface chemistry which is familiar to cells, that *in vivo* thrive on an extracellular matrix made mostly of collagen, elastin, glycoproteins, proteoglycans, laminin and fibronectin; these materials have many advantages to be used as scaffolds (Sachlos & Czernuszka, 2003, as cited Alberts et al., 1994). These polymers can be tailored to produce a wide range of mechanical properties and degradation rates. Synthetic polymers also represent a more reliable source of raw materials with the ability to provoke an immune response in body. Finally synthetic polymers can be react together to combine their unique properties (Willerth et al., 2007; Manzanedo, 2005). Here some of these synthetic polymers used in tissue engineering are described briefly.

- Poly (D, L-lactic acid)

Poly lactic acid (PLA) is a biodegradable polyester attainable by poly-condensation of lactic acid, and amonomeric precursor that can be obtained from renewable resources. Lactic acid is a chiral molecule available in the L and D stereoisomer forms. L-lactic acid occurs in the metabolism of all animals and microorganism, and thus is an absolutely non-toxic degradation product of polylactides. This is proved by the successful application of polylactides as a resorbable medical structure over a period of three decades (kricheldorf, 2001; Onose, 2008).

Since this polymer has biodegradability, biocompatibility, good mechanical properties, and ability to be dissolved in common solvents for processing, it has been successfully employed as matrixes for cell transplantation and tissue regenerations (Kim et al., 2003).

- Poly (lactic-co-glycolic acid)

Poly (glycolic acid) (PGA) and poly (lactic acid) (PLA) are biodegradable synthetic polymers, which can react to form the copolymer poly (lactic-co-glycolic acid) (PLGA). PGA is highly crystalline and has a high melting point and low solubility in organic solvents. An intensive investigation aimed at the improvement of PGA properties was undertaken by many researchers, focusing on the preparation of PGA copolymers with more hydrophilic PLA.

After implantation, the ester bonds that make up the backbone of the polymer can be hydrolyzed, causing the scaffold to degrade into metabolite by-products. These by-products can be absorbed by the body and may cause pH changes around the implantation site. The degradation rate of the scaffolds can be altered by varying the ratio of PGA to PLA in the scaffold.

The presence of an extra methyl group renders lactic acid more hydrophobic. The hydrophobicity of PLA limits the water uptake of films to 2%, and decreases the backbone hydrolysis rate in respect to the one of the PGA homopolymer. Moreover, PLA is more soluble in organic solvents than in PGA. It is important to note that no linear relationship exists between the ratio of glycolic acid to lactic acid and the physico-mechanical properties of their regarding copolymers. The high crystallinity of PGA is rapidly lost in PGA/PLA copolymers. These morphological changes lead to an increase in the hydration and hydrolysis rate and copolymers tend to degrade more rapidly than the homopolymers of PGA or PLA do (Reed et al., 1981).

- Poly-β-hydroxybutyrate

Poly-β-hydroxybutyrate (PHB) is a linear head-to-tail homopolymer of (R)-β hydroxybutiric acid, which forms crystalline cytoplasmic granules in the wide variety of bacteria. This material is biodegradable and biocompatible microbially produced polyester, which after implantation degrades slowly at body temperature and forms a non-toxic metabolite that is secreted in urine (Mosahebi et al., 2001). PHB has been previously used as a wound scaffolding device, designed to support and protect a wound against further damage while promoting healing by encouraging cellular growth on and within the device from the wound surface (Ljungberg et al., 1999).

- Poly-e-caprolactone

Poly-e-caprolactone(PCL), an aliphatic polyester which is bioresorbable and biocompatible, is generally used in pharmaceutical products and wound dressings (Venugopal et al., 2005).

This polymer has low melting point of around 60°C and a glass transition temperature of about −60°C. PCL is degraded by hydrolysis of its ester linkages in physiological conditions (e.g. in the human body) and therefore, has received a great deal of attention for use as an implantable biomaterial (Schnell et al., 2007).

- Poly (ethylene glycol)

Poly ethylene glycol (PEG), also known as poly ethylene oxide (PEO) or polyoxyethylene (POE) is the most commercially important polyethers, which refers to an oligomer or polymer of ethylene oxide resisting protein adsorption and cell adhesion. These characteristics help minimize the immune response after implantation. Additionally, this polymer can also help to seal cell membranes after injury, making it useful for limiting cell death. Hydrophilic PEG hydrogels can be made through a variety of cross-linking schemes to create scaffolds with varying degradation as well as release rates. Further chemistry can be used to modify these gels to add sites for cell adhesion or extracellular matrix (ECM) molecules to allow cells to infiltrate into these scaffolds, extending their potential applications (Willerth et al., 2007).

PEG can be photo-polymerized under mild conditions in the presence of cells to create a hydrogel that is biocompatible and non-toxic. Bioactive molecules such as cell adhesion ligands, growth factors and proteolytic degradation sites have been previously incorporated into PEG hydrogels and shown to influence adhesion, proliferation, migration, and extracellular matrix production of vascular smooth muscle cells.

-　　Poly (glycerol sebacic acid)

Poly glycerol sebacic acid (PGS), also called bio-rubber, is a tough, biodegradable elastomer made from biocompatible monomers. Its main features are good mechanical properties, rubber like elasticity and surface erosion biodegradation. PGS was proved to have similar *in vitro* and *in vivo* biocompatibility to PLGA, a widely used biodegradable polymer. (Manzanedo, 2005; Sundback et al., 2005).

-　　Poly (2-hydroxyethyl methacrylate)

Hydroxyethyl methacrylate (HEMA) is a hydro-soluble monomer, which can be polymerized (under various circumstances) at low temperatures (from -20°C to +10°C). It can be used to prepare various hydrogels and to immobilize proteins or cells. It is widely used in medicine as an appropriate biomaterial. Poly (2-hydroxyethyl methacrylate) (pHEMA) is particularly attractive for biomedical engineering applications. Because of its physical properties and high biocompatibility, this polymer can be easily manipulated through formulated chemistry has been extensively used in medical applications, e.g. contact lenses, kerato prostheses and orbital implants. Furthermore, a pHEMA scaffold could be easily incorporated into the nerve guidance tubes (Flynn et al., 2003).

2.3 Methods used for scaffolds design

Several techniques have been developed to process synthetic and natural scaffold materials into porous structures. These conventional scaffold fabrication techniques are considered as processes that create scaffolds having a continuous, uninterrupted pore structure. An overview of such different techniques is as follows: Electrospinning, Solvent-casting, particulate-leaching, Gas foaming, Fiber meshes/fiber bonding, Phase separation, Melt molding, Emulsion freeze drying, Solution casting and Freeze drying (Mikos& Temenoff, 2003; Sachlos & Czernuszka, 2003).

2.3.1 Electrospinning

Electrospinning is a technique for nano-fibrous scaffold fabrication. Various synthetic or natural polymers can be spun to nano fibers with diameters in nano - to micrometer range. They are characterized by a high surface to volume ratio and thus offer sample substrate for cell attachment.

In this technique, polymers are dissolved into a proper solvent or melted before being subjected to a very high voltage to overcome the surface tension and viscoelastic forces as well as forming different fibers (50 nm - 30 μm) diameters, which feature a morphologic similarity to the extracellular matrix of natural tissue and effective mechanical properties. These nanofibrous scaffolds can be utilized to provide a better environment for cell attachment, migration, proliferation and differentiation when compared with traditional scaffolds (Martins et al., 2007).

In general, the process of electrospinning is mainly affected by (i) system parameters, such as polymer molecular weight, molecular weight distribution and solution properties (e.g. viscosity, surface tension, conductivity); and (ii) process parameters, such as flow rate, electric potential, distance between capillary and collector, motion of collector, etc (Yang et al., 2005).

A study on the fabrication of a scaffold by electrospinning biomaterials such as poly lactic acid (PLA), poly glycolic acid (PGA), poly (ethylene-co-vinyl acetate) (PEVA), and type-I collagen was reported (Bowlin et al., 2001). Moreover, nanostructured electrospun PLGA membranes for anti-adhesion applications were presented (Fang et al., 2001). Electrospinning can even be used to create biocompatible thin films with useful coating designs and surface structures that can be deposited on implantable devices in order to facilitate the integration of these devices with the body (Buchko et al., 1999).

The collagen nanofibers were characterized by a wide range of pore size distribution, high porosity, excellent mechanical strength and high surface area to volume ratios, which are favorable parameters for cell attachment, growth and proliferation. In cell activity assessment, electrospun collagen nanofibers coated with type I collagen or laminin were found to promote cell adhesion and spread of normal human keratinocytes. This may be a consequence of the high surface area available for cell attachment due to their three-dimensional features and restoration of biological and structural properties of natural ECM proteins (Rho et al., 2006).

2.3.2 Freeze-drying

Many of the fabrication technologies for polymers are based on particulate-leaching techniques, heat compression, and extrusion. However, the harsh operating conditions of these processes may limit the incorporation of bioactive proteins, cells and residual amounts of the chemical solvents required may cause toxicity *in vivo*. Freeze-drying is an alternative method to produce porous scaffolds, which do not require additional chemicals relying on the water.

This method, used in hydrogels, forms ice crystals that can be sublimated from the polymer, creating a particular micro-architecture. Because the direction of growth and size of the ice crystals are a function of the temperature gradient, linear, radial, and/or random pore directions and diverse sizes can be produced with this methodology (Stokols et al., 2004). On the other hand, the pore size can be controlled by the freezing rate and pH; a fast freezing rate produces smaller pores (Sachlos et al., 2003).

2.3.3 Molding

Graft implants manufacture is an extraordinary example of micro-architecture. One possible way of making the grafts is the particulate leaching technique (i.e. molding). A mold is made of a medium that may not contaminate the graft and that may melt off at a lower temperature melt point than that of the graft. Already widely used, this method limits design intricacy because heat stress during formation compromises the composition of the graft (Friedman et al., 2002). Another problem with this method is that it would be difficult to make a mold small enough for small animal research. As an example, Moore *et al.* used Poly (lactic-co-glycolic acid) (PLGA) with copolymer ratio 85:15 for their initial experiments. Injection molding with rapid solvent evaporation resulted in scaffolds with a plurality of distinct channels running parallel along the length of the scaffolds (Moore et al., 2006).

2.3.4 Solid free form fabrication

The most promising production method is called solid free-form fabrication (SFF). This approach uses a machine like an inkjet printer to make the graft one layer at time by "printing" one layer on top of the other (Friedman et al., 2002). However, the main problem with this technology is the extremely high cost. The machines have to be permanently modified from their original function. Through these methods one can manage pore size, porosity and pore distribution to produce structures to increase the mass transport of oxygen and nutrients throughout the scaffold.

Although there are several commercial variants of SFF technology, the general process involves producing a computer-generated model using computer-aided design (CAD) software. After expression of cross-sectional layers, data is implemented to the SFF machine to produce the physical model. Through building the layers from bottom to up, each newly formed layer adheres to the previous. Post-processing may be required to remove temporary support structures (Sachlos & Czernuszka, 2003). The methods that use the SSF technologies to fabricate tissue engineering scaffold are: (1) stereolithography (SLA), (2) selective laser sintering (SLS), (3) fused deposition modelling (FDM) and (4) three-dimensional printing (3-DP).

3. Challenges in engineering scaffolds

The major hindrance through tissue engineering constructs is mass transfer. Since scaffolds provide large surface area to volume ratios, it is proved that mass transfer limitation is reduced. Pore interconnectivity directly influences the mass transfer (e.g. oxygen and nutrient supply and removal of toxic metabolites).

During tissue regeneration, permeability of the matrix decreases due to the declined pore size. In addition, the size of most engineered tissues is limited as they do not have their own blood system, and the cells are only supplied by diffusion (Griffith & Naughton, 2002; Kannan et al., 2005). Meanwhile, as only cell layers of 100–200 µm thickness can absorb O_2 by diffusion, oxygen supply is particularly a critical issue. However, since tissue constructions should have larger dimensions, mass transfer limitation can be considered as one of the greatest engineering challenges (Pörtner et al., 2005).

The next challenge is mechanical effect on cells in dynamic systems. Under flow conditions, flow rate needs to be optimized based on (i) nutrient distribution, (ii) effect on assembly of matrix elements and (iii) cellular response to local shear stress (Lawrence et al., 2009). The scaffold architecture affects the local fluid flow velocities of the cell suspension which influences the number of cell-scaffold contacts per time unit and the local cell deposition rates. By changing the scaffolds design, different flow profiles and cell distributions in the scaffolds may be obtained (Melchels et al., 2011).

Although flow of growth medium improves nutrient and waste transport, shear stresses induced by fluid flow could affect the scaffold architecture as well as cellular alignment within the structures (Lawrence et al., 2009 as cited Gray et al., 1988; Huang et al., 2005).The wall shear rate determines the hydrodynamic force, that adhering cells are exposed to, which can be expressed per unit of area as the wall shear stress (in Pa), and is the product of the wall shear rate and the kinematic viscosity of the fluid medium. The critical shear stress for cell detachment has been found to range almost between 1 to 3 Pa that depends on the material on which the cells are cultured. Also the highest cell densities in the scaffolds could be observed in the regions with larger pores, higher fluid flow velocities and higher wall shear rates (Melchels et al., 2011 cited as Isenberg et al., 2006; Macario et al., 2008; Smith et al., 1995).

It has been seen that the pore size is a variable strongly affecting the predicted shear stress level, whereas the porosity only influence the statistical distribution of the shear stress

(Sadir, 2011). Furthermore, it has been showed that porosity and pore size affect pressure drop which is important to be determined during tissue regeneration (Lawrence et al., 2009, cited as Sodian et al., 2000). For the same pore size with decreasing porosities, pressure drop increases, and for the constant porosity with reducing pore size, which could limit the fluid flow and nutrient transport, pressure drop is increased (Sadir, 2011). Here these challenges are investigated by the revision of two case studies.

3.1 Vascular system

The primary functions of blood vessels are the delivery of nutrients and oxygen to the tissues and organs of the body and removal of their respective metabolites for clearance or re-oxygenation (Boland et al., 2004). Diseases of the heart and blood vessels are the most life threatening factor in developed countries (Ikada, 2006). Surgical replacement of vessel segments or bypass surgery is the most common intervention for coronary and peripheral atherosclerotic disease with at least 550,000 bypass performed per year (Chung et al., 2010). Autologous vessels harvested from the patient for bypass surgery include the saphenous vein (SV) from the leg and the internal mammary artery (IMA) from the chest wall. The IMA with elasticity maintains the ability to vasoregulate and is less prone to atherosclerosis (Wise et al., 2011). Apart from the fact that implantation of native vessels is limited by the mismatch of dimensional and mechanical properties (Chung et al., 2010), the acceptable vein is not available in 30 %of patients (Boland et al., 2004).

An artery is composed of three layers having different matrix-tissue compositions. The innermost lumen layer (intima) is composed of endothelial cells (ECs) on extracellular matrixes such as type IV collagen and elastin, providing the necessary antithrombogenic nature for contact with the bloodstream. The middle layer (media) is usually the thickest of the three layers and is composed of multiple layers of smooth muscle cells (SMCs) within a surrounding extracellular matrix (ECM) composed of collagen types I and III, elastin fibers and various proteoglycans. The outermost layer (adventitia) is made of fibroblastic cells on randomly arranged type I collagen. In short the collagens impart tensile strength to the vessel wall while elastin provides the elasticity (Boland et al., 2004) (Thomas et al., 2007).

With respect to this construction, the combined structure is particularly attractive for vascular tissue engineering applications. By designing a multilayered tube, it is now possible to seed or co-culture different cell lines in layers with controlled orientation (Chung et al., 2010).

Tissue engineering uses vascular cells and supporting scaffolds to build functional blood vessels (Tillman et al., 2009). In the domain of cells, both the EC and SMC are critical components of a tissue engineered vascular graft. ECs play an integral role in tissue homeostasis, fibrinolysis, and anti-coagulation while SMCs perform many functions, including vasoconstriction and dilatation, synthesis of various types of collagen, elastin and proteoglycans (Ju et al., 2010). Bare scaffolds without ECs showed abundant platelet adherence, while scaffolds lined with ECs resisted adherence of blood elements. While it is evident that cells play a major role in achieving patent vessels, designing a vascular scaffolds that provide structural support, enabling cells to proliferate and growing into a three-dimensional (3-D) tissue is important (Chung et al., 2010; Tillman et al., 2009)

Developing scaffolds that can withstand the pulsatile nature, high pressure, and high flow rate of the bloodstream (Boland et al., 2004) is especially necessary in cardiovascular applications. The ideal vascular prostheses must be a presentation of its functional characters such as flexibility with kink resistance and biocompatibility (non-thrombogenic)

(Boland et al., 2004). Ability of ECs adherence to form an anti-thrombogenic luminal surface, and SMCs migration, exhibit vasoactive properties, (Williamson et al., 2006 cited as Tiwari et al., 2001; Seifalian et al., 2002), capability of withstanding physiological hemodynamic forces while maintaining structural integrity until mature tissue forms *in vivo* (Ju et al., 2010), and also the compliance is matched with that of the native artery (Boland et al., 2004). Additionally, providing interconnected pores, generating high porosity so as to promote cell–cell and cell–matrix communication and having sufficient mechanical stability are key factors (Chung et al., 2010) for the engineering of blood vessels. Small pore size does not bring about a problem for ECs but would limit the ability of SMCs to colonize the outer portion of neo-vessel. Several approaches have been proposed to generate large sized pores which include the use of the salt leaching technique (Ju et al., 2010, cited as Nam et al., 2007) and coelectrospinning with water-soluble polymers which serve as sacrificial fibers. Researchers indicated that a larger pore size can be achieved by increasing the fiber diameter and that this would facilitate cell infiltration. They showed that the mechanical strength of the scaffolds decreases as fiber diameter increases (Ju et al., 2010).

Of the countless synthetic materials evaluated over the years, expanded poly tetrafluoro ethylene (e-PTFE) and woven or knitted polyethylene terephthalate (PET) fibers have proven to be satisfactory in terms of medium (6-10 mm internal diameter (ID)) and large (>10 mm ID) vascular prosthetics, respectively (Boland et al., 2004). However, clinical success for small diameter (<6 mm) vessels has yet to be demonstrated due to complications such as occlusion, thrombosis and intimal hyperplasia (Chung, 2010).

In addition, biodegradable synthetic polymer scaffolds from polyglycolic acid (PGA), poly-L-lactic acid, polyhydroxyalkanoates such as poly-4-hydroxybutyrate, polycaprolactone - copolylactic acid and polyethylene glycol have been explored. On the other hand, natural polymers such as collagen and fibrin have also been utilized to construct biological vascular grafts, populated and compacted by smooth muscle cells, and exhibiting high tensile strength and flexibility (Zhang et al., 2008).

However, future attempts are addressed as: (i) maufacture of scaffolds with ECM-like nano-fibrous structure; (ii) development of the electrospun structures to manage pore size which improves the ingrowth of large sized cells; (iii) test the structural integrity of these scaffolds during degradation and under dynamic culture conditions (Vaz et al, 2005).

In some cases these materials have been coated with cell-adhesive proteins such as fibronectin, vitronectin and laminin to facilitate EC attachment, which effectively renders the surface to become antithrombogenic However, such modifications can also provide good substrates for platelet adhesion and thrombus formation. EC must be able to resist detachment by the high shear forces exerted by blood flow and turbulence, thereby providing an anti-thrombotic surface. They should also retain vasoactive function but not induce immune reactions to the implant (Williamson et al., 2006, cited as Tiwari et al., 2001).

Based on these results, Ku and Park. suggested that Poly dopamine (PDA) coating generally facilitates the cell adhesion process of culturing human umbilical vein endothelial cells (HUVECs) on electrospun PCL nanofibers. The increase of cell adhesion on this coated nanofibers are attributed to the adsorption/immobilization of serum proteins on PDA ad-layer (Ku & Park, 2010).

As it was mentioned about vascular tissue construction, designing a multilayered tube is attractive to seed or co-culture different cell lines in layers with controlled orientation. Vaz et al. fabricated a bi-layered tubular scaffold of PLA (outer layer) and PCL (inner layer) by multi-layering electrospinning (ME) using a rotating mandrel-type collector. The tensile measurements showed that PLA/PCL scaffolds presented a much lower elongation and a four-fold increase of maximum stress compared to electrospun PCL ones. The PLA/PCL bi-layered scaffold supported the attachment, spread and growth of mouse fibroblasts (Vaz et al, 2005).

Morever, Venugopal et al. proved that PCL nanofibrous matrixes coated with collagen support cell growth or make the three-dimensional structured multilayer of PCL nanofibers and collagen nanofibers suitable for blood vessel engineering. Their report defines the initial adhesion mechanism of SMCs to the sebiocompatible poly (caprolactone) nanofibrous matrixes coated with collagen. These results may be relevant to other cell types (Venugopal et al., 2005c).

Wise et al. reported the production of synthetic human elastin and poly caprolactone multilayered vascular graft which mimics the mechanical properties of the human IMA. They showed that conduits constructed from synthetic elastin (SE) alone had insufficient strength for vascular applications. This graft was systematically modified by addition of PCL, which was selected on the basis of its appropriate mechanical properties and a slow degradation time. On the other hand, they found that PCL interacts with platelets, which can contribute to thrombogenicity. For this reason, the graft lumen was made of pure SE i.e. devoid of PCL to significantly reduce platelet adhesion, in contrast to PCL/collagen hybrids which showed abundant platelet adherence. It should also be mentioned that the outer layer of the graft, a hybrid of SE/PCL modulated the graft mechanic (Wise et al., 2011).

On the other hand, the nanolayer can mimic the ECM, whereas the micro layer provides larger pores which facilitate superior cell infiltration; a micro -and nano-combined structures can be advantageous.

Chung et al. combined electrospinning and melt spinning of poly (L-lactide-co-e-caprolactone) (50:50 PLCL) to fabricate a multilayered tubular construct using a rotating mandrel for both techniques, which makes the nano layer able to adhere completely to the micro layer. They showed that copolymers of PLCL exhibit a range of mechanical properties from rigid solids to elastomers, depending on their composition. Furthermore, it has a slow rate of degradation compared to other bioresorbable polymers such as PGA and PLA (Chung et al., 2010).

Additionally, Surface patterns such as microgrooves have been successfully used to induce both alignment of cell shape and directional cell migration. Uttayarat et al, incorporated both electrospun microfibers and surface microgrooves in the fabrication of 3D synthetic polyurethane (PU) grafts. They found that grooved patterns induced the uniform alignment of endothelial cell monolayers with morphology similar to naturally aligned endothelium under hemodynamic flow. They also extended these findings by demonstrating that a groove depth or a fiber diameter of about 1 μm can guide the alignment of endothelial cells inside tubular PU grafts (Uttayarat et al., 2010).

3.2 Nerve and spinal cord system

The nervous system is classified into the central nervous system (CNS) and the peripheral nervous system (PNS). The CNS comprises the brain, spinal cord, optic, olfactory and auditory systems. Considering these organs, the CNS comprises of a vast number of neurons, astroglia, microglia, and oligodendrocytes.

The spinal cord which is approximately 1 cm thick and 42 cm long has four anatomical divisions; the cervical, thoracic, lumbar, and sacral regions. About 27 %of human spinal cord injuries are lacerations caused by penetrating objects that tear the spinal tissue (Open injuries) resulting in a discontinuity of the cord. The majority of the clinical cases are the result of a temporary compression of the cord that leaves the cord surface intact (closed injuries 73%). Three types of compression injuries are described: massive compression, contusion, and solid cord injury (E.Schmidt et al., 2003; Erschbamer, 2007).

CNS axons do not regenerate appreciably in their native environment. Several glycoproteins in the native extracellular environment (myelin) of the CNS are inhibitory for regeneration. Regeneration in the adult CNS requires a multi-step process. First, the injured neuron must survive, and then the damaged axon must extend its cut processes to its original neuronal targets. According to Horner and Gage investigation, once contact is made, the axon needs to be re-myelinated and functional synapses need to form on the surface of the targeted neurons (Horner and Gage., 2000).

In addition to the nerve graft and other natural tissues, such as autologous muscle and vein grafts, biopolymers can be a practical tool to provide neurotrophic and/or cellular support while simultaneously guiding axonal regeneration (Stokols et al., 2004; Rodriguez et al., 2000). Indeed, numerous natural and synthetic polymers have been used as scaffolds or within scaffolds for peripheral and central nerve regeneration.

Filling of the interior channels with appropriate cell facilitates axon regeneration. The Schwann cell (SC) and its basal lamina are crucial components in the environment through which regenerating axons grow to reach their peripheral targets. They produce myelin, which has important effects on the speed of transmission of electrical signals and are shown to enhance the regeneration of axons in both the peripheral and central nervous systems (Erschbamer, 2007; Alovskaya et al., 2007). Therefore, it seems that application of a nerve grafts (scaffolds) coated with SCs can be an appropriate method for spinal cord regeneration.

Considering the requirements of scaffolds in general and in particular, i.e. in neural tissue engineering, materials appropriate for SC seeding should posse some additional features. In order to successfully design a scaffold that can be used as treatment for SCI, many considerations must be taken into account. The scaffold should lessen glial scar formation, while containing sites for cell adhesion to allow regenerating neurons to extend axons into the injury site (Willerth et al., 2007; Radulescu et al., 2007).

Among natural materials, Matin in 2004 found that implants coated with collagen are more successful than the bare ones. Stokolos et al. chose to fabricate scaffolds with agarose for several reasons. First, when implanted into lesion cavities in the spinal cord as an unstructured solid agarose hydrogel, it did not evoke an immune or inflammatory response and was stable for at least 1 month. Second, it was observed that neither axons nor cells penetrated solid agarose hydrogels, which suggested that walls composed of agarose could effectively delineate pathways for regenerating axons. Third, freeze-drying could be used to fabricate agarose into soft and flexible scaffold. Finally, neurotrophic factors, proven to elicit robust axonal growth could be easily incorporated into these scaffolds.

In other researches, Alvskaya et al. described that by using fibronectin as a substrate in an *in vivo* model of spinal cord repair, the growth of neuritis within the material is accompanied by migration of SCs into the graft and the presence of reactive astrocytes at its surface continued. Within the first 2 weeks of implantation, a number of cells and cellular elements replaced the FN mat as it dissolved. The first cells to infiltrate FN mats were macrophages. The presence of integrin receptors on Schwann cells may be responsible for the extensive infiltration of Schwann cells. The close spatial correspondence between laminin tubules and Schwann cells suggests that they were deposited by the Schwann cells (Alovskaya et al., 2007; King et al., 2006).

Along with natural materials, synthetic polymers have been widely used for tissue engineering. Recently, Patist et al. demonstrated that the implantation of a macro porous PLA tubular scaffold in the transected rat spinal cord elicited a modest axonal regeneration response. These particular scaffolds were prepared by a thermally induced polymer-solvent phase separation process and contained longitudinally oriented macropores connected to each other by a network of micropores (Patist et al., 2004).

Moore et al. described multiple-channel, biodegradable scaffolds that serve as the basis for a model to investigate simultaneously the effects of scaffold architecture, transplanted cells, and locally delivered molecular agents on axon regeneration. PLGA with copolymer ratio 85:15 was used for their experiments. Primary neonatal Schwann cells were distributed in the channels of the scaffold and remained viable in tissue culture for at least 48 h. Scaffolds containing SCs implanted into transected adult rat spinal cords contained regenerating axons at 1 month post-operation. Axon regeneration was demonstrated by three-dimensional reconstruction of serial histological sections (Moore et al., 2006).

Also it is showed that PGS which have similar *in vitro* and *in vivo* biocompatibility to PLGA, had no harmful effect on Schwann cell metabolic activity, attachment, or proliferation, and did not induce apoptosis (Manzanedo, 2005; Sundback et al., 2005).

PHB has been previously used as a wrap-around implant to guide axonal growth after peripheral nerve injury (Ljungberg et al., 1999). Novikova et al. prepared a biodegradable conduit made of PHB fibers which compressed together and running in parallel directions in two perpendicular layers to form a sheet. Implantation of these PHB conduits coated with alginate hydrogel and fibronectin and seeded with SCs has been found to reduce spinal cord cavitation as well as retrograde degeneration of injured spinal tract neurons (Novikov et al., 2002).

PCL is interesting for the preparation of long term implantable devices, owing to its degradation, which is even slower than that of polylactide. Schnell et al. designed biodegradable, aligned poly-e-capro- lactone (PCL) and collagen/PCL (C/PCL) nanofibers as guidance structures were produced by electrospinning and tested in cell culture assays. They compared fibers of 100% PCL with fibers consisting of a 25:75% C/PCL blend. Both types of electrospun fibers supported oriented neurite outgrowth and glial migration from dorsal root ganglia (DRG) explants. SC migration, neurite orientation, and process formation of SCs, fibroblasts and olfactory ensheathing cells were improved on C/PCL fibers, when compared to pure PCL fibers (Schnell et al., 2007).

About PEG, it was showed that focal continuous application of this polymer has minimal toxicity (Cole and Shi, 2005). Duerstock et al. used three-dimensional computer reconstructions of PEG treated and spinal cords to determine whether the pathological character of a 1-month-old injury is ameliorated by application of PEG. In PEG-treated animals, the lesion was more focal and less diffuse throughout the damaged segment of the spinal cord, so that control cords showed a significantly extended lesion surface area (Duerstock and Borgens, 2002).

Furthermore, a pHEMA scaffold could be easily incorporated into the nerve guidance tubes. Flynn et al. developed a method to create longitude in ally oriented channels within (pHEMA) hydrogels for neural tissue engineering applications. They found that these scaffolds have the potential to enhance nerve regeneration after section injuries of the spinal cord by increasing the available surface area and providing guidance to extending axons and invading cells (Flynn et al., 2003).

Several techniques have been developed to process synthetic and natural scaffold materials into porous structures as H. Tabesh et al. reviewed. Among these techniques, creating tissue engineering scaffolds in nano-scale may bring unpredictable new properties to the material-such as mechanical (stronger), physical (lighter), more porous (tunable),optical (color emission), chemical reactivity (more active and less corrosive), electronic properties (more electrically conductive),and magnetic properties (super paramagnetic which are very important in nerve regeneration). Such scaffolds may come up with new functionalities as well-which are unavailable at micro or macro scales (Tabesh et al., 2009).

The process of electrospinning is used for nano-fibrous scaffold fabrication. Electrospinning can even be used to create biocompatible thin films with useful coating designs and surface structures that can be deposited on implantable devices inorder to facilitate the integration of these devices with the body. Silk-like polymers with fibronectin have been electrospun to make biocompatible films used on prosthetic devices aimed to be implanted in the central nervous system (Buchko et al., 1999).

Moreover, an elegant way to produce nanofibrous scaffold using PLLA by a liquid–liquid phase separation method quite similar to natural extracellular matrix (ECM) was developed by a group of scientists. They showed its efficacy in supporting the neural stem cell (NSC) differentiation and neurite outgrowth (Yang et al., 2004).

In addition, a new and facile method for the creation of longitudinally oriented channels in pHEMA gels using a fiber templating technique was described. Biodegradable polycaprolactone (PCL) fibers were extruded and embedded in transparent pHEMA gels, leading to the creation of a pHEMA–PCL composite.

4. Conclusion

In this chapter, efficacious biomaterials (natural and synthetic) for scaffolds in tissue engineering and cell seeding were discussed and also techniques to their fabrication were reviewed. Considering results using such materials and the mentioned criteria for an appropriate scaffold, it is proved that the selection of materials and method of fabrication depend on the cells and their characteristics. The reasons are: scaffold candidates should mimic the structure and biological activity of the native ECM proteins which provide adequate mechanical support and regulate cellular activates. In addition, scaffolds must support and define the three-dimensional structure of the tissue engineered space and maintain the normal state of differentiation within the cellular compartment.

Furthermore the structure of scaffold, pore size and porosity, may affect the mass transfer, shear rate and pressure drop. Mass transfer is the major hindrance in tissue engineering. Although surface area to volume ratios of a scaffold can decrease mass transfer limitations, it is still one of the greatest challenge in tissue engineering. It has been observed that the pore size and shape influence the shear stress level and distribution, while the porosity affects only the distribution. Therefore the wall shear stress is an important parameter in cell adhesion processes.

Two case studies, blood and nerve systems, with regard to their challenges have been investigated. First, for blood system, a scaffold must have the function of native blood and must provide appropriate mechanical, endothelialization and antithrombogenic properties. Therefore choosing a proper biomaterial which provides these characters is prominent. With respect to multilayered construction of blood vessels, the combined structure is particularly attractive for vascular tissue engineering applications. For a better simulation, various types of materials and cells have been used to form different layers of this tissue.

Secondly, for nerve regeneration, it is suggested that blending a synthetic and natural polymer (e.g. poly-e-caprolactone and collagen) is the best choice for Schwann cell seeding to regenerate the spinal cord injuries, considering all results using different materials and the mentioned criteria for an appropriate scaffold. The two suggested polymers have the potential to play role of a scaffold in SCs seeding.

Collagen is a protein of ECM and exists in the basal membrane of the cell. It is easily purified, which can be proposed as a proper substance; however, collagen has less strength to withstand long time, support force adhesion, and degrades enzymatically within short periods. Therefore, using another polymer such as PCL to enhance the stability and mechanical strength of collagen would be crucial. In this sence, an excellent scaffold for Schwann cell adhesion, migration, orientation, and proliferation can be provided.

Also electrospinning is considered as the excellent method for the fabrication of such scaffolds. Additionally, electrospun nanofibers exhibit excellent supports for nerve growth because they can provide large surface area to volume ratios, pore sizes tailored to Schwann cells dimensions, functionalized surfaces, and multiple sites for interaction and attachment, and low mass transfer limitation.

5. References

Alovskaya A.; Alekseeva T.; Phillips J.B.; King V. & Brown R. (2007). Fibronectin, Collagen, Fibrin - Components of Extracellular Matrix for Nerve regeneration. In: *Topics in Tissue Engineerin, Vol. 3,* Ashammakhi Eds.N.; Reis R.L. & Ciellini E. < http://www.oulu.fi/spareparts/ebook_topics_in_t_e_vol3/list_of_contr.html>

Boland E.D.; Matthews J.A.; Pawlowski K.J.; Simpson D.G.; Wnek G.E. & Bowlin G.L. (2004). Electrospinning collagen and elastin: preliminary vascular tissue engineering, *Frontiers in Bioscience,* Vol. 9, 1422-1432.

Bowlin G.L.; Matthews J.A.; Simpson D.G.; Kenawy E.R. & Wnek G.E. (2001). Electrospinning biomaterials. *Journal of Textile Apparel,* Technol Manage, Vol. 1, Special issue: The Fiber Society, Spring 2001 Conference, Raleigh NC.

Buchko C.J.; Chen L.C.; Shen Y. & Martin D.C. (1999). Processing and microstructural characterization of porous biocompatible protein thin films. *Polymer,* Vol. 40, No. 26, pp. 7397-7407.

Chunlin Y.; Hillas P.J.; Buez J.A.; Nokelainen M.; Balan J.; Tang, J.; Spiro R. & Polarek J.W. (2004). The application of recombinant human collagen in tissue engineering, *BioDrugs,* Vol. 18, No. 2, pp. 103_119.

Chung S.; Ingle N. P.; Montero G.A.; Kim S.H. & King M.W. (2010). Bioresorbable elastomeric vascular tissue engineering scaffolds via melt spinning and electrospinning, *Acta Biomaterialia,* Vol. 6, No. 6, pp. 1958–1967, ISSN 1742-7061

Cole A. & Shi R. (2005). Prolonged focal application of polyethylene glycol induces conduction block in guinea pig spinal cord white matter. *Toxicology in Vitro,* Vol. 19, No. 2, pp. 215-220.

Cuy J. (2004). Biomaterials Tutorial: Natural Polymers. University of Washington Engineered Biomaterials. < http://www.uweb.engr.washington.edu/>

Duerstock B. & Borgens R. (2002). Three-dimensional morphometry of spinal cord injury following polyethylene glycol treatment. *The Journal of Experimental Biology*. Vol. 205 (pt 1), pp. 13-24.

Ebner R.; Lackner J.M.; Waldhauser W.; Major R.; Czarnowska E.; Kustosz R.; Lacki P. & Major B. (2006). Biocompatible TiN-based novel nanocrystalline films, *Bulletin of the polish academy of science*, Vol. 54, No. 2, pp. 167-173.

Ellis M.; Jarman-Smith M. & Chaudhuri JB. (2005). Bioreactor Systems for Tissue Engineering: A Four-Dimensional Challenge, In: *Bioreactors for Tissue Engineering: Principles, Design and Operation*, Al-Rubeai M. & Chaudhuri JB., pp. 1–18, Springer, <http://www.springer.com/biomed/book/978-1-4020-3740-5>

Elmstedt N. (2006). Development of biosynthetic conduit for spinal cord and peripheral nerve injury repair, in vitro study. Thesis. Stockholm, Sweden. < http://www.nada.kth.se/utbildning/grukth/exjobb/rapportlistor/2006/>

Erschbamer M. (2007). Experimental spinal cord injury: Development of protection and repair strategies in rats. Thesis. Department of Neuroscience, Karolinska Institutet, Stockholm, Sweden. < http://publications.ki.se/jspui/handle/10616/39259 >

E.Schmidt Ch. & Baier Leach J. (2003). Neural tissue engineering: strategies for repair and regeneration. *Annual review of Biomedical Engineering*. Vol. 5, pp. 293-347.

Fang D.; Xong X.; Chen W.; Cruz S.; Hsiao B. & Chu B. (2001). Nanostructured electrospun poly-D,L-lactideco-glycolide membranes for anti-adhesion applications. *Apparel Technol*, Vol. 1, Special issue: The Fiber Society, Spring 2001 Conference, Raleigh NC.

Flynn L.; Dalton P.D. & Shoichet M.S. (2003). Fiber templating of poly(2-hydroxyethyl methacrylate) for neural tissue engineering. *Biomaterials*, Vol. 24, No. 23, pp. 4265-4272.

Friedman J.A.; Windebank A.J.; Moore M.J.; Spinner R.J.; Currier B.L.; Yaszemski M.I.; Bartolomei J.; Piepmeier J.M.; GHU G.; Fehlings M.G.; Hodge Ch.J. & Wagner F.C. (2002). Biodegradable polymer grafts for surgical repair of the injured spinal cord. *Neurosurgery*, Vol. 51, No. 3, pp. 742-52.

Griffith L. G. & Naughton G. (2002). Tissue engineering–current challenges and expanding opportunities. *Science*, Vol. 295, No. 5557, pp. 1009-1014.

Horner P. & Gage F. (2000). Regenerating the damaged central nervous system. *Nature*, Vol. 407, pp. 963-970.

Hurtado A.; Moon L.D.; Maquet V.; Blits B.; Jerome R. & Oudega M. (2006). Poly (D,L-lactic acid) macroporous guidance scaffolds seeded with Schwann cells genetically modified to secrete a bi-functional neurotrophin implanted in the completely transected adult rat thoracic spinal cord. *Biomaterials*, Vol. 27, No. 3, pp. 430-442.

Ikada Y. (2006). Challenges in tissue engineering, *Journal of Technology Society Interface*, Vol. 3, pp. 589–601

Ju Y.M.; Choi J.S.; Atala A.; Yoo J.J. & Lee S. J. (2010). Bilayered scaffold for engineering cellularized blood vessels, *Biomaterials*, Vol. 31, No. 15, pp. 4313–4321, ISSN: 0142-9612

Ljungberg C.; Johansson-Ruden G.; Bostrom K.J.; Novikov L. & Wiberg, M. (1999). Neuronal Survival Using a Resorbable Synthetic conduit as an Alternative to Primary Nerve Repair, *Microsurgery*, Vol. 19, No. 6, pp. 259_264.

Kannan RY.; Salacinski HJ.; Sales K.; Butler P. & Seifalian A.M. (2005). The roles of tissue engineering and vascularisation in the development of micro-vascular networks: a review. *Biomaterials*, Vol. 26, No. 14, pp. 1857-1875.

King V.R.; Henseler M.; Hunt-Grubbe H.; Brown R. & Priestly J.V. (2006). Cellular and extracellular infiltrates into fibronectin mats implanted into the damaged adult rat spinal cord. *Biomaterials*, Vol. 27, No. 3, pp. 485-496.

Khor E. & Lim L.Y. (2003). Implantable applications of chitin and chitosan, *Biomaterials*, Vol. 24, No. 13, pp. 2339_2349.

Kricheldorf H. R. (2001). syntheses and application of polylactides. *Chemosphere*, Vol. 43, No. 1, pp. 49-54.

Ku S. H. & Park Ch.B. (2010). Human endothelial cell growth on mussel-inspired nanofiber scaffold for vascular tissue engineering, Biomaterials, vol. 31, No. 36, pp. 9431-9437, ISSN: 0142-9612

Lawrence B. J., Devarapalli M. & Madihally S.V. (2009). Flow Dynamics in Bioreactors Containing Tissue Engineering Scaffolds, *Biotechnology and Bioengineering*, Vol. 102, No. 3, pp. 935-947.

Liu W. & Cao Y. (2007). Application of scaffold materials in tissue reconstruction in immunocompetent mammals: our experience and future requirements, *Biomaterials*, Vol. 28, No. 34, pp. 5078–5086.

Madihally S.V. & Matthew H.W.T. (1999). Porous chitosan scaffolds for tissue engineering, *Biomaterials*, Vol. 20, No. 12, pp. 1133_1142.

Manzanedo D. (2005). Biorubber (PGS): evaluation of a novel biodegradable elastomers. Thesis. Massachusetts Institute of Technology, Department of Materials Science and Engineering. < http://hdl.handle.net/1721.1/37687>

Martins A.; Araújo J.V.; Reis R.L. & Neves N.M. (2007). Electrospun nanostructured scaffolds for tissue engineering applications. *Nanomedicine*, Vol. 2, No. 6, pp. 929-42.

Matin S. (2004). Spinal cord regeneration via collagen entubulation. Thesis. Department of Aeronautics and Astronautics, John Hopkins University, Massachuset Institute of Technology. < http://dspace.mit.edu/handle/1721.1/28889>

Melchels F. P.W.; Tonnarelli B.; Olivares A.L.; Martin I.; Lacroix D.; Feijen J.; Wendt D. J. & Grijpma D. W. (2011) . The influence of the scaffold design on the distribution of adhering cells after perfusion cell seeding, *Biomaterials*, Vol. 32, No. 11, pp. 2878-2884, ISSN: 0142-9612

Mikos A. & Temenoff J. (2003). Formation of highly porous biodegradable scaffolds for tissue engineering. *EJB Electronic Journal of Biotechnology*, Vol.3 No.2, (August 15, 2000), ISSN: 0717-3458

Mohd nasir N.F.; mohd zain N.; Graha M. & kardi N.A. (2005). characterization of chitosan – (poly ethylene oxide) blend as haemodiyalysis membrane, *American journal of applied science*, Vol. 2, No. 12, pp. 1578-1583.

Mosahebi. Mohammadi., 2001. Genetic Labelling and Transplantation of Schwann cells to Enhance Pheripheral Nerve Regeneration, Royal Free & University College Medical School, University of London.

Moore M.; Friedmanb J.; Lewellync E.; Mantilaa S.; Krychd A.; Ameenuddinc S.; Knightc Lu L.; Currierd B.; Spinnerd R.; Marshd R.; Windebank A. & Yaszemskia M. (2006). Multiple-channel scaffolds to promote spinal cord axon regeneration. *Biomaterials*, Vol. 27, No. 3, pp. 419-429.

Onose G.; Ciureaa A.V.; Rizeaa R.E.; Chendreanu C.; Anghelescu A.; Haras M. & Brehar F. (2008). Recent advancements in biomaterials for spinal cord injury complex therapeutics. *Digest Journal of Nanomaterials and Biostructures*, Vol. 2, No. 4, pp. 307-314.

Patist C.M.; Mulder M.B.; Gautier S.E.; Maquet V.; Jerome R. & Oudeg M. (2004). Freeze-dried poly (D,L-lactic acid) macroporous guidance scaffolds impregnated with brain-derived neurotrophic factor in the transected adult rat thoracic spinal cord. *Biomaterials*, Vol. 25, No. 9, pp. 1569-82.

Pörtner R.; Nagel-Heyer S.; Goepfert Ch.; Adamietz P. & Meenen M. N. (2005). Bioreactor Design for Tissue Engineering. *Journal of Bioscience and Bioengineering*, Vol. 100, No. 3, pp. 235-245, *ISSN*: 1389-1723.

Radulescu D.; Dhar S.; Young Ch.; Taylor D.; Trost H.; Hayes D. & Evans G. (2007). Tissue engineering scaffolds for nerve regeneration manufactured by ink-jet technology, *Materials Science and Engineering*: C, Vol. 27, No. 3, pp. 534–539.

Reed A.M. & Gilding D. k.; (1981). Biodegradable Polymers For Use in Surgery Poly (lactic acid)/Ploy(glycolic)Homo and Copolymers:2.In vitro Degradation, *Polymer*, Vol. 22, pp. 494_498.

Rho K.; Jeong L.; Lee G.; Seo B.; Park Y.; Hong S.; Roh S.; Cho J.; Park W. & Min B. (2006). Electrospinning of collagen nanofibers: Effects on the behavior of normal human keratinocytes and early-stage wound healing. *Biomaterials*, Vol. 27, No. 8, pp. 1452-1461

Rodriguez F.; Verdu E.; Ceballos D. & Navarro X. (2000). Nerve Guides seeded with autologous Schwann cells Improve Nerve Regeneration, *Experimental Neurology*, Vol. 161, No. 2, pp. 571-584.

Sachlos E. & Czernuszka J.T. (2003). Making tissue engineering scaffolds work Review on the application of solid freeform fabrication technology to the production of tissue engineering scaffolds. *European Cells and Materials*, Vol. 5, pp. 29-40, ISSN: 1473-2262.

Sadir S.; Kadir M.R.A.; Öchsner A. & Harun M.N. (2011). Modeling of Bio Scaffolds: Structural and Fluid

Transport Characterization, World Academy of Science, *Engineering and Technology*, Vol. 74, pp. 621-627.

Schnell E.; Klinkhammer K.; Balzer S.; Brook G.; Klee D. & Dalton P. (2007). Guidance of glial cell migration and axonal growth on electrospun nanofibers of poly-e-caprolactone and a collagen/poly-e-caprolactone blend, *Biomaterials*, Vol. 28, No. 19, pp. 3012-3025.

Sengers B.G.; Taylor M.P.; Please COC. & Oreffo R. (2007). Computational modelling of cell spreading and tissue regeneration in porous scaffolds. *Biomaterials*, Vol. 28, No. 10, pp. 1926–1940.

Stokols Sh. & Tuszynski M. (2004). The fabrication and characterization of linearly oriented nerve guidance scaffolds for spinal cord injury. *Biomaterials*, Vol. 25, No. 27, pp. 5839-5846.

Sundback C.A.; Shyn J.Y.; Wang Y.; Faquin W.C.; Langer R.S.; Vacanti, J.P. & Hadlock T.S. (2005). Biocompability analysis of a poly(glycerol sebacate) as a nerve guide material. *Biomaterials*, Vol. 26, No. 27, pp. 5454-5464.

Tabesh H.; Amoabediny Gh.; Salehi-Nik N.; Heydari M.; Yosefifard M.; Ranaei Siadat S.O. & Mottaghy K. (2009). The role of biodegradable engineered scaffolds seeded with Schwann cells for spinal cord regeneration. *Neurochemistry international*, Vol. 53, No. 2, pp. 73–83, ISSN: 0197-0186

Tillman B.W.; Yazdani S.K.; Lee S.J.; Geary R.L.; Atala A. & Yoo J.J. (2009). The in vivo stability of electrospun polycaprolactone-collagen scaffolds in vascular reconstruction, *Biomaterials*, Vol. 30, No. 4, pp. 583–588, ISSN: 0142-9612.

Thomas V.; Zhang X.; Catledge Sh.A & Vohra Y.K. (2007). Functionally graded electrospun scaffolds with tunable mechanical properties for vascular tissue regeneration, *Biomedical Materials*, Vol. 2, No. 4, pp. 224–232, ISSN: 1748-6041.

Uttayarat P.; Perets A.; Li M.; Pimton P.; Stachelek S.J.; Alferiev I.; Composto R.J.; Levy R.J. & Lelkes P.I. (2010). Micropatterning of three-dimensional electrospun polyurethane vascular grafts, *Acta Biomaterialia*, Vol. 6, No. 11, pp. 4229–4237, ISSN 1742-7061

Vaz C.M.; Tuijl S. V.; Bouten C.V.C. & Baaijens F.P.T. (2005). Design of scaffolds for blood vessel tissue engineering using a multi-layering electrospinning technique, *Acta Biomaterialia*, Vol. 1, No. 5, pp. 575–582, ISSN 1742-7061

Venugopal, J.; Zhang Y.Z. & Ramakrishna S. (2005a). Fabrication of modified and functionalized polycaprolactone nanofibre scaffolds for vascular tissue engineering. *Nanotechnology*, Vol. 16, No. 10, pp. 2138-2142.

Venugopal J.; Zhang Y. & Ramakrishna S. (2005b). Electrospun nanofibres: biomedical applications. *IMechE*, Vol. 218, No. 1, pp. 35-45.

Venugopal J.; Zhang Y. & Ramakrishna S. (2005c). In vitro study of smooth muscle cells on polycaprolactone and collagen nanofibrous matrices. *Cell Biology International*, Vol. 25, No. 10, pp. 861-867.

Willerth S. & Sakayama-Elbert Sh. (2007). Approaches to neural tissue engineering using scaffolds for drug delivery, *Advanced Drug Delivery Reviews*, Vol. 59, No. 4-5, pp. 325–338.

Wise S.G.; Byrom M.J.; Waterhouse A.; Bannon P.G.; Ng M.K.C. & Weiss A.S. (2011) A multilayered synthetic human elastin/polycaprolactone hybrid vascular graft with tailored mechanical properties, *Acta Biomaterialia*, Vol. 7, No. 1, pp. 295–303, ISSN 1742-7061

Williamson M. R.; Black R. & Kielty C. (2006). PCL–PU composite vascular scaffold production for vascular tissue engineering: Attachment, proliferation and bioactivity of human vascular endothelial cells, *Biomaterials*, Vol. 27, No. 19, pp. 3608–3616, ISSN: 0142-9612

Yang L.; Fitie´ C.; van der Werf K.; Bennink M.; Dijkstra P. & Feijen J. (2008). Mechanical properties of single electrospun collagen type I fibers. *Biomaterials*, Vol. 29, No. 8, pp. 955-962.

Yarlagadda PK..; Chandrasekharan M. & Shyan JY. (2005). Recent advances and current developments in tissue scaffolding, *Bio-Medical Materials and Engineering*, Vol. 15, No. 3, pp. 159–177.

Zhang X.; Baughman C.B. & Kaplan D.L. (2008). In vitro evaluation of electrospun silk fibroin scaffolds for vascular cell growth, *Biomaterials*, Vol. 29, No. 14, pp. 2217-2227, ISSN: 0142-9612

8

Optical Detection of Protein Adsorption on Doped Titanium Surface

Raimo Silvennoinen et al.*
Department of Physics and Mathematics, University of Eastern Finland, Joensuu
Finland

1. Introduction

The frequently used biomaterial in hard tissue replacement, such as dental and orthopaedic implants, is titanium. (Ball et al., 1996; Höök et al., 2002a; Huang et al., 2003; Imamura et al., 2008; Jones et al., 2000; Walivaara et al., 1994; Yang et al., 2003) These kind of biomaterial applications made of titanium are satisfactory products, because of their ability to adsorb certain proteins. After implantation, within a few seconds, the biomaterial surface becomes coated with a film of adsorbed proteins mediating the interaction between the implant and the body environment. Since an implant is exposed to blood during implantation, the initial protein layer is mainly composed of plasma proteins. Human plasma fibrinogen (HPF) is the relevant protein, which adsorbs on biomaterial surfaces. HPF partakes in blood coagulation, facilitates adhesion and aggregation of platelets (Cacciafesta et al., 2001; 2000). The structure and composition of the adsorbed protein layer determine the type and extent of the subsequent biological reactions, such as activation of coagulation and immune response and osseointegration (Nygren et al., 1997). Thus, the initially adsorbed protein layer is a factor determining the biocompatibility (Cai et al., 2006; Galli et al., 2001; Garguilo et al., 2004; Hemmersam et al., 2005; Kidoaki & Matsuda, 1999; Ma et al., 2007; Rouahi et al., 2006; Van De Keere et al., 2008; Wang et al., 2003), and also in recent years interest has been focused to preparation of hydrocarbons doped with Ti and used different methods to analyzing of biocompatibility for important proteins (Choukourov et al., 2008; Grinevich et al., 2009; Silvennoinen, Hason, Vetterl, Penttinen, Silvennoinen, Myller, Cernochova, Bartakova, Prachar & Cvrcek, 2010; Silvennoinen, Vetterl, Hason, Tuononen, Silvennoinen, Myller, Cvrcek, Vanek & Prachar, 2008).

The production and application of doped titanium surfaces are under intensive research, and the results have shown the positive views on the adaptation of these materials as a biomaterial, as equal or even better than the bulk titanium. The doping of titanium is performed typically by inserting impurities like N, Nb, Zr, Ta, Al, Cr and V, into the titanium crystal structure. (Archana et al., 2010; Barajas-Ledesma et al., 2010; Cantau et al., 2010; Choi et al., 2010; Czoska et al., 2011; Darriba et al., 2009; Huang et al., 2010; Khaleel et al., 2010; Martin et al., 2006; Shaama, 2005)

*Niko Penttinen, Martti Silvennoinen (*Department of Physics and Mathematics, University of Eastern Finland, Joensuu, Finland*), Stanislav Hason, Vladimír Vetterl (*Institute of Biophysics, v.v.i., Academy of Science of the Czech Republic, Brno, Czech Republic*), Sonia Bartáková, Patrik Prachár, Jiří Vaněk, Vítězslav Březina (*Centre for Dental and Craniofacial Research, Faculty of Medicine, Masaryk University, Brno, Czech Republic*)

Substantial progress has been made in understanding adhesion process on surfaces after or while immersed in liquids. In these research a number of developed techniques such as, X-ray photoelectron spectroscopy (XPS), atomic force microscopy (AFM), ellipsometry, fourier transform infrared attenuated total reflectance (FTIR/ATR), quartz crystal microbalance (QCM), surface plasmon resonance (SPR), dual polarization interferometry, total internal reflectance fluorescence (TIRF), voltammetry and electrochemical impedance spectroscopy (EIS) have been deployed. When measuring in liquid environment the potential amount of usable devices rapidly decreases and none of the listed devices are alone able to describe, in detail, the adsorption process of biomolecules. (Agnihotri & Siedlecki, 2004; Cacciafesta et al., 2001; Cai et al., 2006; 2005; Höök et al., 2002b; Jackson et al., 2000; Jandt, 2001; Roach et al., 2005; Roach, Farrar & Perry, 2006; Roach, Shirtcliffe, Farrar & Perry, 2006; Soman et al., 2008; Sonesson et al., 2007; Swann et al., 2004; Toscano & Santore, 2006; Van De Keere et al., 2008; Vanderah et al., 2004; Wang et al., 2003; Wertz & Santore, 1999; 2001; 2002; Xu & Siedlecki, 2007)

Adsorption of HPF molecules at the rough bulk titanium and modified titanium surfaces increases the scattering and decreases the coherence of the probing laser light beam. These changes in coherence as well as in signal magnitude can be detected with diffractive optical element (DOE) based sensor. DOE sensor is a non-contact optical method, in detection of biomolecules, which can sense changes on the measured biosurface as well as changes in the surface itself. (Silvennoinen, Hason, Vetterl, Penttinen, Silvennoinen, Myller, Cernochova, Bartakova, Prachar & Cvrcek, 2010; Silvennoinen, Peiponen & Myller, 2008; Silvennoinen, Vetterl, Hason, Silvennoinen, Myller, Vanek & Cvrcek, 2010; Silvennoinen, Vetterl, Hason, Tuononen, Silvennoinen, Myller, Cvrcek, Vanek & Prachar, 2008)

HPF protein molecules adsorb stronger on the titanium surface, which is treated by polishing and etching, than at the surface treated only by polishing due to an increase of surface area caused by etching and thus increase the interactions caused by van der Waals forces (Parsegian, 2005; Silvennoinen, Hason, Vetterl, Penttinen, Silvennoinen, Myller, Cernochova, Bartakova, Prachar & Cvrcek, 2010; Silvennoinen, Vetterl, Hason, Tuononen, Silvennoinen, Myller, Cvrcek, Vanek & Prachar, 2008). The effective surface tension, as well as the surface energy related to the topography of surface, is assumed to influence the final interactions of the implant with the surrounding environment. It is also reported that rough surfaces promote better osseointegration than smooth surfaces (Brett et al., 2004; Cochran, 1999; Jansson & Tengvall, 2004; Webster et al., 1999).

Treatment of titanium surface by titanium doped hydrocarbon layer is observed to have strong influence on the adsorption of HPF molecules. Also HPF adsorption is detected on the titanium alloy surface as Ti6Al4V. Adsorption of the elongated HPF on a titanium-based surface is monitored by analyzing permittivity and optical roughness of protein-modified surfaces by using a DOE sensor and a variable angle spectro-ellipsometry (VASE). The biological experiment, when cultivation of osteoblast at the titanium surface was performed, showed that the surface treated by polishing and etching is more proper for the bone cell growth than the surface treated only by polishing. The best adsorption of fibrinogen was observed at the titanium doped hydrocarbon surface prepared by plasma-enhanced chemical vapor deposition, when the optimal ratio was $Ti_{0.38}C_{0.62}$ and $Ti_{0.09}C_{0.91}$. The surface of dental implants treated by this way should increase their biocompatibility, speed up their osseointegration and healing. Also, due to low value of Young modulus and relatively easy forming, even in cold processing, the alloys could be the future material in implantology.

Aging of titanium surface affects to the adsorption behavior of HPF proteins, which may be caused from the consecutive adsorption of carbon and oxygen from air on the surface during the long-term storage of bulk biomaterial. Aging showed decreases in the adhesion magnitudes of titanium doped hydrocarbon surfaces after long term storage, caused by the surface reacting with molecules of air (Silvennoinen, Vetterl, Hason, Tuononen, Silvennoinen, Myller, Cvrcek, Vanek & Prachar, 2008).

2. Characterization of surface

2.1 Diamond stylus

Diamond stylus is one of the standardized and classical methods in measuring the surface roughness. The idea of stylus profilometer is that a thin stylus (with diameter only few microns) is scanned on the measured surface along a line. When a small force is applied to the stylus, the vertical movement can be measured revealing the hills and valleys of the surface ie. the real surface topography. In figure 1 is shown a surface profile measured from a non-polished Si(100) surface.

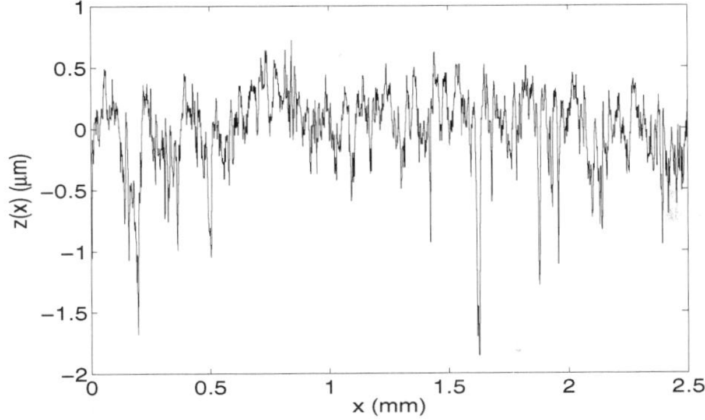

Fig. 1. The measured surface profile $z(x)$ as a function of stylus scan length x. For this surface R_a = 0.23 μm, R_q = 0.31 μm and R_z (peak to valley value) = 2.44 μm.

When the profile curve is recorded, one can calculate various roughness parameters, like average surface as follows

$$R_a = \frac{1}{L} \int_0^L |z(x) - \langle z(x) \rangle|\, dx,\tag{1}$$

where the stylus scan length is L, x is the position along the scan length, $z(x)$ is the vertical position of stylus as a function of x, brackets ($\langle \rangle$) denote the mean value with minimum variance and vertical bar brackets ($||$) denote the absolute value. Another often used parameter is root-mean-square (rms) surface roughness, which can be expressed

$$R_q = \sqrt{\frac{1}{L} \int_0^L (z(x) - \langle z(x) \rangle)^2\, dx}.\tag{2}$$

In general the $R_a \leq R_q$.

2.2 Spectral ellipsometry (VASE)

Spectral ellipsometer can be considered as a special reflectometer, where the only difference is the polarization state of the reflected light wavefront measured from sample surface. The analysis of reflected wavefront data is performed by using the Fresnel laws to solve the optical parameters related to surface such as complex refractive index ($N = n + i\kappa$, where i denotes imaginary unit). The amplitude reflection coefficient (r) of the TM- and TE-polarized light at the angle of incidence (θ_i) can be expressed with the following equations

$$r_{TM} = \frac{(n_2 - i\kappa_2)^2 \cos\theta_i - n_1 \sqrt{(n_2 - i\kappa_2)^2 - n_1^2 \sin^2\theta_i}}{(n_2 - i\kappa_2)^2 \cos\theta_i + n_1 \sqrt{(n_2 - i\kappa_2)^2 - n_1^2 \sin^2\theta_i}} \tag{3}$$

and

$$r_{TE} = \frac{n_1 \cos\theta_i - \sqrt{(n_2 - i\kappa_2)^2 - n_1^2 \sin^2\theta_i}}{n_1 \cos\theta_i + \sqrt{(n_2 - i\kappa_2)^2 - n_1^2 \sin^2\theta_i}}, \tag{4}$$

where n_1 denotes the refractive index of air and $N_2 = n_2 + i\kappa_2$ is the complex refractive index of the reflective material. The respective reflectances for the TM- and TE-polarized light are as follows

$$R_{TM} = r_{TM} r_{TM}^* \quad \text{and} \quad R_{TE} = r_{TE} r_{TE}^*, \tag{5}$$

where the asterisk denotes the complex conjugate. The complex ratio of the amplitude reflectances (r_{TM}/r_{TE}) can be expressed as follows

$$\frac{r_{TM}}{r_{TE}} = \tan\Psi e^{i\Delta}, \tag{6}$$

where $\tan\Psi$ is the amplitude part and $i\Delta$ is the phase part of propagating complex amplitude. Right side of the equation 6 is assumed to be in a form which can be measured experimentally. Utilizing the equation 6, the complex refractive index can be calculated at a known wavelength which was used to measure Ψ and Δ. Example of measured spectral refractive index is presented in figure 2 from vacuum evaporated gold (Au) film surface.

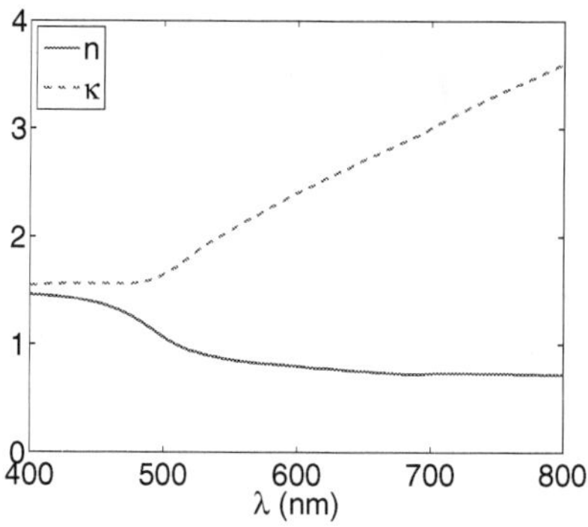

Fig. 2. Measured spectral n and κ ($N = n + i\kappa$) of a gold (Au) film surface measured with VASE.

2.3 Gloss and optical roughness by DOE

DOE sensors have been proven to be accurate in sensing optical changes in two main surface properties as permittivity and roughness also in nanometer scale. Specular reflectance is a function of three variables: complex refractive index/permittivity, angle of incident of the probe beam and surface topography. Standardized specular gloss (G) relates to reflectance and it is normalized to the reflectance of black glass having gloss value of 100 gloss units (GU). The optical roughness, which relates to the surface topography, is sensed as optical path differences which the rough surface generates in reflection. In DOE sensor measurements; (I) the probe beams angle of incidence remains constant (perpendicular to the surface), (II) the complex refractive index of the measured surface is known (measured) and (III) the surface topography acts as a variable. Before DOE measurements, the spectral complex refractive index (N) of each material surface is characterized with variable angle spectro-ellipsometer (VASE). General principle of DOE is to focus coherent light to a 4×4 dot matrix, which takes the optimal shape in its focal plane at wanted wavelength. The formed DOE image is then detected with CCD-camera and saved into personal computers (PC) memory for further analysis (Fig. 3).

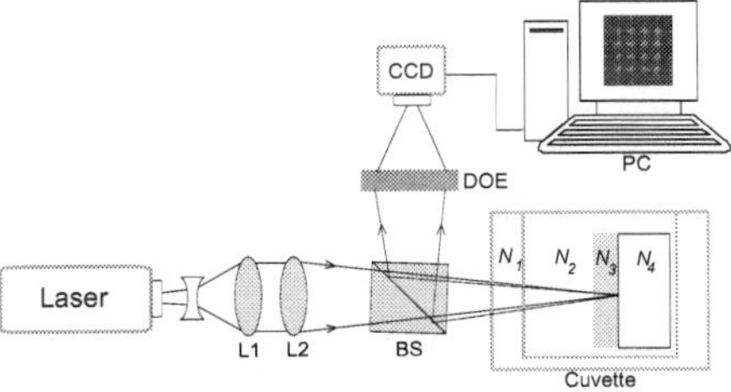

Fig. 3. DOE sensor setup with lenses L1 and L2, beam splitter (BS), sample cuvette, CCD camera, refractive indexes N_1-N_4 and personal computer (PC) for analyzing. DOE image with the 4×4 dot matrix is shown on the screen of PC.

In figures 4 and 5 are presented the gloss and R_{opt} maps scanned from a polished titanium (Ti$_p$) surface with 20 μm probe beam waist diameter. Scanning was done in air from 2mm$\times$2mm surface area.

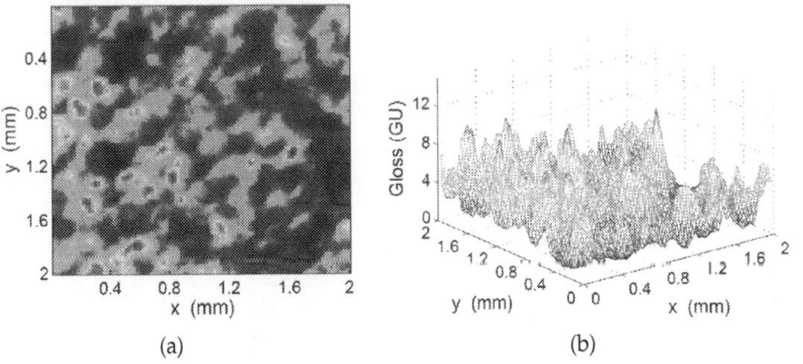

Fig. 4. Gloss from Ti$_p$ surface measured in air. The gloss from the surface is presented (a) as colormap and (b) as meshmap.

The gloss and R_{opt} values can be calculated (see equations 10-13) from the DOE images recorded during the measurements. This calculation is presented in detail in section 3.2. Correlation coefficient between R_{opt} and gloss is ideally -1, because in general when the surface roughness (R_{opt}) increases the reflectance (and G) decreases. Correlation coefficient between the R_{opt} and G values presented in figures 4 and 5 is -0.798.

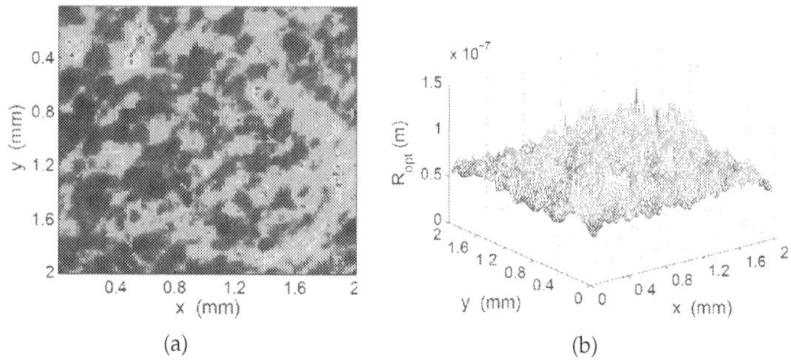

(a)   (b)

Fig. 5. R_{opt} from Ti$_p$ surface measured in air. The R_{opt} from the surface is presented (a) as colormap and (b) as meshmap.

2.4 AFM and SEM

Atomic force microscopy (AFM) bases on the interaction forces of two materials. When the tip of an AFM is in the proximity of a sample surface, these interaction forces (repulsion and attraction) can be detected by measuring the deviations in laser beam that is reflected from the gold sphere on the cantilever in which the tip is attached. Typically, AFM devices can measure the sample surfaces in contact (tip is touching the surface) and non-contact (tip is floating near the surface) mode. In figure 6a is shown an AFM image from a polished silicon Si(100) surface measured in contact mode.

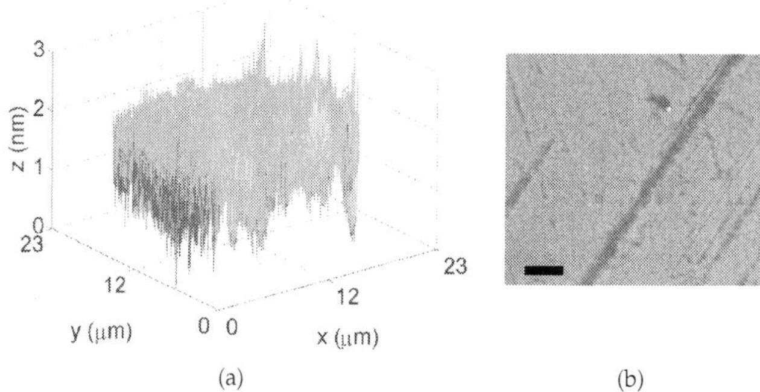

(a)   (b)

Fig. 6. (a) AFM image from a polished silicon Si(100) wafer surface and (b) SEM image from a Ti$_p$ surface, where the bar denotes distance of 1 μm.

Scanning electron microscope (SEM) is constructed in such a way that electrons from filament tip are accelerated, focused and deflected (producing scanning) by coils which have been constructed with knowledge in Coulombian and Lorenzian forces. In figure 6b is shown a SEM image from a polished titanium (Ti_p) surface.

2.5 Spectrophotometry

In spectrophotometry the sample transmittance is measured as a function of wavelength. Thus, the spectrophotometer device includes two equal light pathways for the sample and reference. The probing light is produced from light source like deuterium (UV) and halogen (NIR,VIS) and monochromatized and detected separately with 1 kHz rate for both light paths with a detector (typically one for the UV and one for the VIS/NIR/IR/FIR wavelength regions). Typical parameters, which can be measured by spectrophotometers, are spectral transmittance ($T(\lambda)$), reflectance ($R(\lambda)$) and absorbance ($A(\lambda)$) are presented as follows

$$T(\lambda) = \frac{L_{\lambda t}}{L_{\lambda 0}}, \tag{7}$$

$$R(\lambda) = \frac{L_{\lambda r}}{L_{\lambda 0}}, \tag{8}$$

$$A(\lambda) = \log_e \left(\frac{L_{\lambda 0}}{L_{\lambda t}} \right) = \alpha z = \frac{4\pi\kappa}{\lambda} z, \tag{9}$$

where λ is wavelength, $L_{\lambda t}$, $L_{\lambda 0}$ and $L_{\lambda r}$ are the spectral radiance through, incoming and reflected from the surface respectively, α is the absorbtion coefficient, κ is the extinction coefficient and z is the optical path length in sample. In figure 7 is presented the absorbance of phosphate buffered saline (PBS) and human plasma fibrinogen (HPF) solutions calculated from spectrophotometer measurements.

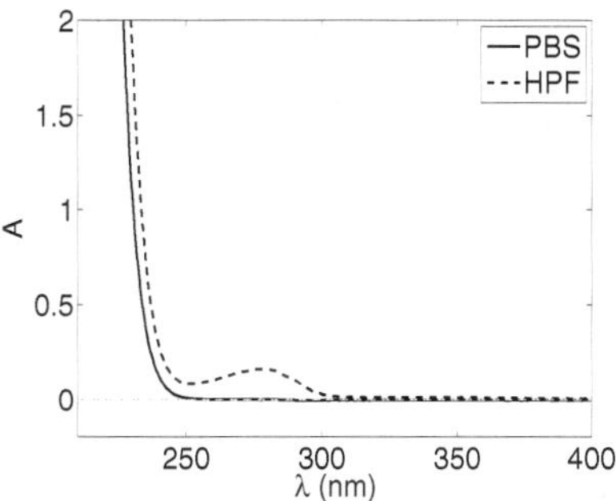

Fig. 7. Calculated absorbance A of (a) PBS and (b) 5 μM HPF solution from spectrophotometer measurement as a function of wavelength λ in nanometers. Optical path length was in solution 10 mm. Black dashed line is the zero absorbance baseline.

3. Sensing of protein molecules

3.1 VASE sensing of protein adsorption

With equation 9 (Beer-Lambert law) one can calculate, from ellipsometry data, the thickness of layer on a sample surface. This requires a smooth sample surface on which the layer (i.e. the proteins) exists. When considering the layer thickness, the sample surfaces complex refractive index without the layer must be measured first and thereafter the measurements are repeated when the layer exists on the sample surface. Then, knowing the complex refractive index N of the formed layer, one can solve the z (double of the layer thickness) from equation 9. As the ellipsometry measurement are done in air, measurements of protein layers should be performed as fast as possible (at least within half an hour), because the drying could have denaturating effects on the proteins.

3.2 Coherent and noncoherent response of DOE sensor

The analysis of DOE image consist of calculating the non-coherent, equation 10, and coherent, equation 11, part of the optical signal, which relate to permittivity and optical roughness, respectively as follows (Silvennoinen, Peiponen & Myller, 2008)

$$I_{NC} = \frac{1}{n_{SW}m_{SW}} \sum_{i_{SW}=1}^{n_{SW}} \sum_{j_{SW}=1}^{m_{SW}} I_{i_{SW},j_{SW}} - I_C, \tag{10}$$

where I_C is the coherent portion of the optical signal being

$$I_C = \frac{1}{n_{pk}m_{pk}} \sum_{i_{pk}=1}^{n_{pk}} \sum_{j_{pk}=1}^{m_{pk}} I_{i_{pk},j_{pk}}. \tag{11}$$

In equations 10 and 11 $I_{i,j}$ denotes the irradiance of reflected probe beam, n_{SW} and m_{SW} are the total numbers of sensor pixel dimensions in signal window (SW), n_{pk} and m_{pk} denotes the respective pixel dimension of each of the 16 peaks (pk) in the DOE image. Gloss G measured in gloss units (GU) is standard measure for optical characterization of a surface. Gloss is a function of three different variables: permittivity (ϵ, connected to N), angle of incidence (θ_i) and topography (roughness). Gloss values are calculated from the DOE image when the coherent response, the CCD values of the peaks, is removed from the CCD values of the DOE image (equation 10). DOE image is produced from the irradiance of the reflected probe beam and because of that, this calculation removes the phase information from this beam, leaving the information (I_{NC}) of the surface reflectance and thus ϵ and N. The calculated I_{NC} for the surface (I_{NCs}) values are normalized with the non-coherent response from black glass I_{NCr}, resulting the gloss value to be as follows

$$G = \frac{I_{NCs}}{I_{NCr}} \times 100. \tag{12}$$

In turn, the optical roughness (R_{opt}) values can be calculated from the DOE image utilizing the equation 11. The I_C values contain the phase information of the reflected probe beam. When the probe light beam is reflected from rough surface, the initial coherence degree of the wave front decreases and distortion appears in DOE image 4×4 peaks (I_C values). Finally, the optical roughness R_{opt} values can be calculated from the following equation

$$R_{opt} = \sqrt{-\log_e \left(\frac{1 - R_s}{1 - R_r} \frac{I_{Cs}}{I_{Cr}} \right) \frac{\lambda}{4\pi}}, \tag{13}$$

where R is reflectance calculated from complex refractive index (N) values measured by variable angle spectro-ellipsometer (VASE), λ is the wave length of the probe beam, and the subscrips s and r denote sample and reference respectively. R_{opt} changes of the investigated surfaces can be determined with accuracy of 0.2 mm (in liquid), which is the detection limit of this one-arm interferometer and reasonable sensitivity to detect nanoscale changes appearing in the bioenvironments (Silvennoinen, Hason, Vetterl, Penttinen, Silvennoinen,

Myller, Cernochova, Bartakova, Prachar & Cvrcek, 2010; Silvennoinen, Peiponen & Myller, 2008; Silvennoinen, Vetterl, Hason, Tuononen, Silvennoinen, Myller, Cvrcek, Vanek & Prachar, 2008).

3.3 DOE sensing of protein adsorption

DOE protein detection was performed by placing the surface vertically in a cuvette in order to minimize the effect of sedimentation. The background electrolyte was phosphate buffered saline (PBS, pH ~ 7.4) prepared with 8 mM Na_2HPO_4, 1.8 mM KH_2PO_4, 140 mM NaCl, and 2.7 mM KCl with the addition of 136 mM sodium citrate. HPF (fraction I, type III) was purchased from Sigma. Optical analysis was done with a 500 nM HPF solution at room temperature. After filling the sample cuvette with water, the baseline measurement for each sample were performed. After the baseline measurement in water, either the background electrolyte or the protein solution was injected in the cuvette.

Human plasma fibrinogen (HPF), found in the circulatory system at a concentration of 2.6 mg/ml is the key structural glycoprotein in blood clotting which, upon thrombin activation, self-assembles forming a fibrin clot. Having a mass of 340 kDa, the rod-like molecule, 46 nm long is a genuine covalent dimer, the two halves having identical sequences that are linked by a central globular domain. Each monomer has three non-identical chains, Aα, Bβ, and γ connected together at the N-terminus by 11 disulfide bridges forming the 'disulfide knot'. The C-terminus of each chain is globular. Those of the α and γ chains extend forming dumbbell shaped ends to the molecule termed the 'D' regions, while the Aα chain globular domains, termed the αC units, interact with each other at the central 'E' region (Wasilewska et al., 2009). Firstly, we study the adsorption ability of HPF proteins on mechanically polished titanium surface by diffractive optical element based sensor (dynamical measurement of adsorption process of protein on surface in electrolyte solution - wet measurement) and ellipsometry (measurement of thickness of adsorbed protein layer on titanium surface after removal the protein-modified surface from protein solution followed by half hour drying in air-dry measurement).

In figure 8 is presented the calculated effects to light reflectance from additional layer (X), PBS electrolyte, air or oxides, on the used polished titanium surface (Ti$_p$) (figure 8a) and alloy surfaces (figure 8b).

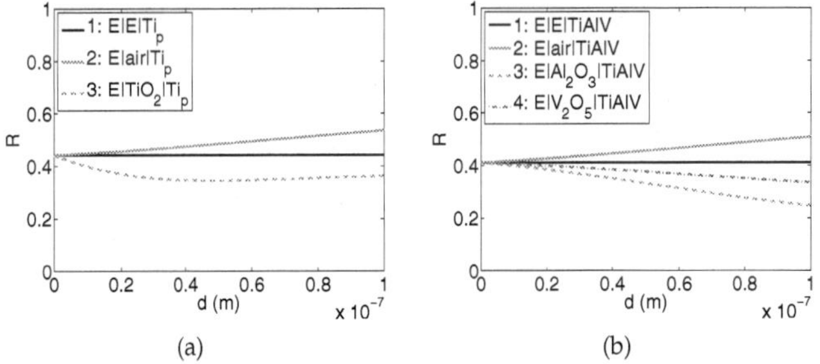

Fig. 8. Calculated reflectance R as a function of layer thickness d on (a) Ti$_p$ and (b) Ti6Al4V surfaces (λ = 632.8 nm). In legend E denotes electrolyte and format E|X|Ti denotes the material layer X between electrolyte and titanium. Complex refractive index for the alloy Ti6Al4V was measured with VASE and other refractive indexes were taken from the book of Palik (1998).

When effective air layer appears on the surface, the reflectance increases. This air layer exists on the surface in liquid in the form of nanobubbles. On the contrary the formation of oxide layer on the surface, for the calculated oxides TiO_2, Al_2O_3 and V_2O_5 decreases the reflectance, and thus the observed gloss values.

Figure 9 presents the temporal responses of gloss and R_{opt} from a polished titanium surface (Ti_p) surface when surface is immersed in water and in PBS without and with HPF molecules.

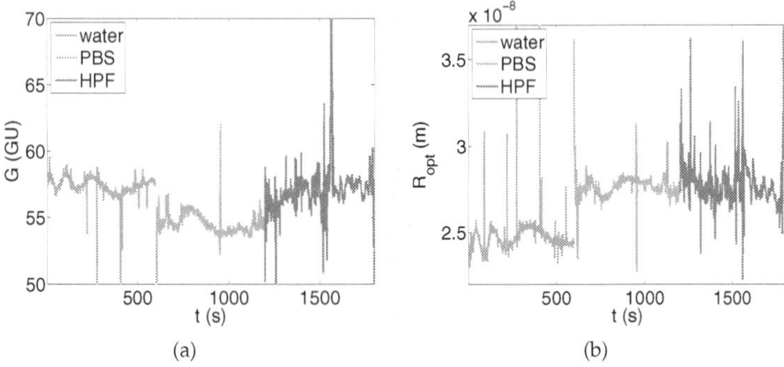

(a)   (b)

Fig. 9. (a) Temporal gloss (G) in GU and (b) optical roughness R_{opt} in meters of a Ti_p surface, when the surface is in water, in PBS without and with HPF-molecules with 500 nM concentration.

When the surface is presented with the PBS, gloss decreases and R_{opt} increases. This is caused by the chemical reaction from PBS, causing surface energy driven formation of oxides and the mobility of initial gas nanobubbles on the surface (see figure 8). Also the formation of additional nanobubbles can occur, but that would require increase in gloss values, like observed in similar case with Ti6Al4V surface in figure 10.

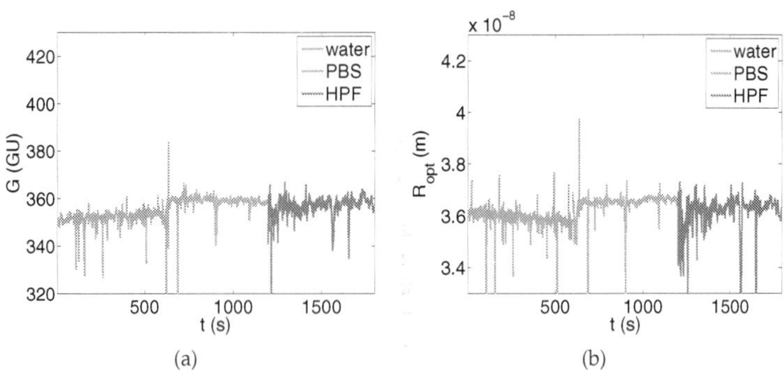

(a)   (b)

Fig. 10. (a) Temporal gloss (G) in GU and (b) optical roughness R_{opt} in meters of a Ti6Al4V surface, when the surface is in water, in PBS without and with HPF-molecules with 500 nM concentration.

After the PBS solution measurement the cuvette is emptied and the HPF solution is injected in the cuvette. The HPF solution typically has higher changes in the signal from water than the PBS, which in turn indicates that the proteins are attaching on the surface. Because of the initial reactions with the PBS solution, the changes that are observed, are the change of gloss and R_{opt} values from water baseline to HPF solution. From the temporal responses of Ti_p and Ti6Al4V like presented in figures 9 and 10 can be noted the stronger reaction of Ti_p compared to the Ti6Al4V, which seems more inert in observed gloss and R_{opt} changes than Ti_p.

3.4 Effect of titanium surface treatment on protein adsorption

In figures 11 and 12 are presented the average values from measurement involving a polished titanium surface (Ti_p) and Ti6Al4V surfaces. All surfaces indicates an initial reaction with the PBS as the gloss decreases, when the PBS without the HPF molecules is injected in the sample cuvette. This is caused by the formation of oxides on the surfaces (see figure 8). At the same time the R_{opt} values have different trends. In the case of the Ti6Al4V alloy, the R_{opt} mean values are higher with the PBS than with the water on surfaces, but on the other hand in the case of titanium, PBS average R_{opt} value is lower compared to the water. This kind of changes in R_{opt} values can be caused by two main reactions of the PBS (i) the mobility of nanobubbles on the surface and additional formation of them and (ii) the surface energy driven formation of oxides. Additional formation of nanobubbles would be seen as an increase in the mean values of the gloss, as the effective layer formed by the bubbles on the surfaces would increase the surface reflection and thus the gloss. However this is not the case observed in the values of the Ti_p and Ti6Al4V surfaces.

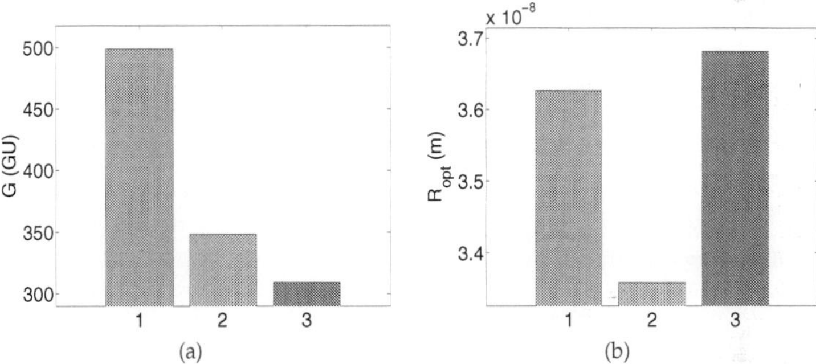

Fig. 11. Average G in GU and optical roughness (R_{opt}) in meters for Ti_p surface. (1) denotes the values from water (2) from PBS and (3) from HPF solution covered surfaces.

When the PBS without the protein is removed, the HPF solution (PBS based) is injected in the sample cuvette. The average signal levels from the HPF measurement are denoted with (3) in the figures 11 and 12. All surfaces indicate lower gloss and higher R_{opt} values than in water, as the proteins are attached on the surface. Also, when comparing the signal value changes from PBS to HPF solution, the changes are similar, which furthermore supports the interpretation that the proteins are attached to the surface. HPF solution is essentially PBS solution with added proteins and when measuring all three types of liquids, in order water, PBS and HPF, one can see the magnitude difference of the adhesion process of samples with DOE sensor data.

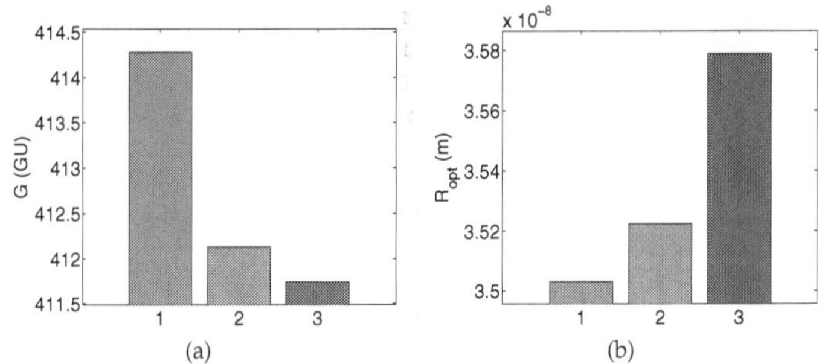

Fig. 12. Average G in GU and optical roughness (R_{opt}) in meters for Ti6Al4V surface. (1) denotes the values from water (2) from PBS and (3) from HPF solution covered surfaces.

As the titanium surface Ti_p is known surface for good biocompatibility, the results (gloss and R_{opt} values) are compared to it. When comparing the results from Ti_p and Ti6Al4V surfaces, one can see that the Ti6Al4V samples are not presenting changes as big as the Ti_p. This would indicate that the Ti6Al4V is not so active as adhesive surface as Ti_p is. The Ti6Al4V behaves in this measurement like an inert surface i.e. indicating unfavorable attachment properties for the HPF molecule.

3.5 Effect of aging of titanium surface on protein adsorption

Effects of aging of the biosurfaces has been studied for titanium-doped hydrocarbon by Silvennoinen, Vetterl, Hason, Tuononen, Silvennoinen, Myller, Cvrcek, Vanek & Prachar (2008). In this study the aging process was observed in 2 years time and the results showed that the long time storage causes impurities to the original surfaces. These impurities was interpreted to be caused by the molecules in air including the carbon and the oxygen. Results showed that the initial adsorption abilities of the measured surfaces were greatly decreased by the long time storage. Thus the aging of biosurface, which have properties to form oxides and other chemical bonds in low energies, should be always considered. For example we measured complex refractive indexes ($\lambda = 632.8$nm) of polished titanium Ti_p and stored titanium Ti_{sto} in aim to compare the possible changes on the surface properties. Complex refractive indexes were $N(Ti_p) = 2.616 + 2.413i$ and $N(Ti_{sto}) = 2.002 + 2.077i$ and respective reflectances were $R(Ti_p) = 0.446$ and $R(Ti_{sto}) = 0.399$. If oxidation of the surface is assumed to be the only reaction on these titanium surfaces, we can estimate from figure 8a the layer thickness of formed native titanium dioxide on this stored surface to be ca. 10 nm. This already would impact on the adhesive properties of the titanium.

4. Alternative optical testing of biocompatibility

4.1 Biological methods in determining the biocompatibility

Cytocompatibility as a basis for biocompatibility testings for several titanium β-alloys (β structure is body-centered cubic structure) has been investigated for example by Bartakova et al. (2009). In this study the possible usage of titanium alloys in dental implants was studied *in-vitro* environment by means of three different tests from which one can determine the biological acceptance of certain material. Test of the cell-spread on standard substrate (coverglass or bottom of culture flask) with (1) inoculation and (2) cell monolayer (both tests used time-lapse capturing of pictures) and (3) assay tests showing possible chromosome aberrations, mutagenesis and neoplastic cell forming. Differences for methods (1) and (2) is that in (1) the cells are inoculated on the surface having lower cell density than in (2) in which the cell density is around 70 %. The results are evaluated by optically or with scanning electron microscopy which can show morphological anomalies on cell membrane surface.

4.2 Cytocompatibility of dental implants alloys

In cell area dilatation test (spreading test) it is possible to use any heteronuclear cell line, for example HeLa, L929, PtK, CHO and other. Aim of this test is to show time for full spreading (dilatation) on standard substrate. Standard substrate is a coverglass, or bottom of culture flask. Positive control is a laboratory standard, negative control is a glass surface or inner surface of culture flask.

Experiment shows an ability of cell membrane to spread and evaluates some disturbance in a cell attached mechanisms on the surface. When the environment is a toxic matter, the cell area dilatation (spreading) is very low and this is a marker for damage of cell membranes. When cells didn't spread, the material was interpreted to be toxic. Zone between 50 and 70 % of cell spread is kept tolerant and 80 and 99 % is interpreted as a cytocompatible material when cultivation time is kept constant. Example of light microscope image of cell area dilation tests shown in figures 13 and 14 for Ta and Cu surfaces respectively.

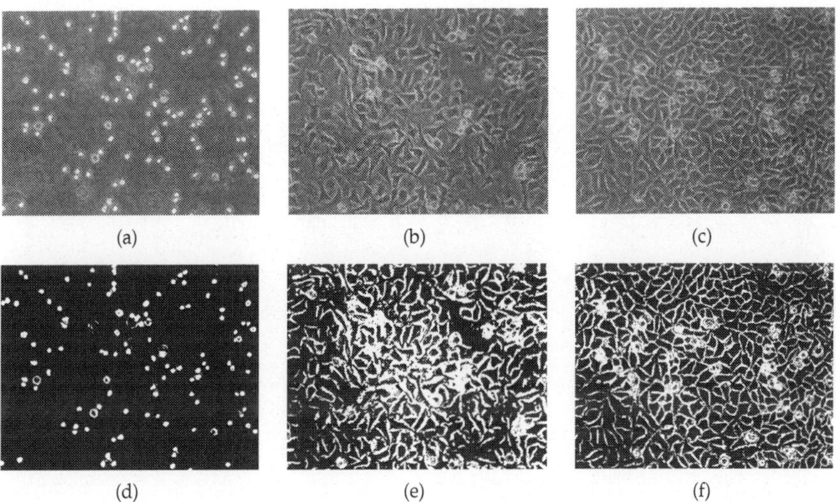

(a)   (b)   (c)

(d)   (e)   (f)

Fig. 13. Evolution of cell spread on Ta surface. Light microscope images from the surface after (a,d) 0 min, (b,e) 120 min and (c,f) 500 min of inoculation. (d), (e) and (f) shows the corresponding processed images.

To make the area occupied by the cells more visible, we have processed the optical microscope images (see figures 13 and 14) by chroma tone threshold processing.

From figures 13 and 14 one can observe the difference of higher (figure 13) and lower (figure 14) cell area dilation. Higher cell division rate on Ta surface indicates that the surface can hold cells on it while the cells maintain their natural division abilities, thus the Ta surface could be considered as a cytocompatible surface. The Cu surface shows toxic reaction to the cells (see figure 14) as the cells are not propagating on the surface and some of the cells shows lost of their biological function. From the images b and c in the figure 14, the dying cells can be observed to have black nucleus from the consequence of increased light absorbtion of lost cell chromosomes.

Optical analysis of Ti_p and Ti6Al4V from the gloss and R_{opt} data in the section 3.4 indicated that the Ti6Al4V surface is not showing strong reaction in the HPF adsorption, but polished titanium surface (Ti_p) is observed to adsorb the HPF molecules. In the study by Bartakova et al. (2009), the Ti6Al4V alloy was recognized to be biologically tolerant, but not as good as Ti_p (biocompatible), thus the two studies seem to be consistent with each other.

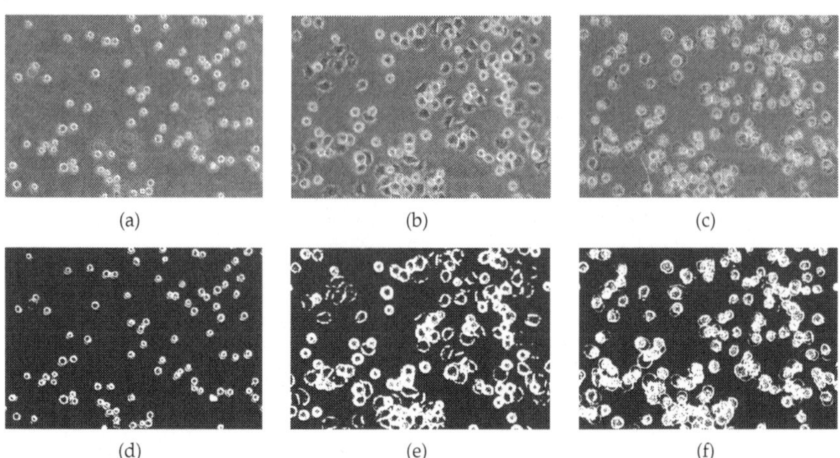

Fig. 14. Evolution of cell spread on Cu surface. Light microscope images from the surface after (a,d) 0 min, (b,e) 40 min and (c,f) 180 min of inoculation. (d), (e) and (f) shows the corresponding processed images.

5. Conclusion

Doped titanium alloys can show both mechanical and biocompatibility properties that exceed the ones of titanium, which has been the often used biomaterial in bioapplications. Study results reported here indicated that many dopant for titanium can be used, but the initial biological reaction should be observed before using the doped titanium in applications. Before the optical detection the surface needs to be characterized. The characterization of the surface is not only basis for the optical measurement, but also helps one to understand more about the biophysical reactions that occurs with the biomolecules on the surface. Results presented here shows that optical detection of biomolecules is a capable non-contact method for determining the reaction of the biomolecules on a surface *in-vitro*. Especially the diffractive optics based elements as the DOE, which can measure multiple surface parameters simultaneously, could give promising new aspects in development of future bio-optoelectronics.

6. Acknowledgement

This work was supported by the Grant Agency of the Czech Republic (P205/10/2378) and the Academy of Sciences of the Czech Republic (KAN200040651), the Ministry of

Education, Youth and Sport of the Czech Republic (1M0528), and institutional research plans (AVOZ50040507 and AVOZ50040702) and the Academy of Finland (131539/31.03.2009).

7. References

Agnihotri, A. & Siedlecki, C. (2004). Time-dependent conformational changes in fibrinogen measured by atomic force microscopy, *Langmuir* 20(20): 8846–8852.

Archana, P., Jose, R., Jin, T., Vijila, C., Yusoff, M. & Ramakrishna, S. (2010). Structural and electrical properties of nb-doped anatase TiO_2 nanowires by electrospinning, *Journal of The American Ceramic Society* 93(12): 4096–4102.

Ball, V., Bentaleb, A., Hemmerle, J., Voegel, J. & Schaaf, P. (1996). Dynamic aspects of protein adsorption onto titanium surfaces: Mechanism of desorption into buffer and release in the presence of proteins in the bulk, *Langmuir* 12(6): 1614–1621.

Barajas-Ledesma, E., Garcia-Benjume, M., Espitia-Cabrera, I., Bravo-Patino, A., Espinoza-Beltran, F., Mostaghimi, J. & Contreras-Garcia, M. (2010). Photocatalytic activity of Al_2O_3-doped TiO_2 thin films activated with visible light on the bacteria escherichia coli, *Materials Science and Engineering B* 174(1-3, Sp. Iss. SI): 74–79.

Bartakova, S., Prachar, P., Kudrman, J., Brezina, V., Podhorna, B., Cernochova, P., Vanek, J. & Strecha, J. (2009). New titanium beta-alloys for dental implantology and their laboratory-based assays of biocompatibility, *Scripta Medica* 82(2): 76–82.

Brett, P., Harle, J., Salih, V., Mihoc, R., Olsen, I., Jones, F. & Tonetti, M. (2004). Roughness response genes in osteoblasts, *Bone* 35(1): 124–133.

Cacciafesta, P., Hallam, K., Watkinson, A., Allen, G., Miles, M. & Jandt, K. (2001). Visualisation of human plasma fibrinogen adsorbed on titanium implant surfaces with different roughness, *Surface Science* 491(3): 405–420.

Cacciafesta, P., Humphris, A., Jandt, K. & Miles, M. (2000). Human plasma fibrinogen adsorption on ultraflat titanium oxide surfaces studied with atomic force titanium oxide surfaces studied with atomic force, *Langmuir* 16(21): 8167–8175.

Cai, K., Bossert, J. & Jandt, K. (2006). Does the nanometre scale topography of titanium influence protein adsorption and cell proliferation?, *Colloids and Surfaces B-biointerfaces* 49(2): 136–144.

Cai, K., Muller, M., Bossert, J., Rechtenbach, A. & Jandt, K. (2005). Surface structure and composition of flat titanium thin films as a function of film thickness and evaporation rate, *Applied Surface Science* 250(1-4): 252–267.

Cantau, C., Pigot, T., Dupin, J. & Lacombe, S. (2010). N-doped TiO_2 by low temperature synthesis: Stability, photo-reactivity and singlet oxygen formation in the visible range, *Journal of Photochemistry and Photobiology A-chemistry* 216(2-3, Sp. Iss. SI): 201–208.

Choi, J., Park, H. & Hoffmann, M. (2010). Combinatorial doping of TiO_2 with platinum (Pt), chromium (Cr), vanadium (V), and nickel (Ni) to achieve enhanced photocatalytic activity with visible light irradiation, *Journal of Materials Research* 25(1, Sp. Iss. SI): 149–158.

Choukourov, A., Grinevich, A., Slavinska, D., Biederman, H., Saito, N. & Takai, O. (2008). Scanning probe microscopy for the analysis of composite ti/hydrocarbon plasma polymer thin films, *Surface Science* 602(5): 1011 – 1019.

Cochran, D. (1999). A comparison of endosseous dental implant surfaces, *Journal of Periodontology* 70(12): 1523–1539.

Czoska, A., Livraghi, S., Paganini, M., Giamello, E., Di Valentin, C. & Pacchioni, G. (2011). The nitrogen-boron paramagnetic center in visible light sensitized N-B Co-doped TiO_2.

experimental and theoretical characterization, *Physical Chemistry Chemical Physics* 13(1): 136–143.

Darriba, G., Errico, L., Eversheim, P., Fabricius, G. & Renteria, M. (2009). First-principles and time-differential gamma-gamma perturbed-angular-correlation spectroscopy study of structural and electronic properties of Ta-doped TiO_2 semiconductor, *Physical Review* 79(11).

Galli, C., Coen, M., Hauert, R., Katanaev, V., Wymann, M., Groning, P. & Schlapbach, L. (2001). Protein adsorption on topographically nanostructured titanium, *Surface Science* 474(1-3): L180–L184.

Garguilo, J., Davis, B., Buddie, M., Kock, F. & Nemanich, R. (2004). Fibrinogen adsorption onto microwave plasma chemical vapor deposited diamond films, *Diamond and Related Materials* 13(4-8): 595–599.

Grinevich, A., Bacakova, L., Choukourov, A., Boldyryeva, H., Pihosh, Y., Slavinska, D., Noskova, L., Skuciova, M., Lisa, V. & Biedermanl, H. (2009). Nanocomposite ti/hydrocarbon plasma polymer films from reactive magnetron sputtering as growth support for osteoblast-like and endothelial cells, *Journal of Biomedical Materials Research Part A* 88(4): 952–966.

Hemmersam, A., Foss, M., Chevallier, J. & Besenbacher, F. (2005). Adsorption of fibrinogen on tantalum oxide, titanium oxide and gold studied by the QCM-D technique, *Colloids and Surfaces B-biointerfaces* 43(3-4): 208–215.

Höök, F., Vörös, J., Rodahl, M., Kurrat, R., Boni, P., Ramsden, J., Textor, M., Spencer, N., Tengvall, P., Gold, J. & Kasemo, B. (2002a). A comparative study of protein adsorption on titanium oxide surfaces using in situ ellipsometry, optical waveguide lightmode spectroscopy, and quartz crystal microbalance/dissipation, *Colloids and Surfaces B-biointerfaces* 24(2): 155–170.

Höök, F., Vörös, J., Rodahl, M., Kurrat, R., Böni, P., Ramsden, J., Textor, M., Spencer, N., Tengvall, P., Gold, J. & Kasemo, B. (2002b). A comparative study of protein adsorption on titanium oxide surfaces using in situ ellipsometry, optical waveguide lightmode spectroscopy, and quartz crystal microbalance/dissipation, *Colloids and Surfaces B: Biointerfaces* 24(2): 155–170.

Huang, L., Ning, C., Ding, D., Bai, S., Qin, R., Li, M. & Mao, D. (2010). Wettability and in vitro bioactivity of doped TiO_2 nanotubes, *Journal of Inorganic Materials* 25(7): 775–779.

Huang, N., Yang, P., Leng, Y., Chen, J., Sun, H., Wang, J., Wang, G., Ding, P., Xi, T. & Leng, Y. (2003). Hemocompatibility of titanium oxide films, *Biomaterials* 24(13): 2177–2187.

Imamura, K., Shimomura, M., Nagai, S., Akamatsu, M. & Nakanishi, K. (2008). Adsorption characteristics of various proteins to a titanium surface, *Journal of Bioscience and Bioengineering* 106(3): 273–278.

Jackson, D., Omanovic, S. & Roscoe, S. (2000). Electrochemical studies of the adsorption behavior of serum proteins on titanium, *Langmuir* 16(12): 5449–5457.

Jandt, K. (2001). Atomic force microscopy of biomaterials surfaces and interfaces, *Surface Science* 491(3): 303–332.

Jansson, E. & Tengvall, P. (2004). Adsorption of albumin and IgG to porous and smooth titanium, *Colloids and Surfaces B-biointerfaces* 35(1): 45–51.

Jones, M., Mccoll, I., Grant, D., Parker, K. & Parker, T. (2000). Protein adsorption and platelet attachment and activation, on TiN, TiC, and DLC coatings on titanium for cardiovascular applications, *Journal of Biomedical Materials Research* 52(2): 413–421.

Khaleel, A., Shehadi, I. & Al-shamisi, M. (2010). Structural and textural characterization of sol-gel prepared nanoscale titanium-chromium mixed oxides, *Journal of Non-crystalline Solids* 356(25-27): 1282–1287.

Kidoaki, S. & Matsuda, T. (1999). Adhesion forces of the blood plasma proteins on self-assembled monolayer surfaces of alkanethiolates with different functional groups measured by an atomic force microscope, *Langmuir* 15(22): 7639–7646.

Ma, W., Ruys, A., Mason, R., Martin, P., Bendavid, A., Liu, Z., Ionescu, M. & Zreiqat, H. (2007). DLC coatings: Effects of physical and chemical properties on biological response, *Biomaterials* 28(9): 1620–1628.

Martin, E., Manceur, A., Polizu, S., Savadogo, O., Wu, M. & Yahia, L. (2006). Corrosion behaviour of a beta-titanium alloy, *Bio-medical Materials and Engineering* 16(3): 171–182.

Nygren, H., Tengvall, P. & Lundstrom, I. (1997). The initial reactions of TiO_2 with blood, *Journal of Biomedical Materials Research* 34(4): 487–492.

Palik, E. (1998). *Handbook of Optical Constants of Solids (Volumes I, II and III)*, Academic Press.

Parsegian, V. (2005). *Van der Waals Forces: A Handbook for Biologists, Chemists, Engineers, and Physicists*, Cambridge University Press, Cambridge.

Roach, P., Farrar, D. & Perry, C. (2005). Interpretation of protein adsorption: Surface-induced conformational changes, *Journal of the American Chemical Society* 127(22): 8168–8173.

Roach, P., Farrar, D. & Perry, C. (2006). Surface tailoring for controlled protein adsorption: Effect of topography at the nanometer scale and chemistry, *Journal of the American Chemical Society* 128(12): 3939–3945.

Roach, P., Shirtcliffe, N., Farrar, D. & Perry, C. (2006). Quantification of surface-bound proteins by fluorometric assay: Comparison with quartz crystal microbalance and amido black assay, *Journal of Physical Chemistry B* 110(41): 20572–20579.

Rouahi, M., Champion, E., Gallet, O., Jada, A. & Anselme, K. (2006). Physico-chemical characteristics and protein adsorption potential of hydroxyapatite particles: Influence on in vitro biocompatibility of ceramics after sintering, *Colloids and Surfaces B-biointerfaces* 47(1): 10–19.

Shaama, F. (2005). An in vitro comparison of implant materials cell attachment, cytokine and osteocalcin production, *West Indian Medical Journal* 54(4): 250–256.

Silvennoinen, R., Hason, S., Vetterl, V., Penttinen, N., Silvennoinen, M., Myller, K., Cernochova, P., Bartakova, S., Prachar, P. & Cvrcek, L. (2010). Diffractive-optics-based sensor as a tool for detection of biocompatibility of titanium and titanium-doped hydrocarbon samples, *Applied Optics* 49(29): 5583–5591.

Silvennoinen, R., Peiponen, K.-E. & Myller, K. (2008). Specular gloss, *Elsevier, Amsterdam* .

Silvennoinen, R., Vetterl, V., Hason, S., Silvennoinen, M., Myller, K., Vanek, J. & Cvrcek, L. (2010). Optical sensing of attached fibrinogen on carbon doped titanium surfaces, *Advances in Optical Technologies* 2010(22): 7.

Silvennoinen, R., Vetterl, V., Hason, S., Tuononen, H., Silvennoinen, M., Myller, K., Cvrcek, L., Vanek, J. & Prachar, P. (2008). Sensing of human plasma fibrinogen on polished, chemically etched and carbon treated titanium surfaces by diffractive optical element based sensor, *Optics Express* 16(14): 10130–10140.

Soman, P., Rice, Z. & Siedlecki, C. (2008). Measuring the time-dependent functional activity of adsorbed fibrinogen by atomic force microscopy, *Langmuir* 24(16): 8801–8806.

Sonesson, A., Callisen, T., Brismar, H. & Elofsson, U. (2007). A comparison between dual polarization interferometry (DPI) and surface plasmon resonance (SPR) for protein adsorption studies, *Colloids and Surfaces B-Biointerfaces* 54(2): 236–240.

Swann, M., Peel, L., Carrington, S. & Freeman, N. (2004). Dual-polarization interferometry: an analytical technique to measure changes in protein structure in real time, to determine the stoichiometry of binding events, and to differentiate between specific and nonspecific interactions, *Analytical Biochemistry* 329(2): 190–198.

Toscano, A. & Santore, M. (2006). Fibrinogen adsorption on three silica-based surfaces: Conformation and kinetics, *Langmuir* 22(6): 2588–2597.

Van De Keere, I., Willaert, R., Hubin, A. & Vereeckent, J. (2008). Interaction of human plasma fibrinogen with commercially pure titanium as studied with atomic force microscopy and x-ray photoelectron spectroscopy, *Langmuir* 24(5): 1844–1852.

Vanderah, D., La, H., Naff, J., Silin, V. & Rubinson, K. (2004). Control of protein adsorption: Molecular level structural and spatial variables, *Journal of the American Chemical Society* 126(42): 13639–13641.

Walivaara, B., Aronsson, B., Rodahl, M., Lausmaa, J. & Tengvall, P. (1994). Titanium with different oxides - *in-vitro* studies of protein adsorption and contact activation, *Biomaterials* 15(10): 827–834.

Wang, X., Yu, L., Li, C., Zhang, F., Zheng, Z. & Liu, X. (2003). Competitive adsorption behavior of human serum albumin and fibrinogen on titanium oxide films coated on LTI-carbon by IBED, *Colloids and Surfaces B-biointerfaces* 30(1-2): 111–121.

Wasilewska, M., Adamczyk, Z. & Jachimska, B. (2009). Structure of fibrinogen in electrolyte solutions derived from dynamic light scattering (DLS) and viscosity measurements, *Langmuir* 25(6): 3698–3704.

Webster, T., Siegel, R. & Bizios, R. (1999). Osteoblast adhesion on nanophase ceramics, *Biomaterials* 20(13): 1221–1227.

Wertz, C. & Santore, M. (1999). Adsorption and relaxation kinetics of albumin and fibrinogen on hydrophobic surfaces: Single-species and competitive behavior, *Langmuir* 15(26): 8884–8894.

Wertz, C. & Santore, M. (2001). Effect of surface hydrophobicity on adsorption and relaxation kinetics of albumin and fibrinogen: Single-species and competitive behavior, *Langmuir* 17(10): 3006–3016.

Wertz, C. & Santore, M. (2002). Fibrinogen adsorption on hydrophilic and hydrophobic surfaces: Geometrical and energetic aspects of interfacial relaxations, *Langmuir* 18(3): 706–715.

Xu, L. & Siedlecki, C. (2007). Effects of surface wettability and contact time on protein adhesion to biomaterial surfaces, *Biomaterial* 28(22): 3273–3283.

Yang, Y., Cavin, R. & Ong, J. (2003). Protein adsorption on titanium surfaces and their effect on osteoblast attachment, *Journal of Biomedical Materials Research Part A* 67a(1): 344–349.

Part 2

Biomaterials for Dental, Bone and Cartilage Tissues Engineering and Treatment

9

Biomaterials and Biotechnology Schemes Utilizing TiO$_2$ Nanotube Arrays

Karla S. Brammer[1], Seunghan Oh[2], Christine J. Frandsen[1] and Sungho Jin[1]
[1]Materials Science & Engineering, University of California,
San Diego, La Jolla, California
[2]Department of Dental Biomaterials, College of Dentistry,
Wonkwang University, Iksan,
[1]USA
[2]South Korea

1. Introduction

Ti and Ti alloys are corrosion resistant, light, yet sufficiently strong for utilization as load-bearing and machinable orthopaedic implant materials. They are one of the few biocompatible metals which osseo-integrate, provides direct chemical or physical bonding with the adjacent bone surface without forming a fibrous tissue interface layer. For these reasons, they have been used successfully as orthopaedic and dental implants (Ratner 2004). To impart even greater bioactivity to the Ti surface and enhance integration properties, surface treatments such as surface roughening by sand blasting, formation of anatase phase TiO$_2$ (Uchida et al. 2003), hydroxyapatite (HAp) coating, or chemical treatments (Ducheyne et al. 1986; Cooley et al. 1992) have been employed. However, these treatments are generally on the micron scale. Webster et al. (Webster et al. 2001; Webster, Siegel, and Bizios 1999) reported that it is even more advantageous to create nanostructured, in particular in the less than 100nm regime, surface designs for significantly improved bioactivity at the Ti implant interface and for enhanced cell adhesion. Since then, advances in biomaterial surface structure and design, specifically on the nanoscale, have improved tissue engineering in general. This chapter is a report on titanium dioxide (TiO$_2$, or Titania) nanotube surface structuring for optimization of titanium (Ti) implants utilizing nanotechnology.

The main focus will be on the unique 3-D tube-shaped nanostructure of TiO$_2$ and its effects on creating profound impacts on cell behavior. We will also shed light on the effects of changing the nanotube diameter size and optimizing the geometry for enhanced cell behavior. This work focuses on the tissue specific areas of cartilage and bone. Specifically, we will discuss how the desired cell behavior and functionality are enhanced on surfaces with TiO$_2$ nanotube surface structuring. Here we reveal how the TiO$_2$ surface nano-configurations are advantageous in various tissue engineering and regenerative medicine applications, for osteo-chondral, orthopedic, and osteo-progenitor implant applications discussed here and beyond. This chapter will also shed light on future applications and the direction of nanotube surface structuring.

2. Electrochemical anodization

In general, the mechanism of TiO$_2$ nanotube formation in fluorine-ion based electrolytes is said to occur as a result of three simultaneous processes: the field assisted oxidation of Ti metal to form titanium dioxide, the field assisted dissolution of Ti metal ions in the electrolyte, and the chemical dissolution of Ti and TiO$_2$ due to etching by fluoride ions, which is enhanced by the presence of H$^+$ ions (Shankar et al. 2007). TiO$_2$ nanotubes are not formed on the pure Ti surface but on the thin TiO$_2$ oxide layer naturally present on the Ti surface. Therefore, the mechanism of TiO$_2$ nanotubes formation is related to oxidation and dissolution kinetics. Schematic diagram of the formation of TiO$_2$ nanotubes by anodization process is shown in Figure 1. For a description of the process displayed in Figure 1, the anodization mechanism for creating the nanotube structure is as follows:

a.   Before anodization, a nano scale TiO$_2$ passivation layer is on the Ti surface.

b.   When constant voltage is applied, a pit is formed on the TiO2 layer.

c.   As anodization time increases, the pit grows longer and larger, and then it becomes a nanopore.

d.   Nanopores and small pits undergo continuous barrier layer formation. (e) After specific anodization time, completely developed nanotubes are formed on the Ti surface.

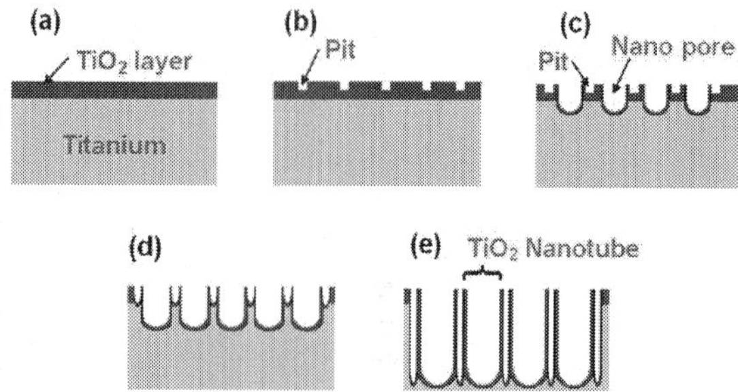

Fig. 1. Schematic illustration of TiO$_2$ nanotube formation.

Furthermore, based on the mechanism of nanotube formation, it is inherent that the nano-tubular structure formation depends on both the intensity of applied voltage and the concentration of fluorine ions in the electrolyte solution. It is well-known that by increasing the applied voltage, larger diameter nanotubes can be formed. This aspect of diameter manipulation using applied voltage will be further emphasized and the effects on cell function and fate is also discussed.

3. Nanotube size effects

The Jin lab was the first to demonstrate that TiO$_2$ nanotubes can significantly accelerate osteoblast (bone cell) adhesion and proliferation at the biomaterial/tissue interface and enhance bone mineral formation. The TiO$_2$ nanotubes are formed as vertically aligned configuration, with an average diameter of ~100 nm, a height of ~300 nm, and a wall thickness of ~10 nm. According to published research (Oh S 2006), nanotube arrays on titanium surfaces induced proliferation of osteoblasts by as much as 300 – 400% compared to non-modified titanium surfaces. In other research groups studying nanoporous materials, major accomplishments have been made in the generation of geometrically defined surfaces with the fabrication of Al and Si nanostructured surfaces. There is rapidly increasing evidence that the lateral spacing of features on the nanoscale can impact and change cell behaviour (Boyen et al. 2002; Cavalcanti-Adam et al. 2006; Popat et al. 2006). Therefore, in order to optimize the lateral spacing of the TiO$_2$ nanotube system, by changing the geometry of the nanotubes, four different pore sizes (30, 50 70, and 100nm in diameter) were created (Figure 2) for examination of cartilage chondrocyte cells, bone osteoblast cells, and osteo-progenitor mesenchymal stem cells.

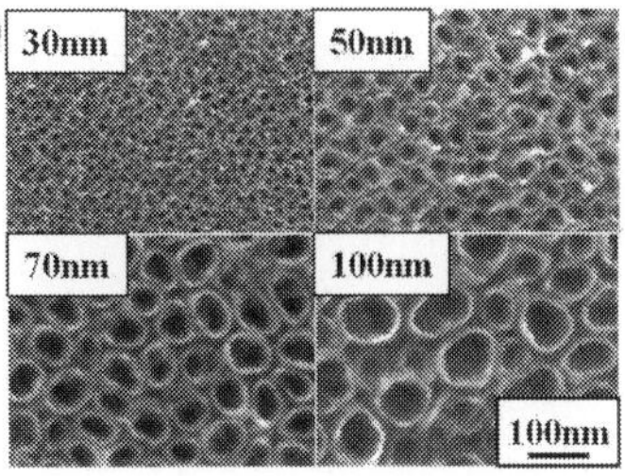

(b)

Sample	Applied Voltage (V)	Estimated Inner Pore Size (nm)	Roughness Ra (nm)	Contact Angle (°)
Ti	NA	NA	9.7	54
30nm	5	27.0	13.0	11
50nm	10	46.4	12.7	9
70nm	15	69.5	13.5	7
100nm	20	99.6	13.2	4

Fig. 2. Physical characterization of different size nanotube surfaces. (a) SEM micrographs of self-aligned TiO$_2$ nanotubes with different diameters. The images show highly ordered nanotubes with four different pore sizes between 30-100nm created by controlling the voltage from 5-20V. (b) Table with the applied voltage parameter, estimated inner pore size from SEM images, average roughness (Ra) and surface contact angle measurements for Ti and 30-100nm TiO$_2$ nanotube surfaces.

In terms of current biologically active implants, enhanced surface roughness is one of the important factors in providing the proper cues for a positive cell response to implanted materials. However, much of the research related to the effect of macro and micro-roughness on cellular responses and tissue formation are inconclusive due to the non-uniformity of macro and micro-roughness stemming from crude fabrication methods like polishing, sand blasting, chemical etching and so on. An important aspect of our nanotube system shown in the SEM images (Figure 2) is that the nano-topography can feature a more defined, reproducible and reliable roughness than micro and macro-topography for enhanced bone cell function *in vivo*. Although, the heights of the nanotube walls increase proportionally to the increasing diameter, there is no evidence of changes in surface roughness between the different sized nanotubes based on atomic force microscopy (AFM) data (Figure 2 (b)). As expected, the nanotube surfaces have a slightly higher roughness over flat Ti, but between the nanotubes, there appears to be no difference. The AFM data was performed because it is a somewhat standard surface analysis technique as it is useful for coarser or microscale roughness measurements, say for other convential coatings, but for the nanotube dimensions it may not always represent the true roughness when the probe tip radius is not substantially finer than the nanotube dimensions such as in the TiO$_2$ nanotube case. The wall thickness, pore diameter, nanotube spacing, etc can be as small as ~10 nm, while the AFM probe tip diameter can be as large as 30 - 50 nm.

Furthermore, it can be assumed that the surface area on the nano-scale may be affected based on the various sizes and the surface area probably increases proportionally with increasing nanotube size. It is expected that the surface area to be 3 times higher on the 100nm diameter nanotubes compared to the 30nm diameter nanotubes, respectively. Additionally, the contact angle describing the wettability of the surface is enhanced, more hydrophilic, on the nanotube surfaces (showing contact angles between 4-11°), which can been advantageous for enhancing protein adsorption and cell adhesion.

3.1 Protein adhesion properties based on pore size

Cells respond to the amount and area of proteins that are available for binding. In fact, cells do not see a naked material, *in vivo* or in *in vitro* culture. At all times, the material is conditioned by the components of the fluid in which the material is immersed, whether it is serum, saliva, cervicular fluid or cell culture media. As the cell begins to adhere and spread on the nanotubes, there will be a dissimilar protein density and extra cellular configuration based on the nanotube diameter. The behaviour of protein adsorption on the nanotube surfaces are shown in Figure 3. On the 30nm diameter nanotubes there is a large number and thorough distribution of protein nanoparticles covering the whole surface of the nanotubes after just 2 hours of incubation in culture media. However, proteins on 100 nm TiO_2 nanotubes can only adhered sparsely at the top wall surface owing to the presence of large empty nanotube pore spaces. This inherent protein adsorption property of the nanotubes based on poresize is hypothesized to influence cell shape and fate. It is shown in the next sections that the changes in poresize even in such a small range of dimensions (30-100nm) will have huge impacts on downstream cell morphology and behavior.

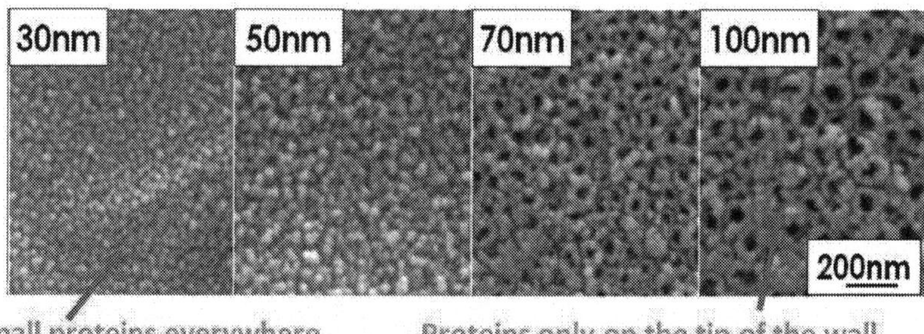

Fig. 3. SEM micrographs of flat Ti and 30, 50, 70, 100nm diameter TiO_2 nanotube surfaces after 2 hours of culture showing protein adsorption from media.

4. Osteo-chondral applications of TiO_2 nanotube constructs

Artificial cartilage prepared from cultured chondrocytes offers promise as a treatment for cartilage defects (Fedewa et al. 1998), but connecting this artificial soft tissue to bone in the attempts to restore the defected cartilage is difficult. One strategy employed in this section is to develop a dually functional substrate that supports the growth and attachment of cartilage tissue on one extremity and encourages osseointegration, a direct structural and functional connection to living bone, on the other. This substrate should be an engineered interface between artificial cartilage and native bone (Zhang, Ma, and Francis 2002).

In recent studies, Ti has emerged as a candidate material in cartilage tissue formation as well. It has been demonstrated that a micrometer porous substrate of Ti-6Al-4V provided conditions that favored cartilage tissue formation by influencing cell attachment, spreading and the amount and composition of cartilaginous tissue that forms (Spiteri, Pilliar, and Kandel 2006; Bhardwaj et al. 2001; Ciolfi et al. 2003). Not only porosity, but also surface geometry and topography have been found to have positive effects on the behaviour of chondrocytes (Bhardwaj et al. 2001).

Nanoscale topographic effects have been illustrated in nanostructured poly-lactic-co-glycolic acid (PLGA)/nanophase Titania (TiO$_2$) composites, which have elicited an enhanced chondrocyte response compared to surfaces with a conventional or micrometer topography (Savaiano and Webster 2004). We have recently reported on our hypothesis that the nanotopographical cues, from porous nanotubular structured substrates made of TiO$_2$, already being an osseointegrating biomaterial (Bjursten 2009; Oh et al. 2006), may also be a candidate for providing an alternative way to positively influence cartilage formation and the cellular behaviour of cartilage chondrocytes.

4.1 Up-regulated chondrocyte synthesis of extracellular matrix components

The dimensions of the nanotubes were varied in order to determine if the size of the nanotube diameters would play a role in the chondrocyte behaviour. For this comparative chondrocyte cell culture study, a commercially pure Ti surface, without surface modification was used as a control, as it commonly used as implant material.

It is well known that chondrocytes, the primary cells of cartilage, are extremely active cells. They produce a large amount of extracellular matrix (ECM) that is critical for the mechanical properties and joint lubrication characteristics of cartilage. In the SEM micrographs in Figure 4, the nanotubes substrates appear that they are inducing a positive response from the chondrocytes because the cells begin synthesizing abundant ECM deposition and fibril organization. In the SEM observations of chondrocytes a striking difference in the production of ECM fibrils between the flat Ti without a nanostructure vs. TiO$_2$ nanotube surfaces is revealed. Fibrils are abundant and extending from all areas of the chondrocyte cell creating a dense network of ECM on the nanotube substrates.The flat Ti most likely lacks surface structuring cues for signaling ECM fibril production and organization. One possibility is that ECM protein formation into dense fibrils on the surface may be "nano-inspired" to form on the nanotube structure because of the precise dimensions or fine scale cues of the top surface (tip of the vertical wall) of TiO$_2$ nanotubes having a physically confined geometry which could aid in fibril formation. It was demonstrated previously that the nanotubes produced bio-active nanostructured formations of sodium titanate nanofibers directly on the top of TiO$_2$ nanotube walls when the nanotubes were exposed to NaOH solution (Oh S 2005). ECM proteins once secreted, in this study, may also self-assemble according to the top-wall surface geometric nanocues.

In the lower panel of Figure 4, immunofluorescent images for collagen Type II (red color) are illustrated for flat Ti vs. TiO$_2$ nanotubes. Both surfaces stained positive for collagen type II, but there was large networks of connected bundles expressed across the surface of the nanostructure.

When a morphological analysis was conducted, it was determined that nanotubes induce a more spherical chondrocyte cell shape (data not shown). Fibroblastic shaped cells were observed on the flat controls. The percentage of round shaped, spherical cells was significantly lower for chondrocytes on the polystyrene, Ti, and the smallest diameter (30nm) nanotube substrates compared to the larger diameter 50, 70, and 100nm TiO$_2$

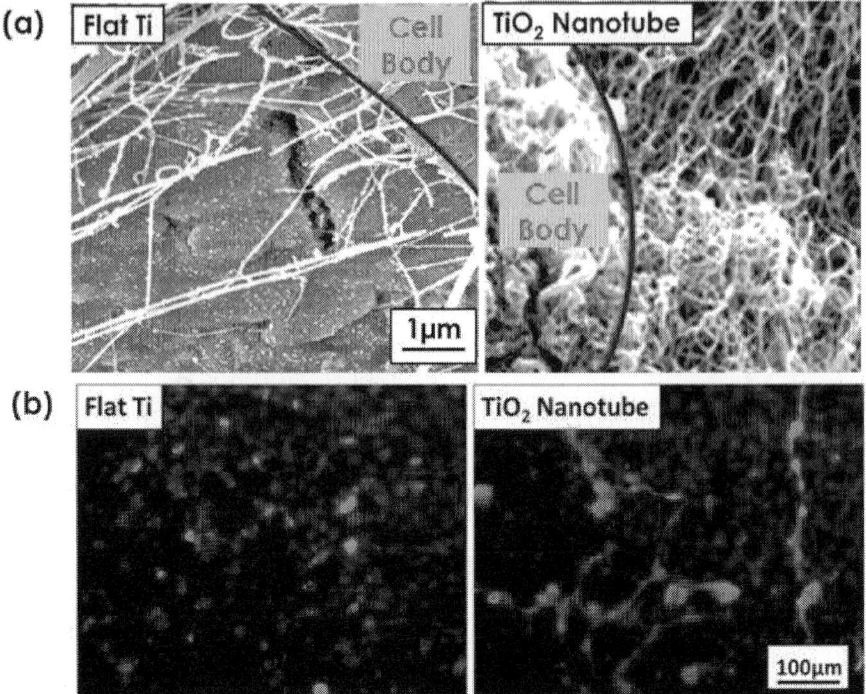

Fig. 4. Extracellular matrix (ECM) production on experimental surfaces Ti vs. TiO$_2$ nanotubes. High magnification SEM observations of chondrocytes reveal a striking difference in the production of ECM fibrils between the flat Ti without a nanostructure vs. TiO$_2$ nanotube surfaces. Fibrils are abundant and extending from all areas of the chondrocyte cell creating a dense network of ECM on the nanotube substrates. (b) Immunofluorescent images of collagen type II (red) ECM fibrils produced by chondrocytes on flat Ti and nanotube surfaces (100nm diameter shown in this image).

nanotube surfaces respectively. It was also determined that all diameter nanotube surfaces were significantly higher than flat Ti which probably indicates that the cell shape was influenced by the presence of the nanostructure itself. The nanotube geometry seen in Figure 2 most likely aids in preserving the chondrocyte spherical morphology because of the distinct structure of the surface. Cells may be localized atop the pores, anchored possibly at the tip of the nanotube walls and confined by the tube contour. The chondrocytes on the flat Ti seem to be spread along the surface probably because the necessary structuring cues and nanopores needed for shape confinement are absent. To further describe the chondrocyte shape phenomenon found the experimental surfaces, a schematic is shown in Figure 5.

It was formerly assumed that focal contacts should be a specific length in order to promote adhesion and that the maximum overall contact of the cell with its substrate was most favourable (Ohara and Buck 1979). Yet, more recent studies suggest that cell-flattening or

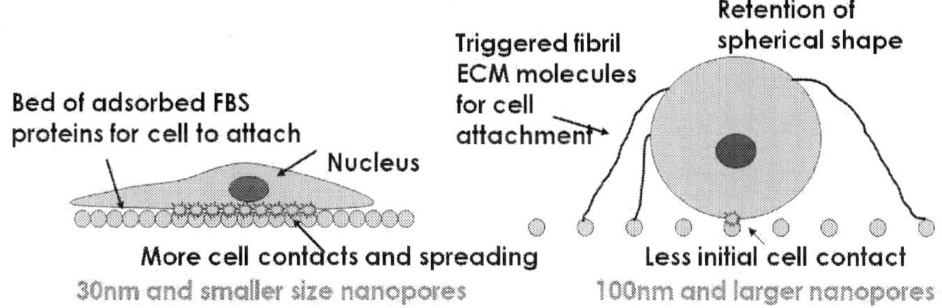

Fig. 5. Chondrocyte cell adhesion and spreading schematic determined by the size of the nanotube diameter and focal attachment sites.

spreading is not always compatible with differing cell types, particularly in the case of chondrogenesis (Solursh 1989; Benya and Shaffer 1982; Solursh 1982; Zanetti and Solursh 1984). Thus, the type of focal adhesion and its geometry can influence the shape the cell assumes, ultimately influencing the phenotypic expression. It has been reported that the dedifferentiation of chondrocytes in culture is usually associated with changes in cell morphology, from a rounded to a spread one (Costa Martinez et al. 2008). The results reported here suggest that creating pores by fabricating nanotubes on Ti surfaces provides a more favorable environment for the retention of the rounded morphology and the prevention of chondrocyte spreading, reducing the risk of a loss of phenotype. It should be noted that although chondrocyte cells retain this type of spherical morphology in response to the nanotube pores, different cell types will differ in size, shape, function, and how they operate on the nanotube surfaces. It is well known that different cell types elicit their own unique responses to environmental cues. The chondrocyte cells with their spherical morphology are much different than mesenchymal stem cells and osteoblast cells, described in later sections, and therefore will adhere differently to the topography and form different morphologies.

Because chondrocytes are very dynamic cells that produce and maintain the cartilaginous matrix, which consists mainly of collagen and proteoglycans, it is important to test the biochemical ECM production on the different experimental surfaces. Therefore, to further evaluate the response of BCCs for this comparative report, the glycosaminoglycan (GAG) secretion in the media was also studied and shown in Figure 6.

Naturally, aggrecan draws water into the tissue and swells against the collagen network, thereby resisting compression and allowing for proper joint movement (Muir 1995). An interesting concept worthy to note is that the up-regulation of GAG chains indicative of increased aggrecan production observed on the larger sized nanotube pores could imply that because there are increased storage volume capabilities as pore size increases it triggers a higher rate of production because the molecule retention ability of the cellular environment has been inflated.

It was determined that there was a correlation in the increased GAG secretion in the media and the reduction of fibroblastic shaped cells. Specifically, TiO₂ nanotubes with diameters in the range of 50-100nm had significantly higher levels of both round, spherical shaped cells (more phenotypic) and GAG secretion over flat Ti and small 30nm TiO_2 nanotube surfaces.

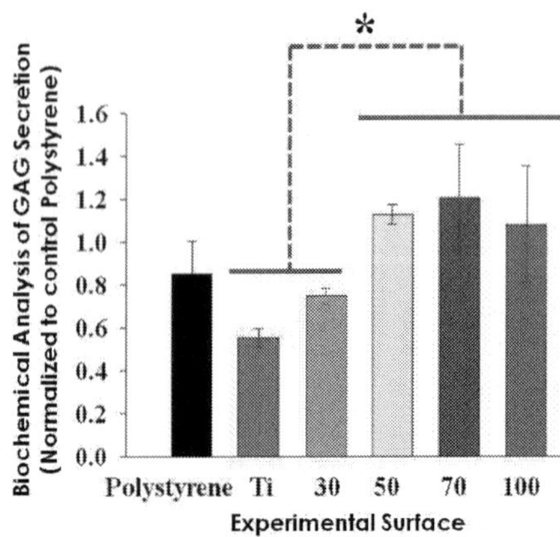

Fig. 6. Glycosaminoglycan (GAG) secretion in the media.

In this study, the larger diameters (50nm-100nm) nanotubes were revealed to be most suitable for chondrocyte culture *in vitro*. It would be interesting to investigate nanotube diameter sizes larger than 100nm by other fabrication means so as to elucidate the possible effect of large pore size and find an ultimate productivity saturation limit; this would certainly allow more light to be shed on the beneficial nature of nanotopography.

4.2 Conclusions and considerations for further osteo-chondral development

As Ti is the well accepted orthopaedic implant material, the results obtained are very encouraging and suggest that the use of nanotube structures could up-regulate production of extracellular matrix by chondrocytes. In clinical applications the TiO_2 nanotube surface can be utilized as an *in-vitro* culture surface to enhance chondrocyte cell behavior and extracellular matrix production during patient-specific *in-vitro* chondrocyte expansion, which can then be transplanted to the defective cartilage areas. In addition, the TiO_2 nanotube surface exhibits significantly augmented, mechanically and chemically strong osseo-integration with existing bones with a minimal chance of bone loosening evidenced by *in vitro* data [22], and our preliminary *in vivo* animal data indicating a strong new bone integration on the nanotube surface with reduced soft tissue trapping (data not shown). Therefore, a Ti implant with all surfaces covered with the nanotubes can be potentially utilized, for some specific types of articular cartilage injuries, to serve with dual function of accelerated osseointegration to the existing articular bone surface at the bone-facing contact interface while the exposed nanotube surface can accelerate the cartilage tissue regeneration by providing positive surface nanostructuring effects on chondrocytes, as illustrated in Figure 7.

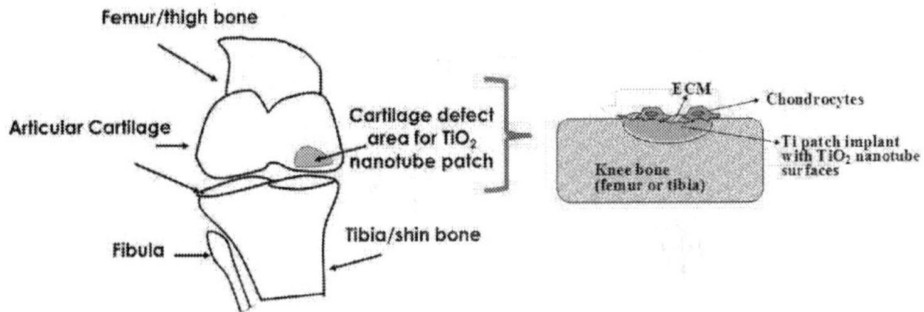

Fig. 7. Schematic illustration of TiO₂ nanotubes for dual function of osseointegration and enhanced chondrocyte function and ECM production.

5. Orthopedic implant applications of TiO₂ nanotube constructs

As mentioned earlier, while a thin TiO₂ passivation layer on the Ti surface can impart improved bioactivity and better chemical bonding to the bone (Feng et al. 2003), other techniques have been developed to further enhance the bioactivity of a pure Ti surface, such as direct coating of bioactive materials like hydroxyapatite and calcium phosphate (Puleo et al. 1991; Salata 2004; Satsangi et al. 2003). However, even though these surface modified layers have good bioactivity and high surface area, they tend to delaminate at the interface between the implant and the bone due to the relatively large, micrometer-regime thickness of the coated layer on Ti (Ong et al. 1992), presumably due to the stress accumulation commonly seen in a thick coating of foreign material. This ultimately leads to implant failure. In order to overcome this problem, some plasma spray Ca-Si based ceramic coatings have been developed but still have roughness and layer thickness in the micrometer range. For the purposes of this study, the focus is on nanoscale thickness surface coatings. Therefore, developing an implant bioactive surface layer having high surface area for enhanced bonding yet thin enough, in the nanometer range per se, to minimize delamination would be desirable.

Recent reports indicate that modifying Ti surfaces with TiO₂ nanotubes for orthopedic applications significantly enhances the mineral formation (Oh et al. 2005), adhesion of osteoblasts *in vitro* (Oh et al. 2006), and strongly adherent bone growth *in vivo* (Bjursten 2009), showing better bone bonding characteristics than conventional micro-roughened Ti surfaces by sandblasting. One physical advantage of the TiO₂ nanotube surface system is that it is composed of and created directly from the native underlying Ti constituent, unlike the foreign ceramic and spray coatings on Ti or Ti alloyed surfaces mentioned previously. As well, the nanotube layer is at most ~300nm tall (for the purposes of this work) which in the scheme of things is a much thinner layer and this nanometer length scale eliminates the tendency of delamination prevalent in thick micrometer layers.

Because orthopedic implants encounter two types of cells, osteoblast cells in bone tissue and osteo-progenitor cells also know as mesenchymal stem cells (MSCs) present in bone marrow, it is advantageous to look at the differentiation potential of orthepeadic implant surfaces in order to initiate a mature population of bone building cells for enhanced osseointegration. One key principle in terms of orthepaedic implant technologies, to initiate the differentiation of bone in the absence of chemical factors, hormones, or any other synthetic, possibly toxic chemicals traditionally used by biochemists in *in vitro* differentiation. The two cell types are illustrated and described in Figure 8.

Mesenchymal stem cells develop into osteoblast cells

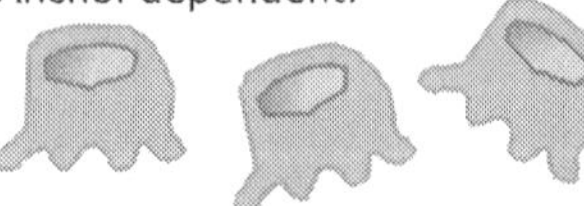

- Bone marrow derived.
- Pluripotent cells with the capacity to differentiate into different cell types:
  - Osteoblast, chondrocytes, adipocytes, etc.
- Fibroblastic.

- Mature, adult cells specific to bone tissue.
- Bone building cell, deposits minerals to make up bone matrix component.
- Anchor dependent.

Fig. 8. Schematic illustration showing osteo-progenitor cells developing into osteoblast cells. A description of the two cell types is also portrayed in list format.

Stem cells, which have the potential to differentiate into multiple cell types, provide great promises in advances in regenerate medicine. The differentiation of stem cells into the appropriated lineages is temporal- and spatial-specific, with the surrounding microenvironment playing critical roles in governing the stem cell fate. For generation of osteoblasts, the MSCs need to be guided to selectively differentiate to osteoblasts, rather than differentiating into other types of cells. The use of nanostructures and surface topographical features have recently been shown to have positive effects on specific differentiation of mesenchymal stem cells (Dalby, Andar et al. 2008), illustrating that the surface topography alone can stimulated osteogenic differentiation.

5.1 Osteogenic functionality depends on nanotube diameter

In terms of the dimensions of the nanotubes in our osteoblast (bone cell) and mesenchymal stem cell (osteo-progenitor cell) studies, we have reported a unique variation in cell behavior even within a narrow range of nanotube diameters from 30-100nm (Brammer et al. 2009; Oh et al. 2009) and it seems that a similar trend is established for both cell types.

When cells were grown on four different diameter nanotubes, shown in Figure 2, osteoblast functionality in terms of bone forming ability, or alkaline phosphatase activity (ALP), and mesenchymal stem cell (MSC) osteogenesis (bone cell differentiation) in terms of osteogenic gene expression (osteopontin (OPN), osteocalcin (OCN), and alkaline phosphatase (ALP)) were most prominent on the large 100 nm diameter nanotubes, Figure 9 (a and b). During periods of active bone growth, ALP activity levels are elevated in osteoblast cells so it is beneficial to design an implant surface that would enhance the ALP activity to initiate the formation of new bone. As well, it is critical to design a surface that is capable of allowing the attachment of MSCs and promote osteogenic differentiation of cells for delivering a mature osteoblastic cell population capable of rapidly forming bone. On the nanotube surfaces, a reoccurring trend was revealed that as we increased the diameter of the nanotubes, there was an increase in osteogenic biochemical activity and relative gene expression, Figure 9 (c).

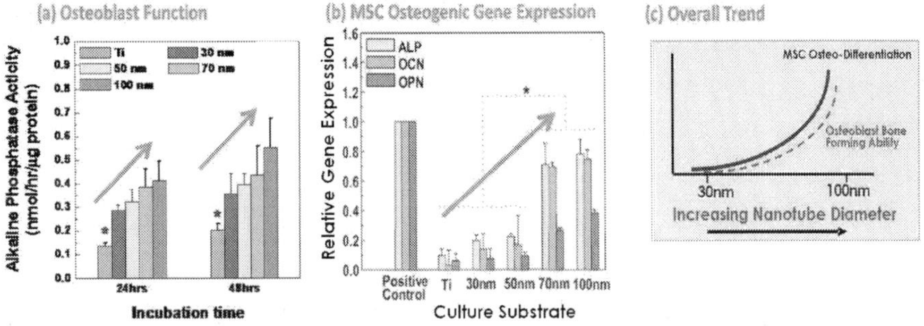

Fig. 9. Comparative graphs showing the influence of TiO₂ nanotube diameter on osteogenic cell behavior. (a) Osteoblast functionality in terms of alkaline phosphatase activity or bone forming ability. (b) Osteogenic gene expression. RNA levels of alkaline phosphatase (ALP), osteocalcin (OCN), and osteopontin (OPN) are given for mesenchymal stem cells grown on different size diameter nanotubes. (c) Overall all trend for diameter effect on osteogenic cell behavior. As the nanotube diameter increases the osteogenic cell behavior is enhanced.

In a recent review article, Bettinger et al. claimed that the most palpable effect of nanotopography on cells is the distinct changes in cell geometry or shape (Bettinger, Langer, and Borenstein 2009). In fact, on the nanotube substrates with varying diameters, it was revealed that the osteoblast and mesenchymal stem cells have reacted to the nanostructures by changing shape. As the nanotube diameter increases we found a clear trend of increasing cell elongation, Figure 10. The stretching aspect ratio was as great as 12:1 (length:width) on the 100 nm diameter nanotubes. It can be assumed that the initial cell stretching and elongated shape of the adhering cells on the large nanotubes impacted the cytoskeletal (actin) stress. This is supported by the general notion that nanostructures (and the adhered protein configuration for example, Figure 3) act as an extracellular matrix which imposes physical forces and morphological changes to the cell (Chen 2008). It is probable that cells must elongate their bodies to find a protein deposited surface, extending across larger areas and thus eventually forming an exceedingly elongated shape on the 100nm diameter nanotubes (because of the sparse distribution, Figure 3). Thus altering the density of extracellular matrix (ECM) attachment sites or initial protein adhesion density affects the shape of adhered cells. It has been reported that the focal attachments made by the cells with their substrate determine cell shape which, when transduced via the cytoskeleton to the nucleus, result in expression of specific phenotypes (Boyan et al. 1996). In our results, we hypothesize that increasing nanotube diameters, changes the focal adhesion sites of the osteoblast and mesenchymal stem cells, increases the cell elongation, and increases osteogenic potential.

Interestingly, the cell nuclei also exhibit a somewhat similar trend of increased elongation with increasing nanotube diameter, with the 100nm TiO₂ nanotubes showing the most significantly elongated nuclear shape (by ~20-25%) (data not shown). It can be hypothesized that the nucleus organelle elongation on the TiO₂ nanotube surfaces is in part due to the

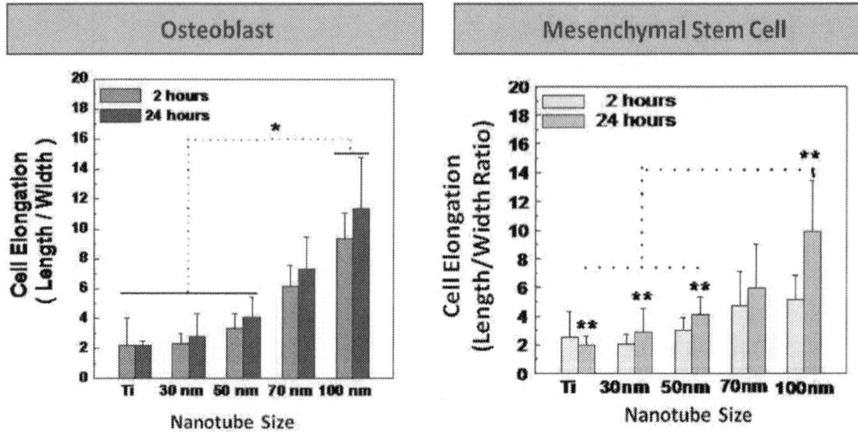

Fig. 10. Quantification of cell elongation for osteoblast and mesenchymal stem cells on nanotubes with different diameters ranging from 30-100nm. There is a trend of increased elongation with increasing nanotube diameter.

gross elongated cytoskeletal morphology of the cell. It has been reported that cell shape maintained by the cytoskeletal assembly may also facilitate nuclear shape distortion which may promote DNA synthesis by releasing mechanical restraints to DNA unfolding, changing nucleocytoplasmic transport rates, or alternating the distribution and function of DNA regulatory proteins that are associated with the nuclear protein matrix (Maniotis, Chen, and Ingber 1997). A change in the nuclear structure has an effect on the 3-dimensional internal organization (Getzenberg et al. 1991). It appears that the osteoblasts and mesenchymal stem cells are adapting to the nanotube substrate nanotopography by organizing both external and internal shapes.

A concept developed by Dalby et al. (Dalby et al. 2008) suggests that MSC osteo-differentiation is determined by mechanotransductive pathways that were stimulated because the cell was under tension caused by way the cell was adhering and the shape it assumed due to the underlying nanostructure surface. Cell morphology/spreading dominates cell fate. McBeath et al. showed that commitment of stem cell differentiation to specific lineages is dependent upon cell shape (McBeath et al. 2004). In a single cell experiment with micropatterned surfaces the critical role of cell spread/shape in regulating cell fate was determined.

Nonetheless there is a need to better understand the mechanism by which such nanosurfaces direct MSC osteogenesis, and to optimize the culture conditions in order to maximize MSC expansion and differentiation. For bone growth, it requires cell proliferation and selective differentiation, and these processes are found to occur at different but discrete nanosurface topography conditions such as variations in nanotube diameter.

5.2 Conclusions and considerations for further orthopaedic considerations

Establishing possible connections between mechanical properties of MSCs in differentiation is valuable to the field of mechanobiology. It can be speculated that natural forces that MSCs encounter in a physical environment does not need a strong cytoskeleton, however upon osteogenic differentiation, in which differentiation of MSCs become part of a larger bone structure that functions to provide both form and strength, the supporting structure of the cells are enhanced to enable function and withstand load bearing wear that bones endure. Understanding the physical characteristics of MSCs during differentiation may aid in the development of new biomaterials, which can potentiate the necessary mechanics of the cells for the advancement of tissue engineering.

In terms of the dimensions of the nanotubes in the osteoblast (bone cell) and mesenchymal stem cell (osteo-progenitor cell) studies, it was reported that a unique variation in cell behaviour even within a narrow range of nanotube diameters (Brammer et al. 2009; Oh 2009).The results of the previous research can be simply summarized: osteogenic functionality, both biochemical activity in osteoblasts and internal gene regulation of osteo-progenitor cells, were altered by the size/diameter of TiO₂ nanotubes, as the nanotube diameter increased, the osteogenic function also increased. Such a trend can be utilized for improvement and control of the bone forming functionality for advanced orthopaedic implant technologies.

In these studies however, TiO₂ nanotubes having a 1: 3 diameter: height aspect ratio was used, which was determined by the electrochemical anodization conditions including electrolyte solution, voltage, time, etc. While this current-state-of-the-art self assembly process of TiO₂ anodization does not easily allow fabrication of TiO₂ nanotubes with the diameter larger than ~100nm with the electrolyte used in this study, it would be interesting to study the effect of even larger diameter TiO₂ nanotubes, possibly using a modified chemical process, on osteogenic cells.

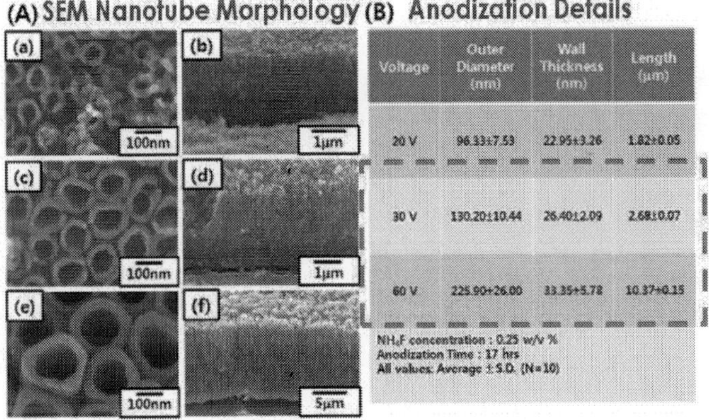

Voltage	Outer Diameter (nm)	Wall Thickness (nm)	Length (μm)
20 V	96.33±7.53	22.95±3.26	1.82±0.05
30 V	130.20±10.44	26.40±2.09	2.68±0.07
60 V	225.90+26.00	33.35+5.78	10.37+0.15

NH₄F concentration : 0.25 w/v %
Anodization Time : 17 hrs
All values: Average ± S.D. (N=10)

Fig. 11. Large diameter nanotubes prepared in 0.25 w/v% NH₄F with various applied voltages. (A) SEM micrographs showing nanotube morphology by top view (a, c, e) and cross-sectional view (b, d, f). (B) Chart describing the effect of applied voltage on the physical nature of the nanotube dimensions.

Other methods for making large diameter (>100nm) TiO₂ nanotubes using aqueous organic electrolytes with a future potential use as orthopaedic implant surfaces have been explored. Previously the anodization electrolyte method included aqueous dilute hydrofluoric acid. In an alternative fabrication method ammonium fluoride (NH$_4$F) in an ethylene glycol solution as an electrolyte can be investigated. Figure 11 illustrates the SEM figures of TiO$_2$ nanotubes prepared by 0.25 w/v% NH$_4$F electrolyte at various anodization voltage for 17 hrs. The table in Figure 11 indicates the variations in applied voltage and the resultant physical dimensions of the fabricated nanotubes. By controlling the applied voltage, the diameter of the nanotubes could also be controlled. Nanotubes with diameters from ~130-225nm have been successfully prepared. Future work should include the use of large size nanotubes to find the most optimal stem cell osteogenesis and bone cell/tissue function.

6. Future direction and applications

In the future, nanostructured ceramics will be created that demonstrate even higher degrees of integration between materials and biology (Narayan et al. 2004). Future ceramic nanostructures may possess even more precisely tailored grain and/or pore sizes in order to obtain specific tissue interactions. For instance, loading nanoporous structures can provide an implant with biological functionalities (Cowan et al. 2003) are key prospects in deriving therapeutic based nanostructures that integrate and for wound healing, bone repair, and cardio vascular restoration, to name just a few. For example, silver and zinc-containing zeolites, marketed under the trade name AgION®, are currently being assessed for use in wound dressings. Figure 12 shows the idea of a "nanodepot" using the nanopores of the TiO$_2$ nanotubes for loading biological agents.

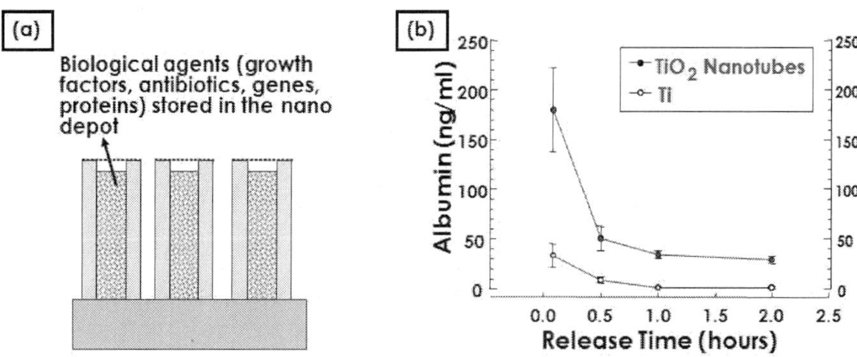

Fig. 12. (a) Schematic illustration of "nanodepot" concept in TiO$_2$ nanotubes. (b) Example of albumin protein elution from inner pores of TiO$_2$ nanotube structures.

In the previous sections it has been shown that the physical environment of nanotopography has positive effects on cell behaviour, yet direct comparisons of nanotopographic surface chemistry has not been fully explored. For instance the possibility of nanotubes made of different materials, i.e. carbon, gold-Pd, zirconia, or tantalum for instance.

To date, a large part of the interest has remained on titanium oxide (TiO$_2$) nanotubes because it is well known that titanium (Ti) is a biocompatible orthopedic material which provides an excellent osseointegrative surface. However, little notice has been given to zirconium oxide (ZrO$_2$) nanotubes, which are formed via the similar self-assembled mechanism as TiO$_2$ nanotubes, through an electrochemical anodization process (Berger et al. 2008). Zirconium (Zr) is similar to titanium in that it possesses a thin passivation oxide layer which makes it highly resistant to corrosion in bodily fluids (Oliveira et al. 2005). In fact, while the corrosion resistance and biocompatibility of certain Zr alloys are as good as those of Ti alloys, the mechanical properties have been found to be superior to those of the commonly used Ti-6Al-4V alloy (Kobayashi et al. 1995). Furthermore, a recent study by Bauer and co-workers demonstrated that mesenchymal stem cells react in the same manner to ZrO$_2$ nanotubes, AuPd-coated TiO$_2$ nanotubes, and as-formed TiO$_2$ nanotubes (Bauer et al. 2009). Their results indicate that the cell response is chiefly due to nanotopographical cues instead of a specific surface chemistry pertaining only to TiO$_2$.

There is a vast parametric space in which to explore novel nanostructures, nanopore dimensions, and material compositions for optimizing and advancing implant designs for desired tissue interactions.

7. Acknowledgment

The authors acknowledge financial support of this research by Iwama Fund at UC San Diego and UC Discovery Grant No. ele08-128656/Jin. This work was also partially supported by Postdoctoral Fellowship (for S. Oh) from the California Institute for Regenerative Medicine (CIRM).

8. References

Bauer, S., J. Park, J. Faltenbacher, S. Berger, K. von der Mark, and P. Schmuki. "Size Selective Behavior of Mesenchymal Stem Cells on Zro(2) and Tio(2) Nanotube Arrays." *Integr Biol (Camb)* 1, no. 8-9 (2009): 525-32.

Benya, P. D., and J. D. Shaffer. "Dedifferentiated Chondrocytes Reexpress the Differentiated Collagen Phenotype When Cultured in Agarose Gels." *Cell* 30, no. 1 (1982): 215-24.

Berger, S., J. Faltenbacher, S. Bauer, and P. Schmuki. "Enhanced Self-Ordering of Anodic Zro2 Nanotubes in Inorganic and Organic Electrolytes Using Two-Step Anodization." *Physica Status Solidi-Rapid Research Letters* 2, no. 3 (2008): 102-04.

Bettinger, C. J., R. Langer, and J. T. Borenstein. "Engineering Substrate Topography at the Micro- and Nanoscale to Control Cell Function." *Angew Chem Int Ed Engl* 48, no. 30 (2009): 5406-15.

Bhardwaj, T., R. M. Pilliar, M. D. Grynpas, and R. A. Kandel. "Effect of Material Geometry on Cartilagenous Tissue Formation in Vitro." *J Biomed Mater Res* 57, no. 2 (2001): 190-9.

Bjursten, L. M., Rasmusson, L., Oh, S., Smith, G.C., Brammer, K.S., Jin, S. "Titanium Dioxide Nanotubes Enhance Bone Bonding in Vivo." *J Biomed Mater Res* 88A, no. 92 (2009): 1218-1224.

Boyan, B. D., T. W. Hummert, D. D. Dean, and Z. Schwartz. "Role of Material Surfaces in Regulating Bone and Cartilage Cell Response." *Biomaterials* 17, no. 2 (1996): 137-46.

Boyen, H. G., G. Kastle, F. Weigl, B. Koslowski, C. Dietrich, P. Ziemann, J. P. Spatz, S. Riethmuller, C. Hartmann, M. Moller, G. Schmid, M. G. Garnier, and P. Oelhafen. "Oxidation-Resistant Gold-55 Clusters." *Science* 297, no. 5586 (2002): 1533-6.

Brammer, K. S., S. Oh, C. J. Cobb, L. M. Bjursten, H. van der Heyde, and S. Jin. "Improved Bone-Forming Functionality on Diameter-Controlled Tio(2) Nanotube Surface." *Acta Biomater* 5, no. 8 (2009): 3215-23.

Cavalcanti-Adam, E. A., A. Micoulet, J. Blummel, J. Auernheimer, H. Kessler, and J. P. Spatz. "Lateral Spacing of Integrin Ligands Influences Cell Spreading and Focal Adhesion Assembly." *Eur J Cell Biol* 85, no. 3-4 (2006): 219-24.

Chen, C. S. "Mechanotransduction - a Field Pulling Together?" *J Cell Sci* 121, no. Pt 20 (2008): 3285-92.

Ciolfi, V. J., R. Pilliar, C. McCulloch, S. X. Wang, M. D. Grynpas, and R. A. Kandel. "Chondrocyte Interactions with Porous Titanium Alloy and Calcium Polyphosphate Substrates." *Biomaterials* 24, no. 26 (2003): 4761-70.

Cooley, D. R., A. F. Van Dellen, J. O. Burgess, and A. S. Windeler. "The Advantages of Coated Titanium Implants Prepared by Radiofrequency Sputtering from Hydroxyapatite." *J Prosthet Dent* 67, no. 1 (1992): 93-100.

Costa Martinez, E., J. C. Rodriguez Hernandez, M. Machado, J. F. Mano, J. L. Gomez Ribelles, M. Monleon Pradas, and M. Salmeron Sanchez. "Human Chondrocyte Morphology, Its Dedifferentiation, and Fibronectin Conformation on Different Plla Microtopographies." *Tissue Eng Part A* 14, no. 10 (2008): 1751-62.

Cowan, M. M., K. Z. Abshire, S. L. Houk, and S. M. Evans. "Antimicrobial Efficacy of a Silver-Zeolite Matrix Coating on Stainless Steel." *J Ind Microbiol Biotechnol* 30, no. 2 (2003): 102-6.

Dalby, M. J., A. Andar, A. Nag, S. Affrossman, R. Tare, S. McFarlane, and R. O. Oreffo. "Genomic Expression of Mesenchymal Stem Cells to Altered Nanoscale Topographies." *J R Soc Interface* (2008).

Ducheyne, P., W. Van Raemdonck, J. C. Heughebaert, and M. Heughebaert. "Structural Analysis of Hydroxyapatite Coatings on Titanium." *Biomaterials* 7, no. 2 (1986): 97-103.

Fedewa, M. M., T. R. Oegema, Jr., M. H. Schwartz, A. MacLeod, and J. L. Lewis. "Chondrocytes in Culture Produce a Mechanically Functional Tissue." *J Orthop Res* 16, no. 2 (1998): 227-36.

Feng, B., J. Weng, B. C. Yang, S. X. Qu, and X. D. Zhang. "Characterization of Surface Oxide Films on Titanium and Adhesion of Osteoblast." *Biomaterials* 24, no. 25 (2003): 4663-70.

Getzenberg, R. H., K. J. Pienta, W. S. Ward, and D. S. Coffey. "Nuclear Structure and the Three-Dimensional Organization of DNA." *J Cell Biochem* 47, no. 4 (1991): 289-99.

Kobayashi, E., S. Matsumoto, H. Doi, T. Yoneyama, and H. Hamanaka. "Mechanical Properties of the Binary Titanium-Zirconium Alloys and Their Potential for Biomedical Materials." *J Biomed Mater Res* 29, no. 8 (1995): 943-50.

Maniotis, A. J., C. S. Chen, and D. E. Ingber. "Demonstration of Mechanical Connections between Integrins, Cytoskeletal Filaments, and Nucleoplasm That Stabilize Nuclear Structure." *Proc Natl Acad Sci U S A* 94, no. 3 (1997): 849-54.

McBeath, R., D. M. Pirone, C. M. Nelson, K. Bhadriraju, and C. S. Chen. "Cell Shape, Cytoskeletal Tension, and Rhoa Regulate Stem Cell Lineage Commitment." *Dev Cell* 6, no. 4 (2004): 483-95.

Muir, H. "The Chondrocyte, Architect of Cartilage. Biomechanics, Structure, Function and Molecular Biology of Cartilage Matrix Macromolecules." *Bioessays* 17, no. 12 (1995): 1039-48.

Narayan, Roger, Prashant Kumta, Charles Sfeir, Dong-Hyun Lee, Daiwon Choi, and Dana Olton. "Nanostructured Ceramics in Medical Devices: Applications and Prospects." *JOM Journal of the Minerals, Metals and Materials Society* 56, no. 10 (2004): 38-43.

Oh S, Daraio C, Chen L-H, Pasanic TR, Finones RR, Jin S. . "Significantly Accelerated Osteoblast Cell Growth on Aligned Tio2 Nanotubes." *Journal of Biomedical Materials Research Part A* 78A, no. 1 (2006): 97-103.

Oh S, Finones R, Daraio C, Chen L-H, Jin S. "Growth of Nano-Scale Hydroxapatite Using Chemically Treated Titanium Oxide Nanotubes." *Biomaterials* 23 (2005): 2945-54.

Oh, S., K. S. Brammer, Y. S. Li, D. Teng, A. J. Engler, S. Chien, and S. Jin. "Stem Cell Fate Dictated Solely by Altered Nanotube Dimension." *Proc Natl Acad Sci U S A* 106, no. 7 (2009): 2130-5.

Oh, S., Brammer, K.S. , Li, Y. S. Julie, Teng, D. , Engler, A.J. , Chien, S., and Jin, S. . "Stem Cell Fate Dictated Solely by Altered Nanotube Dimension." *Proceedings of the National Academy of Sciences* (2009).

Oh, S., C. Daraio, L. H. Chen, T. R. Pisanic, R. R. Finones, and S. Jin. "Significantly Accelerated Osteoblast Cell Growth on Aligned Tio2 Nanotubes." *J Biomed Mater Res A* 78, no. 1 (2006): 97-103.

Oh, S. H., R. R. Finones, C. Daraio, L. H. Chen, and S. Jin. "Growth of Nano-Scale Hydroxyapatite Using Chemically Treated Titanium Oxide Nanotubes." *Biomaterials* 26, no. 24 (2005): 4938-43.

Ohara, P. T., and R. C. Buck. "Contact Guidance in Vitro. A Light, Transmission, and Scanning Electron Microscopic Study." *Exp Cell Res* 121, no. 2 (1979): 235-49.

Oliveira, N. T., S. R. Biaggio, R. C. Rocha-Filho, and N. Bocchi. "Electrochemical Studies on Zirconium and Its Biocompatible Alloys Ti-50zr At.% and Zr-2.5nb Wt.% in Simulated Physiologic Media." *Journal of Biomedical Materials Research Part A* 74, no. 3 (2005): 397-407.

Ong, J. L., L. C. Lucas, W. R. Lacefield, and E. D. Rigney. "Structure, Solubility and Bond Strength of Thin Calcium Phosphate Coatings Produced by Ion Beam Sputter Deposition." *Biomaterials* 13, no. 4 (1992): 249-54.

Popat, K. C., R. H. Daniels, R. S. Dubrow, V. Hardev, and T. A. Desai. "Nanostructured Surfaces for Bone Biotemplating Applications." *J Orthop Res* 24, no. 4 (2006): 619-27.

Puleo, D. A., L. A. Holleran, R. H. Doremus, and R. Bizios. "Osteoblast Responses to Orthopedic Implant Materials in Vitro." *J Biomed Mater Res* 25, no. 6 (1991): 711-23.

Ratner, B. D. *Biomaterials Science : An Introduction to Materials in Medicine.* 2nd ed. Amsterdam ; Boston: Elsevier Academic Press, 2004.

Salata, O. "Applications of Nanoparticles in Biology and Medicine." *J Nanobiotechnology* 2, no. 1 (2004): 3.

Satsangi, A., N. Satsangi, R. Glover, R. K. Satsangi, and J. L. Ong. "Osteoblast Response to Phospholipid Modified Titanium Surface." *Biomaterials* 24, no. 25 (2003): 4585-9.

Savaiano, J. K., and T. J. Webster. "Altered Responses of Chondrocytes to Nanophase Plga/Nanophase Titania Composites." *Biomaterials* 25, no. 7-8 (2004): 1205-13.

Shankar, K., G. K. Mor, H. E. Prakasam, O. K. Varghese, and C. A. Grimes. "Self-Assembled Hybrid Polymer-Tio2 Nanotube Array Heterojunction Solar Cells." *Langmuir* 23, no. 24 (2007): 12445-9.

Solursh, M. "Cartilage Stem Cells: Regulation of Differentiation." *Connect Tissue Res* 20, no. 1-4 (1989): 81-9.

Solursh, M. "Cell Interactions During in Vitro Limb Chondrogenesis." *Prog Clin Biol Res* 110 Pt B (1982): 139-48.

Spiteri, C. G., R. M. Pilliar, and R. A. Kandel. "Substrate Porosity Enhances Chondrocyte Attachment, Spreading, and Cartilage Tissue Formation in Vitro." *J Biomed Mater Res A* 78, no. 4 (2006): 676-83.

Uchida, M., H. M. Kim, T. Kokubo, S. Fujibayashi, and T. Nakamura. "Structural Dependence of Apatite Formation on Titania Gels in a Simulated Body Fluid." *J Biomed Mater Res A* 64, no. 1 (2003): 164-70.

Webster, T. J., L. S. Schadler, R. W. Siegel, and R. Bizios. "Mechanisms of Enhanced Osteoblast Adhesion on Nanophase Alumina Involve Vitronectin." *Tissue Eng* 7, no. 3 (2001): 291-301.

Webster, T. J., R. W. Siegel, and R. Bizios. "Osteoblast Adhesion on Nanophase Ceramics." *Biomaterials* 20, no. 13 (1999): 1221-7.

Zanetti, N. C., and M. Solursh. "Induction of Chondrogenesis in Limb Mesenchymal Cultures by Disruption of the Actin Cytoskeleton." *J Cell Biol* 99, no. 1 Pt 1 (1984): 115-23.

Zhang, K., Y. Ma, and L. F. Francis. "Porous Polymer/Bioactive Glass Composites for Soft-to-Hard Tissue Interfaces." *J Biomed Mater Res* 61, no. 4 (2002): 551-63.

10

Cartilage Tissue Engineering Using Mesenchymal Stem Cells and 3D Chitosan Scaffolds – *In vitro* and *in vivo* Assays

Natália Martins Breyner, Alessandra Arcoverde Zonari,
Juliana Lott Carvalho, Viviane Silva Gomide,
Dawidson Gomes and Alfredo Miranda Góes
Universidade Federal de Minas Gerais (Federal University of Minas Gerais),
Institute of Biologic Science, Department of Biochemistry and Immunology,
Brazil

1. Introduction

Cartilage tissue has only one cell type, the chondrocyte, wich is immerse in extracellular matrix composed mainly by collagen type II. Because of such properties, cartilage tissue doens't heal spontaneously after a lesion, which with time becomes progressive and chronic. Cartilage lesions may be caused by automobile and sport accidents, as well as by normal wear due to age, and usually generate severe pain and difficulty of mobility in patients. Therefore, cartilage disease is a common type of lesion to which everyone is susceptible and represents a very important public health problem in the world (Willians et al, 2006).

Initial therapies to treat cartilage lesions included replacement surgery with artificial or natural organs and tissue grafts. Artificial and natural organ transplants and tissue grafts, on the other hand, are able to fully replace organs or tissues, but require continuous and permanent immune therapy to reduce immunological response to graft and to increase the longevity of transplanted tissue. Therefore, although major progresses were done in the field of cartilage tissue regenerative medicine during the years, current therapies still present limitations. Moreover, no adequate cartilage substitute has been developed. Thus, most of the severe injuries related to cartilage are still unrecoverable or not adequately treated. Therefore, these methods are helpful but need modification to develop better novel or alternative therapies (Ikada et al, 2006 and Tabata et al, 2009).

In such context emerges tissue engineering, which has been defined by Langer and Vacanti as: "an interdisciplinary field of research that applies the principles of engineering and the life sciences towards the development of biological substitutes that restore, maintain, or improve tissue function" (Salgado et al, 2004).

Tissue Engineering or Bioengineering is based on three elements: (i) cells; (ii) scaffolds and (iii) signalling molecules. These elements integrate themselves and promote the new tissue development (Langer and Vacanti, 1993; Ikada et al, 2006 and Chiang et al, 2009). In order to mimic tissue structure, tissue engineering also requires 3 dimensional cell cultures, which, in contrast to traditional bidimensional cell culture, has only been developed recently. Nowadays, it is beyond dispute that this cell culture strategy presents many advantages, including continuous exchange of nutrients and oxygen, metabolite removal and mechanical and chemical stimuli. All these factors allow and facilitate cell differentiation and proliferation (Ikada et al, 2006 and Tabata et al, 2009).

Used as scaffolds to 3D cultures, biomaterials studied in tissue engineering can be derived from natural or synthetic sources and may belong to one of three classes: metals, ceramics, or polymers. Once transplanted, biomaterials can be reabsorbed in vivo and replaced by new tissue (Ikada et al, 2006 and Tabata et al, 2009).

1.1 Cells

Tissue engineering strategy demands high numbers of cells, therefore, ideal cell sources for tissue engineering application must be easily isolated, expandable to higher passages, be non-immunogenic and have a protein expression pattern similar to the tissue regenerated (Salgado et al, 2004).

Chondrocytes derived from autologous tissue constitutes the most obvious choice to be used in tissue engineering, for their absence of immunogenicity and possibility of limited expansion in vitro. However this methodology suffers from many limitations, such as the generation of a second site of cartilage lesion, as well as the limited amount of cells obtained at the end of the procedure.

As an alternative, stem cells present a great therapeutic potential due to their capacity of differentiation to many cell lineages. These cells are able to self-renew and proliferate for long periods *in vitro* (Zuk et al, 2002 and Mountford et al, 2008).

Stem cells are divided into two great classes: adult and embryonic stem cells and also divided based on their differentiation potential. even though they may also be described based on their differentiation potential. According to this latter classification, the zygote and the cells produced by its first two divisions are considered totipotent, or capable of generating any cell of the embryo as well as the trophoblast. Continuing the embryo development, at the fifth day the embryo is constituted of two cell types, which compose the trophectoderm and the inner cell mass. Cells from the inner cell mass (ICM) are also called embryonic stem cells and are classified as pluripotent, for their capacity of generating the three embryo germ lines. ICM cells are not totipotent because they lack the capacity to generate extra-embryonary tissues. Later in the development, present in fully differentiated tissues, there are multipotent stem cells, which present more limited differentiation potential, being restricted to generate cells from the same embryonary origin as the tissue where they are found. However, according to the literature, multipotent stem cells may present a broader differentiation capacity then initially expected (Friedenstein et al, 1966; Owen et al, 1988; Zuk et al, 2002 and Conrad et al, 2004).

In 2007 yet a another type of stem cell was generate in vitro, the induced Pluripotency Stem Cell (iPSC). This new type of cells is produced by reprogramming adult cells, such as fibroblasts to a pluripotent state similar to that observed in embryonic stem (ES) cells, by retroviral transduction of some genes (Nanog, Oct4, Sox2, c-Myc, Klf 4 and Lin 28). The forced expression of such genes was capable of giving differentiated adult cells pluripotent differentiation capacity akin to the embryonic stem cell. This technique was termed cellular reprogramming (Takahashi et al, 2006; Yu et al, 2008 and Yamanaka et al, 2009).

It is important to aknowledge each stem cell type properties, for all stem cell types present inherent advantages and disadvantages, depending on their application.

Among adult stem cells there are:

- *Mesenchymal stem cells* (MSCs), which take part of the mesenchyme of varied tissues such as the bone marrow, adipose tissue, brain, dental pulp and skin, and are capable of differentiating into many cell lineages. MSCs present great potentials to the treatment of several diseases due to their low immunogenicity, immunomodulatory properties, the possibility of autologous transplantation, easy isolation and *in vitro* proliferation possibility.
- The bone marrow was the first source of MSCs described in the literature, and still remains the more thoroughly studied stem cell type. Also present in the bone marrow, there are:
- *Hematopoietic stem* cells (HSCs), which differentiate into all the hematopoietic and lymphoid cells from the blood. Therefore, HSCs are studied due to their roles in leukemia and other blood diseases. Usually, the treatment of such diseases include the substitution of the sick bone marrow to a healthy one, and in accordance to such fact, studies involving HSCs are mainly focused on how HSCs behave in different live organisms. Autologous grafts, or the implantation into a genetically similar live organism, may be performed in order to treat blood related diseases, as well as heterologous implantations, or grafting into genetically different live organisms. Presently, these different graft types show paradox behaviours. Heterologous grafts cause immune rejection in the host, requiring the host to be continuously submitted to immunossuppression. This therapy can lead to patient death due to the absence of an immune response to opportunist pathogens, however, this treatment is still commonly used today. In cases where the patient's conditions are good, cells can be extracted from the patient himself. This method is named autologous transplantation and is not susceptible to host rejection (Friedenstein et al, 1966; Owen et al, 1988, Conrad et al, 2004; Davila et al, 2004 and Gregory et al, 2005).

Even though stem cells derived from bone marrow have been well studied, they do not constitute the ideal mesenchymal stem cell source, due to the limited extent of MSC isolation (low extractable quantity of tissue) and donor discomfort. Therefore, new alternative sources have been proposed, including the adipose tissue. Adipose tissue is an excellent tissue to obtain great quantities of mesenchymal stem cells and it presents low discomfort when compared to bone marrow. According to Zuk, 2002, adipose tissue is a viable source to obtain mesenchymal stem cells, and these cells present similar characteristics to bone marrow MSCs.

Since the discovery of so many MSC sources, the International Society for Stem Cell Therapy postulated that a cell will only be considered a MSC if it presents 3 characteristics: 1. Being able to attach to cell culture surface; 2. Specific surface antigen expression; 3. Multipotent differentiation potential (osteoblast, adipocytes, chondroblasts) (Dominici et al, 2006). MSCs must then be capable of differentiating into cartilage, bone and muscular cell lineages, self-renewing and proliferating *in vitro*.

Embryonic stem cells (ESCs) are derived from the inner cell mass of the blastocist and present great moral, religious and ethical barriers due to their isolation technique, which leads to embryo destruction. They constitute a very promising stem cell type considering the tissue engineering field, for their pluripotent differentiation potential and unlimited proliferation capacity. Besides their ethical issues, ESCs also present the possibility of when injected *in vivo*, to produce teratomas. Thus, even though ESC present endless wonderful possibilities to be used in several science fields, more studies are necessary to ensure their safety and efficiency to be used in humans (Takahashi et al, 2006). In 2009, the biotech company Geron received the FDA approval to start the first human clinical trial of embryonic stem cell-based therapy in order to assess the safety of using differentiated embryonic stem cells to treat spine cord injuries. In 2011, they finally injected the first cells in patients and are waiting for results.

With respect to the potential of these stem cells, researchers have developed methods to trace these cells in both live and post-mortem stages. This is very important, because there are many routes and ways to introduce stem cells in an organism. If we can determine where these cells are going, wheter they can stay inside the 3D scaffold and differentiate or if they can stimulate others cells to migrate and graft, many unanswered questions will be addressed. Therefore, researchers have developed tracing techniques to locate injected cells in the organism. In basic science models, genetically modified transgenic organisms that express fluorescent proteins (FP) have been used to generate cells expressing such markers. Green fluorescent protein, the first FP generated, was derived from fluorescent seaweeds found in the US. After the isolation and characterization of the protein, its gene was introduced in mice and many other animals, so that those animals fluoresce when exposed to UV radiation. Cells taken from these transgenic animal and introduced into other non-transgenic animal (of similar lineages) do not present rejection problems and allow for trafficking of these cells (Ogawa et al, 2004). Many other tracing strategies were also developed including: radioisotopes, DNA and mitochondria dyes, as well as fluorescent microbeads. (Ogawa et al, 2004).

1.2 Tissue engineering

Tissue engineering or Bioengineering, as definided by Langer and Vancanti, constitute a innovation in regenerative medicine and is based on three elements (i) cells; (ii) scaffold and (iii) growth factors. Scaffolds can be bi-dimensional and three-dimensional, bi-dimensional structures allow us to observe only cell behavior with reference to medium composition, cell-cell interaction, cell viability and cell differentiation. However, three-dimensional structures allow us more physiologically realistic factors including dynamic fluids rich in O_2, mechanical forces, and cell adhesion but this interaction is three-dimensional and can be modify cell behaviour. For instance, nowadays it is beyond dispute that scaffolds are sources of instructive signals for cell differentiation, migration, proliferation and orientation, and of paramount role in phenotype maintenance. Therefore, many studies have searched for great biomaterials that can be used as a surrogate for extracellular matrix (ECM) tissue (Ikada et al, 2006; Chiang et al, 2009; Tabata et al, 2009 and Mingliang et al, 2011).

One of the main goals of Tissue engineering is to create a scaffold that can mimic ECM due to better cells, and micro-environment interactions. This interaction permits cells adhesion, migration, proliferation, differentiation and long-term viability (Bacakova et al, 2004). To produce a new organ or tissue we need scaffolds that are biodegradable and biocompatible. These structures need stable and appropriate porosity and architecture to permit formation of a vascular net able to give nutrients and O_2. These scaffolds should be gradually degraded to be occupied by new tissue formed by the interaction among the 3D scaffold, stem cells and growth factors.

To construct our 3D structure, we use chitosan and gelatin. Chitosan is derived from chitin presents in arthropods, including shrimp and crab. It is a polysaccharide very similar to glycosaminoglycans present in cartilage ECM. It is acid soluble, forms like-gel solutions and is water insoluble. Therefore, chitosan is available in nature and easily manipulated beyond its seemed ECM polysaccharides state (Roughley et al, 2006 and Dong et al, 2010).

Gelatin is derived from collagen, mainly proteins presents in cartilage ECM. Cartilage tissue is composed of type II colagen. In spite of this, gelatin is not composed of type II collagen, it present RGD motifs like all types collagen and this motif is able promote cell adhesion and differentiation and/or promote phenotype maintenance. Gelatin is water and acid soluble and is able to mix to chitosan gels (Tortelli and Cancceda, 2009 review). Both biomaterials have properties similar to cartilage ECM and they are biocompatible and biodegradable and are able to form porous where fluid can pass.

The connection between chitosan and gelatin is termed reticulates. Reticulates have a property to make chemical connections between molecules. These connections are stable and require the maintenance of stable scaffold architecture. In this study, we used two reticulates: Genipin and Glutaraldehyde. These reticulates are used beyond stable architecture, to increase degradation time *in vivo*. The importance of that controllable degradation is that it guarantees new tissue formation (e.g. ECM secretion) by differentiated cells. This way, the scaffold is able to provide a temporary matrix for developing cells, such as a support for cell attachment and tissue neomorphogenesis (Bacakova et al, 2004 and Mironov et al, 2009).

Glutaraldehyde is the reticulate most used in tissue engineering; it helps the 3D matrix creation through freeze and freezing drying. This process creates pores inside the scaffold and these pores are favorable to cell development, adhesion, proliferation and differentiation beyond the exchange of metabolites and food (Hofmann et al, 2009). Despite being most used, glutaraldehyde presents high levels of citotoxicity and limited reactivity with acetylates molecules. Therefore, a new approach is needed to find new reticulate that overcome all prior difficulties.

There is a new reticulate that had been studied, genipin. Genipin is derived from vegetable (Gardennia jasminoides, ELLIS) and it presents good capacity to increase mechanical properties for biomaterial-based protein. It forms pores and delays degradation that favors new tissue formation (Al amar et al, 2009 and Beier et al, 2009). This work verified which reticulates are better for our goal, the design of cartilage tissue.

The ultimate goal of tissue engineering is to design and fabricate close-to-natural functional human organs suitable for regeneration, repair and replacement of damaged, injured or lost human organs. Without tissue engineering, living functional human organs can be produced only during natural embryonic development. Therefore, according of Miranov and colleagues (2009) one of the most logical and obvious ways to look for possible alternatives to solid biodegradable scaffold-based tissue engineering approaches is to understand how tissue and organs are formed during normal embryonic development. Organ printing (one biomedical application of rapid prototyping) is an emerging transforming biomimetic technology that has potential for surpassing traditional solid scaffold-based tissue engineering (Miranov et al, 2009).

1.3 Signalling molecules

As the third pillar to tissue engineering, besides cells and scaffolds, it is important to deal with media constitution. All biochemical molecules present in culture media are able to stimulate cells. These stimuli differentiate all cells in culture, so theses molecules are important in creating new tissue or regenerate damage tissue. Due to stem cell differentiation capacities, they are cultivated in special medium that stimulates them during the differentiation process. Here we describe signaling molecules that stimulate chondrogenic differentiation of stem cells, mainly type II collagen secretion (Raghunath et al, 2005).

There are key molecules to chondrocyte differentiation: TGF-β and dexamethasone (Raghunath et al, 2005; Betre et al, 2006; James et al, 2007; Melrose et al, 2008 and Mueller et al, 2008). According to Lee and colleagues (2004) and Melrose and colleagues (2008), TGF-β induces the synthesis of type II collagen through Sox-9 pathway. Mueller and colleagues (2008) agree that dexamethasone causes chondrocyte hypertrophy because it induces type X collagen synthesis from cells. If this process occurs, the neotissue will suffer mineralization and will lose its properties (e.g. smoothness). However, most studies that aim to achieve chondrogenic differentiation use such molecules in differentiation medium, as the beneficial aspects outweigh disadvantages and such molecules seem to necessary to the chondrogenic differentiation process (Otto et al, 2004; Medrado et al, 2006; Huang et al, 2006 and Koay et al, 2007).

Still considering the example given, one viable option to obtain cartilage tissue in vitro and adequate to tissue engineering application is the combination of chitosan and gelatin, reticulated either by glutaraldehyde or genipin, seeded with mesenchymal stem cells. Here we show *in vitro* analysys performed to verify stem cell behavior in control (no differentiation stimuli) and differentiation medium. We tested whether differentiated cells in 3D scaffolds maintained differentiated phenotype *in vivo*.

2. Materials and methods

All animals were used and sacrificed in accordance to CETEA – UFMG (Ethical Committee Animals Experiments) # 153/2006. We used 30 rats (Lewis, male and female, 4 to 6 weeks old) from Physiology Department of UFMG and 5 rats (Lewis transgenic GFP - Lew-Tg e-GFP, 4 to 6 weeks old) from Missouri University (USA).

2.1 Cells and characterization by flow cytometry

Mesenchymal stem cells were obtained from rat adipose tissue (Lewis and Lewis eGFP). Rats were killed with anesthetic overdose and adipose tissue was removed from the abdominal region and it was maintained in conical tubes with DMEM supplemented 10% serum bovine fetal (SBF). After a few minutes, this tissue was digested with collagenase type II for 60 minutes, in 37°C and 5% CO_2. Every 15 minutes this solution was manually shaken. After this procedure, this solution was centrifuged for 10 minutes (1400rpm). The pellet was recovered and cultivated in DMEM with 10%SBF for 3 days. On the third day, the medium was collected, centrifuged to recover the non adherent cells and adhesive cells were cultivated in DMEM + 10% SBF. When this culture became confluent (80 to 90%), cells were trypsinized and expanded to new culture flasks (Zuk et la, 2002).

These cells were phenotyped by flow cytometry and used for differentiation studies at the 4th pass by flow cytometer. This procedure used anti-CDs antibodies to label markers present on the cell surface. The CDs are markers present in mesenchymal stem cells, hematopoietic stem cells and other cell types. We used CD54, CD91 and CD73 as MSC markers and CD45 as HSC (Zuk et al, 2002 and Ucelli et al, 2006). All cells are fixed with formaldehyde (2%) and analyzed by FACScalibur (USA). For control, we use only secondary antibody and selected the gate for cells to be analyzed we used no marker cells. Around 20,000 events (minimum) were used for fluorescence capture in Cell Quest software. All data were analyzed by WinMid 2.8 software. This procedure was performed according Zuk et al, 2002. To conduct flow cytometry we used 1×10^6 cells and stained with antibodies (anti-CDs) isolated before we used secondary antibody (FITC for Lewis and PE for Lewis eGFP).

2.2 Chondrogenic medium

Chondrogenic medium was based in protocol by Medrado and cols (2006), Huang and cols (2006), Koay and cols (2008) and Breyner and cols (2010), from these we used 10µg/L recombinant TGF-β3 (Bioclone), 10^{-7} M dexamethasone (Sigma) with 1% SBF.

2.3 Cell differentiation

Mesenchymal stem cells were cultivated in T75 flasks with control medium (DMEM + 10%SBF) and chondrogenic medium for 1, 3, 6 and 9 weeks. To verify to differentiation process immunofluorescence was perfomed using antibodies for collagen II, CD54, CD90,CD45 and CD73, osteocalcin (Zuk et al, 2002 and Huang et al, 2006).

2.4 Immuno-fluorescence

Mesenchymal stem cells (1×10^5) cultivate in normal medium and chondrogenic medium for 1, 3, 6 and 9 weeks were used for immunofluorescence. To conduct this experiment, we used cells from eGFP rats. Due to green GFP fluorescence, these cells were able to fluoresce under confocal microscopy (green) and we used others markers with red fluorescence. Initially, when we used to identify chondrogenic differentiation, we used markers from msenchymal stem cells (CD90, CD54and CD73), hematopoietic stem cells (CD45), cartilage-specific (type II collagen) and bone-specific (osteopontin). Cells were seeded in glass laminules. By 48 h, all cells were fixed with formaldehyde (2%) and stained with each antibody. Each well was washed with cold PBS 0,15M and the secondary antibodies (polyclonal anti-rat IgG made in rabbit, Molecular Probes) were applie and the cells were observed in confocal microscope (Soliman et al, 2008).

2.5 3D scaffolds

We developed two kinds of 3D scaffolds with similar chitosan (85% deacetylated, Sigma) and gelatin (Sigma) ratios. The difference between scaffolds is the reticulate use: glutaradehyde (0,1%, Sigma) or genipin (0,1%, Challenge Bioproducts). In order to solubilise chitosan we diluted it in acetic acidic (0,5mM) and gelatin was diluted in water. Misture of chitosan to gelatine was performed maintaining the ratio of 3 parts of chitosan solution to 1 part gelatin solution. Immediately, we shook these solutions and distributed 1mL/well in 24 wells plate. These plates were shaken in mechanic shaker overnight, protect from light. On the next day we added 1mL reticulate to each well. After 60min, these plates were frozen at -20°C overnight and then were transferred to -80°C. Plates were freeze-drying for 8 hours. The cilinders (3D scaffolds) were sterililzed (120°C) and used for cell culture (Guo et al, 2006 and Yamane et al, 2005). The scaffolds were analyzed at the Department of Metallurgical and Materials Engineering, Federal University of Minas Gerais. The matrices were covered with gold (Sputter Coater - SPI Supplies) for 90 sec at 13mA. The images were obtained by means of scanning electron microscope (JEOL 6360LV), at 15kV and 750mA, to qualitatively assess the pore interconnectivity and size.

2.6 Cell and 3D scaffold

GFP cells were used because of the need to trace these cells; they were processed the same way as the MSC characterized by IMF. These cells were cultured in a 3D scaffold with chondrogenic medium for 3 weeks and surgically grafted in the rat subcutaneous dorsal region. After 6 weeks, all animals were dead and the samples were analyzed by IMF to verify presence of collagen type II. This assay was conducted because of the need to know whether differentiated cells can dedifferentiate after implantation (Janune et al, 2006).

2.7 Statiscal analyzes

All data are presented as an average ± standard deviation (SD). To test the significance of the observed differences between the study groups, a statistical evaluation was carried out using a one-way ANOVA. A value of P< 0.05 was considered to be statistically significant.

3. Results

3.1 In vitro assay
3.1.1 Cell characterization

Cells were characterized by Immunofluorescence using GFP, MSC membrane surface markers CD90, CD73 and CD54, HSC surface marker, CD45 antibodies to show the absence of cell contamination with another source of stem cells. All cells were derived from GFP-Lewis rats and verified with an anti-GFP primary antibody and PE (phicoerithrine) conjugate secondary antibody. Our results showed that all cells were positively for GFP (red, Fig. 1). For other markers, we used secondary antibodies with FITC (fluorecein isothiacyanate, Fig 1) and we observed that, cells were negative for CD45 and positive for CD90, CD73 e CD54. Nuclei were marked with DAPI (blue, Fig. 1).

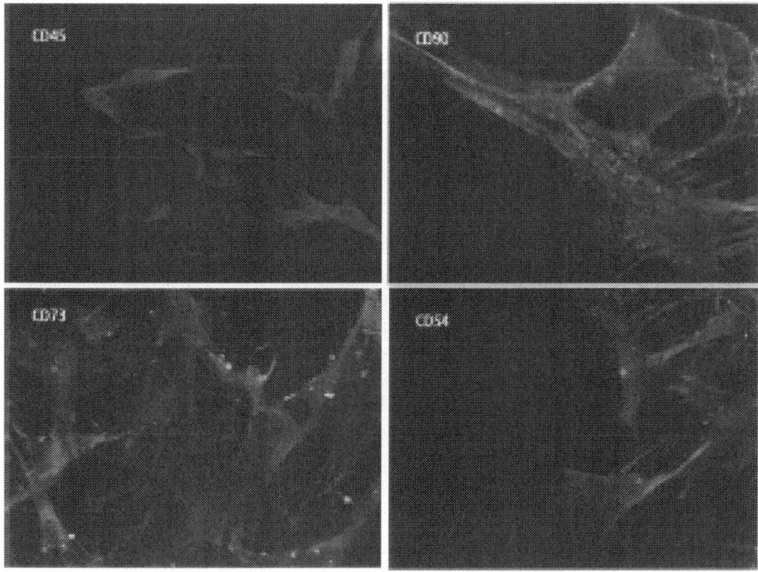

Fig. 1. Characterization of mesenchymal stem cells derived from adipose tissue of GFP-Lewis rats. A- CD45, B-CD90,C CD73 and D-CD54. GFP: red, surface marker: green and nucleus: blue.

3.1.2 Cell differentiation

In order to assess mesenchymal stem cell phenotype changes Cells differentiation was characterized by Immunofluorescence using membrane surface markers, CD90, CD73 and

CD54, MSC markers and CD45 a HSC marker. The goal of this experiment was to observe wether the cellular phenotype changed during the 9 weeks in chondrogenic medium. It is important to note that GFP cells are able to fluoresce without secondary antibody. Therefore anti-GFP were not used , only anti-CDs antibodies and secondary antibodies conjugated to PE to not GFP used. We observed that when cells were cultivated in chondrogenic medium, they were phenotypical alterated. Differentiated cells were positive for CD73 and Collagen II (red, Fig. 2) and they were negative for CD90, CD45, CD54 and osteocalcine (Fig. 2).

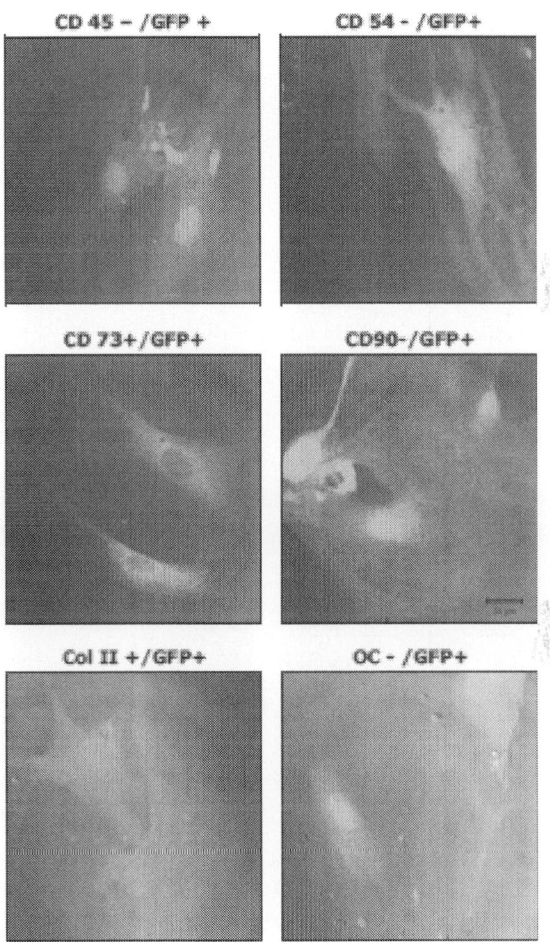

Fig. 2. Cells differentiated with chondrogenic medium for 9 weeks. CD45, CD54, CD73, CD90, Collagen II and osteocalcine. Cells were GFP labeled.

3.2 Biomaterials
3.2.1 Scaffold development and analysis

The characterization of the biomaterials developed began at macroscopic aspects. In in one hand glutaraldehyde – reticulated scaffolds presented yellow color, genipin-reticulated scaffolds were dark blue. Microscopically, glutaraldeyde-reticulated matrices presented round pores with sizes ranging between 100 - 500 μm (Fig. 3A and 3C). On the other hand, genipin-reticulated matrices also presented pores with sizes between 100 - 500 μm, however these pores had the appearance of being more fine and fragile (Fig. 3 B and 3D).

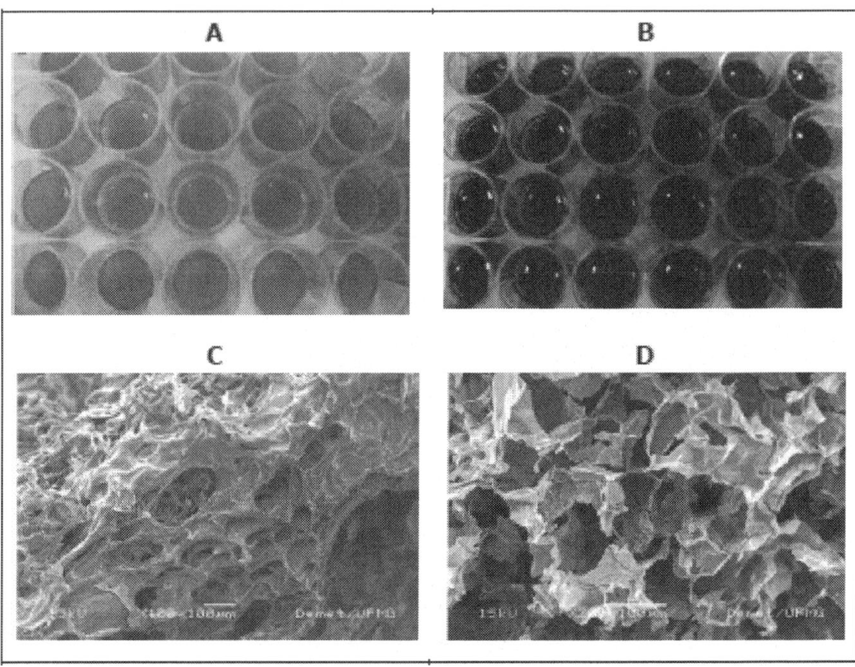

Fig. 3. 3D Scaffolds. (A) and (C) 3D scaffold with glutaraldehyde reticulates. (A) Without magnification and (C) SEM X100. (B) and (D) 3D scaffold with genipin reticulate. (B) Without magnification and (D) SEM X200.

3.2.2 Scaffold and cells

In order to assess the biomaterials' citotoxicity, we seeded cells on the scaffold and verified if the cells were viable after 1 week. To verify this we performed a stablished citotoxicity assay called MTT, which verifies cell viability through the assessment of mitochondria function. It was possible then to verify that cells were kept viable in both scaffolds viable (Fig. 4). The cells colonized both scaffolds as shown in figure 5. Rat MSC extended pseudopodes to link onto wall scaffolds.

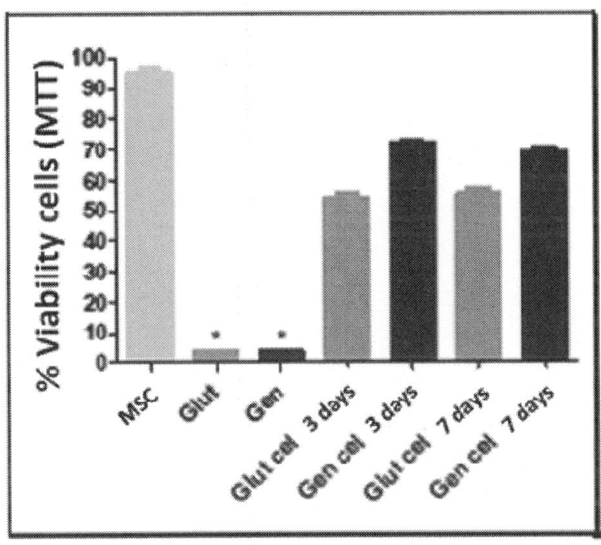

Fig. 4. Cells Viability (MTT). The graph shows cell viability when cultivated on scaffolds for 1 week.

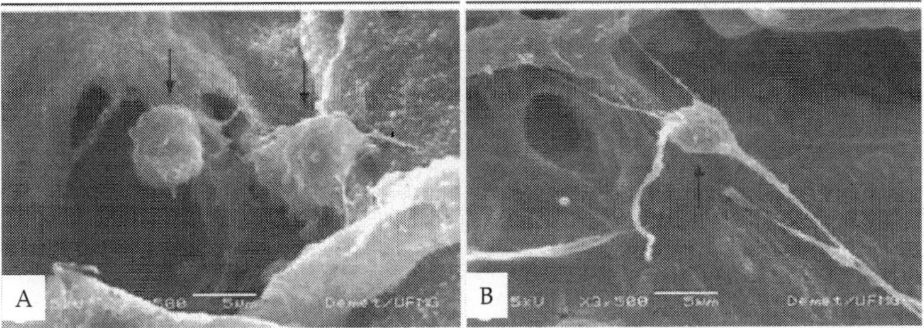

Fig. 5. Scanning electronic microcopy. (A) and (B) Cells attached to scaffolds.

3.3 *In vivo* assay

After 3 weeks of culture in chondrogenic medium in respective scaffolds, we grafted those constructs (association of scaffolds and cells) subcutaneous onto the dorsal region of Lewis rats. Those rats were sacrificed with anesthestic overdose on the 3rd week after implantation. These samples were analyzed by immunofluorescence and we verified that collagen type II was present in both samples. This result showed that once cells differentiated in the scaffold they do not dedifferentiate. The cells were able to maintained the differentiated phenotype inside the scaffolds (Fig. 6).

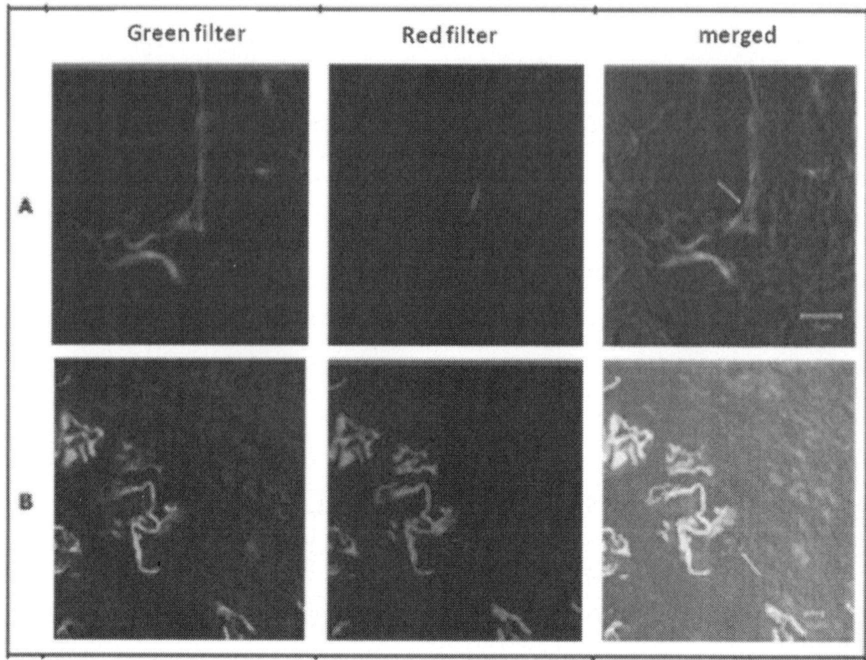

Fig. 6. Collagen II staining *in vivo*. Grafts of scaffolds with differentiated cells. (A) glutaraldehyde scaffold (B) Genipin scaffold.

4. Discussion

An ideal scaffold for been used in cartilage tissue engineering should be biodegradable, tissue compatible and display some degrees of rigidity and mechanical flexibility. It should have a three-dimensional configuration that provides a favorable environment for proliferation of chondrocytes and stem cells, and for cell migration and differentiation (Puppi et al, 2010 and Hutmacher et al, 2000). Furthermore, engineered bio-interfaces covered with biomimetic motifs, including short bioadhesive ligands, are a promising material-base strategy for tissue repair in regenerative medicine.

A 3D culture system that promoted the chondrogenesis of MSCs was established in this study. It was observed that the combination of chitosan and gelatin scaffolds provided a supporting environment for the chondrogenesis of rat MSCs. The MSCs formed cartilage-like tissue formation *in vivo* and *in vitro* via stimulation with common combinations of bioactive substances such as transforming growth factor (TGF-β) and dexamethasone. The chondrogenic differentiation of MSCs is typically detected by the formation of cell spheres in culture expression type II collagen in the extracellular matrix, surface markers alteration and the confirmation of typical gene expressions profiles by PCR analysis as determined by Breyner and colleagues, 2010.

Thus, the chitosan-gelatin scaffold used in our work mimic the natural environment leading to increased ECM synthesis and promoting differentiation of MSCs to chondrocytes.

Other publications have promoted variations in the matrix composition and GAG fine structure among the scaffolds used for cartilage tissue engineering in order to improve articular chondrocyte culture and chondrogenesis of progenitor cells (Mouw et al, 2005 and Melhorn et al, 2007). Our work, however, demonstrates that the 3D structure and chemical composition of the chitosan scaffolds and chondrogenic medium promoted MSC activation, proliferation and differentiation into chondrocytes, as was detected by a decrease in ALP production, an increase in collagen type II production and a lack of osteocalcin, a known osteogenic marker (Breyner et al, 2010; Huang et al, 2008 and Medrado et al, 2006).

The 3D chitosan-gelatin structure is perhaps an indication that the attachment of MSCs to chitosan matrix could improve cell differentiation after matrix deposition as seen in the development of chondrocytes. One of the advantages of 3D systems is the substantial surface area to volume ratio can maximize cell-material contact when compared to monolayer culture systems.

An ideal scaffold for cartilage tissue engineering should be biocompatible, non-cytotoxic and have favorable structural features for cell attachment and proliferation (Lefebvre et al, 1997 and Mingliang et al, 2011). This study showed that cells attached, proliferated and secreted extracellular matrix in the two 3D porous scaffolds used. After 3 days in culture the viability and cell number in the 3D scaffold culture group had increased and was higher than that of the cells in monolayer culture treated with glutaraldehyde or genipin. The percentage of cells in different stages was determined by the MTT metabolization assay. Strong cell attachment and proliferation demonstrated that there was no cytotoxicity in either scaffolds used. The results agreed well with previous studies showing that initial cell adhesion was largely influenced of RGD in gelatin bound to chitosan (Dong et al, 2010). It was reported that immobilization of RGD peptide onto a scaffold enabled the adhesion of stem cells to the scaffold and inhibited the immediate matrix-induced cell aggregation (Re'em et al, 2010). It allowed better access for cells to nutrients, oxygen and chondrogenic inducer. TGF-β1 is the main chondrogenic inducer during MSCs chondrogenesis (Barry et al, 2001). Studies suggested that RGD interaction with α5 and β1 integrin subunits enhanced TGF-β1 secretion, and RGD-dependent integrin activation should be linked to modulation of TGF-β1 activity (Ortega-Velasquez et la, 2003). In this study, the homogeneous spread of MSCs and abundant matrix secretion in scaffold indicated that TGF-β1 had efficiently induced MSCs chondrogenesis. These results may indicate that the 3D scaffold culture is superior to the monolayer culture because it is more effective in promoting ECM secretion or expression.

The immunoflurescence staining also revealed type II collagen accumulation between scaffolds seeded with MSCs after *in vivo* implantation. It is also interesting to note that the 3D culture system not only enhanced chondrogenesis but also increased the cell proliferation of MSCs. This may be a positive effect from chitosan-gelatin combination and TGF-β added to the chondrogenesis medium.

Generally, cell attachment and spreading can occur in a serum-containing environment regardless of surface coating because many factors regulating cell adhesion and spreading, such as fibronectin, vitronectin, and cytokines are found in serum (Underwood et al, 2001). However, during *in vitro* chondrogenic differentiation of MSCs, defined chondrogenic differentiation inducing medium was used without serum. Therefore, cell adhesion was the vital step for MSCs chondrogenic differentiation. However, cell density strongly influences MSCs differentiation and cell-matrix secretion (Hui et al, 2008).

5. Conclusion

Rat mesenchymal stem cells are able to differentiate when they are cultivated in chondrogenic medium. These cells can colonize scaffolds and differentiate inside them. Therefore, when these constructs were subcutaneously grafted in the rat dorsal region we verified that cells maintained a differentiated phonotype after 6 weeks. All together, we concluded that chitosan and gelatin are good candidates for scaffolds used to differentiate stem cells with chondrogenic treatment.

6. References

Williams DF. (2006). To engineering is to create: the link between engineering and regeneration. *Trends in Biotechnol* 24 (1): 4-8.

Ikada Y.(2006). Challenges in tissue engineering. *J.R.Soc Interface* 3:589-601.

Tabata Y.(2006). Biomaterial technology for tissue engineering applications. *J. R. Soc. Interface* 6: S311-S324.

Langer R & Vacanti JP. (1993). Tissue Engineering. *Science* 260: 920-926.

Chiang H & Jiang CC. (2009). Repair of articular cartilage defects: review and perspectives. *J Formos Med Assoc* 108: 87-102.

Zuk PA; Zhu M; Ashjian P; De Ugarte DA; Huang JI; Mizuno H; Alfonso ZC; Fraser JK; Benhaim P; Hedrick MH.(2002). Human adipose tissue is a source of multipotent stem cells. *Molecular Biology of the Cell* 13: 4279-95.

Mountford JC.(2008). Human embryonic stem cells: origins, characteristics and potential for regenerative therapy. *Transfusion Medicine* 18: 1-12.

Friedenstein AJ; Gorskaja JF; Kalugina NN. (1976). Fibroblast precursors in normal and irradiated mouse hematopoietic organs. *Experimental Hematol* 4:267-274.

Owen M & Friedenstein JA. (1988). Stromal stem cells: marrow derived osteogenic precursors. *Ciba Found Symp* 126: 42-60.

Conrad C & Huss R. (2004). Adult stem cell lines in regenerative medicine and reconstructive surgery. *Journal of Surgical Research* 124:201 – 208.

Takahashi K & Yamanaka S. (2006). Induction of pluripotent stem cells from mouse embryonic and adult fibroblast cultures by defined factors. *Cell* 126: 663-76.

Yu J and Thomson JA. (2008). Pluripotent stem cell lines. *Genes and dev* 22: 1987-97.

Yamanaka S. (2009). Elite and stochastic model for induced pluripotent stem cell generation. *Nature* 460: 49-52.

Davila JC; Cezar G; Thiede M; Strom S; Miki T; Trosko J. (2004). Use and application of stem cells in toxicology. *Toxicological Science* 79: 214-223.

Gregory CA; Prockop DJ; Spees J L. (2005). Non hematopoietic bone marrow stem cells: Mollecular control of expansion and differentiation. *Experimental Cell Research* 306: 330 – 335.

Ogawa R; Mizuno H; Hyakosuko H; Watanabe A; Migita M; Shimada T. (2004). Chondrogenic and osteogenic differentiation of adipose derived stem cells isolated from GFP transgenic mice. *J Nippon Med Sch* 71 (4): 240-1.

Bacakova L; Filova E; Rypacek; Svorcik V; Stary V. (2004). Cell adhesion on artificial materials for tissue engineering. *Physiol Res* 53: 35-45.

Roughley P; Hoemann C; DesRosiers E; Mwale F; Antoniou F; Alini M. (2006). The potencial of chitosan-based gels containing intervertebral disc cells for nucleous pulposus supplementation. *Biomaterials* 27 (3): 388-96.

Tortelli F & Cancedda R. (2009). Three dimensional cultures of osteogenic and chondrogenic cells: a tissue engineering approach to mimic bone and cartilage *in vitro. Eur Cells and Materials* 17:1-14.

Miranov V; Visconti, RP; Kasyanov V; Forgacs G; Drake CJ; Markwald RR. (2009). Organ printing: tissue spheroids as building blocks. *Biomaterials* 30, 2164-74.

Hoffmann B; Seitz D; Mencke A; Kokott A; Ziegler G. (2009). Glutaraldehyde and oxidized dextran as crosslinker reagents for chitosan-based scaffold for cartilage tissue engineering. *J Mater Sci: Mater Med* 1-9.

Al-Ammar A; Drummond JL; Bedran-Russo AKB. (2009). The use of collagen cross-linking agents to enhance dentin bond strength. *J Biomed Mater Res B Appl Biomater*, 91 (1):419-24.

Beier JP; Klumpp D; Rudisile M; Dersh R; Wendorff JH; Bleiziffer O; Arkudas A; Polykandriotis E; Horch RE; Kneser U. (2009). Collagen matrices from sponge to nano: new perspective for tissue engineering of skeletal muscle. *BMC Biotechnology* 9 (34): 1-14.

Raghunath J; Salacinsky HJ; Sales KM; Butler PE and Seifalian AM. (2005). Advanced cartilage tissue engineering: the application of stem cell technology. *Current opinion in biotechnology* 16:503-509.

Betre H; Ong SR; Guilak F; Chilkoti A; Fermor B; Setton LA. (2006). Chondrocytic differentiation of human adipose-derived adult stem cells in elastin-like polypeptide. *Biomaterials* 27: 91-99.

James CG; Ulici V; Tuckermann J; Underhill TM; Beier F. (2007). Expression profiling of dexamethosone-treated primary chondrocytes identifies targets of glucocorticoid signaling in endochondral bone development. *BMC Genomics* 8: 1-27.

Melrose J; Chuang C; Whitelock J. (2008). Tissue engineering of cartilages using biomatrices. *J Chem Technol Biotechnol* 83: 444-63.

Mueller MB & Tuan RS. (2008). Functional characterization of hypertrophy in chondrogenesis of human mesenchymal stem cells. *Arthritis & Rheumatism* 58:1377-88.

Lee JW; Kim YH; Kim SH; Han SH; Hahn SB. (2004). Chondrogenic differentiation of mesenchymal stem cells and its clinical applications. *Yonsei Medical Journal* 45:41-47.

Otto WR and Rao J. (2004). Tomorrow's skeleton staff: mesenchymal stem cell and the repair of bone and cartilage. *Cell Prolif* 37: 97-110.

Medrado GC; Machado CB; Valerio P; Sanches MD; Goes AM. (2006). The effect of a chitosan-gelatin matrix and dexamethasone on the behavior of rabbit mesenchyma stem cells. *Biomed Mater* 13: 155-161.

Hwang NS; Kim MS; Sampattavanich S; Baek JH; Zhang Z; Elisseeff J. (2006). Effects of three dimensional culture and growth factors on the chondrogenic differentiation of murine embryonic stem cells. *Stem cells* 24: 284-291.

Koay EJ; Hoben GMB; Athaunasiou KA. (2007). Tissue engineering with chondrogenically differentiated human embryonic stem cells. *Stem cells* 25: 2183-2190.

Ucelli A; Moretta L and Pistoia V. (2008). Mesenchymal stem cells in health and disease. *Nature* 8:726-37.

Breyner NM; Hell RC, Carvalho LR, Machado CB, Peixoto Filho IN; Valerio P; Pereira MM, Goes AM. (2010). Effect of three dimensional chitosan porous scaffold on the differentiation of mesenchymal stem cells into chondrocytes. *Cell Tissue and Organs*. 191, 119-28.

Soliman EM; Rodrigues MA, Gomes DA; Sheung N; Yu J; Amaya MJ, Nathanson MH, Dranoff JA. (2009). Intracellular calcium signals regulate growth of hepatic stellate cell via specific effects in cell cycle progression. *Cell calcium*. 68: 284-92.

Guo T; Zhao J; Chang J; Ding Z; Hong H; Chen J; Zhang J. (2006). Porous chitosan-gelatin scaffold containing plasmid DNA encoding transforming growth factor-β1 for chondrocytes proliferation. *Biomaterials* 27: 1095-1103.

Yamane S; Iwasaki N; Majima T; Funakoshi T; Masuko T; Harada K; Minami A; Monde K; Nishimura S. (2005).   Feasibility of chitosan-based hyaluronic acid hybrid biomaterial for a novel scaffold in cartilage tissue. *Biomaterials* 26: 611-19.

Janune DDJ; Cestari TM; De Oliveira RC; Taga EM; Taga R e Granjeiro JM. (2006). Avaliação da resposta tecidual ao implante de osso bovino misto medular em subcutâneo de ratos. *Innovations implant journal.* 2: 21-28.

Puppi D, Chiellini F, Piras AM, Chiellini E. (2010). Polymeric materials for bone and cartilage repair. *Prog Polym Sci* 35:403-40.

Hutmacher DW. (2000) Scaffolds in tissue engineering bone and cartilage. *Biomaterials* 21:2529-43.

Mouw, J.K., N.D. Case, R.E. Guldberg, A.H. Plaas, M.E. Levenston. (2005). Variations in matrix composition and GAG fine structure among scaffolds for cartilage tissue engineering. *Osteoarthritis Cartilage* 13: 828–836.

Mehlhorn, A.T., P. Niemeyer, K. Kaschte, L Muller, G. Finkenzeller, D. Hartl, N.P. Sudkamp, H. Schmal. (2007). Differential effects of BMP-2 and TGF-beta1 on chondrogenic differentiation of adipose derived stem cells. *Cell Prolif* 40: 809–823.

Lefebvre V, Huang W, Harley VR, Goodfellow PN, De Crombrugghe B. (1997). SOX9 is a potent activator of the chondrocyte-specific enhancer of the pro alpha1 (II) collagen gene. *Mol Cell Biol* 17:2336-46.

Mingliang You, Gongfeng Peng, Jian, Ping Maa, Zhihui Wang, Weiliang Shu, Siwu Peng, Guo-Qiang Chen. (2011). Chondrogenic differentiation of human bone marrow mesenchymal stem cells on polyhydroxyalkanoate (PHA) scaffolds coated with PHA granule binding protein PhaP fused with RGD peptide. *Biomaterials* 32, 2305-2313.

Dong Y, Li P, Chen CB, Wang ZH, Ma P, Chen GQ. (2010). The improvement of fibroblast growth on hydrophobic biopolyesters by coating with polyhydroxyalkanoate granule binding protein PhaP fused with cell adhesion motif RGD. *Biomaterials* 31:8921-30.

Re'em T, Tsur-Gang O, Cohen S. (2010) The effect of immobilized RGD peptide in macroporous alginate scaffolds on TGF [beta] 1-induced chondrogenesis of human mesenchymal stem cells. *Biomaterials* 31:6746-55.

Ortega-Velázquez R, Díez-Marqués ML, Ruiz-Torres MP, González-Rubio M, Rodríguez-Puyol M, et al. (2003). Arg-Gly-Asp-Ser peptide stimulates transforming growth factor-b1 transcription and secretion through integrin activation. *FASEB J*;17:1529-31.

Underwood PA, Bean PA, Mitchell SM, Whitelock JM. (2001). Specific affinity depletionof cell adhesion molecules and growth factors from serum. *J Immunol Methods* 247:217-24.

Hui TY, Cheung KMC, Cheung WL, Chan D, Chan BP. (2008). In vitro chondrogenic differentiation of human mesenchymal stem cells in collagen microspheres: influence of cell seeding density and collagen concentration. *Biomaterials* 29:3201-12.

11

Magnetoelectropolished Titanium Biomaterial

Tadeusz Hryniewicz[1], Ryszard Rokicki[2] and Krzysztof Rokosz[1]
[1]Politechnika Koszalińska, Division of Surface Electrochemistry
[2]Electrobright, Macungie PA
[1]Poland
[2]USA

1. Introduction

The high standard and bio-electrochemical stability of any metallic biomaterial components are the main conditions for their safe implantation into human body. The most critically indispensable properties for metallic biomaterials are their corrosion resistance, inertness, low toxicity, and durability. The work aims at analyzing the titanium biomaterial characterization after electrolytic polishing in a magnetic field, named as the magnetoelectropolishing (MEP), in comparison with the material finish after a conventional electropolishing (EP).

Titanium and titanium alloys gradually became the main biomedical materials used presently in orthopaedic applications. The leaching of metallic ions such as nickel during the corrosion process from other biomaterials such as: 316L stainless steel, L-605 cobalt-chromium alloy, or Nitinol, has caused considerable concerns due to the allergies, inflammations, etc. It appears to be non existent when CP titanium is employed. Titanium is also insensitive to oxygen concentration and by this titanium ions release by this mechanism is not applicable in this case. Also it is well known that when titanium is exposed to body fluids, its surface undergoes spontaneous modification by Ca^{2+} and PO_4^{3-} ions and prolong exposure leads to formation of hydroxyapatite layer, which is indispensable to bone-implant osseointegration.

High quality metal alloys of titanium are commonly used for orthopaedic prostheses as bone plates, nails, screws, etc. Fortunately, the oxides and hydroxides of titanium have extremely low solubilities – so a passive oxide film readily and spontaneously forms over the titanium's surface. In spite of its very high corrosion resistance the spontaneously formed oxide film consists of some inclusion and discontinuity spots, which can cause the problems in integration at the bone-implant interface. To overcome these problems, several surface treatments are employed: chemical etching, plasma treatment, ion implantation, electrochemical or wet chemical hydroxyapatite precipitation following either hydrothermal treatment or sintering, anodizing, electropolishing, etc. The electropolishing process seems to be the best way to eliminate these problems. By dissolving the existing imperfect oxide, electropolishing process creates the base for formation the more perfect homogeneous oxide over the base titanium metal. As it was shown in many works, including also previous ours, it is the presence of this oxide film that is responsible for titanium's excellent corrosion resistance, which enables it to be used in surgical applications.

2. Titanium biomaterial characteristics

Titanium and some of its alloys Ti6Al4V, TI6AL7Nb, Nitinol, are classified as biologically inert biomaterials and are commercially used in orthopaedic and dental application. This is a result of their outstanding corrosion resistance and inertness. The inertness of those materials is due to titanium oxide (TiO_2) which spontaneously covers them after exposure to ambient air or water according to the following reactions:

$$Ti + O_2 \rightarrow TiO_2$$

$$Ti + 2H_2O \rightarrow TiO_2 + 4H^+ + 4e^-$$

The stability of TiO_2 can be compromised only by complexing species such as HF or H_2O_2 and can lead to its dissolution. Without those species TiO_2 is thermodynamically stable in the wide range of pH = 2-12 (Schenk, 2001) and by this CP-Ti is totally corrosion resistant in the presence of neutral physiological solutions. In the case of Al-containing alloys (Ti6Al4V, Ti6Al7Nb) acidification of solutions make them more prone to dissolution than CP-Ti (Ruzickova et al., 2005).

In present work our research concentrates on CP Titanium as on precursor of another titanium alloys. Along excellent corrosion resistance, inertness to human body and excellent osseointegration titanium offers more suitable properties for implantable biomaterials. Some of those properties include: fatigue resistance, strength, density, elastic module, etc. The density of around 4.5 g.cm^{-3} makes it almost two times lighter than cobalt-chromium or 316L stainless steel alloys. As pure element CP titanium also excludes possibility of leaching harmful elements which could be detrimental to surrounding tissues as can be in case of 316L stainless steel and Nitinol where possibility of leaching nickel is reality. The biggest disadvantage of CP-Ti is its relatively low wear resistance and by this it should not be used in devices where contact wear is unavoidable as for example modular interface corrosion between Co-Cr heads and Ti stem in total hip replacement prosthesis (Singh &, Dahotre, 2007; Salvati et al., 1995).

2.1 Mechanism of passive film formation and growth

Even that the passive film on titanium is only in nanometers range it creates very protective barrier against biological environment of human body. Its protectiveness depends on several features: morphology, homogeneousness, thickness, kind and quantity of foreign chemicals species incorporated in it. All of those properties of passive film determine the tunneling rate end speed of ions moving through the film as well as its dissolution by surrounding fluids.

According to (Cabrera & Mott, 1948) high field mechanism for oxide film formation and growth theory the main prerequisite is adsorption of oxygen on bare titanium surface which creates oxide monolayer (Fig. 1). The next step is electron tunneling from titanium to monolayer of adsorb oxygen which by adding electrons became electron traps on the outer surface of the oxide. As the number of electron traps increases the potential drop across the films grows. The drop of potential creates the electric field across the passive film which lowers the activation energy necessary for further ions transport though passive film. The oxide on titanium is classified as N-type semiconductor which means that anion transport through film is dominant way of film grow and is due to oxygen ions movement toward titanium metal. The thickening of oxide film increases the activation energy necessary for further oxygen ions transport and limits further passive film formation. The only way for further grow of passive film in this point is to increase the potential drop across the film which simultaneously increases the electric field.

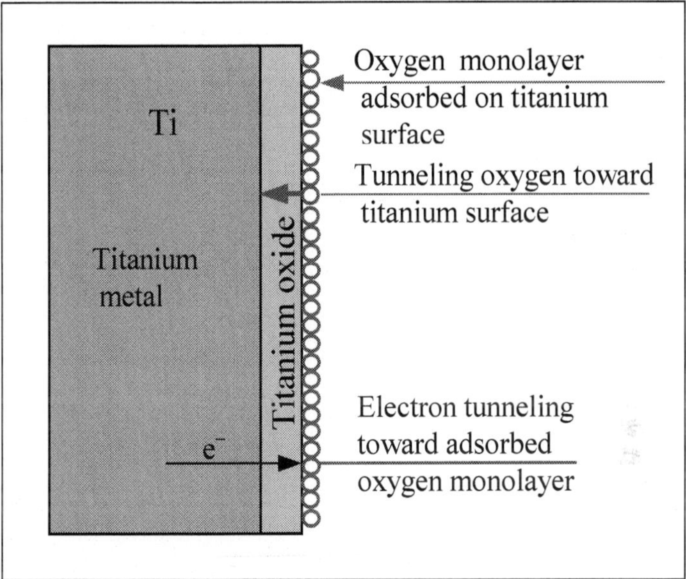

Fig. 1. Oxide growth on titanium.

2.2 Passive film components

The composition of titanium oxides depends on conditions and chemical media in which oxide is created. There are three main forms of titanium oxide which can exist on titanium separately or simultaneously in different proportions. The three main forms of titanium oxides are rutile-tetragonal crystals, anatase – also tetragonal crystals, but of more amorphous form and brookite-orthorombic crystals. To lesser extent Ti_2O_3, Ti_3O_5 and some hydrated oxides can also form on titanium surface. The naturally created oxide is mainly in TiO_2 form, transparent, not visible to eye and less than 10 nanometer thick (Alekseeva, 1964).

2.3 The processes altering the titanium surface

To improve corrosion resistance, biocompatibility, osseointegration, cleanability, overcome galling and seizing problems many processes were invented in last several decades including: acid etching, DC anodizing, AC spark anodizing (Rokicki, 1992), electropolishing, thermal oxidation, conversion coating (fluoride-phosphate), alloying of surface layer with palladium by thermal decomposition, laser irradiation, phothocatalytic treatment, plasma spraying for hydroxyapatite incorporation, and most recently proposed magnetoelectropolishing process (Rokicki, 2009). From all of those processes the most popular is DC anodization. This is very easy to perform not complicated electrochemical process which gives very attractive colorized finish to titanium in following colour succession: yellow, purple, dark blue, sky blue, greenish, golden-purple (mosaic), violet-greenish (mosaic), and gray. The colour succession is due to thickening of oxide by rising voltage. This process is very often utilized as finish of standard micro-rough CP Ti orthopaedic locking compression plates (yellow colour) because this finish gives very high osseointegration. This is most probably attributed to anatase form of titanium oxide which predominantly covers anodized surface of titanium (Gopal et al., 2003; Simka et al., 2011). Another recently more demanded finish for titanium is electropolished finish. The electropolishing dissolves existing natural oxide with all its imperfections and creates new more corrosion resistant oxide mainly in the form of rutile (Rokicki, 1990) according to following reactions:

1.   dissolution and transfer of titanium ions into solution

$$Ti = Ti^{4+} + 4e^-$$

2.   evolution of the oxygen from the anode surface

$$4OH^- = O_2 + 2H_2O + 4e^-$$

3.   formation of the passive film on the anode surface

$$Ti + 2OH^- = Ti_xO_y + H_2O + 2e^-$$

The oxide created by electropolishing process is very homogeneous with few dislocation sites. The advantage of this oxide over oxides created naturally on mechanically polished or chemically etched titanium surface was shown in experiment performed by the author over two decades ago (Rokicki, 1990). In that experiment two samples of titanium, one electropolished another chemically etched, were anodized to the same colour (golden-yellow which is first colour obtained during anodization). After one year of exposure to ambient atmosphere the colour on chemically etched titanium sample changed colour to purple (thickening of the passive film field). The colour on electropolished sample stays unchanged (and stayed unchanged to this day). Above experiment indicates that oxide on electropolished titanium resists further oxidation by ambient atmosphere. In the case of chemically etched titanium surface the oxide was very unstable and underwent further oxidation by ambient air. This fact supports the Cabrera & Mott (1948) theory of titanium oxide growth by oxygen movement toward titanium surface through existing oxide (Fig. 1).

3. Magnetoelectropolishing of titanium

Magnetoelectropolishing (MEP) appears to be an important and effective process for obtaining modified surface properties (Rokicki, 2009; Hryniewicz et al., 2008; Hryniewicz et al., 2009). In our studies, for comparison the electrolytic polishing was performed both in the absence and in the presence of a magnetic field. The experiments were carried out with the use of wires, commercial endodontic files, and flat samples (plates). For the MEP experiments, a constant external magnetic field below 500 mT was applied to the electropolishing (EP) system by neodymium ring magnets. For both processes, conventional EP and MEP, the same type of an acidic electrolyte was used, which was mixture of sulfuric, hydrofluoric and nitric acids. We use nitric acid addition to well known H_2SO_4/HF electrolyte composition for electropolishing titanium as a precaution from possibility of hydrogen adsorption during electropolishing. This method was successfully applied by Higuchi & Sato, (2003). Decreased hydrogen concentration in the stainless steel samples after MEP in comparison with EP, and MP ones, was recently reported by Hryniewicz et al., (2011).

The bath was unstirred during the process carried out with absence of externally applied magnetic field. During magnetoelectropolishing the stirring was self imposed by Lorentz Force as a result of interaction of electric and magnetic fields. For comparison, also Ti samples after mechanical polishing using an abrasive paper of the grit size up to 1000 were used.

4. Towards improving the titanium biomaterials

Much attention has been concentrated on improving the properties of titanium biomaterial (Schenk, 2001; Rokicki, 1992; Gopal et al., 2003; Simka et al., 2011; Rokicki, 1990; Hryniewicz et al., 2009; (2) Hryniewicz et al., 2009; Schultz & Watkins, 1998; Virtanen et al., 2008; Buly et al., 1994; La Budde et al., 1994; Burstein et al., 2005; Virtanen & Curty, 2004; Burstein & Souto, 1995; Mickay & Mitton, 1995; Khan et al., 1999; Hanawa et al., 1998; Budzynski et al., 2006; Hayes et al., 2010). Osseointegration of a metallic implant into bone or adaptation in soft tissue involves many complex physiological reactions related both to the material itself and to the living host. It was thus realized that during implantation a hydrated oxide layer grows on titanium, suggested to be due to the metabolic activity at the site of implantation (Virtanen et al., 2008). A prerequisite for clinical success of orthopaedic and dental implants is a strong and long-lasting connection between the implant and bone. Surface roughness has been suggested as one important factor for establishing clinically reliable bone attachments (Buly et al., 1994; La Budde et al., 1994; Burstein et al., 2005; Virtanen & Curty, 2004; Burstein & Souto, 1995). Implant-related factors, mechanical loading, surgical technique, implant site and patient variables influence bonding between implants and bone (Buly et al., 1994). Several authors suggest methods to modify the surface structure of titanium implants, which all may lead to altered chemical and mechanical properties of the metal surface (La Budde et al., 1994; Mickay & Mitton, 1995). Textured implant surfaces can be produced by several methods, all of which provide different characteristics of the biomaterial surface. It is not clear what influence these characteristics have on the bone response after implantation, or to what extent the response depends on geometry and degree of the surface roughness (Khan et al., 1999). Results from in vitro studies suggest a positive correlation between surface roughness and cellular attachment and osteoblast-like cell activity (Duisabeau et al., 2004).

5. Comparison of surface roughness

Definitions of roughness parameters are given in accordance to Polish Standard: PN-EN ISO 4287 (1999),

R – parameter calculated from the roughness profile

Ra – is the arithmetic mean of the sum of roughness profile values

$$Ra = \frac{1}{l}\int_0^l |z(x)|\,dx \tag{1}$$

where $|z(x)|$ – absolute ordinate value inside the elementary measuring length, with $l = lp$, and lp – elementary length in x direction (on average line) used for identification of unevenesses characterizing profile under evaluation,
R_z – the height of roughness acc. to 10 points as the measure of the range of roughness values in the profile. It is generally determined as the mean of 5 single measuring lengths of the roughness profile it corresponds to the mean peak-to-valley height,
Rt – the total height of the primary roughness profile defined as the difference between height of the highest profile peak and the depth of the lowest profile valley of the respective profile within the evaluation length ln ($l = ln$).
A computerized HOMMEL TESTER T800 system of Hommelwerke GmbH for roughness measurement was used for the study of surface roughness. The comparison of roughness studies results after both conventional electropolishing (EP) and magnetoelectropolishing (MEP) were carried out on the Ti samples.
Surface roughness measurements on CP Ti Grade 2 samples were performed both on wires and plates after standard electropolishing (EP), and magnetoelectropolishing (MEP), with mechanically abrasive polishing (MP) samples serving as a reference. Comparison of the R_z, R_zISO, Rt, and Ra results obtained after MP, EP, and MEP are presented in Fig. 2. Dependent on the treatment proposed, a decreasing surface roughness is observed with MEP roughness being the least (Figs. 2a,b,c).
The maximum height of scale limited surface Sz is the sum of the largest peak height value and the largest pit depth value within a definition area. The arithmetic mean height Sa is the arithmetic mean of the absolute of the height within a definition area

$$Sa = \frac{1}{A}\int_A |z(x,y)|\,dx dy \tag{2}$$

with A being the definition area (Standard ISO 25178-2, 2008).
The interferometric roughness studies were performed on CP Ti Grade 2 strips after MP, EP, and MEP (Fig. 3). The obtained results are even more pronounced with detailed data given in Table 1. Even if Sa surface roughness parameter measured after EP is very low (of 88%) in comparison with Sa after MP, this parameter of Ti sample surface after MEP is still reduced over 12%.
Our previous studies performed on MP, EP, and MEP sample surfaces of different metallic biomaterials indicated decreasing roughness, evaluated both by 2D standard roughness measurements (Ra, R_z, Rt), as well as 3D interferometry measurements (Sa, Sz), (Hryniewicz & Rokosz, 2009) as well as presented in our works elsewhere (Hryniewicz & Rokicki, 2007; (2) Hryniewicz & Rokicki, 2007; Hryniewicz et al., 2007). Consequently, regarding reduction in hydrogenation, the results obtained by SIMS are in agreement with the sample surface roughness data concerning their mode of treatment (Hryniewicz et al., 2011).
Some of the studies on surface roughness performed by other technique (Atomic Force Microscopy, AFM) on Nitinol (Fig. 4) also confirm improvement of MEP treated samples against EP ones (Rokicki et al., 2008; Hryniewicz & Rokicki, 2008).

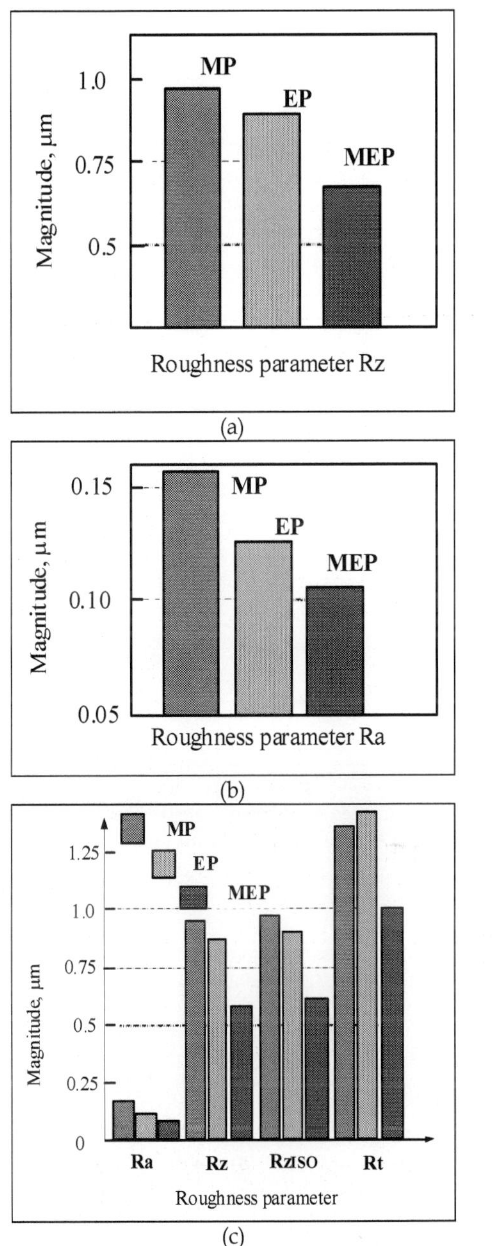

Fig. 2. Surface roughness parameters measured on a CP Ti Grade 2 wire: (a) R_z, (b) Ra, (c) comparison of Ra, R_z, R_zISO, Rt.

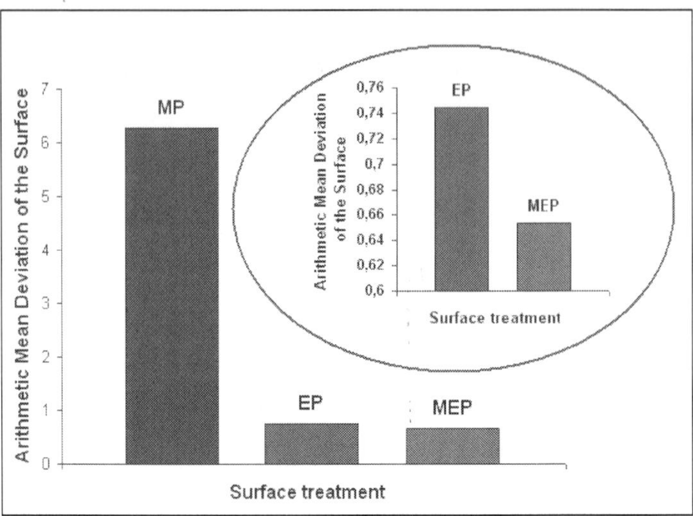

Fig. 3. *Sa* surface roughness parameter measured on a CP Ti Grade 2 strip after MP, EP, and MEP.

Treatment	Sa
MP	6.28
EP	0.745
MEP	0.654

Table 1. *Sa* data of interferometry studies performed on MP, EP, and MEP samples of CP Ti Grade 2.

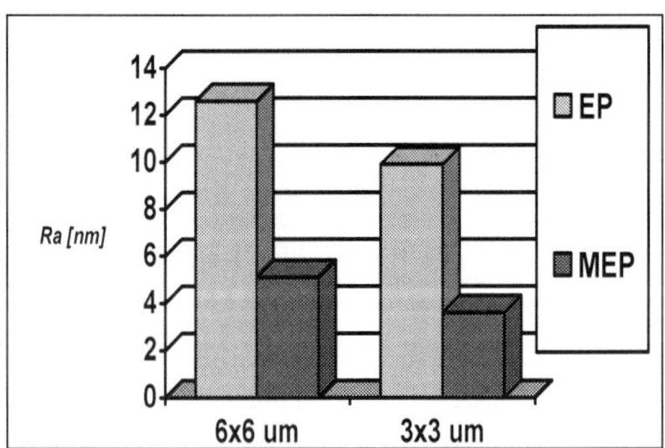

Fig. 4. Average surface roughness *Ra* parameter of Nitinol sample after EP, and MEP, measured by AFM (Rokicki et al., 2008).

6. Wettability and microscopic studies

The orthopaedic titanium implants can be divided in two groups. One group of implant consists of devices which after implantation will stay permanently in the body (spine implants). Second group covers implants which are temporarily implanted and after healing is over those devices are removed from the body (fracture fixation devices as bone plates, nail etc). The one of most desired property for permanent implants are their complete and strong osseointegration. For temporary devices the strong osseointegration is not desirable because it complicates removal procedure, which can lead to unnecessary blood loss, infection or even fracture. To accomplish this goal the metallic implants require different surface finish which was shown in recent work of Hayes et al., (2010).

It is well known that wettability of metallic implant surfaces plays great role in their bio integrity with surrounding tissues in titanium case most importantly with bones. In our study electropolished and magnetoelectropolished surfaces had shown very different wettability (Rokicki & Hryniewicz, 2008). The simple experiment with drop of water dragged upon the EP and MEP surface have shown very different behaviour. Dragged droplet of water on EP surface was leaving the trace behind; this indicates that surface is very hydrophilic. Contrary dragged droplet of water on magnetoelectropolished surface did not leave any traces and this indicates very hydrophobic surface property. Taking under consideration finding of Hayes et al. (2010) that electropolished titanium surface leads to lesser bone integration we hypothesize that more hydrophobic in our case magnetoelectropolished titanium surface should lead to better and more fuller osseointegration which will be very desirable for permanent orthopaedic implants.

7. Corrosion measurements

A metal corrodes if the electrode potential $E = E_{corr}$ is the corrosion potential and flowing current $i_a = -i_c = i_{corr}$ in the created cell is the corrosion current. The relationship between current density and potential of anodic and cathodic electrode reactions under charge transfer control is given by the Butler-Volmer equations:

$$i_a = i_{oa} \exp\left[2.303 \frac{E - E_{oa}}{b_a} \right] \tag{3}$$

$$i_c = i_{oc} \exp\left[2.303 \frac{E - E_{oc}}{b_c} \right] \tag{4}$$

where:
b_a, b_c – Tafel constants
i_{oa}, i_{oc} – exchange current densities for the anodic and cathodic processes, respectively
E_{oa}, E_{oc} – equilibrium potentials for the anodic and cathodic processes, respectively.
Summary current density:

$$i = i_a + i_c \tag{5}$$

then

$$i = i_{corr}\left(\exp 2.303\frac{\Delta E}{b_a} - \exp 2.303\frac{\Delta E}{b_c}\right) \tag{6}$$

where

$$\Delta E = E - E_{corr} \tag{7}$$

$$\left(\frac{\partial i}{\partial E}\right)_{E=E_{corr}} = R_t^{-1} = 2.303 i_{corr}\left(\frac{1}{b_a} + \frac{1}{b_c}\right) \tag{8}$$

and

$$i_{corr} = \frac{b_a b_c}{2.303(b_a + b_c)}\left(\frac{\partial i}{\partial E}\right)_{E=Ecorr} \tag{9}$$

The corrosion rate in the anodic sites on the metal surface is proportional to the current intensity. The current intensity flowing through the current cell results in change of potentials of both anodic (oxidation) and cathodic (reduction) reactions (Hryniewicz et al., 2009). The cathodic potential shifts into negative direction and anodic potential shifts into positive direction.

The GPES (General Purpose Electrochemical Software) provides a convenient interface for making Tafel plots, calculating Tafel slopes and corrosion rates. First we specify the anodic and cathodic Tafel region. Once the regions are selected the GPES software automatically calculates the Tafel slopes and the corrosion currents. A correct estimate of the Tafel slopes is possible only if the linear Tafel region covers at least one decade in current. Some additional data are the density of investigated material, equivalent mass, and the studied surface area. Having these the algorithm serves to calculate corrosion current density, polarization resistance, and the corrosion rate CR.

The corrosion studies of conventionally electropolished (EP) and magnetoelectropolished (MEP) titanium samples below oxygen evolution regime were carried out in a Ringer's solution (Table 2) at 25 °C (Figs. 5, and 6). Ringer's (*Solution Ringeri* by Fresenius Kabi) and Hank's solutions are the main artificially created human fluids generally recognized by the researchers to carry out the corrosion studies on biomaterials.

Solution components	g/dm³	Ions	mEq/dm³	mmol/dm³
Sodium chloride	8.60	Na⁺	147.16	147.16
Potassium chloride	0.30	K⁺	4.02	4.02
Calcium chloride	0.48	Ca²⁺	4.38	2.19
		Cl⁻	156.56	156.56

Table 2. Ringer's solution composition used for the corrosion study.

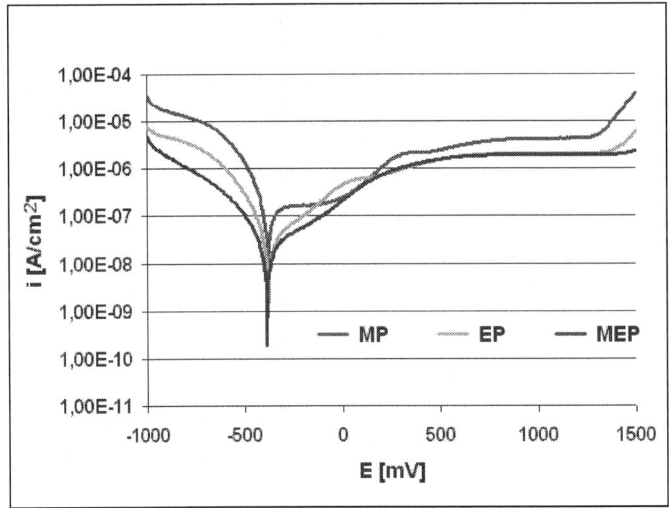

Fig. 5. Polarization cures of Ti samples in Ringer's solution after: MP – abrasive polishing, EP – standard electropolishing, MEP – magnetoelectropolishing.

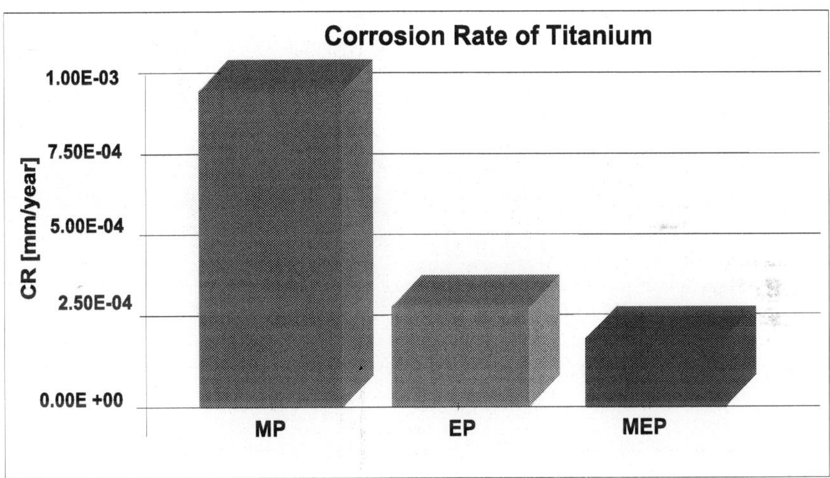

Fig. 6. Comparison of corrosion rates of Ti samples in Ringer's solution after: MP – abrasive polishing, EP – standard electropolishing, MEP – magnetoelectropolishing.

In the corrosion studies, polarization curves obtained on Ti samples in Ringer's solution (Fig. 5) indicated a significant differentiation in their courses dependent on the treatment mode considered. Calculated corrosion rates CR (Fig. 6) reveal even more clearly the decrease in CR after electropolishing (EP), and further reduction in CR is observed after MEP process.

The microscopic observations of the Ti samples after EP, MEP and MP were performed to compare the results of the study ((2) Hryniewicz et al., 2009). The pictures which were taken by Neophot and SEM very well characterize visually the Ti surface after each of the treatment performed. The example images of the wire samples are presented in Fig. 7. Changes in surface topography after MP, EP, and MEP may be viewed, showing the smoothest surface after MEP (Fig. 7c). Fig. 7a presents the typical surface pattern after abrasive pretreatment, whereas Ti image after EP (Fig. 7b) shows multiple irregularities and cavities. The reason of increased fatigue resistance of titanium biomaterial after MEP was due to different plastic behaviour of the same Ti wire after two electropolishing treatments, EP and MEP (Hryniewicz et al., 2009).

8. XPS surface film composition studies

Surface film composition on wire samples of CP Ti (commercial purity titanium) Grade 2 after a standard electropolishing (EP), magnetoelectropolishing (MEP), compared with mechanically polished (MP) samples was studied and the results presented in consecutive Figs. 8, 9, 10, and in Table 3. In Fig. 8, the high resolution XPS (X-ray Photoelectron Spectroscopy) spectra of CP-Ti samples after MP, EP, and MEP, are compared. Characteristic peaks of titanium and oxygen may be observed, with a carbon peak coming from environment, occurring always in this kind of studies.

In Figs. 9a-f, the results of surface film composition measured on titanium samples are presented. Figs. 9a,b show the results after MP, Figs. 9c,d – after EP, and Figs. 9e,f – after MEP. In Figs. 9 (a, c, e) spectra are referred to titanium (Ti^0, Ti^{2+}, Ti^{3+}, Ti^{4+}), and in Figs. 9 (b, d, f) spectra are referred to oxygen group (O^{2-}, OH^-, H_2O).

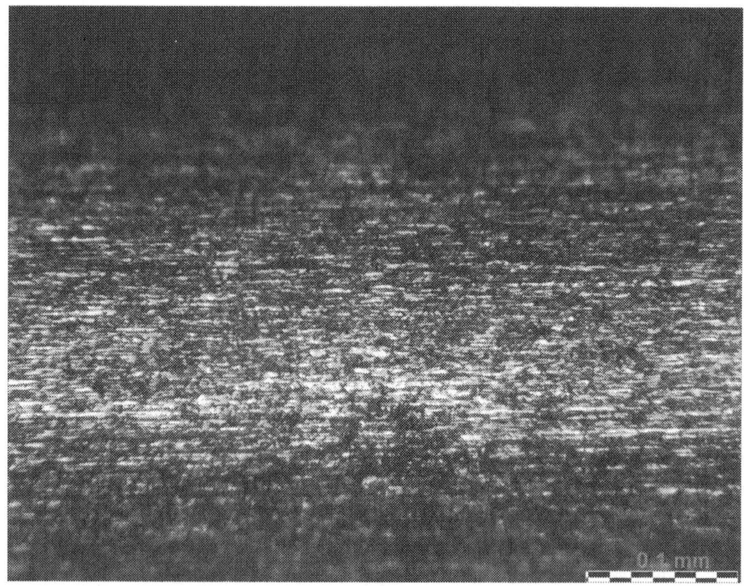

(a)

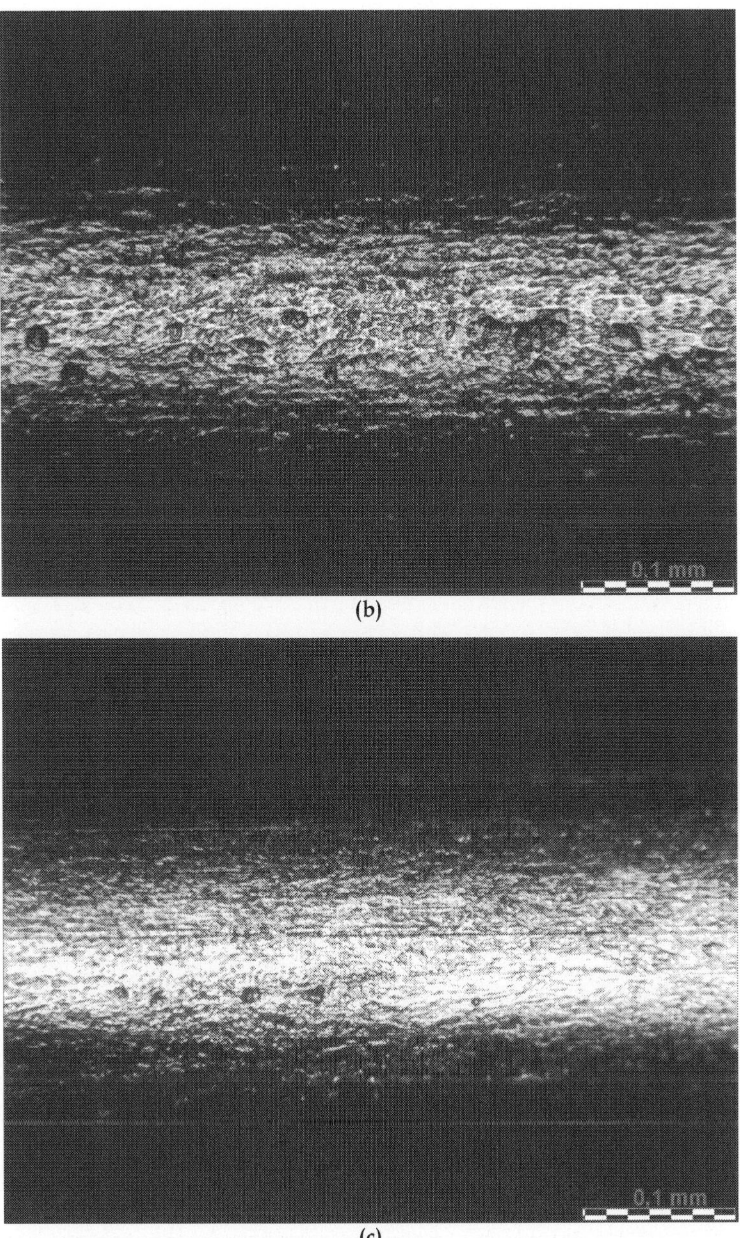

Fig. 7. Micrographs of Ti wire sample surfaces after: (a) MP – abrasive polishing, (b) EP – standard electropolishing, (c) MEP – magnetoelectropolishing.

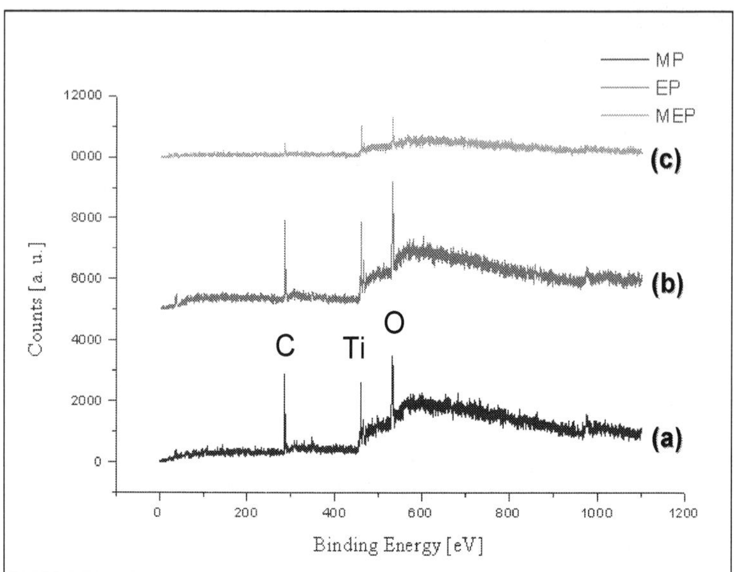

Fig. 8. High resolution XPS spectra of Ti samples (wires) comparison obtained on surface: (a) as-received MP, (b) after EP, (c) after MEP.

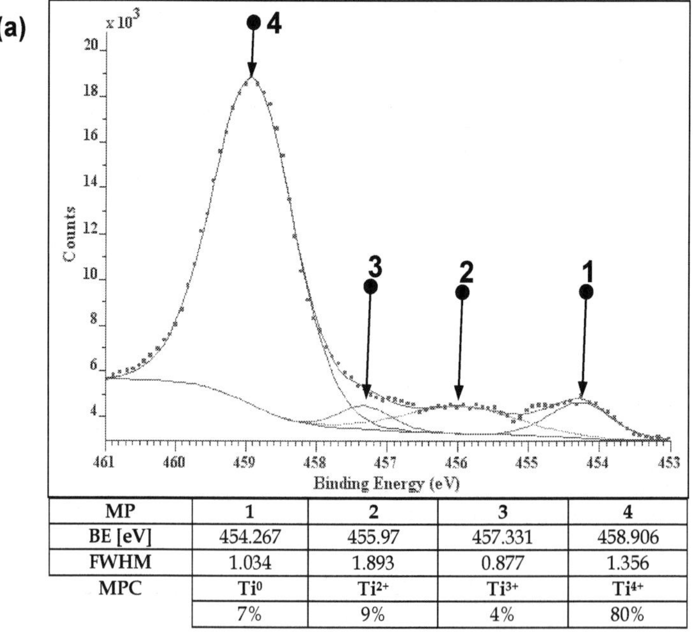

MP	1	2	3	4
BE [eV]	454.267	455.97	457.331	458.906
FWHM	1.034	1.893	0.877	1.356
MPC	Ti^0	Ti^{2+}	Ti^{3+}	Ti^{4+}
	7%	9%	4%	80%

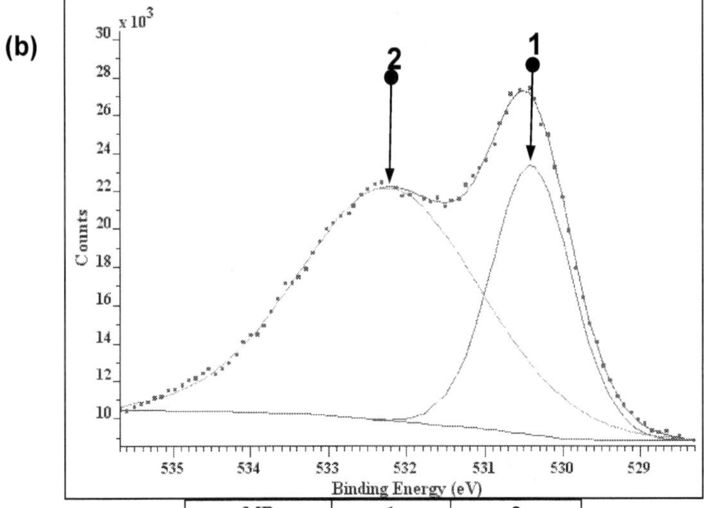

MP	1	2
BE [eV]	530.4	532.226
FWHM	1.215	2.732
MPC	O^{2-}	H_2O
	34%	66%

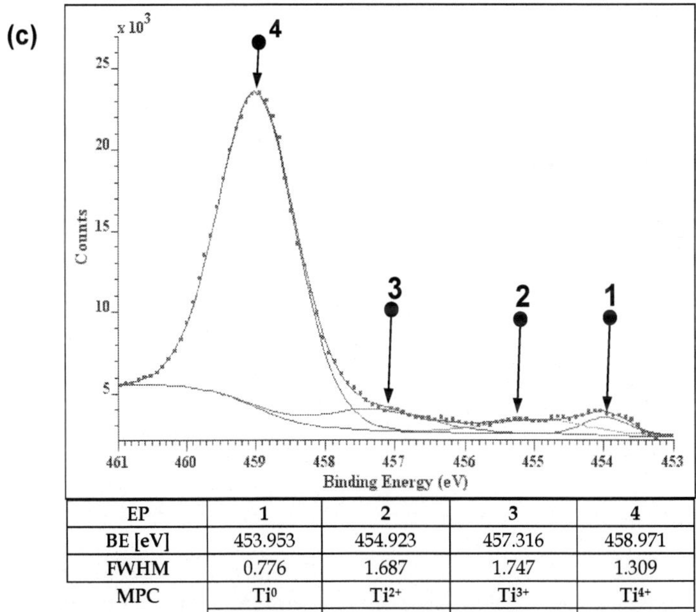

EP	1	2	3	4
BE [eV]	453.953	454.923	457.316	458.971
FWHM	0.776	1.687	1.747	1.309
MPC	Ti^{0}	Ti^{2+}	Ti^{3+}	Ti^{4+}
	3%	5%	8%	84%

(d)

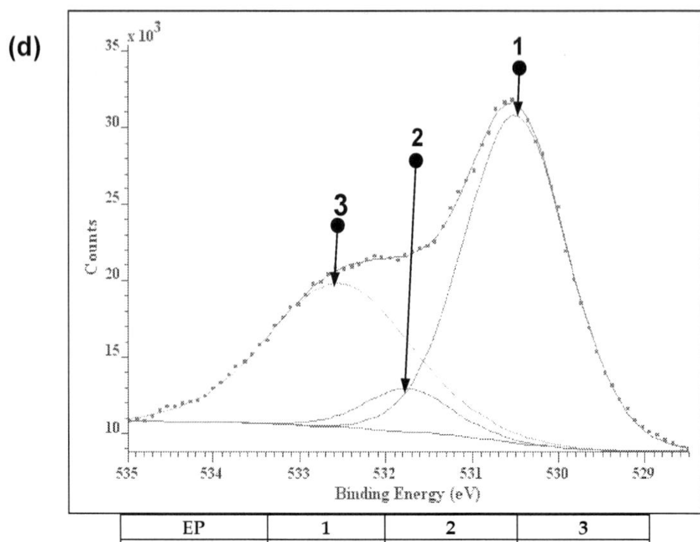

EP	1	2	3
BE [eV]	530.498	531.738	532.544
FWHM	1.408	1.172	1.962
MPC	O^{2-}	OH^-	H_2O
	58%	8%	34%

(e)

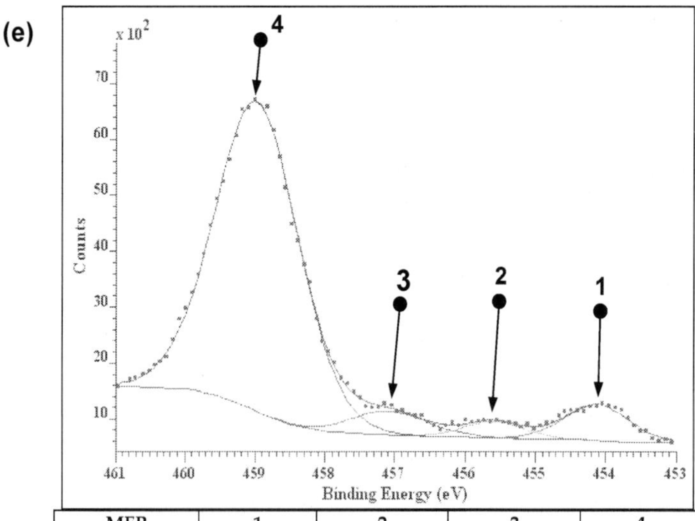

MEP	1	2	3	4
BE [eV]	454.126	455.589	457.092	458.978
FWHM	1.057	0.964	1.256	1.391
MPC	Ti^0	Ti^{2+}	Ti^{3+}	Ti^{4+}
	7%	3%	6%	84%

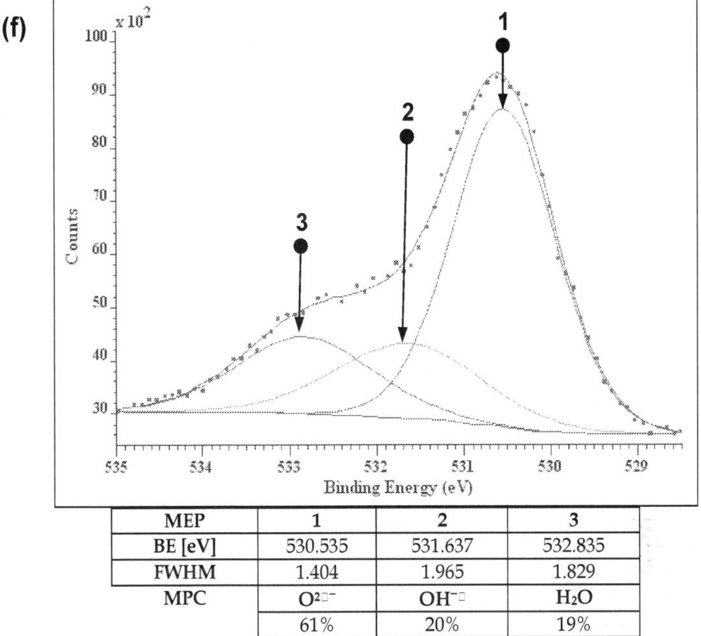

MEP	1	2	3
BE [eV]	530.535	531.637	532.835
FWHM	1.404	1.965	1.829
MPC	O^{2-}	OH^-	H_2O
	61%	20%	19%

Fig. 9. Surface film composition measured by XPS of the treated CP Ti Grade 2 for 3 samples (wires) after: (a, b) mechanical polishing MP, (c, d) electrochemical polishing EP, (e, f) magnetoelectropolishing MEP; (a, c, e) spectra referred to titanium (Ti^0, Ti^{2+}, Ti^{3+}, Ti^{4+}), (b, d, f) spectra referred to oxygen group (O^{2-}, OH^-, H_2O). Under description:

- **BE** – binding energy
- **FWHM** – an expression of the extent of a function, given by the difference between the two extreme values of the independent variable at which the dependent variable is equal to half of its maximum value
- **MPC** – most probably compound

Each of the presented plots in Fig. 9 has been connected with the underlying Table 3 revealing all important data, indicating the most probable compound (MPC) with the relevant percentage in Ti surface film investigated.

Treatment	Ti^0	Ti^{2+}	Ti^{3+}	Ti^{4+}
MP	7%	9%	4%	80%
EP	5%	5%	8%	84%
MEP	7%	3%	6%	84%

Table 3. XPS data of the treated CP Ti Grade 2 for 3 samples (strips).

The most interesting spectra referred to titanium (Ti^0, Ti^{2+}, Ti^{3+}, Ti^{4+}), concerning Fig. 9 (a, c, e), have been gathered and reported in Table 3. Visual presentation of the results is given in

Fig. 10. In Fig. 10a the comparison with full presentation is given, and in Fig. 10b there are shares excluding Ti^{4+} reported. Depending on the surface treatment used, characteristic differentiation in composition of surface film is apparent.

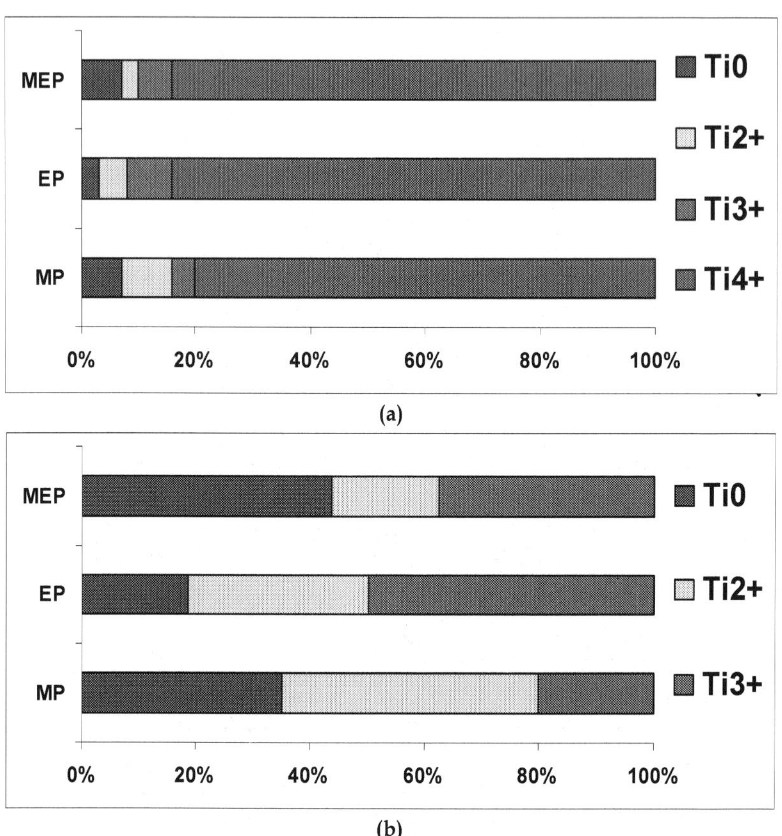

Fig. 10. Surface film composition by XPS of the treated CP Ti Grade 2 for 3 samples (strips): (a) shares with full representation, (b) shares excluding Ti^{4+}.

9. Fatigue resistance

Our extensive studies over metallic biomaterials show that the kind of finish affects also some mechanical properties. The experiments carried out on samples of CP-titanium and Ti-alloys (Nitinol) indicate that electropolishing (EP) greatly affects the resistance to bending in advantage of this process against abrasive polishing (MP), and the results obtained on samples after MEP are better than those recorded after a standard EP process (Fig. 11). Our research study results performed on fatigue resistance according to the Polish Standard (1975) have been also confirmed by another team of authors on finished by us commercial endodontic files (Praisarnti et al., 2010).

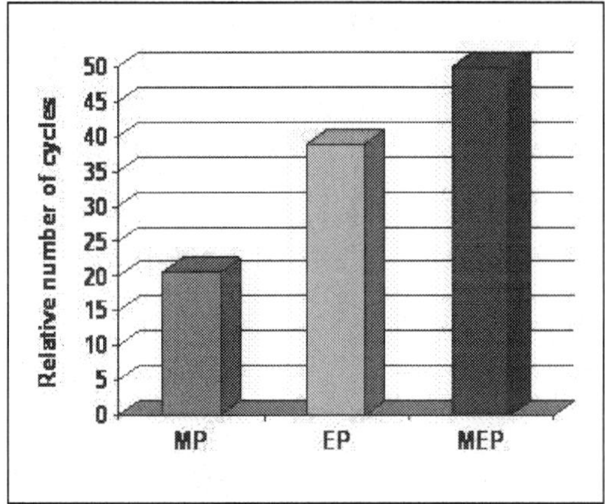

Fig. 11. Relative resistance to bending of Ti samples after: MP – abrasive polishing, EP – standard electropolishing, MEP – magnetoelectropolishing.

10. Conclusion

Selection of titanium and titanium alloys with relatively high strength for critical high-temperature, corrosive environment application in the twentieth century was limited to those commercial alloys developed and designed specifically for aerospace use. Most commercial titanium and Ti alloys were not fully resistant to localized attack and/or stress corrosion in chloride environments.

Titanium (and its alloys), though long known, has found favour in recent years as biomaterial. As a biomaterial, titanium is widely used for surgical and medical implants, both skeletal and dental. The state of the surface finish is critical, and numerous studies have shown the benefits of electropolishing, with advantageous using of magnetic fields.

The application of externally applied magnetic fields to the electropolishing process provides the super-critical refinement of surface properties to the new high level required for medical implant devices. Improved corrosion resistance of magnetoelectropolished titanium surface is caused by more homogenized amorphous mixture of titanium oxides and hydroxides compared with very crystalline titanium oxide mainly in rutile form on conventionally electropolished surface.

The following final conclusions may be drawn after investigation of titanium biomaterial:

a. a new improved technology effective for titanium biomaterial has been offered

b. the surface roughness after MEP is much less than after a standard EP; all detected roughness parameters after MEP were better than those after EP

c. very good corrosion characteristics have been obtained with the best performance corrosion potential after MEP; the highest corrosion resistance has been obtained on samples after MEP (CR=1.84×10^{-4} mm/year), the worst one after MP (CR=9.63×10^{-4} mm/year), and CR=2.67×10^{-4} mm/year after EP (Figs. 5 and 6)

d. Ti biomaterial samples after MEP appeared to be the most resistant to fracture, of about 27% better than the same Ti samples after EP, and much better than after abrasive polishing MP (Fig. 11).

It was found that titanium biomaterial properties may be well improved after using a new technology utilizing the magnetic field during electrolytic polishing. This study shows that both titanium and titanium alloys for medical applications may be considerably improved.

11. Acknowledgment

Authors of this paper highly acknowledge Prof. Steinar Raaen, Department of Physics, NTNU Trondheim, Norway, for making available the XPS apparatus used for surface film composition studies.

12. References

Schenk, R. (2001). The corrosion properties of titanium and titanium alloys, In: *Titanium in Medicine,* Brunette D.M., Tengvall P., Textor M., Thoms P., (Eds.), 145-170, Berlin, Springer-Verlag

Ruzickova, M.; Hildebrand, H. & Virtanen, S. (2005). On the Stability of Passivity of Ti-Al Alloys in Acidic Environment. *Zeitschrift für Physikalische Chemie*, Vol.219(11), pp. 1447–1459

Singh, R. &, Dahotre, N.B. (2007). Corrosion degradation and prevention by surface modification of biometallic materials. *J. Materials Science: Materials in Medicine*, Vol.18(5), pp. 725-751

Salvati, E.A.; Lieberman, J.R.; Huk, O.L. & Evans, B.G. (1995). Complications of femoral and acetabular modularity. *Clin. Orthop. Related Res.*, Vol.319, pp. 85–93

Cabrera, N. & Mott, N.F. (1948). Theory of Oxidation of Metals. *Rept. Progr. Phys.*, Vol.12, pp. 163-181

Andreeva, V.V. (1964). Behavior and nature of thin oxide films on some metals in gaseous media and in electrolyte solutions. *Corrosion,* Vol.20, p. 351

Rokicki, R. (1992). Conversion Coating of Titanium and its Alloy Using Alternating Current. *Metal Finishing,* April, pp. 9-10

Rokicki, R. (2009). Apparatus and method for enhancing electropolishing utilizing magnetic field. US Patent No 7632390, http://www.patengenius.com/patent/7632390.html

Gopal, J.; Muraleedharan, P.; George, P. & Khatak H.S. (2003). Investigations of the antibacterial properties of an anodized titanium alloy. *Trends Biomater. Artf. Organs*, Vol.17(1), pp. 13-18

Simka, W.; Nawrat, G.; Maciej, A.; Nieużyła, Ł. & Michalska, J. (2011). Badania procesu pasywacji anodowej tytanu. *Przemysł Chemiczny*, Vol.90(1), pp. 98-102

Rokicki, R. (1990). The passive oxide film on electropolished titanium. *Metal Finishing*, Vol. 88(2), pp. 65-66

Hryniewicz, T.; Rokicki, R. & Rokosz, K. (2009). Corrosion and Surface Characterization of Titanium Biomaterial after Magnetoelectropolishing. *Surface and Coatings Technology*, Vol. 203(10-11), pp. 1508-1515, published online 2008, available from http://dx.doi.org/10.1016/j.surfcoat.2008.11.028)

Hryniewicz, T., Rokicki, R. & Rokosz, K. (2008). Co-Cr alloy corrosion behaviour after electropolishing and "magnetoelectropolishing" treatments. *Materials Letters,* Vol.62, pp. 3073-3076

Hryniewicz, T.; Rokosz, K. & Rokicki, R. (2009). Surface investigation of NiTi rotary endodontic instruments after magnetoelectropolishing, *MRS Proceedings of 18th*

International Materials Research Congress, Vol. 1244E, pp. 21-32, ISBN 978-1-60511-221-3, 9. Biomaterials, Cancun, Mexico, 16-20 August 16-20, 2009, Available from http://www.mrs.org/s_mrs/sec_subscribe.asp?CID=24949&DID=280254

Hryniewicz, T.; Konarski, P.; Rokosz, K. & Rokicki, R. (2011). SIMS analysis of hydrogen content in near surface layers of AISI 316L SS after electrolytic polishing under different conditions. *Surface & Coatings Technology*, Vol.205, pp. 4228-4236 doi:10.1016/j.surfcoat.2011.03.024

Schultz, R.W. & Watkins, H.B. (1998). Recent developments in titanium alloy application in the energy industry. *Materials Science and Engineering A*, Vol.243, pp. 305-315

Virtanen, S.; Milosev, I.; Gomez-Barrena, E.; Trebse, R.; Salo, J. & Konttinen, Y.T. (2008). Special modes of corrosion under physiological and simulated physiological conditions. *Acta Biomaterialia*, Vol.4, pp. 468-476

Buly, R.L.; Huo, M.H.; Salvati, E.; Brien, W. & Bansal, M. (1994). Titanium wear debris in failed cemented total hip arthroplasty. An analysis of 71 cases. *J. Arthroplasty*, Vol.9, pp. 315-323

La Budde, J.K.; Orosz, J.F.; Bonfiglio, T.A. & Pellegrini Jr, V.D. (1994). Particulate titanium and cobalt-chrome metallic debris in failed total knee arthroplasty. A quantitative histologic analysis. *J. Arthroplasty*, Vol.9, pp. 291-304

Burstein, G.T.; Liu, C. & Souto, R.M. (2005). The effect of temperature on the nucleation of corrosion pits on titanium in Ringer's physiological solution. *Biomaterials*, Vol.26 pp. 245-256

Virtanen, S. & Curty, C. (2004). Metastable and stable pitting corrosion of titanium in halide solutions. *Corrosion*, 60643-60649

Burstein, G.T. & Souto, R.M. (1995). Observations of localized instability of passive titanium in chloride solution. *Electrochimica Acta*, Vol.40, pp. 1881-1888

Mickay, P. & Mitton, D.B. (1985). An electrochemical investigation of localized corrosion on titanium in chloride environments. *Corrosion*, Vol.41, pp. 52-62

Khan, M.A.; Williams, R.L.; Williams, D.F. (1999). Conjoint corrosion and wear in titanium alloys, *Biomaterials*, Vol.20, 765-772

Hanawa, T.; Asami, K.; Asaoka, K. (1998). Repassivation of titanium and surface oxide film regenerated in simulated bioliquid. *J. Biomed. Mater. Res.*, Vol.40, pp. 530-538

Budzynski, P.; Youssef, A.A. & Sielanko, J. (2006). Surface modification of Ti-6Al-4V alloy by nitrogen ion implantation. *Wear*, Vol.26(11-12), pp. 1271-1276

Duisabeau, L.; Combrade, P. & Forest B. (2004). Environmental effect on fretting of metallic materials for orthopaedic implants. *Wear*, Vol.256, pp. 805-816

Higuchi, T. & Sato, K. (2003). Hydrogen Absorption in Electropolishing of Niobium. *Hydrogen in Materials & Vacuum Systems : First International Workshop on Hydrogen in Materials and Vacuum Systems. AIP Conference Proceeding*, Vol.671, pp. 203-219

Hayes, J.S.; Seidenglanz, U.; Pearce, A.I.; Pearce, S.G.; Archer, C.W. & Richards, R.G. (2010). Surface Polishing Positively Influences Ease of Plate and Screw Removal. *European Cells and Materials*, Vol.19, pp. 117-126

Polish Standard: PN-EN ISO 4287 (1999). Specyfikacje geometrii wyrobów – Struktura geometryczna powierzchni: metoda profilowa – Terminy, definicje i parametry struktury geometrycznej powierzchni

Standard ISO 25178-2 (2008). Geometrical product specification (GPS) – Surface texture: Areal – Part 2: Terms, definitions and surface texture parameters

Hryniewicz, T. & Rokicki, R. (2007). Progress in Surface Treatment of Biomaterials. *Proc. of 4th International Congress on Precision Machining ICPM 2007* (eds. S. Adamczak, K. Stępień), Vol. 2, Sandomierz-Kielce, September 2007, pp. 159-164

Hryniewicz, T. & Rokicki, R. (2007). On the surface modification of stainless steels. Proceedings of 15th Annual Intl. Conference on Composites/Nano-Engineering, ICCE-15, 15-21 July, 2007, Haikou, Hainan Island, China, ed. by David Hui, pp. 345-346

Hryniewicz, T.; Rokicki, R. & Rokosz, K. (2007). Magnetoelectropolishing for metal surface modification. *Transactions of the Institute of Metal Finishing*, Vol.85(6), pp. 325-332

Hryniewicz, T. & Rokosz, K. (2009). On the Wear Inspection and Endurance Recovery of Nitinol Endodontic Files. *PAK (Pomiary Automatyka Kontrola)*, Vol.55(4), pp. 247-250

Rokicki, R.; Hryniewicz, T. & Rokosz, K. (2008). Modifying Metallic Implants with Magnetoelectropolishing. *Medical Device & Diagnostic Industry*, Vol.30(1), pp. 102-111 (INVITED PAPER)

Hryniewicz, T. & Rokicki, R. (2008). Improved surface properties of Nitinol after magnetoelectropolishing. *Proceedings of 16th Annual Intl.Confer. on Composites/Nano-Engineering, ICCE-16*, 20-26 July, 2008, Kunming, China (City of Eternal Spring), ed. by David Hui, Extended Abstract, (Poster Session)

Rokicki, R. & Hryniewicz, T. (2008). Nitinol™ Surface Finishing by Magnetoelectro-polishing. *Transactions of the Institute of Metal Finishing*, Vol.86(5), pp. 280-285

Polish Standard PN-75/M-80002: Technological testing of wires and rods (1975)

Praisarnti, C.; Chang J.W.W. & Cheung G.S.P. (2010). Electropolishing Enhances the Resistance of Nickel-Titanium Rotary Files to Corrosion-Fatigue Failure in Hypochlorite. *Journal of Enfdodontics*, Vol.36(8), 1354-1357

Specyfikacje geometrii wyrobów – Struktura geometryczna powierzchni: metoda profilowa – Terminy, definicje i parametry struktury geometrycznej powierzchni

12

Low Modulus Titanium Alloys
for Inhibiting Bone Atrophy

Mitsuo Niinomi
Institute for Materials Science, Tohoku University,
Japan

1. Introduction

Metallic biomaterials constitute approximately 70% - 80% of all implant materials, and thus represent a very important class of biomaterials. Among the various metallic biomaterials such as stainless steels and Co-Cr alloys, titanium and its alloys exhibit the best biocompatibility. Consequantly, titanium and its alloys have been the fucus of attension for biomedical materials, especially for use in load bearing implants such as artificial hip joints, bone plates and screws, spinal instruments, and dental implants.

The Young's moduli of metallic biomaterials are generally much higher than that of the human cortical bone (hereafter, bone) (Niinomi, 2002a). If the Young's modulus of a load bearing implant made of metallic biomaterials is higher than that of the cortical bone, bone atrophy occurs because of the stress shielding between the implant and bone (Sumitomo, 2008). Stress shielding causes loosening of the implants such as artificial joints or bone re-fracture after extraction of the implants. Therefore, the Young's modulus of the metallic biomaterials must be equal to that of the natural bone. Titanium and its alloys possess lower Young's moduli than those of other mettalic biomaterials such as stainless steels and Co-Cr alloys. β-type titanium alloys, in particular, having a single β phase exhibit much smaller Young's moduli compared with α- and (α + β)-type titanium alloys (Niinomi, 2002b). As such, the β-type titanium alloys are particulary promissing for biomedical applications. A number of β-type titanium alloys composed of non-toxic and non-allergic elements with low Young's modulus have been developed and even more are currently under development (Niinomi, 2011a). Much of the current research in the biomaterials field has been directed at lowering the Young's modulus of the β-type titanium alloys for biomedical applications.

The mechanical biocompatibility factors such as fatigue strength, fretting fatigue strength, tensile properties, wear resistance, fracture toughness, etc., including the Young's modulus are very important factors for the practical application of the alllloys in biomaterials as well as the biological biocompatibilities (Niinomi, 2008a). Among the mechanical biocompatibility factors, the endurance, i.e., the fatigue strength is one of the most important factors (Niinomi, 20007). The development of metallic biomaterials with improved fatigue strength and a simaltaneously low Young's modulus is desirable for biomedical applications.

The effect of the Young's modulus of the metallic implant should be clarified by animal experiments prior to medical applications, and recently, these kinds of studies have begun to be implemented (Niinomi, 2002b).

A recent report highlited the contrasting requirements of patients versus surgeons for metallic implants. The rquirements of the patients dictate that the implants have a Young's modulus similar to that of the bone, whereas while the surgeons require a high Young's modulus for inhibiting springback both during and after the operation procedure (Nakai, 2011a). Titanium alloys that simultaneously satisfy the demands of both patients and surgeons are thus necessiated.

There is a demand to remove the implant when the bone fracture is healed, in which case, the adhesion of the implant to the bone must be weak enough to inhibit the refracture of the bone. This requires titanium alloys having poor bone conductivity, but excellent biocompatibility (Zhao, 2011).

This chapter introduces low modulus β-type titanium alloys and various methods of lowering the Young's modulus of the β-type titanium alloy, improving the strength and fatigue strength of the β-type titanium alloy while maintaining a low Young's modulus, the evaluation of the effect of the Young's modulus on bone atrophy using rabbits, titanium alloys with variable Young's moduli, and removable titanium alloys are described.

2. Low modulus β–type titanium alloys for biomedical applications

A number of β-type titanium alloys with low Young's modulus have been developed for use in the human body. The titanium alloys composed of safe alloying elements developed to date include the following; Ti-13Nb-13Zr, Ti-12Mo-6Zr-2Fe, Ti-15Mo, Ti-15Mo-5Zr-5Sn, Ti-15Mo-5Zr-3Al, Ti-16Nb-10Hf, Ti-15Mo-2.8Nb-0.2Si, Ti-30Ta, Ti-35Zr-10Nb, Ti-8Fe-8Ta, Ti-8Fe-8Ta-4Zr, Ti-35Nb-7Zr-5Ta, Ti-29Nb-13Ta-4.6Zr (TNTZ), and Ti-Nb-Sn system alloys (Niinomi, 2011a). Shape memory and super-elastic Ti-Nb based alloys have been also been developed; Ti-Nb, Ti-Nb-O, Ti-Nb-Sn, Ti-Nb-Al, Ti-22Nb-(0.5-2.0)O (at%),Ti-Nb-Zr, Ti-Nb-Zr-Ta, Ti-Nb-Zr-Ta-O, Ti-Nb-Ta-Zr-N, Ti-Nb-Mo, Ti-22Nb-6Ta(at%), Ti-Nb-Au, Ti-Nb-Pt, Ti-Nb-Ta, and Ti-Nb-Pd system alloys. Ti-Mo based alloys have been developed; Ti-Mo-Ga, Ti-Mo-Ge, Ti-Mo-Sn, Ti-Mo-Ag, Ti-5Mo-(2-5)Ag (at%), Ti-5Mo-(1-3)Sn (at%), in addition to Ti-Sc-Mo system alloys. The Ti-Ta based alloys are Ti-50Ta, Ti-50Ta-4Sn, and Ti-50Ta-10Zr. Other alloys such as Ti-7Cr-(1.5, 3.0, 4.5)Al super elastic and shape memory alloys, Gum Metal (Ti-25at% (Ta, Nb, V) + (Zr, Hf, O)), and Ti-9.7Mo-4Nb-2V-3Al super elastic alloys have also been developed (Niinomi, 2011b).

3. Further decreases in Young's modulus

Improvements in the static strength of biomaterials such as the tensile strength can be achieved by employing strengthening mechanisms including work hardening, grain refinement strengthening, precipitation strengthening, and dispersion strengthening. One of the best ways to increase tensile strength while maintaining a low Young's modulus is to introduce a number of dislocations into the alloy system by conventional severe cold working techniques such as severe cold rolling and swaging, and by special severe cold working techniques such as high pressure torsion (HPT), accumulated roll-bonding (ARB) and equal channel angular pressing (ECAP) (Yilmazer, 2009).

Figure 1 (Niinomi, 2010a) shows the relationships between the tensile properties and working ratio of Ti-29Nb-13Ta-4.6Zr (TNTZ) while Fig. 2 (Niinomi, 2010a) shows the relationship between the Young's modulus and working ratio of TNTZ after subjecting TNTZ to severe cold working by general severe cold rolling or swaging in both cases.

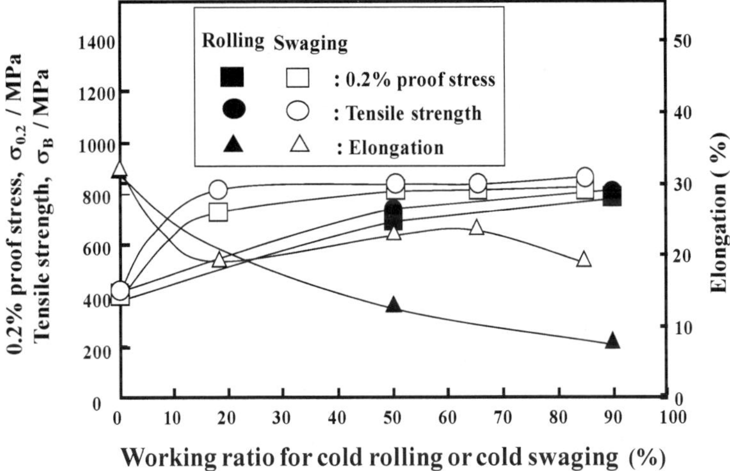

Fig. 1. Tensile properties of TNTZ subjected to cold rolling or cold sawaging as a function of working ratio.

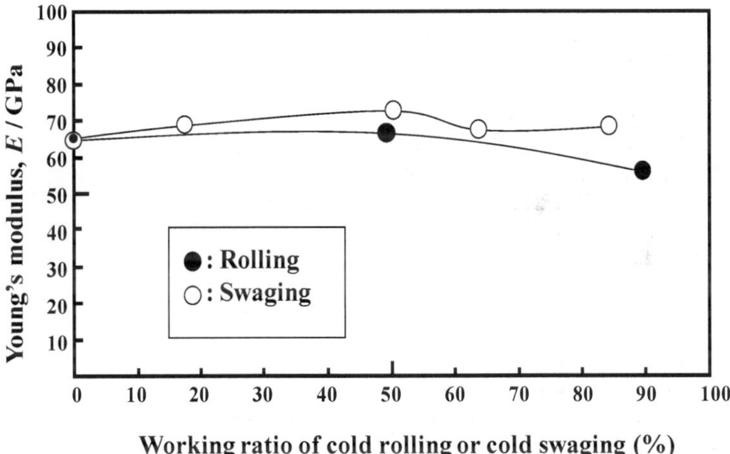

Fig. 2. Young's modulus of TNTZ subjected to cold rolling or cold swaging as a function of working ratio.

The tensile strength and 0.2% proof strength of cold rolled and swaged TNTZ increase with increasing working ratio up to approximately 20 % and then become almost constant. The ductility (elongation) of cold rolled TNTZ decreases with increasing cold working ratio, but that of swaged TNTZ decreases with increasing cold working ratio up to approximately 20 % and then becomes constant while maintaining high elongation. The tensile and 0.2 %

proof stress of TNTZ subjected to cold rolling and swaging are nearly equal to those of Ti-6Al-4V extra-low interstitial alloy (Ti-6Al-4V ELI); having a tensile strength of around 800 MPa) at high cold working ratio with good elongation.

The Young's modulus of TNTZ subjected to cold rolling or cold swaging is almost constant with increasing working ratio. The Young's modulus of TNTZ subjected to cold rolling tends to decrease when the working ratio is high because the trend of the formation of the texture becomes significant.

3.1 Lowering the Young's modulus by control of the crystal direction

Anisotorophy of the mechanical properties of β-type titanium alloy, TNTZ, has been reported to be significantly larger than those of other metallic materials such as carbon steel, S45C. Figure 3 (Niinomi, 2008b) shows the tensile strain (ε) versus lattice strain (ε_l) of TNTZ and ferrite in S45C, both of which have the bcc structure. Strains were calculated from the (110), (200), and (211) planes of the β phase of TNTZ and ferrite in S45C from the XRD profiles, obtained from in-situ X-ray analysis under tensile loading. The degree of the change in lattice strain with tensile strain for S45C is smaller than that for TNTZ. The relationship between the lattice strain and tensile strain obtained from the (100), (200), and (211) planes is nearly the same for each plane in S45C, but varies significantly for TNTZ. This illustrates that the anisotropy of the mechanical properties of TNTZ is large. Based on this trend in mechanical properties, the Young's modulus of TNTZ is considered to exhibit large anisotprophy. The single crystal of TNTZ, which grows to a certain direction, is expected to exhibit a low Young's modulus.

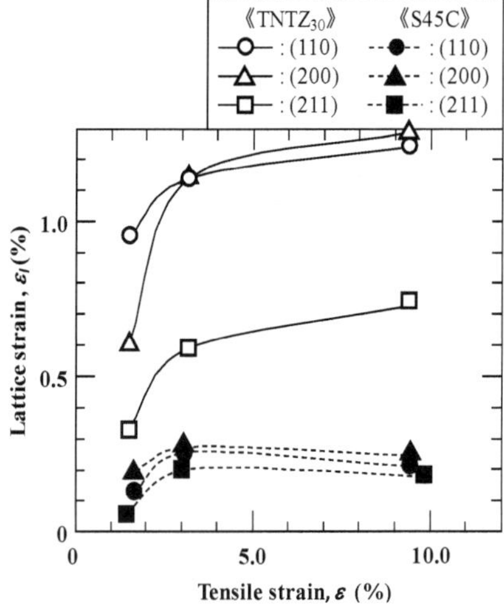

Fig. 3. Relationships between tensile strain and lattice strain calculated using several diffraction angles of TNTZ₃₀ (Ti-30Nb-10Ta-5Zr) and carbon steel (S45C).

Figure 4 (Tane, 2008) shows the orientation dependence of the Young's modulus in Ti-29Nb-Ta-Zr and Ti-25Nb-Ta-Zr single crystals between the <100> and <110> directions, which were calculated by coordinate conversion of c_{ij}. The symbol θ denotes the angle from the

<100> direction on the <110> zone axis. The Young's modulus of Ti-29Nb-Ta-Zr is lower than that of Ti-25Nb-Ta-Zr in all directions, but there are many common features independent of the alloy composition. The Young's modulus of both single crystals shows anisotropy as a function of θ; the Young's modulus of both crystals in the <100> direction, E_{100}, is approximately two times lower than the Young's modulus in the <111> direction, E_{111}, where E_{100} and E_{111} are the lowest and highest among all of the Young's moduli in all the directions, respectively. The lowest Young's modulus, E_{100}, of the Ti-29Nb-Ta-Zr single crystal is quite low at a value of only about 35 GPa, which is comparable to that of cortical bone. This level may be effective in the suppression of stress shielding in bone. Therefore, metallic single crystals of titanium alloys may be applicable as biomaterials; these may be referred to as "single crystal biometals".

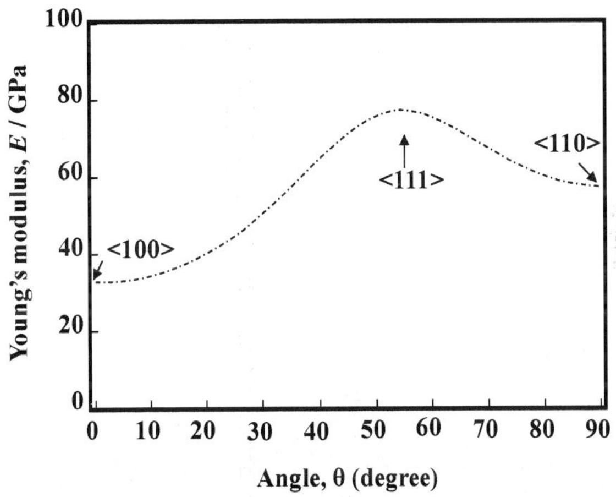

Fig. 4. Young's modulus of single-crystal Ti-29Nb-13Ta-4.6Zr in directions between [100] and [110].

3.2 Lowering the Young's modulus through structural design

Introduction of porosity into titanium and its alloys is a very effective method for further reducing the Young's moduli of titanium and its alloys. Introduction of porosity into titanium may affect a drastic reduction of the Young's modulus, and offer control of the Young's modulus by variation of the porosity. The relationship between the Young's modulus and porosity of porous titanium made from titanium powders with different particle diameters have been compared with the Young's modulus of bulk titanium in a recent report. According to that report, at a porosity of approximately 30%, the Young's modulus is nearly equal to that of cortical bone. The use of a titanium alloy with an even lower Young's modulus than titanium may allow for the achievement of a Young's modulus equal to that of cortical bone at lower porosity compared with the case of titanium. Pores of the proper size also enhance the bone conductivity. On the other hand, however, increasing the porosity of titanium results in a drastic decrease in its strength. At a porosity of approximately 30%, which leads to the Young's modulus equal to that of cortical bone,the 0.2% proof stress is below 100 Mpa (Oh, 2002).

This decrease in the strength of porous titanium can be prevented by combining with a biocompatible polymer. Penetration of the polymer into the porous titanium can be achieved by pressing. In the pressing method, HMDP (high molecular density polyethylene) is pressed into porous titanium.

Another proposed method for penetration of a polymer into the titanium pores (Nakai, 2010) involves firstly using the monomer of PMMA. The porous titanium (pTi) is first immersed into a monomer solution of PMMA leading to penetration of the monomer into the pores of titanium. The PMMA monomer in the porous titanium is then subjected to polymerization by heating. By combination with PMMA, the strength of porous titanium increases as shown in Fig. 5 (Nakai, 2010). The tensile strength of PMMA infiltrated porous titanium is greater than that of porous titanium, whereas the Young's modulus of PMMA infiltrated porous titanium is nearly equal to that of porous titanium as shown in Fig. 6 (Nakai, 2010). The tensile strength of PMMA infiltrated porous titanium increases by silane coupling treatment while the Young's modulus remains unchanged as shown in Figs. 5 and 6.

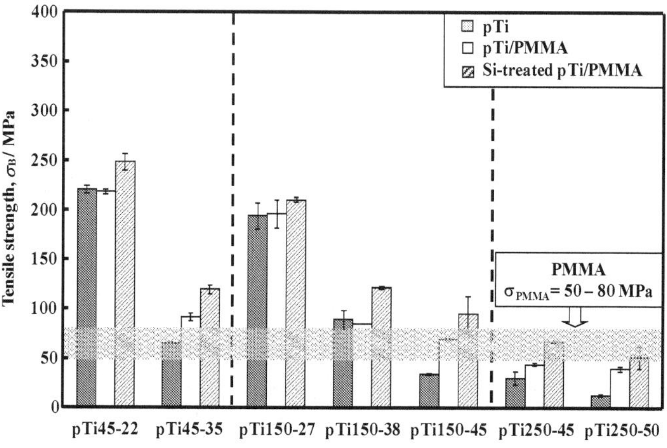

Fig. 5. Tensile strengths of pTi, pTi/PMMA, and Si-treated pTi/PMMA.

Another advantage offered by porous titanium and polymer composites is the ease with which bio-functionalities may be added given that the surface of porous titanium can be covered with polymers. Instead of PMMA, biodegradable PLLA can also be filtrated into the pores of porous titanium by modifying the process for PMMA filtration. Fig. 7 (Nakai, 2011b) shows the compressive 0.2% proof stress of porous titanium and PLLA infiltrated porous titanium as a function of porosity in the range of 5%–45%. In this figure, the compressive 0.2% proof stress of PLLA obtained experimentally is also shown for comparison. The compressive 0.2% proof stress of PLLA infiltrated porous titanium is higher than those of porous titanium independent to porosity. This result indicates that the PLLA filling can improve the compressive 0.2% proof stress of porous titanium at any degree of porosity. In particular, the increase in compressive 0.2% proof stress due to PLLA filling is relatively large for porosities higher than or equal to 35%. The compressive 0.2% proof stress of PLLA obtained is around 80–120 MPa, which is higher than that of PMMA (around 50–80MPa) (Honda, 1961; Imai and Brown, 1976).

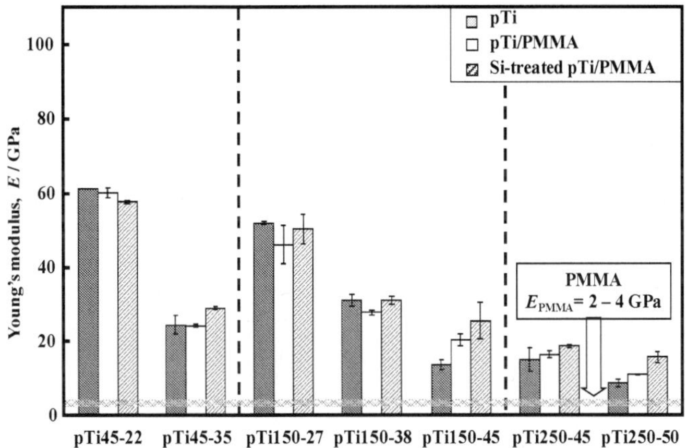

Fig. 6. Young's moduli of pTi, pTi/PMMA, and Si-treated pTi/PMMA.

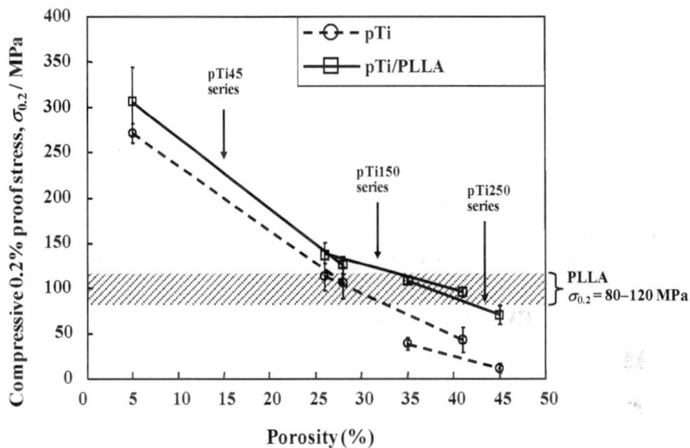

Fig. 7. Compressive 0.2% proof stresses of porous titanium (pTi) and porous titanium filled with poly-L-lactic acid (pTi/PLLA).

Figure 8 (Nakai, 2011b) shows the compressive Young's modulus of porous titanium and PLLA infiltrated porous titanium as a function of porosity in the range of 5%–45%. In this figure, the compressive Young's modulus of PLLA obtained experimentally is also shown for comparison. The compressive Young's modulus of porous titanium decreases with increasing porosity. The compressive Young's modulus is higher for PLLA infiltrated porous titanium than for porous titanium with a relatively high porosity of ≥35%.

Since PLLA is biodegradable, an agent to enhance the bone conductivity can be added to the PLLA in the porous titanium, and the agent can then be released into the body fluid. The released agent is expected to enhace the bone conductivity of the porous titanium.

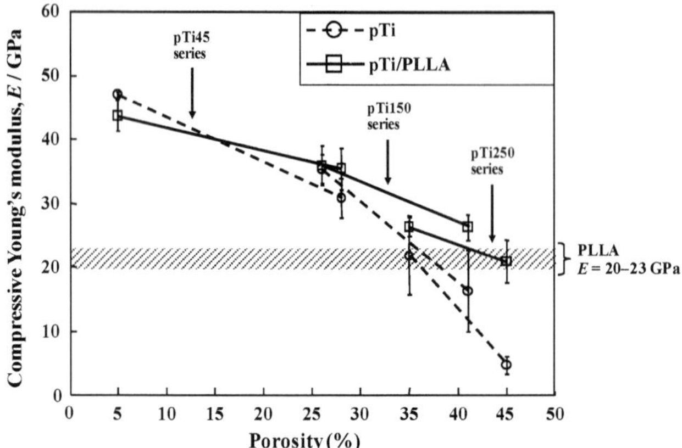

Fig. 8. Compressive Young's moduli of porous titanium (pTi) and porous titanium filled with poly-L-lactic acid (pTi/PLLA).

3.3 Strengthening or increasing endurance while maintaining a low Young's modulus

As already shown in Figs.1 and 2, the tensile strength and 0.2% proof strength of TNTZ both increase with an increase in the cold working ratio, and at high cold working ratio, these parameters become almost equal to those of Ti-6Al-4V ELI (having a tensile strength of around 800 MPa). Good elongation is also achieved when TNTZ is subjected to both cold rolling and cold swaging while the Young's modulus is kept low for both types of treatments.

However, the dynamic strength, i. e., the fatigue strength of TNTZ cannot be improved by severe cold working as shown in Fig. 9 (Akahori, 2003). Therefore, work hardening is not effective for improving the fatigue strength of TNTZ. Precipitation strengthening and dispersion strengthening are expected to improve the fatigue strength of TNTZ.

ω-phase precipitation significantly increases the strength and the Young's modulus of TNTZ as compared to α-phase precipitation, although the ω-phase enhances the brittleness of the alloy. Therefore, a small amount of ω-phase precipitation is expected to improve the fatigue strength of TNTZ while maintaining a low Young's modulus. For this purpose, short-time aging at fairly low temperatures, which enhances the precipitation of small amounts of the ω-phase, is effective.

Figure 10 (Nakai, 2011c) shows the relationship between tensile strength and the Young's modulus of TNTZ subjected to various thermomechanical treatments. In this figure, the terms CR, AT3.6, AT10.8, and AT86.4 indicate TNTZ subjected to severe cold rolling at a reduction ratio of 87 %, and aged at 573 K for 3.6 ks, 10.8 ks, and 86.4 ks, respectively. After severe cold rolling at a reduction ratio of 87 % (the samples are reffered to as (aging treatment) AT samples). The data for TNTZ subjected to aging treatments at 573 K, 598 K, 673 K, and 723 K for various times after solution treatment are also presented. TNTZ aged at 573 K, 673 K, and 723 K for various times after solution treatment showsYoung's moduli greater than or equal to 80 GPa. TNTZ aged at 573 K or 598 K has much higher Young's moduli (around 100-120 GPa). The strengths of these samples are scattered across a larger range than those of TNTZ subjected to aging treatments at other temperatures. These effects

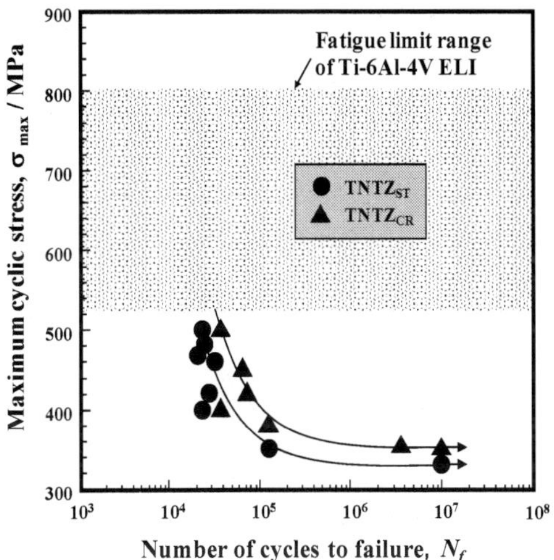

Fig. 9. S-N curves of TNTZ subjected to solution treatment (TNTZ$_{ST}$) and severe cold rolling with a reduction ratio of around 87 % (TNTZ$_{CR}$) along with fatigue limit range of Ti-6Al-4V ELI in air.

are the result of the presence of a large amount of ω-phase; TNTZ containing the ω- phase often exhibits inproved strengths (around 1400 MPa), when the samples do not fail in the elastic deformation range during tensile testing. However, in other cases, the brittleness becomes too high to attain plastic deformation during tensile testing, resulting in relatively low tensile strengths (around 800–900 MPa). In contrast, the TNTZ samples subjected to aging treatment (AT) , except for AT86.4, have Young's moduli below 80 GPa because of the small amount of ω-phase formed as a result of the short aging time. Among the AT samples, AT3.6 and AT10.8 exhibit an excellent balance between high strength and low Young's modulus (numbers 4 and 5 in Fig.10). Therefore, further examination of their fatigue properties was performed.

Figure 11 (Nakai, 2011c) shows the relationship between fatigue strength and Young's modulus of the TNTZ subjected to various thermomechanical treatments. In this figure, the abbreviations of the data are same as used in Fig.10. AT3.6 is classified as being in the low fatigue strength group. The AT10.8 falls in the intermediate level of fatigue strength among the samples, but possesses the highest fatigue strength among the samples having a Young's modulus less than 80 GPa.Therefore, it is possible to effectively control the proper precipitation of the-ω phase, evidenced by the fact that short-time aging at relatively low temperatures improves the fatigue strength of TNTZ while maintaining a low Young's modulus.

The addition of a small amount of ceramics particles such as TiB$_2$ and Y$_2$O$_3$ into the titanium matrix is also expected to improve the fatigue strength of β-type titanium alloys while maintaining a low Young's modulus. Figure 12 (Niinomi, 2011c) shows the Young's modulus and fatigue limit of TNTZ with TiB$_2$ or Y$_2$O$_3$ additions, subjected to severe cold rolling; as a function of B or Y concentration along with those of TNTZ subjected to severe cold rolling (TNTZ$_{CR}$) and solution treatment (TNTZ$_{ST}$), and Ti-6Al-4V ELI (Ti64 ELI). The fatigue limit of TNTZ is improved with 0.1 mass% and 0.2 mass% B concentration or 0.2 mass% and 0.5 mass% Y concentration while maintaining a very low Young's modulus.

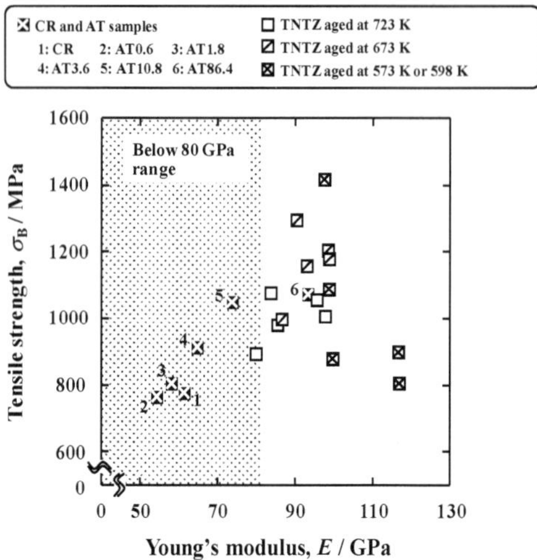

Fig. 10. Relationship between tensile strength and Young's modulus of TNTZ subjected to various thermomechanical treatments.

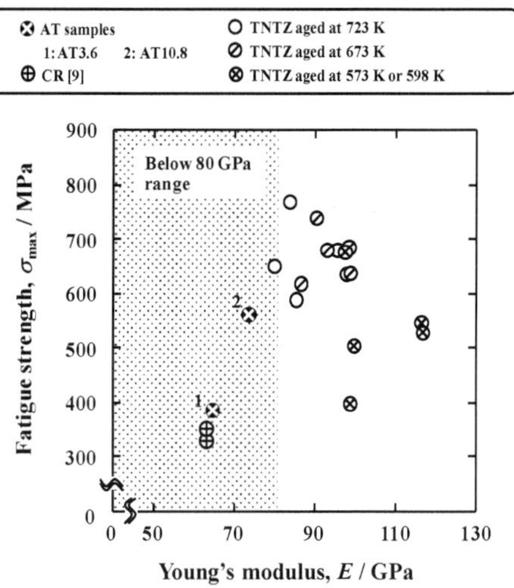

Fig. 11. Relationship between fatigue strength and Young's modulus of TNTZ subjected to various thermomechanical treatments.

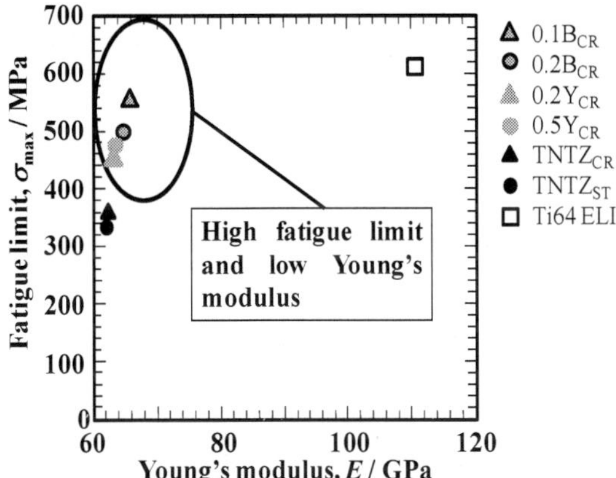

Fig. 12. Relationships between Young's modulus and fatigue limit of TNTZ with B concentrations of 0.1 and 0.2 mass% (0.1B$_{CR}$ and 0.2B$_{CR}$,), and Y concentrations of 0.2, 0.5, and 1.0 (0.2Y$_{CR}$, 0.5Y$_{CR}$, 1.0Y$_{CR}$,) subjected to severe cold rolling, TNTZ subjected to severe cold rolling (TNTZ$_{CR}$), TNTZ subjected to solution treatment (TNTZ$_{ST}$), and Ti-6Al-4V ELI (Ti64 ELI).

4. Bone remodeling and Young's modulus

It is very important to prove that alloys for implants have a Young's modulus similar to that of bone, which will inhibit bone atrophy and induce good bone remodeling as stated above. It is known that the geometry of the implant is another factor used to control the Young's modulus, but the effect of geometry is not treated in this chapter. Studies on bone plates made of low Young's modulus β-type titanium alloy (TNTZ), and conventional and practical (α + β)-type titanium alloy (Ti-6Al-4V ELI), and stainless steel (SUS 316L) in fracture models made into the tibiae of rabbits have been conducted. The Young's moduli of TNTZ, Ti-6Al-4V ELI, and SUS 316L stainless steel used for intramedullary rods, which were measured by three point bending tests, were 58, 108, and 161 GPa, respectively. In that study, an increase in the diameter of the tibia and the double-wall structure in the intramedullary bone tissue were reported to be observed only for the case of the bone plate made of TNTZ as shown in Fig. 13 (Niinomi, 2010a, b). Figure 13 shows that the inner wall bone structure is the original (old) cortical bone whereas the outer wall bone structure is newly formed bone. This is the possible result of bone remodeling with a bone plate having a low Young's modulus.

Furthermore, understanding of the Young's modulus level that is most effective in inhibiting bone atrophy and bone remodeling is necessary. Figure 14 (Niinomi, 2010a) shows the profiles of the extracted bone plates made of TNTZ subjected to solution treatment (TNTZ-ST), TNTZ subjected to aging after solution treatment (TNTZ-AT), and SUS 316 L stainless steel (SUS 316L) attached to the tibiae of the rabbits at 52 weeks after implantation. The Young's moduli of TNTZ-ST, TNTZ-AT, and SUS 316L measured by three-point bending tests were 58 GPa, 78 GPa, and 161 GPa, respectively. The upper surface and sides of each bone plate are covered by newly formed bone, but a fairly large amount of

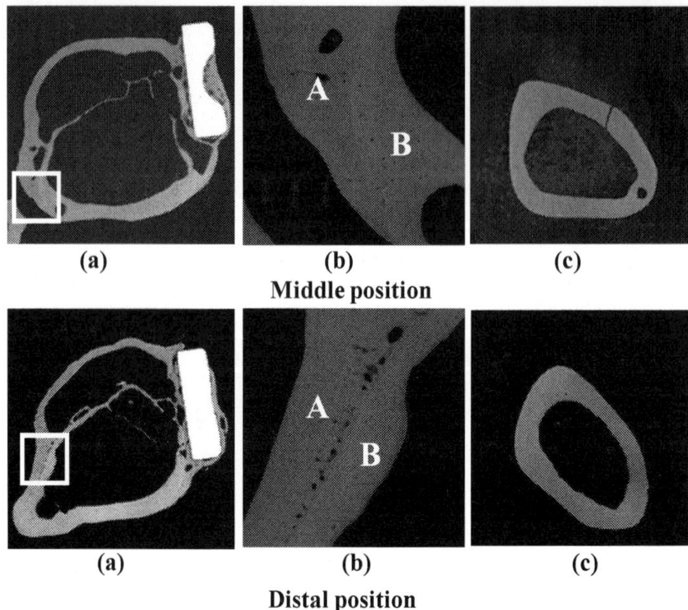

(a)   (b)   (c)

Middle position

(a)   (b)   (c)

Distal position

Fig. 13. CMRs of cross sections of fracture models implanted with and without bone plates made of TNTZ at middle position and distal position   at 48 weeks after implantation : (a) cross section of fracture model, (b) parts of (a), namely high magnification CMR of branched parts of bones formed outer and inner sides of tibiae, and (c) cross sections of un-implanted tibiae.

newly formed bone can be observed on the heads of the screws made of TNTZ-ST and TNTZ-AT; these have been encircled. Figure 15 (Niinomi, 2010a) shows the optical micrographs of the bone state beneath the bone plates made of TNTZ-ST, TNTZ-AT, and SUS 316L. Bone atrophy can be observed for all the cases, but it is more exident with a titanium having a lower Young's modulus will be advantageous for inhibiting bone atrophy leading to better bone remodeling.

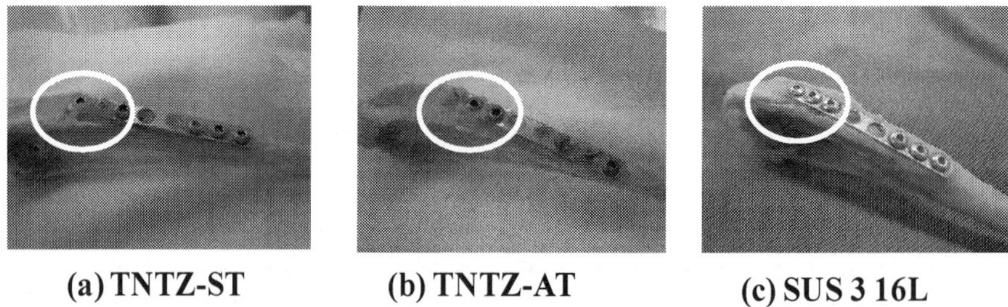

(a) TNTZ-ST   **(b) TNTZ-AT**   **(c) SUS 3 16L**

Fig. 14. Profiles of extracted bone plates made of (a) TNTZ-ST, (b) TNTZ-AT, and (c) SUS 316L stainless steel fixed to tibiae of rabbits at 52 weeks after implantation.

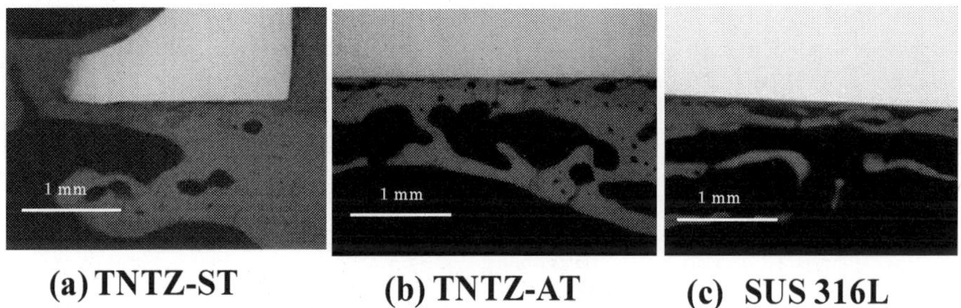

(a) TNTZ-ST   **(b) TNTZ-AT**   **(c) SUS 316L**

Fig. 15. Optical photographs of bones formed around extracted bone plates made of (a) TNTZ-ST, (b) TNTZ-AT, and (c) SUS 316L stainless steel fixed to tibiae of rabbits at 52 weeks after implantation.

5. Variable Young's modulus titanium alloys

While using low modulus titanium alloys, some surgeons specializing in spinal diseases, such as scoliosis, spondylolisthesis, and spine fracture, pointed out that the amount of spring-back in the implant rods should be small so that the implant offers better handling ability during surgeries. The implant rods undergo bending when they are manually handled by surgeons within the small space inside the patient's body for in-situ spine contouring. It is considered that the amount of spring-back depends on both the strength and the Young's modulus of the implant rod. If two implant rods having the same strength but with different Young's moduli are used, the implant rod having lower Young's modulus shows greater spring-back. Implant rods made of low modulus titanium alloys exhibit a lower Young's modulus, resulting in greater spring back. Thus, a low Young's modulus, which is one of the key features of β-type titanium alloys such as TNTZ as a metallic biomaterial, is obviously a desirable property for patients but becomes an undesirable property for surgeons. Titanium alloys, which satisfy the requirements of both surgeons and patients with regard to the Young's modulus of the implant rod, are currently being developed (Nakai, 2011a).

The amount of spring back is considered to be small for an alloy having a higher Young's modulus than for an alloy having a low Young's modulus. Therefore, a low Young's modulus β-type titanium alloy having a variable Young's modulus that becomes high only at the deformed part may reduce spring-back while simultaneously satisfying the low Young's modulus condition. This concept is called "self-adjustment of Young's modulus" .

In general, the Young's modulus of metals and alloys does not change upon deformation. However, in the case of certain metastable β-type titanium alloys, non-equilibrium phases such as α'-, α"-, and ω-phases appear in the β matrix during deformation. If the Young's modulus of the deformation-induced phase is higher than that of the original β-phase, the Young's modulus of only the deformed part of the implant rod increases, whereas that of the non-deformed part remains low. In orthopedic operations performed for the treatment of spinal diseases, the implant rod is bent by the surgeons so that it corresponds to the curvature of the spine. Therefore, if a suitable titanium alloy is employed as the implant rod material, spring-back can be suppressed by the deformation-induced phase transformation that occurs during bending in the course of operation, while a low Young's modulus can be retained for patients. In general, the Young's modulus of the ω-phase is much greater than those of the α-, α′-, α″-, and β-phase. Among these phases, the ω- , α′-, and α″-phase can be induced by deformation in β-type titanium alloys with certain chemical compositions.

Ti-12Cr has been reported to be one of the candidate alloys with self-adjustable Young's modulus for biomedical applications. Figure 16 (Nakai, 2011a) shows the Young's moduli of Ti-12Cr subjected to solution treatment (Ti-12Cr-ST) and severe cold rolling (Ti-12Cr-CR) along with those of TNTZ subjected to solution treatment (TNTZ-ST) and severe cold rolling (TNTZ-CR). Ti-12Cr-ST exhibits a low Young's modulus of ~70 GPa; this value is comparable to that of TNTZ-ST, which has been developed as a biomedical β-type titanium alloy having a low Young's modulus. TNTZ-CR also shows a low Young's modulus almost equal to that of TNTZ-ST. Thus, cold rolling leads to negligible change in the Young's modulus of TNTZ. However, in the case of Ti-12Cr, the Young's modulus increases upon cold rolling and that of Ti-12Cr-CR is >80 GPa. The deformation-induced ω phase was detected in Ti-12Cr, but no induced phase was detected in TNTZ. Therefore, the increase in Young's modulus of Ti-12Cr is probably due to the deformation-induced ω phase transformation.

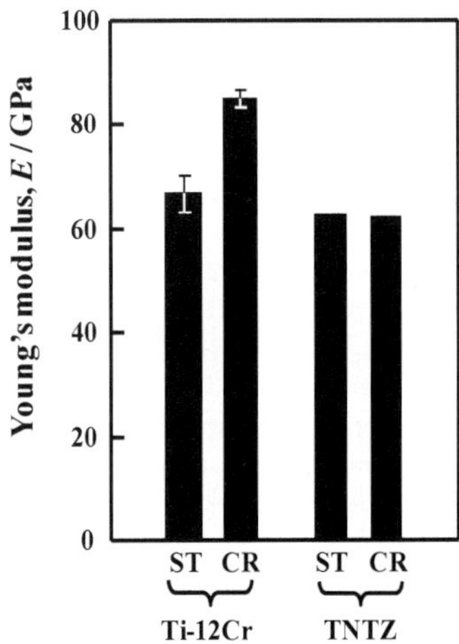

Fig. 16. Comparison of Young's moduli of the designed alloy (Ti-12Cr) and those of an alloy with a low Young's modulus (TNTZ): Ti-12Cr subjected to solution treatment (ST) and cold rolling at a reduction ratio of 10 % (CR), and TNTZ subjected to solution treatment (ST) and severe cold rolling at a reduction ratio of 87 % (CR).

Figure 17 (Nakai, 2011a) shows the tensile properties of Ti-12Cr-ST, Ti-12Cr-CR, TNTZ-ST, and TNTZ-CR. The tensile strengths of both Ti-12Cr-ST and TNTZ-ST show an increase, but the elongation due to cold rolling tends to decrease. This trend is probably caused by the occurrence of work hardening. Furthermore, the tensile strengths of Ti-12Cr-ST and Ti-12Cr-CR may be higher than those of TNTZ-ST and TNTZ-CR, respectively. Moreover, the elongations of Ti-12Cr-ST and Ti-12Cr-CR are >10% and ~10%, respectively. High strength is an essential requirement from the viewpoint of practical application, although such high strength could lead to undesirable spring-back. Therefore, the fundamental composition of Ti-12Cr makes it one of the preferred candidates for use in spinal fixation devices as a biomedical titanium alloy with the ability to self-adjust its Young's modulus.

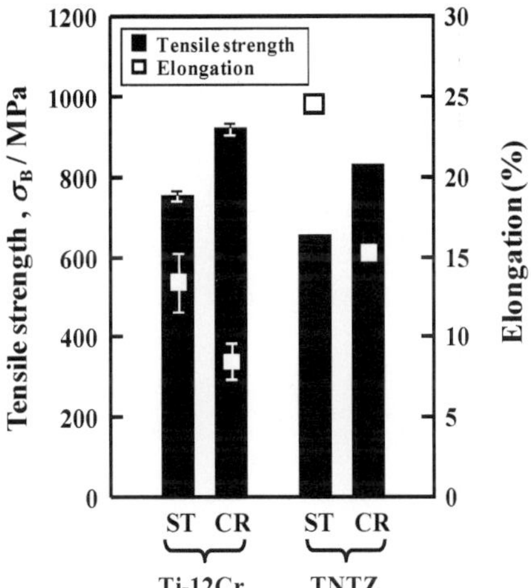

Fig. 17. Comparison of tensile properties of the designed alloy (Ti-12Cr) and those of an alloy with a low Young's modulus (TNTZ): Ti-12Cr subjected to solution treatment (ST) and cold rolling at a reduction ratio of 10 % (CR), and TNTZ subjected to solution treatment (ST) and severe cold rolling at a reduction ratio of 87 % (CR).

6. Low Young's modulus titanium alloys for reconstructive implant devices

In the case of some types of internal fixation devices implanted into the bone marrow such as femoral, tibia, and humeral marrow, in the case of screws used for bone plate fixation (Kobayashi, 2007), and in the case of implants used for children, which otherwise would grow into the bone, it is essential to remove the internal fixation device after surgery owing to certain specific indications; these indications include significant local symptoms such as palpable hardware, wound dehiscence/exposure of hardware, or athletes returning to contact sport (Kambouroglou, 1998) (Coook, 1985). The assimilation of removable internal fixation devices into the bone due to precipitation of calcium phosphate might cause refracture of the bone during the removal of the fixation device. Therefore, in these cases, it is essential to prevent the adhesion of the alloys with the bone tissues. Hence, considering this requirement, it is essential to inhibit the precipitation of calcium phosphate. It is reported that Zr, which is a non-toxic and allergy-free element, has the ability to prevent precipitation of calcium phosphate (Kawahara, 1963), and Ti alloys with Zr contents exceeding 25 mass% prevent the formation of calcium phosphate, which is the main component of human bones (Narushima, 2005). Thus, Ti-30Zr-Mo has been proposed as a low Young's modulus titanium based biomaterial for use in removable implants.

Figure 18 (Zhao, 2011) shows the Young's moduli of Ti-30Zr-xMo (x = 5, 6, and 7) subjected to solution treatment and those of the alloys considered for comparison. The Young's modulus of Ti-30Zr-xMo is lower than that of the alloys considered for comparison except TNTZ. The minimumYoung's modulus is obtained for Ti-30Zr-6Mo with a value of around 60 GPa, and TNTZ also shows a low Young's modulus.

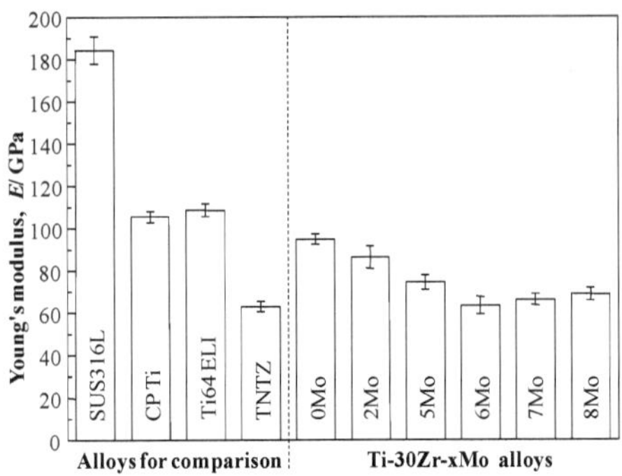

Fig. 18. Young's moduli of Ti-30Zr-xMo (x = 0, 2, 5, 6, 7, and 8 mass%) alloys subjected to solution treatment and the alloys considered for comparison.

In orthopedic applications, ideal biomedical implant materials are required to have high strength and a low Young's modulus. The elastic admissible strain, defined as the strength-to-modulus ratio, is a useful parameter considered in orthopedic applications. The higher the elastic admissible strain, the more suitable are the materials for such applications (Williams , 1971). Figure 19 (zhao, 2011) shows the distribution of the as-solutionized Ti-30Zr-xMo and the alloys considered for comparison in the plot of elastic admissible strain against elongation. Ti-30Zr-6Mo and -7Mo exhibit larger elongation and higher elastic admissible strain than the other Ti-30Zr-xMo and SUS316L, CP Ti, Ti64 ELI, and TNTZ.

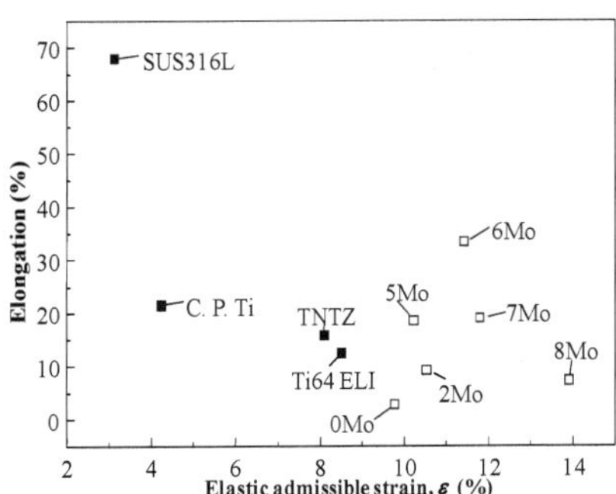

Fig. 19. Distribution of solutionized Ti-30Zr-xMo (x = 0, 2, 5, 6, 7, and 8 mass%) alloys and the alloys (SUS 316 L stainless steel, TNTZ and commercially pure titanium (C.P.Ti) considered for comparison in a plot of elastic admissible strain against elongation.

Figure 20 (Zhao, 2011) shows the density of cells cultured for 24 h in the presence of Ti-30Zr-7Mo and the alloys considered for comparison. Ti-30Zr-7Mo has the highest value of cell density. Therefore, Ti-30Zr-6Mo and Ti-30Zr-7Mo show promising potential to be new candidates for use in biomedical applications.

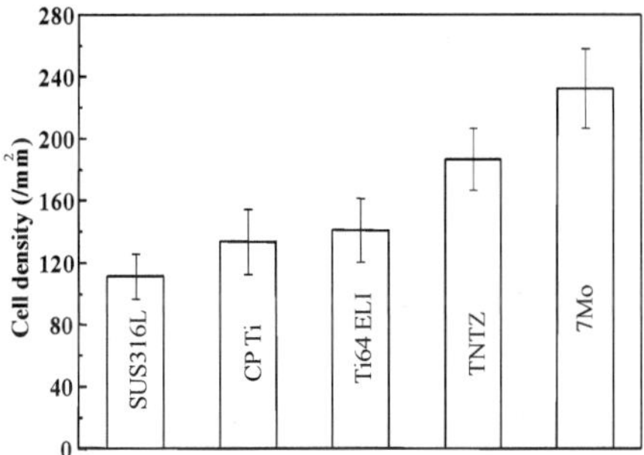

Fig. 20. Density of cells cultured in 7Mo and the alloys (SUS 316 L stainless steel, commercially pure titanium (CPTi), Ti-6Al-4V ELI (Ti64ELI), and TNTZ) considered for comparison.

Figure 21 (Zhao, 2011) shows Young's moduli of Ti-30Zr-5Mo, Ti-30Zr--6Mo, and Ti-30Zr--7Mo subjected to solution treatment (referred to as 5Mo-ST, 6Mo-ST, and 7Mo-ST, respectively) and subjected to cold rolling at a reduction ratio of 10 % (referred to as 5Mo-CR, 6Mo-CR, and 7Mo-CR respectively). In the ST samples, with increasing Mo content, the Young's modulus initially decreases from 75 GPa in 5Mo-ST to 63 GPa in 6Mo-ST and then increases slightly to 66 GPa in 7Mo-ST. The Young's moduli of the ST alloys are lower than those of conventional biomedical alloys such as SUS316L stainless steel (SUS 316L), commercial pure Ti (CP Ti), and Ti–6Al–4V extra-low interstitial alloy (Ti64 ELI). The change in Young's modulus after cold rolling varies with the Mo content: the Young's modulus of 5Mo-CR decreases drastically to 59 GPa from 75 GPa (after ST), and the Young's modulus of 6Mo-CR deceases to 61 GPa from 63 GPa (after ST). However, the Young's modulus of 7Mo-CR increases to 73 GPa from 66 GPa (after ST). Therefore, Ti-30Zr-7Mo is expected to be a Young's modulus self-adjustable titanium alloy for biomedical applications.

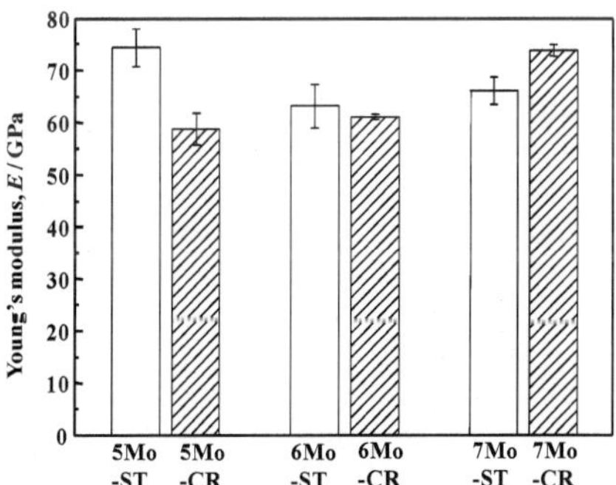

Fig. 21. Young's moduli of Ti–30Zr–(5, 6, 7)Mo alloys subjected to solution treatment (5Mo-, 6Mo-, 7Mo-ST) and cold-rolling (6Mo-, 6Mo-, 7Mo-CR).

7. Summary

The removal of metallic biomaterials from implanted bone tissue is fairly new concept because tight adhesion between the metallic biomaterial and bone is currently one of the targets of biomaterials researchers.
Nowadays, conflicting properties of a low Young's modulus to inhibit bone atrophy and high Young's modulus to inhibit spring-back are simultaneously desired in biomaterials. i.e., Young's modulus self-adjustment ability is required in the metallic biomaterials. In order to satisfy these demands, metastable β-type titanium alloys that exhibit deformation-induced transformation are prospective candidates materials can be selectively changed from low to high at the deformation point. However, the degree of the change in the Young's modulus is currently insufficient to satisfy the biomaterial demand and thus, further investigation of these kinds of titanium alloys is warranted.

8. Acknowledgements

The authors thank Professor T. Hattori of Meijo University, Nagoya, Japan, and Miss. X. Zhao of Institute for Materials Research, Tohoku University, Sendai, Japan for their contributions to the experiments. This study was supported in part by the Global COE Program "Materials Integration International Center of Education and Research, Tohoku University", Ministry of Education, Culture, Sports, Science and Technology (MEXT) (Tokyo, Japan) and The New Energy and Industrial Technology Development Organization (NEDO) (Tokyo, Japan), the collaborative project between Tohoku University and Kyusyu University on "Highly-functional Interface Science: Innovation of Biomaterials with Highly-functional Interface to Host and Parasite", MEXT (Tokyo, Japan), The Light Metal Educational Foundation, Inc. (Osaka, Japan), the cooperative research program of Institute for Materials Research, Tohoku University (Sendai, Japan), and the cooperative research program of the Advanced Research Center of Metallic Glasses, Institute for Materials Research, Tohoku University (Sendai, Japan).

9. References

Akahori, T., Niinimi, M., Ishimizu, K., Fukui, H., & Suzuki, A. (2003). Effect of Thermomechanical Processing on Fatigue Characteristics of Ti-29Nb-13Ta-4.6Zr, *J. Jpn. Inst. Metals*, Vol. 67, No. 11, pp. 652-660.

Cook, S. D., Renz, E. A., Barrzak, R., Thomas. K. A., Harding, A. F., Haddad R.J, Jr. & Mllicic, M. (1985). Clinical and Metallurgical Analysis of Retrieved Internal Fixation Devices, *Clin. Orthop. Relat. R.*, Vol. 194, pp. 236–247.

Kambouroglou, G. & Axelrod, T. (1998). Complications of the AO/ASIF Titanium Distal Radius Plate System (π plate) in Internal Fixation of the Distal Radius: A Brief Report, *J. Hand Surg.*, Vol. 23, pp. 737–741.

Kawahara, H., Ochi, S., Tanetani, K., Kato, K., Isogai, M., Mizuno, Y., Yamamoto, H., & Yamaguchi, A. (1963). Biological Test of Dental Materials. Effect of Pure Metals upon the Mouse Subcutaneous Fibroblast. Strain L Cell in Tissue Culture, *J. Jpn. Soc. Dent. Apparat. Mater.*, Vol. 4, pp. 65–75.

Kobayashi, E., Ando, M., Tsutsumi, Y., Doi, H., Yoneyama, T., Kobayashi, M., & Hanawa, T. (2007). Inhibition Effect of Zirconium Coating on Calcium Phosphate Precipitation of Titanium to Avoid Assimilation with Bone, *Mater. Trans.* , Vol. 48, pp. 301–306.

Nakai, M., Niinomi, M., Akahori, T., Tsutsumi, H., Itsuno, S., Haraguchi, N., Itoh, Y., Ogasawara, T., Onishi, T., & Shindoh., T. (2010). Development of Biomedical Porous Titanium Filled with Medical Polymer by In-Situ Polymerization of Monomer Solution Infiltrated into Pores, *J. Mech. Behav. Biomed. Mater.*, Vol. 3, pp. 41-50.

Nakai, M., Niinomi, M., Zhao, X. F., & Zhao, X. L. (2011a). Self-Adjustment of Young's Modulus in Biomedical Titanium Alloys during Orthopaedic Operation. *Mater. Lett.*, Vol. 65, pp. 688-690.

Nakai, M., Niinomi, M., & Ishii, D. (2011b). Mechanical and Biodegradable Properties of Porous Titanium Filled with Poly-L-lactic Acid by Modified In-Situ Polymerization Technique, *J. Mech. Behav. Biomed. Mater.*, in press.

Nakai, M., Niinomi, M., & Oneda, T. (2011c). Improvement in Fatigue Strength of Biomedical β-type Ti–Nb–Ta–Zr Alloy while Maintaining Low Young's Modulus through Optimizing ω-phase Precipitation, submitted to *Metal. Mater. Trans. A*.

Narushima, N. (2005). Titanium and Titanium Alloys, *J. Jpn. Soc. Biomat.*, Vol. 23, pp. 86–94.

Niinomi, M. (2002a). Recent Metallic Materials for Biomedical Applications, *Metall. Mater. Trans. A*, Vol. 33A, No.3, pp. 477-486.

Niinomi, M., Hattori, T., Morikawa, K., Kasuga, T., Suzuki, A., Fukui, H. & and Niwa, S. (2002b). Development of Low Rigidity β-type Titanium Alloy for Biomedical Applications, *Materi. Trans.*, Vol. 43, No. 12, pp. 2970-2977.

Niinomi, M. (2007). Fatigue Characteristics of Metallic Biomaterials, *Int. J. Fatigue*, Vol. 29, pp. 992-1000.

Niinomi, M. (2008a). Mechanical Biocompatibilities of Titanium Alloys for Biomedical Applications. *J. Mech. Behav. Biomed.l Materi.*, Vol. 1, No. 1, pp. 30-42.

Niinomi, M., Akahori, T., & Nakai, M. (2008b). In Situ X-Ray Analysis of Mechanism of Nonlinear Super Elastic Behavior of Ti-Nb-Ta-Zr System Beta-Type Titanium Alloy for Biomedical Applications, *Mater. Sci. Eng. C*, Vol. 2, pp. 406-413.

Niinomi, M. (2010a). Trend and Present State of Titanium Alloys with Body Centered Structure for Biomedical Applications, *Bulletin* of *The Iron and Steel Inst. Jpn.*, Vol. 15, No. 11, pp. 661-670.

Niinomi, M., & Hattori, T. (2010b). Effect of Young's Modulus in Metallic Implants on Atrophy and Bone Remodeling, *Interface Oral Health Science 2009*, Sasano, T., Suzuki, O., Stashenko, P., Sasaki, K., Takahashi, N., Kawai, T., Taubman, M. A., & H. C. Margolis,H. C., (Eds.), Springer, pp. 90-99.

Niinomi, M. & Nakai, M. (2011a). Titanium-based Biomaterials for Preventing Stress Shielding between Implant Devices and Bone, *Int. J. Biomaterials*, in press.

Niinomi, M., (2011b). Shape memory, Super Elastic and Low Young's Modulus Alloys, submitted to *Biomaterials for Spinal Surgery*, Ambrosio, L. & Tanner, K. E., (Eds.), Woodhead Publishing Ltd., Cambridge, UK.

Niinomi, M., Nakai, M., Song, X., & Wang, L. (2011c). Improvement of Mechanical Strength of a β-type Titanium Alloy for Biomedical Applications with Keeping Young's Modulus Low by Adding a Small Amount of TiB_2 or Y_2O_3, *Proc. PFAMXIX*

(Processing and Fabrication of Advanced Materials), Vol. 2, Bhattacharyya, D., Lin, R. J. T., & Srivatsan, T. S. (Eds.), pp. 817-827.

Oh, I. H., Nomura, N., & and Hanada, S. (2002). Microstructures and Mechanical Properties of Porous Titanium Compacts Prepared by Powder Sintering, *Mater. Trans.*, Vol. 43, pp. 443-446.

Sumitomo, N., Noritake, K., Hattori T., Morikawa, K., Niwa, S, Sato, K. & Niinomi, M. (2008). Experiment Study on Fracture Fixation with Low Rigidity Titanium Alloy - Plate Fixation of Tibia Fracture Model in Rabbit, *J. Mater. Sci.: Mater. in Medicine*, Vol. 19, No. 4, pp. 1581-1586.

Tane, M., Akita, S., Nakano, T., Hagihara, K., Umakoshi, Y., Niinomi, M., & and Nakajima, H. (2008). Peculiar Elastic Behavior of Ti-Nb-Ta-Zr Single Crystals, *Acta Mater*, Vol. 56, pp. 2856-2863.

Williams, J. C., Hickman, B. S., & D.H. Leslie, D. H. (1971). The Effect of Ternary Additions on the Decomposition of Detestable Beta Phase Titanium Alloys, *Metall. Mater. Trans.*, Vol. B 2, pp. 477–484.

Yilmazer, H., Niinomi, M., Akahori, T., Nakai, M., & Tsutsumi, H., (2009). Effects of Severe Plastic Deformation and Thermo-mechanical Treatments on Microstructures and Mechanical Properties of β-type Titanium Alloys for Biomedical Applications, *Proc. PFAMXIII*, pp. 1401-1410.

Zhao, X. L., Niinomi, M., Nakai, M., Miyamoto, G., & Furuhara, T. (2011). Microstructures and Mechanical Properties of Metastable Ti-30Zr-(Cr, Mo) Alloys with Changeable Young's Modulus for Spinal Fixation Applications. *Acta Biomater.*, Vol. 7, pp. 3230-3236.

Zhao, X. L., Niinomi, M., Nakai, M., Ishimoto, T., & Nakano, T. (2011), Development of High Zr-containing Ti-based Alloys with Low Young's Modulus for use in Removable Implants, Mater. Sci. Eng. A, in press.

13

Mesopore Bioglass/Silk Composite Scaffolds for Bone Tissue Engineering

Chengtie Wu and Yin Xiao
Queensland University of Technology
Australia

1. Introduction

In the past 20 years, mesoporous materials have been attracted great attention due to their significant feature of large surface area, ordered mesoporous structure, tunable pore size and volume, and well-defined surface property. They have many potential applications, such as catalysis, adsorption/separation, biomedicine, etc. [1]. Recently, the studies of the applications of mesoporous materials have been expanded into the field of biomaterials science. A new class of bioactive glass, referred to as mesoporous bioactive glass (MBG), was first developed in 2004. This material has a highly ordered mesopore channel structure with a pore size ranging from 5–20 nm [1]. Compared to non-mesopore bioactive glass (BG), MBG possesses a more optimal surface area, pore volume and improved *in vitro* apatite mineralization in simulated body fluids [1,2]. Vallet-Regí et al. has systematically investigated the in vitro apatite formation of different types of mesoporous materials, and they demonstrated that an apatite-like layer can be formed on the surfaces of Mobil Composition of Matters (MCM)-48, hexagonal mesoporous silica (SBA-15), phosphorous-doped MCM-41, bioglass-containing MCM-41 and ordered mesoporous MBG, allowing their use in biomedical enineering for tissue regeneration [2-4]. Chang et al. has found that MBG particles can be used for a bioactive drug-delivery system [5,6]. Our study has shown that MBG powders, when incorporated into a poly (lactide-co-glycolide) (PLGA) film, significantly enhance the apatite-mineralization ability and cell response of PLGA films. compared to BG [7]. These studies suggest that MBG is a very promising bioactive material with respect to bone regeneration. It is known that for bone defect repair, tissue engineering represents an optional method by creating three-dimensional (3D) porous scaffolds which will have more advantages than powders or granules as 3D scaffolds will provide an interconnected macroporous network to allow cell migration, nutrient delivery, bone ingrowth, and eventually vascularization [8]. For this reason, we try to apply MBG for bone tissue engineering by developing MBG scaffolds. However, one of the main disadvantages of MBG scaffolds is their low mechanical strength and high brittleness; the other issue is that they have very quick degradation, which leads to an unstable surface for bone cell growth limiting their applications.

Silk fibroin, as a new family of native biomaterials, has been widely studied for bone and cartilage repair applications in the form of pure silk or its composite scaffolds [9-14]. Compared to traditional synthetic polymer materials, such as PLGA and poly(3-

hydroxybutyrate-*co*-3-hydroxyvalerate) (PHBV), the chief advantage of silk fibroin is its water-soluble nature, which eliminates the need for organic solvents, that tend to be highly cytotoxic in the process of scaffold preparation [15]. Other advantages of silk scaffolds are their excellent mechanical properties, controllable biodegradability and cytocompatibility [15-17]. However, for the purposes of bone tissue engineering, the osteoconductivity of pure silk scaffolds is suboptimal. It is expected that combining MBG with silk to produce MBG/silk composite scaffolds would greatly improve their physio-chemical and osteogenic properties for bone tissue engineering application. Therefore, in this chapter, we will introduce the research development of MBG/silk scaffolds for bone tissue engineering.

2. Preparation, characterization, physio-chemistry and biological property of MBG/silk composite scaffolds

In the section, we will introduce the novel development of MBG/silk composite scaffolds prepared by two methods for bone tissue engineering. One is that we will use silk-modified MBG scaffolds to enhance mechanical, biological and drug-delivery properties for bone regeneration application; the other is to incorporate MBG powders into silk scaffolds to improve their physio-chemistry and *in vivo* osteogenesis.

2.1 Silk-modified MBG scaffolds
2.1.1 Composition optimization and characterization of silk-modified MBG scaffolds

To prepare and optimize MBG scaffolds, a series of MBG scaffolds with varied composition (molar composition: 100Si; 90Si-5Ca-5P; 80Si-15Ca-5P and 70Si-25Ca-5P) have been prepared by co-template method of P123 (EO_{20}-PO_{70}-EO_{20}) and polyurethane sponges (PUS), in which P123, as the template of mesopore formation, creates well-ordered mesoporous channels (around 5 nm) and PUS, as the template of large pores, produces hierarchically large pores (around 200-400 µm) (Figure 1). Our study has shown that MBG with the composition of 80Si-15Ca-5P has optimized bioactivity among four scaffolds [18]. Therefore, in this study, MBG (80Si-15Ca-5P) was selected for the further study.

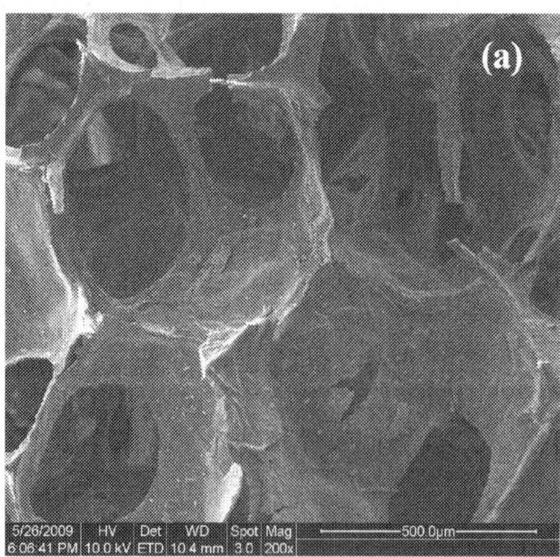

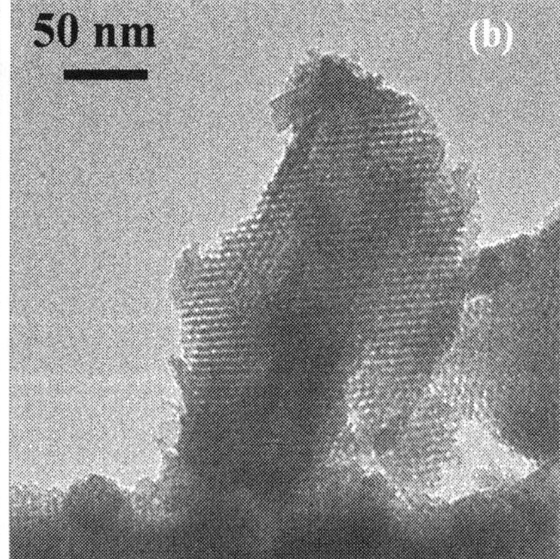

Fig. 1. SEM (a) and TEM (b) images for the prepared MBG scaffolds.

Silk-modified MBG scaffolds were prepared by dip-coating silk fibroin solution (wt. 2.5% and 5%) on the surface of scaffold pore walls with the cross linking of ethanol. After modification, a smooth silk film had formed on the surface of pore wall (Fig. 2b). Silk-modified MBG scaffolds showed a more uniform and continuous pore network (Fig. 2b) compared to unmodified MBG scaffolds with numerous collapsed and un-continuous pore networks due to their brittle nature (Fig. 2c). Silk-modified MBG scaffolds had a highly porous structure with the large-pore size of 400μm and maintained high porosity (95%). These characteristics indicate that silk-modified MBG scaffolds satisfy the requirements of pore structure architecture for cell and blood vessel ingrowth and nutrient supply [8].

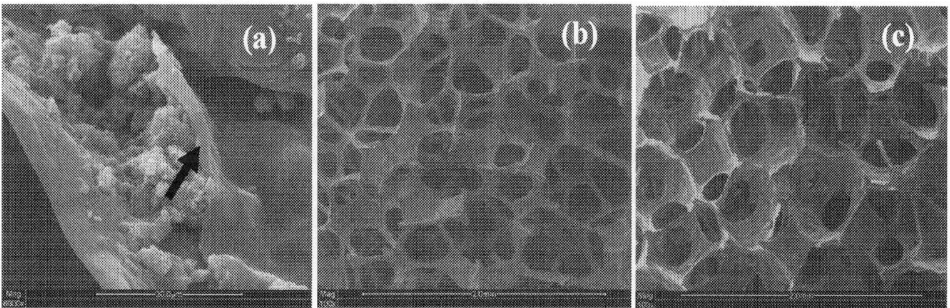

Fig. 2. SEM images for the formed silk layer (see arrow) (a), silk-modified MBG scaffolds (b) and un-modified MBG scaffolds (c) [19].

2.1.2 Physio-chemistry and biological property of silk-modified MBG scaffolds

The concentration of silk fibroin plays an important role to influence the compressive strength of MBG scaffolds. The compressive strength of pure MBG scaffolds was estimated to be 60kPa. Silk modification significantly improved the compressive strength of MBG scaffolds, which increases to 120kPa for 2.5%Silk-MBG scaffolds and 250kPa for 5.0%Silk-MBG scaffolds, a 100 and 300% increase, respectively (Fig. 3). There are two possible explanations as to why silk-modification improves the mechanical properties of the MBG scaffolds [19]. (1) Silk modification may induce a more uniform and continuous pore network within the MBG scaffolds, which, due to their natural brittleness, would otherwise have collapsed pore networks and micro-defects (micropores) and which contributes to their low mechanical strength; or (2) silk, which has greater mechanical strength than any other traditional polymer [15], may form an intertexture within the MBG scaffolds, linking the inorganic phase together and, in effect, reinforce the scaffolds [20].

By comparison, the compressive strength of hydroxyapatite andβ-Tricalcium phosphate is only 30kPa [21] and 50kPa [22], when their porosity is greater than 90%. The compressive strength of spongy bone (not the strut) is in the range of 0.2–4 MPa [23]. Therefore, the silk-modified MBG scaffolds fall within this range and therefore mimic that of cancellous bone.

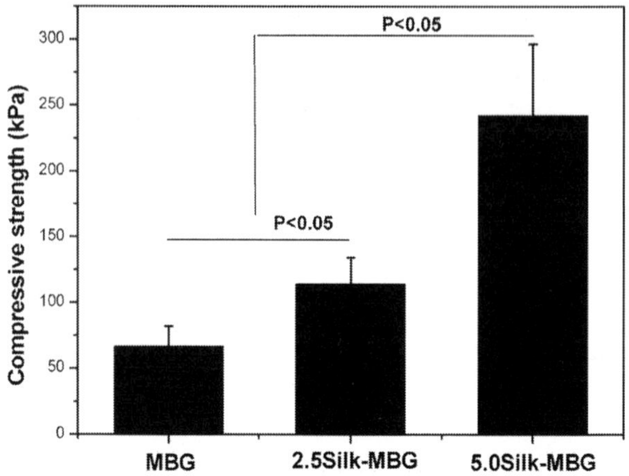

Fig. 3. The effect of silk concentration on the mechanical strength of MBG scaffolds [19].

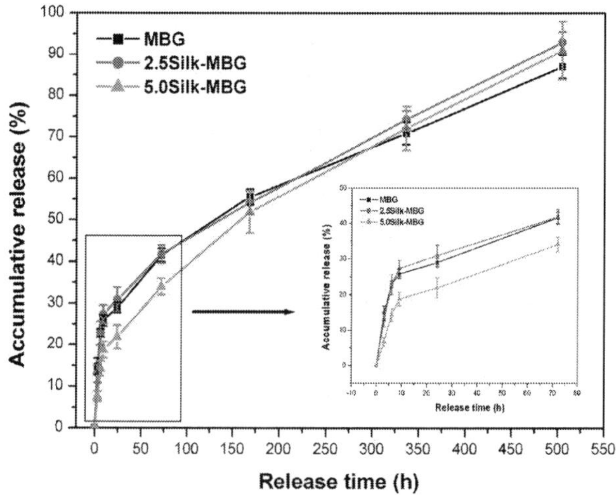

Fig. 4. Drug release from silk-modified and unmodified of MBG scaffolds [19].

Drug delivery represents another major challenge for scaffolds applicable for bone tissue engineering. In traditional scaffolds it was very difficult to combine the function of drug delivery due to the absence of a nano-pore structure. Due to the existence of well-ordered mesoporous structure in the MBG scaffolds, they can be used for the drug carrier. Dexamethasone can be easily loaded in the matrix of scaffolds by a simple soaking method.

The 5.0%Silk-MBG scaffolds had a decreased burst release compared to MBG scaffolds, and maintained a sustained release (Fig. 4). The most likely explanation for this is that the 5% silk solution forms a relatively dense silk film on the surface of pore walls which slows the drug release; however, over time the silk will begin to degrade and its effect on the release kinetics will therefore abate. This was evident by the fact that there was no discernible difference of the drug-release rate of the 5.0%Silk-MBG scaffold compared to the 2.5%Silk-MBG and the MBG scaffolds (Fig. 4). Further study has shown that after the drug release in phosphate buffer solution (PBS), a thin Ca-P layer of micro-particles was found to have been deposited on the surface of pore walls (Fig. 5). The formed Ca-P layer, one the one hand, will have an inhibitory effect on drug release [24]; On the other hand, it is indicated that silk-modified MBG scaffolds maintained the bioactivity of the surface chemistry as the Ca-P formation ability was regarded as one of important factor for bioceramics according to Kokubo's view [25].

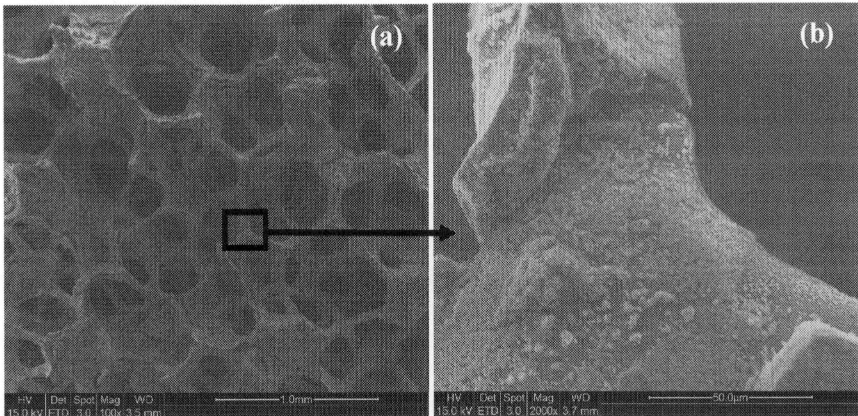

Fig. 5. SEM images for silk-modified MBG scaffolds after drug release in PBS.

The biological properties of silk-modified MBG scaffolds were further investigated by evaluating the attachment, proliferation, differentiation and bone cell-relative gene (alkaline phosphatase activity (ALP) and osteocalcin (OCN)) expression of bone marrow stromal cells (BMSC). After 1 day of culture, BMSCs were attached to the surface of the pore walls in all three types of scaffolds (Fig. 6). There was no obvious difference in the cell numbers and morphology of the attached cells among the scaffold types after one day. After 7 days, however, the density of BMSCs on 2.5%Silk-MBG and 5.0Silk-MBG scaffolds was higher than that of pure MBG scaffolds, the BMSCs on the 5.0%Silk-MBG scaffolds eventually reaching confluence. High magnification images show that BMSCs on 5.0%Silk-MBG scaffolds had a more spread out morphology than those on pure MBG scaffolds and that the cells had close contact with the pore walls of 5.0%Silk-MBG scaffolds (Fig. 6).

Silk modification significantly improved the proliferation and ALP activity of BMSCs on MBG scaffolds. At day 1, the number of cells on the silk-modified MBG scaffolds were comparable with that of pure MBG scaffolds (Fig. 7a), whereas by day 7, the number of cells on the silk-modified MBG scaffolds was significantly higher than that on the pure MBG scaffolds (p<0.05). ALP activity was used as an early marker of BMSC differentiation on scaffolds. After 7 days of culture, the ALP activity of the cells on all three scaffold types was roughly equal (Fig 7b); however, after 14 days the ALP activity of BMSCs in the 5.0%Silk-MBG scaffolds was greater than that of the other two scaffold types (Fig. 7b). The osteoblastic differentiation was further assessed by the mRNA expression of ALP and OCN using RT-qPCR method. There was an upregulation of the osteogenic marker genes of OCN and ALP after the modification of MBG scaffolds by silk (Fig. 8).

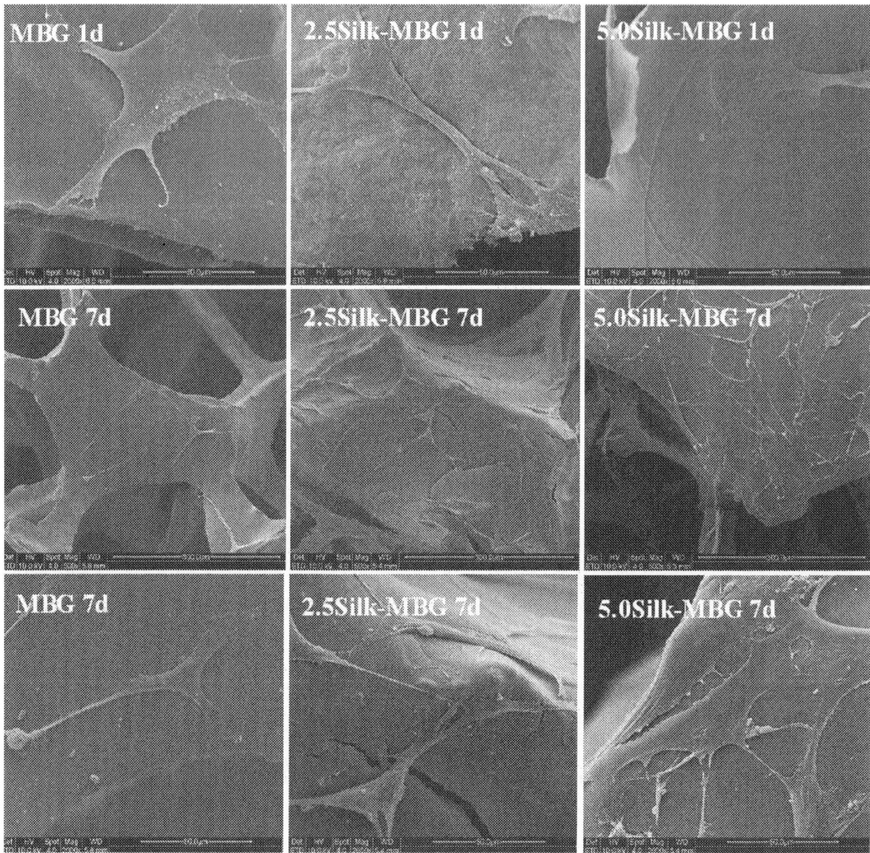

Fig. 6. The morphology of bone marrow stromal cells on the silk-modified and unmodified MBG scaffolds after 1 and 7 days of culture [19].

From the results above, it is indicated that that silk modification of MBG scaffolds had a positive effect on the attachment, proliferation and differentiation of BMSCs [19]. Generally, the ionic environment and material surface are two main factors which influence the interaction of cells and biomaterials [7,26-29]. In this study, silk modification did not create any significant effect on either ionic release or weight loss of MBG scaffolds. It has been speculated that the ionic environment and pH value of culture media may not be the most important factors affecting the cell response. Therefore, there is a reasonable hypothesis that

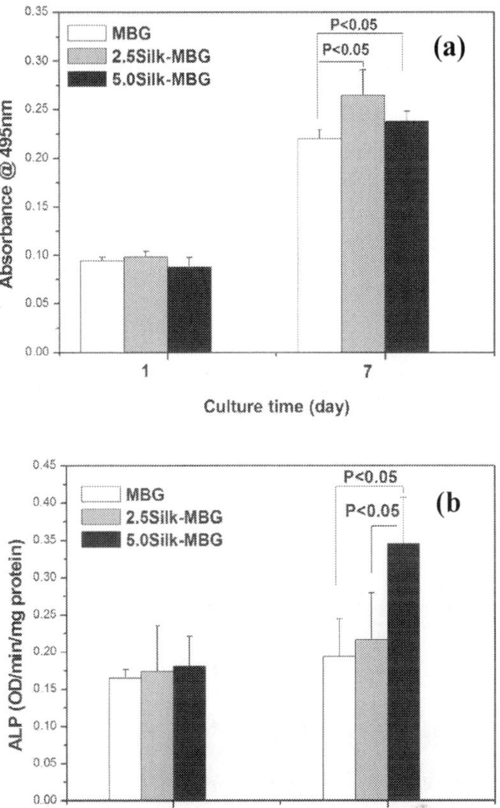

Fig. 7. The proliferation (a) and ALP activity of bone marrow stromal cells on the silk-modified and unmodified MBG scaffolds [19].

the silk itself may be responsible for enhancing BMSC proliferation and differentiation. It is known that, generally, the stable surface of biomaterials enhances cell attachment and proliferation, compared to the unstable surface of biomaterials which have a higher rate of dissolution [28]. MBG scaffolds have a high rate of degradation leaving their surface relatively unstable, and this may affect cell growth. The silk used in this study is the relatively stable fibroin, which consists of a β-sheet structure. When applied to the MBG scaffolds this silk may provide a relatively stable surface interface to support BMSC proliferation and differentiation. It has been reported that silk-functionalized titanium surfaces can enhance osteoblast functions [30,31], and also that silk modification of poly (D,L-lactic acid) improves ostoblast differentiation [32]. Although the mechanisms underlying this stimulatory effect on cell functions remains unclear, the explanation may be related to the interaction with specific chemical groups in silk, such as amino acids [15].

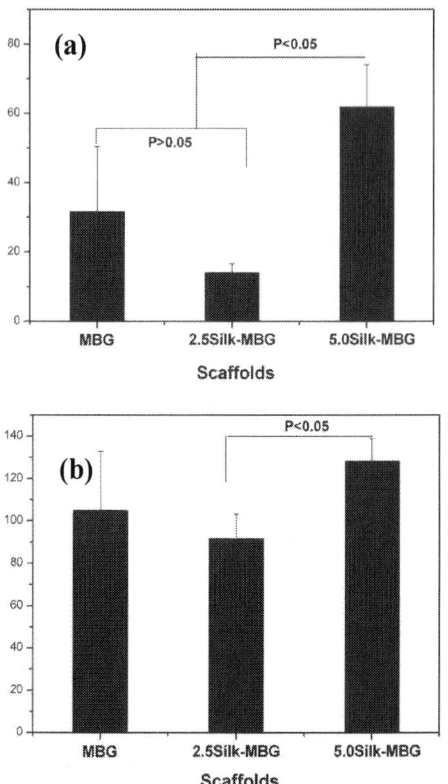

Fig. 8. The bone-relative gene expression of OCN (a) and ALP (b) of bone marrow stromal cells on the silk-modified and unmodified MBG scaffolds [19].

2.2 MBG powders-incorporated silk scaffolds
2.2.1 Preparation, characterization and physio-chmistry of MBG-incorporated silk scaffolds

MBG powders (molar composition: 80Si-15Ca-5P) with a particle size lower than 45µm were synthesized by an evaporation-induced self-assembly process according to the publications [7]. The obtained MBG powders possess well-ordered mesoporous structure (see Fig. 9a). Non-mesoporous bioglass (BG) powders with same composition were synthesized for the control materials (Fig. 9b). The surface area and pore volume of MBG are about 400m^2/g and 0.5cm^3/g, respectively, which are obviously higher than those of BG (57 m^2/g for surface area, 0.09 cm^3/g for pore volume).

Porous MBG/silk scaffolds with 10% MBG (w/w) were fabricated using a freeze-drying method. Pure silk and BG/silk scaffolds were prepared by same method for the control materials. The silk, MBG/silk and BG/silk scaffolds were highly porous (Fig. 10), with near identical porosities, 78%, 76% and 76%, respectively. The pure silk scaffolds had a flat pore

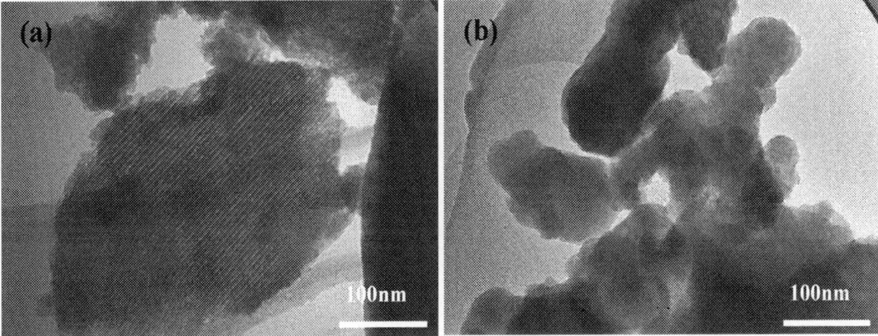

Fig. 9. TEM images of MBG (a) and BG (b) powders [7].

morphology (Fig. 10a), whereas the MBG or BG composite scaffolds had a more open pore morphology (Fig. 10b and c) compared to the silk scaffolds. The pore size of the pure silk scaffolds ranged from several tens to one hundred micrometers; the pore size of the composite scaffolds is larger than that of pure silk scaffolds [33].

The compressive strength and modulus of MBG/silk scaffolds were 420kPa and 0.70MPa, respectively, figures that were comparable with those of pure silk scaffolds and greater than those of BG/silk scaffolds (300kPa for compressive strength and 0.5MPa for modulus). It is speculated that the incorporation of BG particles into silk scaffolds may destroy the inner structure of silk and lead to the detrimental effect of the mechanical strength of silk scaffolds. Although MBG particles may also destroy the inner structure of silk, however, MBG has high surface area and pore volume, and parts of silk solution may enter into the nanopores of MBG during preparation, which leads to a strong bond between MBG particles and silk after freeze-drying. Thus, the incorporation of MBG into silk will not decrease the mechanical strength of silk scaffolds [33].

The apatite-mineralization ability and ion release of scaffolds were carried out using acellular simulated body fluids (SBF). The morphology of the three scaffold species, after soaking in SBF, is shown in Figure 11. There was no apatite particles deposit visible on the pore wall surfaces for pure silk and BG/silk scaffolds (Fig. 11a and b). However, a layer of apatite microparticles formed on the pore wall of MBG/silk scaffolds (Fig. 11c) and at higher magnification apatite was seen as nano-sized particles (Fig. 11d). EDS analysis revealed the ratio of Ca/P of the apatite to be 2.3 [33]. Apatite mineralization of silicate materials, such as $CaSiO_3$ ceramics, 45S5 bioglass, etc. is thought to be an important phenomenon in the chemical interactions between the implant materials and the bone tissue, which ultimately affects the *in vivo* osteogenesis of the bone grafting materials [34-36]. In this study, MBG/silk scaffolds had an obvious apatite mineralization in SBF, whereas neither BG/silk nor pure silk scaffolds induced apatite mineralization. This suggests that MBG/silk scaffolds have an improved *"in vitro* bioactivity", a term that has been used in previous studies [25,37,38].

There was a sustained release of Si ions from both the MBG/silk and BG/silk scaffolds, even across an extended period of soaking and the MBG/silk scaffolds had a faster rate of Si ion release than BG/silk scaffolds. The pH value of SBF with MBG/silk scaffolds stayed within a range of 7.25–7.5 throughout the 6 weeks of soaking. The pH values of the pure silk and

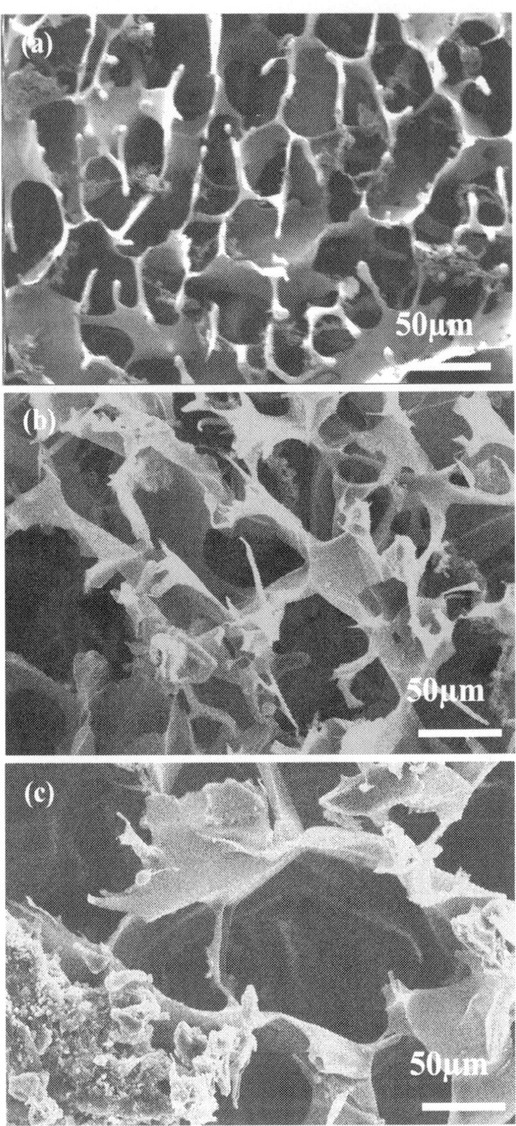

Fig. 10. Surface morphology of porous silk (a), MBG/silk (b) and BG/silk (c) scaffolds.

BG/silk scaffolds resulted in a slight decreased in SBF, varying from 7.1 to 7.4 [33]. Therefore, it is very obvious that the incorporation of MBG powders into silk scaffolds significantly improved their physio-chmestry. Further study for the effect of in vivo osteogenesis has been further investigated in the following section.

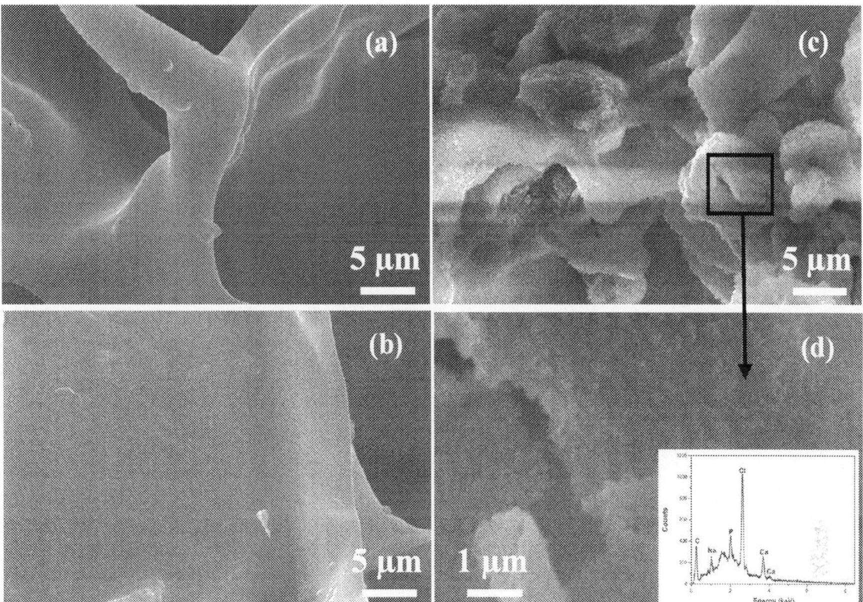

Fig. 11. SEM and EDS analysis silk (a), BG/silk (b) and MBG/silk (c and d) scaffolds after soaking in simulated body fluids for 7 days [33].

2.2.2 The *in vivo* osteogenesis of MBG-incorporated silk scaffolds

To further investigate the in vivo osteogenesis of MBG-incorporated silk scaffolds, the scaffolds were implanted into calvarial defects in adult severe combined immunodeficient (SCID) mice and the degree of *in vivo* osteogenesis was evaluated by micro-computed tomography (µCT), hematoxylin and eosin (H&E) and immunohistochemistry (type I collagen) analyses.

Both MBG/silk and BG/silk scaffolds clearly showed better bone repair ability than pure silk scaffolds. The defects implanted with MBG/silk scaffolds had been completely filled with new bone mineral tissues (Fig. 12a). The BG/silk scaffolds also induced new bone formation in the defects (Fig. 12b). However, the skull defects implanted with pure silk scaffolds revealed little mineralized tissues around the border and no new bone formation at all in the middle of the defects (Fig. 12c). Quantitative analysis from μCT data showed that the mineralized tissue volume for MBG composite was a little higher than that of BG composite. The volume of mineralized tissue for silk, MBG/silk and BG/silk scaffolds was 2.5, 7.0 and 6.1 mm³, respectively (Fig. 12d) [33].

New bone filled most of the MBG/silk scaffolds from the edge to the center and formed a continuous plate of bone area (Fig. 13a and b). Most of MBG/silk scaffolds had been degraded (Fig. 13a). In the BG/silk scaffolds the majority of the new bone was located in the periphery, with some bone islands forming centrally. There was only limited degradation of the BG/silk scaffolds (Fig. 13c and d). In the skull defects implanted with pure silk scaffolds there was no evidence of bone formation.

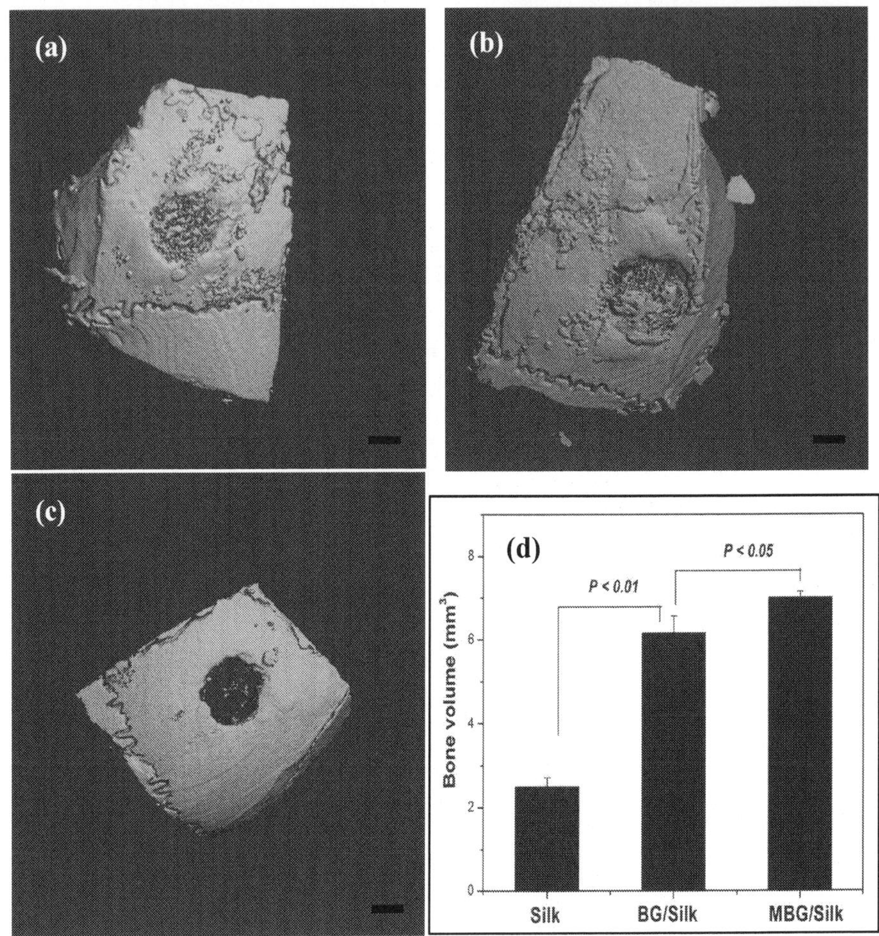

Fig. 12. Micro-CT analysis for the *in vivo* bone formation of MBG/silk (a), BG/silk (b), silk (c) scaffolds, and new bone volume (d) after implanted in calvarial defects of SCID mice for 8 weeks [33].

Immunohistochemical analysis revealed type I collagen (COL1) expression in the *de novo* bone in both MBG/silk and BG/silk scaffolds (Fig. 14); there was certainly slightly strong COL1 expression in the bone matrix of the MBG/silk scaffolds (Fig. 14a and b) and this expression was discernibly stronger compared to that in the BG/silk scaffolds (Fig. 14c and d) [33].

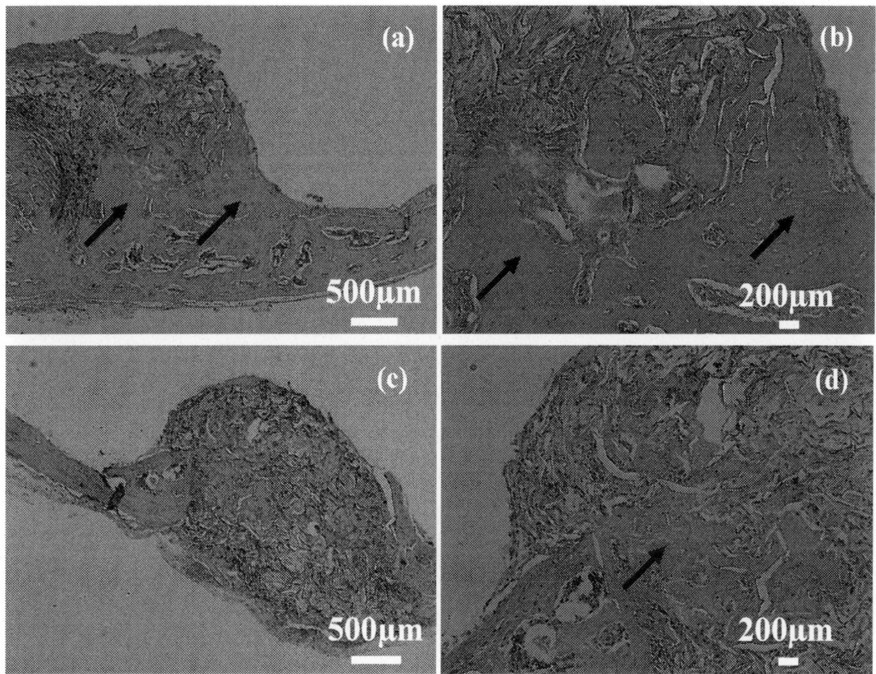

Fig. 13. The *in vivo* bone formation was evaluated by hematoxylin and eosin (H & E) staining. (a) and (b): MBG/silk; (c) and (d): BG/silk. (b) and (d) are higher magnification images. Arrows point to new formed bone [33].

There are three reasons that best explains why MBG/silk scaffolds have improved new-bone formation, compared to BG/silk scaffolds [33]: (1) apatite mineralization plays an important role in bone repair and studies suggest that apatite mineralization of 45S5 bioglass [36], A–W bioactive glass ceramics [39] and CaSiO₃ ceramics [34,40], is the direct factor influencing the *in vivo* osteogenesis potential of these materials. In the present study, we show that MBG/silk has a better apatite-mineralization ability than does BG/silk, leading us to draw the tentative conclusion that this may be one of the most important factors to improve new bone formation. (2) The faster rate of dissolution and Si ion release of the MBG/silk scaffolds compared to BG/silk scaffolds may enhance new-bone formation; this is supported by a study that showed that CaSiO₃ ceramics has significantly faster rate of degradation than does β-tricalcium phosphate ceramics and leads to an improved *in vivo* osseointegration [34]. It has been reported that Si ions may be associated with the initiation of pre-osseous tissue mineralization, both in periosteal or in endochondral ossification, in the early stages of calcification [41,42]. *In vitro* studies have confirmed that silicon released from the materials results in a significant up-regulation of osteoblast proliferation and gene expression [26,43,44]. The faster rate of degradation may in fact provide the space and environment for matrix deposition and tissue growth [45], and, at the same time, the quicker release Si ions from MBG/silk scaffolds may stimulate the viability of osteoblast around the defects, to the benefit of *in vivo* osteogenesis. (3) One cannot overlook the beneficial role that the stable pH environment of MBG/silk scaffolds has on *in vivo* osteogenesis [46,47].

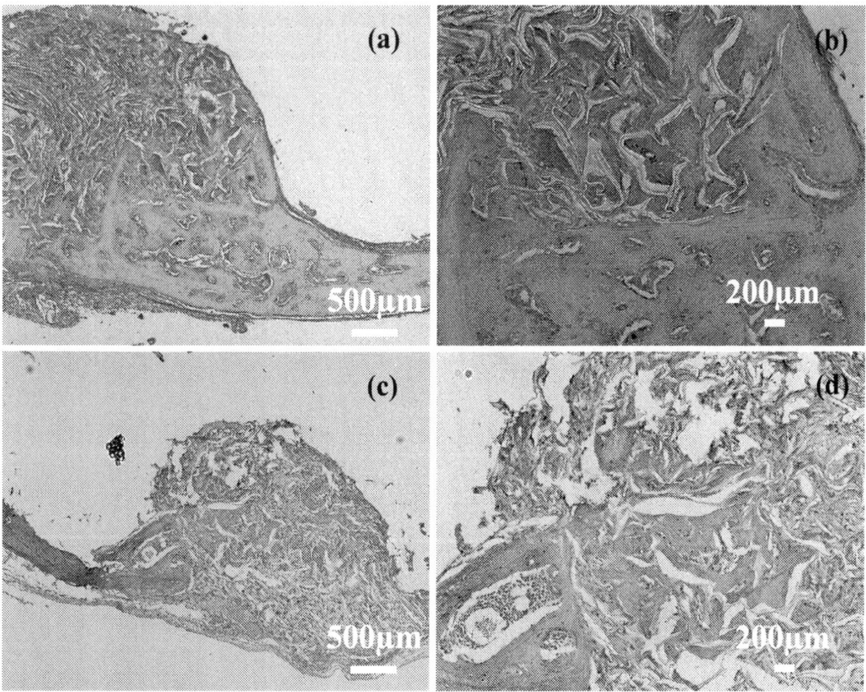

Fig. 14. Immunohistochemical analysis by Collagen I staining on the new bone tissues. (a) and (b): MBG/silk; (c) and (d): BG/silk. (b) and (d) are higher magnification images [33].

3. Conclusion

In summary, we have successfully developed scaffolds containing MBG and silk components for the purpose of bone tissue engineering. Two approaches in scaffold fabrication have been investigated, namely silk surface-coating for MBG scaffolds and MBG integrated silk scaffold.

Porous silk-modified MBG scaffolds with high porosity and large-pore size have been prepared by coating silk on the pore walls surfaces of MBG scaffolds. Silk modification improves the continuity of pore network and mechanical strength of MBG scaffolds, resulting enhanced BMSC proliferation, differentiation and in vivo bone formation.

We have prepared MBG powders-incorporated silk scaffolds and found that MBG can significantly improve the *in vitro* bioactivity and *in vivo* osteogenesis of silk scaffolds. MBG/silk scaffolds have the improved physio-chemistry and new-bone formation ability compared to BG/silk scaffolds.

Our study indicates that MBG/silk composite scaffolds, in the forms of silk-modified MBG scaffolds and MBG powder-integrated silk scaffolds, are very promising potential biomaterials for bone repair and regeneration.

4. Acknowledgment

The authors would like to acknowledge the support of Vice-chancellor Fellowship (241402-0120/07) and BlueBox (241402-0126/07) of Queensland University of Technology and the NHMRC Australia-China Fellowship Scheme,. We also thank Dr Yufang Zhu and A/Prof Yufeng Zhang's for their contributions in this work.

5. References

[1] Zhao, D., Feng, J., Huo, Q., Melosh, N., Fredrickson, G.H., Chmelka, B.F. and Stucky, G.D. (1998) Triblock copolymer syntheses of mesoporous silica with periodic 50 to 300 angstrom pores. *Science*, 279, 548-552.

[2] Lopez-oriega A, Arcos D, Izquierdo-Barb I, Sakamoto Y, Terasaki O and Vallet-Regi, M. (2006) Ordered mesoporous bioactive glasses for bone tissue regeneration. *Chem Mater*, 18, 3137-3144.

[3] Vallet-Regi, M. (2006) Revisiting ceramics for medical applications. *Dalton Transactions*, 5211-5220.

[4] Vallet-Regi, M.A., Ruiz-Gonzalez, L., Izquierdo-Barba, I. and Gonzalez-Calbet, J.M. (2006) Revisiting silica based ordered mesoporous materials: medical applications. *J Mater Chem*, 16, 26-31.

[5] Xia, W. and Chang, J. (2006) Well-ordered mesoporous bioactive glasses (MBG): a promising bioactive drug delivery system. *J Control Release*, 110, 522-530.

[6] Xia W and Chang, J. (2008) Preparation, in vitro bioactivity and drug release property of well-ordered mesoporous 58S bioactive glass. *J Non-Cryst Solids*, 15, 1338-1341.

[7] Wu, C., Ramaswamy, Y., Zhu, Y., Zheng, R., Appleyard, R., Howard, A. and Zreiqat, H. (2009) The effect of mesoporous bioactive glass on the physiochemical, biological and drug-release properties of poly(DL-lactide-co-glycolide) films. *Biomaterials*, 30, 2199-2208.

[8] Hutmacher, D.W. (2000) Scaffolds in tissue engineering bone and cartilage. *Biomaterials*, 21, 2529-2543.

[9] Park, S.H., Gil, E.S., Kim, H.J., Lee, K. and Kaplan, D.L. Relationships between degradability of silk scaffolds and osteogenesis. *Biomaterials*, 31, 6162-6172.

[10] Wang, Y., Rudym, D.D., Walsh, A., Abrahamsen, L., Kim, H.J., Kim, H.S., Kirker-Head, C. and Kaplan, D.L. (2008) In vivo degradation of three-dimensional silk fibroin scaffolds. *Biomaterials*, 29, 3415-3428.

[11] Yun, H.S., Kim, S.E. and Hyeon, Y.T. (2007) Design and preparation of bioactive glasses with hierarchical pore networks. *Chem Comm*, 2139-2141.

[12] Nazarov, R., Jin, H.J. and Kaplan, D.L. (2004) Porous 3-D scaffolds from regenerated silk fibroin. *Biomacromolecules*, 5, 718-726.

[13] Mandal, B.B. and Kundu, S.C. (2009) Cell proliferation and migration in silk fibroin 3D scaffolds. *Biomaterials*, 30, 2956-2965.

[14] Zhang, Y., Wu, C., Friis, T. and Xiao, Y. (2010) The osteogenic properties of CaP/silk composite scaffolds. *Biomaterials*, 31, 2848-2856.

[15] Vepari, C. and Kaplan, D.L. (2007) Silk as a Biomaterial. *Prog Polym Sci*, 32, 991-1007.

[16] Zhao, J., Zhang, Z., Wang, S., Sun, X., Zhang, X., Chen, J., Kaplan, D.L. and Jiang, X. (2009) Apatite-coated silk fibroin scaffolds to healing mandibular border defects in canines. *Bone*, 45, 517-527.

[17] Altman, G.H., Diaz, F., Jakuba, C., Calabro, T., Horan, R.L., Chen, J., Lu, H., Richmond, J. and Kaplan, D.L. (2003) Silk-based biomaterials. *Biomaterials*, 24, 401-416.

[18] Zhu, Y., Wu, C., Ramaswamy, Y., Kockrick, E., Simon, P., Kaskel, S. and Zreiqat, H. (2008) Preparation, characterization and in vitro bioactivity of mesoporous bioactive glasses (MBGs) scaffolds for bone tissue engineering. *Micropor Mesopor Mat*, 112, 494-503.

[19] Wu, C., Zhang, Y., Zhu, Y., Friis, T. and Xiao, Y. (2010) Structure-property relationships of silk-modified mesoporous bioglass scaffolds. *Biomaterials*, 31, 3429-3438.

[20] Wu, C., Ramaswamy, Y., Boughton, P. and Zreiqat, H. (2008) Improvement of mechanical and biological properties of porous CaSiO3 scaffolds by poly(D,L-lactic acid) modification. *Acta Biomater*, 4, 343-353.

[21] Kim, H.W., Knowles, J.C. and Kim, H.E. (2005) Hydroxyapatite porous scaffold engineered with biological polymer hybrid coating for antibiotic Vancomycin release. *J Mater Sci Mater Med*, 16, 189-195.

[22] Wu, C., Chang, J., Zhai, W. and Ni, S. (2007) A novel bioactive porous bredigite (Ca(7)MgSi (4)O (16)) scaffold with biomimetic apatite layer for bone tissue engineering. *J Mater Sci Mater Med*, 18, 857-864.

[23] Gibson LJ, A.M. (1999, 429–452.) *Cellular solids: structure and properties. 2nd ed.* Pergamon, Oxford.

[24] Jongpaiboonkit, L., Franklin-Ford, T. and Murphy, W.L. (2009) Mineral-Coated Polymer Microspheres for Controlled Protein Binding and Release. *Adv Mater*, 21, 1960-1963.

[25] Kokubo, T. and Takadama, H. (2006) How useful is SBF in predicting in vivo bone bioactivity? *Biomaterials*, 27, 2907-2915.

[26] Xynos, I.D., Edgar, A.J., Buttery, L.D., Hench, L.L. and Polak, J.M. (2000) Ionic products of bioactive glass dissolution increase proliferation of human osteoblasts and induce insulin-like growth factor II mRNA expression and protein synthesis. *Biochem Biophys Res Commun*, 276, 461-465.

[27] Wu, C., Ramaswamy, Y., Soeparto, A. and Zreiqat, H. (2008) Incorporation of titanium into calcium silicate improved their chemical stability and biological properties. *J Biomed Mater Res A*, 86, 402-410.

[28] John, A., Varma, H.K. and Kumari, T.V. (2003) Surface reactivity of calcium phosphate based ceramics in a cell culture system. *J Biomater Appl*, 18, 63-78.

[29] Zreiqat, H., Valenzuela, S.M., Nissan, B.B., Roest, R., Knabe, C., Radlanski, R.J., Renz, H. and Evans, P.J. (2005) The effect of surface chemistry modification of titanium alloy on signalling pathways in human osteoblasts. *Biomaterials*, 26, 7579-7586.

[30] Zhang, F., Zhang, Z., Zhu, X., Kang, E.T. and Neoh, K.G. (2008) Silk-functionalized titanium surfaces for enhancing osteoblast functions and reducing bacterial adhesion. *Biomaterials*, 29, 4751-4759.

[31] Cai, K., Hu, Y. and Jandt, K.D. (2007) Surface engineering of titanium thin films with silk fibroin via layer-by-layer technique and its effects on osteoblast growth behavior. *J Biomed Mater Res A*, 82, 927-935.

[32] Cai, K., Yao, K., Lin, S., Yang, Z., Li, X., Xie, H., Qing, T. and Gao, L. (2002) Poly(D,L-lactic acid) surfaces modified by silk fibroin: effects on the culture of osteoblast in vitro. *Biomaterials*, 23, 1153-1160.

[33] Wu, C., Zhang, Y., Zhou, Y., Fan, W. and Xiao, Y. (2011) A comparative study of mesoporous-glass/silk and non-mesoporous-glass/silk scaffolds: physiochemistry and in vivo osteogenesis. *Acta Biomater*, 7, 2229-2236.

[34] Xu, S., Lin, K., Wang, Z., Chang, J., Wang, L., Lu, J. and Ning, C. (2008) Reconstruction of calvarial defect of rabbits using porous calcium silicate bioactive ceramics. *Biomaterials*, 29, 2588-2596.

[35] Hench, L.L. (1991) Bioceramics: from concept to clinic. *J Am Ceram Soc*, 74, 1487-1510.

[36] Hench, L.L. (1998) Biomaterials: a forecast for the future. *Biomaterials*, 19, 1419-1423.

[37] Hench, L.L. and Wilson, J. (1984) Surface-active biomaterials. *Science*, 226, 630-636.

[38] Wu, C., Zhang, Y., Ke, X., Xie, Y., Zhu, H., Crowford, R. and Xiao, Y. (2010) Bioactive mesopore-bioglass microspheres with controllable protein-delivery properties by biomimetic surface modification. *J Biomed Mater Res A*, In Press.

[39] Kokubo T. (1990) Surface chemistry of bioactive glass-ceramics. *J Non-Cryst Solids* 120, 138-151.

[40] Xue, W., Liu, X., Zheng, X. and Ding, C. (2005) In vivo evaluation of plasma-sprayed wollastonite coating. *Biomaterials*, 26, 3455-3460.

[41] Carlisle, E.M. (1970) Silicon: a possible factor in bone calcification. *Science*, 167, 279-280.

[42] Schwarz, K. and Milne, D.B. (1972) Growth-promoting effects of silicon in rats. *Nature*, 239, 333-334.

[43] Valerio, P., Pereira, M.M., Goes, A.M. and Leite, M.F. (2004) The effect of ionic products from bioactive glass dissolution on osteoblast proliferation and collagen production. *Biomaterials*, 25, 2941-2948.

[44] Wu, C., Chang, J., Ni, S. and Wang, J. (2006) In vitro bioactivity of akermanite ceramics. *J Biomed Mater Res A*, 76, 73-80.

[45] Burg, K.J., Porter, S. and Kellam, J.F. (2000) Biomaterial developments for bone tissue engineering. *Biomaterials*, 21, 2347-2359.

[46] Verrier, S., Blaker, J.J., Maquet, V., Hench, L.L. and Boccaccini, A.R. (2004) PDLLA/Bioglass composites for soft-tissue and hard-tissue engineering: an in vitro cell biology assessment. *Biomaterials*, 25, 3013-3021.

[47] Putnam, D. (2008) The heart of the matter. *Nat Mater*, 7, 836-837.

14

To Build or Not to Build:
The Interface of Bone Graft
Substitute Materials in Biological
Media from the View Point of the Cells

Andrey Shchukarev[1], Maria Ransjö[2] and Živko Mladenović[2]
[1]Department of Chemistry, Umeå University
[2]Department of Odontology, Umeå University
Sweden

1. Introduction

The phenomena at the biomaterial - biological media interface are crucial in maintaining optimal cellular functions, initiating biomineralization and bone tissue regeneration. In contrast to "inert" implants substituting hard and soft tissues, bioactive ceramics used as bone graft substitutes are known to possess a reactive surface with respect to the body liquids and similar model solutions (Hench & Wilson, 1993). It is the interface that cells are approaching and "deciding" to build or not to build a new bone.

At interfacial equilibrium, an electrical double layer (EDL) is formed at the interface. The key feature that determines the EDL structure is the surface charge developed due to chemical interactions between bone graft substitute material and biological media. Three scenarios are possible depending on biomaterial surface chemistry, composition of the media, and the media pH. Positively or negatively charged surfaces encourage electrostatic (long-range forces) attraction of oppositely charged species. For example, ions, protonated/deprotonated aminoacids, peptides, proteins and even, probably, cells. For the particles with zero surface charge, short-range (chemical binding) interactions are preferable. Since body fluids have relatively constant composition and stable buffered pH, it is mainly the bulk and surface chemistry of biomaterial that affect the surface charge development. Electrostatic attraction of inorganic cations and anions from biological media is particularly important to initiate biomineral surface nucleation. The surface charge of biomaterial plays a pivotal role in adsorption of organic species from the media. The adsorption can change hydrophobic/hydrophilic properties of the surface if preferential accumulation of lipids, proteins, and vitamins from one side, or sugars, small organic acids and aminoacids from another side occurs. The adsorbed organic molecules will determine if the cells can attach to the bioactive surface with functional groups on the cell membrane. Moreover, the surface charge will influence the interface pH and bone graft substitute dissolution/re-precipitation processes. The extent of biomaterial dissolution is very important considering the rate of bone regeneration and biomaterial substitution by new bone. Another important factor which can significantly improve initial adhesion of the cell to the bone graft substitute materials is the surface morphology. Considering the cell size (μm) and the interface (EDL) thickness (nm), it should be the surface morphology which certainly maintains comfortable hosting for the cell to adhere, to fine tune their micro environment, and to communicate with other cells. The main parameters to consider are particles size, shape, porosity, surface (macro and micro) roughness, and availability of natural and artificial scaffolds. Initial morphology is not kept during the formation of new bone; however it can contribute at the first stage of the tissue regeneration process.

"Biomaterial in biological media" system can be separated to three different parts: solid phase (macro), solution phase (macro), and solid-solution interface (nano). Continuous phases are easy to investigate, but the most important information about the interface is almost inaccessible. The formation of new bone can be experimentally observed already at micro level using histological sections of biopsies from patients; changes in composition of biological media are routinely monitoring by standard chemical tools. However, specific characteristics necessary for a potential interfacial "nano-reactor" and the key elements which initiate and control cellular activity are still unknown. To understand these phenomena at a molecular level we need supplementary physico-chemical information about the surface of bone graft substitute material and the immediate interface between the material and the biological media. This information will allow material scientists and bone biologists to improve existing or to create a new tissue substitutes with attractive interface properties, leading the cell towards the decision – "To build or not to build" the bone.

2. Biomaterials and bone cell interactions

Although bone is unique in terms of being able to repair itself (e.g. fracture healing), bone defects caused by trauma and pathological conditions will not always heal spontaneously (Feng & McDonald, 2011; Nakahama, 2010). Autologous bone graft is considered the best treatment option, but has the limitation of donor sites, and the interest for bone graft substitute materials is increasing (Porter et al., 2009). Ideally, bone formation induced by the biomaterial will mimic the physiological process where osteoblasts differentiate from mesenchymal stem cells and produce extracellular proteins that serve as a template for the biomineralization process. It is crucial that the biomaterial does not promote an imbalanced bone remodeling process with uncoordinated activities between bone forming osteoblasts and bone resorbing osteoclasts. Osteoclasts are multinucleated cells with a hematopoietic origin formed by fusion of mono nuclear cells mainly regulated by osteoblasts and stromal cells (Raggatt & Partridge, 2010).

Cell differentiation in general is regulated by complex interacting signaling pathways which eventually lead to altered gene expression and subsequent changes in cell behavior. The interaction between cell surface molecules and extracellular structures is one of the key mechanisms regulating the differentiation and activities of bone cells. Mesenchymal stem cells and stromal cells, including osteoblasts, as well as osteoclasts, are depending on attachment to extra cellular matrix (ECM) proteins. Cellular interactions with adhesion proteins/ECM proteins occur via transmembrane integrins receptors which recognize proteins containing the Arg-Gly-Asp (RGD) amino acid sequence. The RGD sequence can be found within proteins such as fibronectin and vitronectin which are components in ECM and interestingly also among the serum proteins which are adsorbed to implanted surfaces (Kundu & Putnam, 2006; Raggatt & Partridge, 2010; Wilson et al., 2005). Vitronectin is reported to promote osteogenic differentiation of mesenchymal stem cells via integrin-mediated signals (Kundu & Putnam, 2006). The mitogen activated protein (MAP)-kinase cascade is an important downstream intracellular signaling transduction pathway in integrin-mediated signals. MAP kinase activation leads to extracellular-regulated kinase (ERK) activation via a pathway of downstream protein effectors ultimately regulating gene transcription and cell activities (Kundu & Putnam, 2006; Yee et al., 2008) (Illustrated in Fig. 1. based on: Geiger et al., 2009; Kundu & Putnam, 2006; Lu & Zreiqat, 2010; Milani et al., 2010; Raggatt & Partridge, 2010; Stevens & George, 2005; Wang et al., 2009; Wilson et al., 2005; Yee et al., 2008)

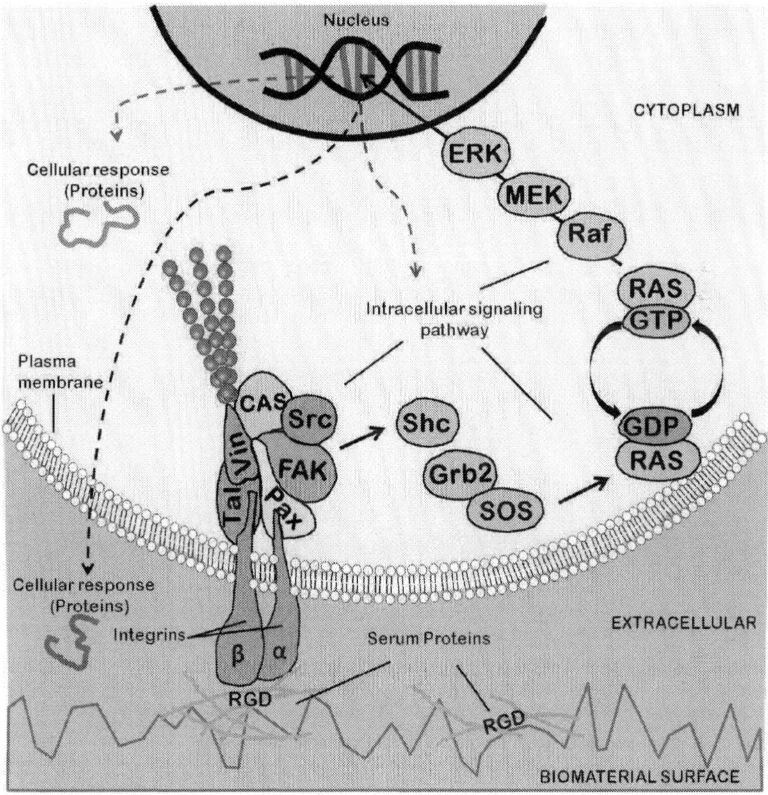

Fig. 1. RGD-integrin mediated cell attachment to biomaterial surface which leads to intra cellular signaling and change of cell functions. Figure was produced using Servier Medical Art.

It has also been proposed the inorganic components of bone could support the attachment and interactions between cells and bone (Feng, 2009). Moreover, it has recently been demonstrated that bone-specific extracellular matrix proteins could promote the osteogenic differentiation of embryonic stem cells (Evans et al., 2010). In addition to the control of cell activities by matrix proteins and signaling molecules, it is becoming evident that nanoscale alterations in topography will influence osteoblast differentiation and formation of a mineralized matrix (Diniz et al., 2002; Stevens & George, 2005). The mechanisms by which biomaterials support bone formation are largely unknown but will certainly be determined by cell-surface interactions and the interface between the biomaterial and cells. Bone graft materials may regulate cell adhesion and functions through chemical and physical properties, dissolution and precipitation reactions, surface charge, protein adsorption and surface topography. A schematic illustration of the hypothetical time dependent development of the biomaterial-tissue interface leading to bone formation is presented in Fig. 2.

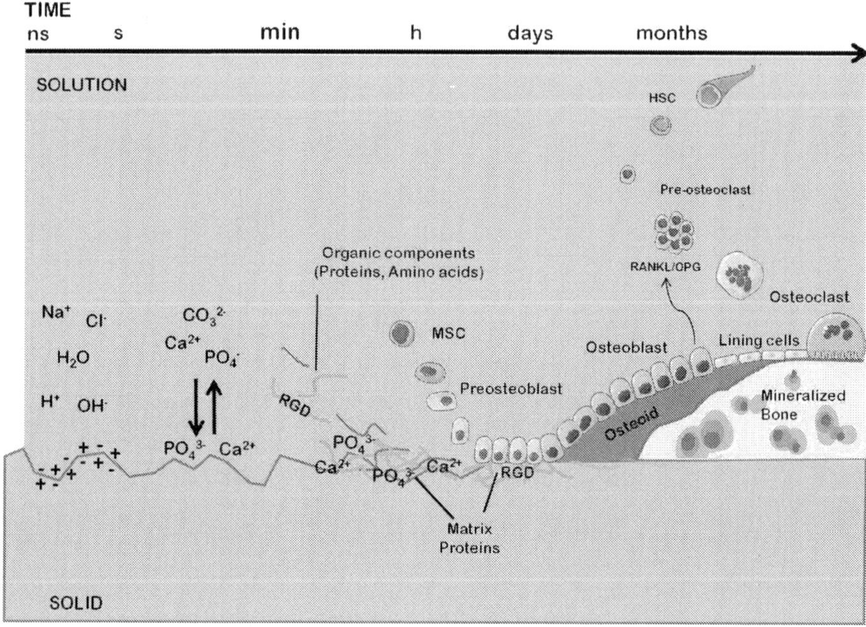

Fig. 2. Time dependent development of the biomaterial-tissue interface leading to bone formation. Figure was produced using Servier Medical Art.

When the bone graft substitute has been implanted in the patient, the biomaterial and surrounding tissue fluids will start interacting within nanoseconds (ns). After minutes (min) to hours (h), depending on material and solution properties, equilibrium will be reached, organic components will be attracted. The development of a biomaterial-tissue interface will be started and this interface will facilitate (pre)osteoblast attachment. A possible attachment mechanism could be the integrin-RGD sequence binding leading to activation of intracellular signaling and change of cellular functions. Activated and differentiated osteoblasts will then produce the bone matrix (osteoid), which needs to be mineralized to become bone. The mature mineralized bone will undergo bone remodeling involving both osteoclast and osteoblast activity in response to physiological changes in the environment. (Fig. 2 based on: Hamilton & Brunette, 2007; Hing, 2004; Hoppe et al., 2011; Jell & Stevens, 2006; Raggatt & Partridge, 2010; Roach et al., 2007; Shekaran & Garcia, 2011; Zambuzzi et al., 2011).

3. Bone graft substitute materials and biological media

Biomaterials have different origins and may be grouped into natural materials, metals, ceramics and polymers. Chemical and physical properties of the material might differ between the bulk and the surface. The material bulk will determine the mechanical and other physical characteristics, while the surface is responsible for the interaction with the surrounding environment. In the present study, three commercially available biomaterials (described below) were used as model systems to analyze the surface reactions which may promote cell adhesion and bone formation. The materials investigated have different origin and are already used in the clinic to repair dentoalveolar bone defects in patients. Frios® Algipore® (Algipore), is a biological fluoro-hydroxyapatite derived from calcifying marine algae. Bio-Oss® (Bio-Oss), is a deproteinized sterilized cancellous bovine bone chemically and structurally comparable to mineralized human bone. Bioactive glass 45S5 (45S5), a synthetic material, is a four-component glass. Bulk chemical composition and some important properties of these three bone graft substitutes are summarized in Table 1.

Biomaterial properties			
Biomaterial	Algipore	Bio-Oss	45S5
Particle size(μm)	500–1000	250–1000	70–700
Chemical composition	$CaCO_3$ transformed into $Ca_5(PO_4)_3OH_xF_{1-x}$[†]	n/a (deproteinized bone)	45wt.% SiO_2, 24.5wt.% CaO, 24.5wt.% Na_2O & 6wt. % P_2O_5 [‡]
Surface area (m^2/g)	10.4	85.5	0.16
Pore Volume (cm^3/g)	0.06	0.42	0.0009
Average pore size (nm)	23.6	19.5	23.3

Table 1. Biomaterial properties for Algipore ([†]Ewers, 2005), Bio-Oss & Bioactive glass 45S5 ([‡] Hench, 2006).

A great advantage is that histological studies in patients treated with these biomaterials have been published. It is clear from the clinical studies that new bone is formed around all three materials. However, the materials react very different when they are placed in patients. Algipore is resorbed and replaced by newly formed bone (Ewers, 2005), Bio-Oss is neither resorbed by osteoclasts nor degraded (it can still be seen after several years) but new bone is formed in close contact around the Bio-Oss particle (Mordenfeld et al., 2010). 45S5 glass particles corrode and bone formation seems to start inside the particle which is completely replaced by bone after approximately 16 months (Tadjoedin et al., 2000). The ability of these three materials to promote bone formation in vivo taken together with the different material properties provides the concept for our experimental investigation of their surfaces in equilibrium with biological medium. The formation of new bone at the biomaterial surface implies a reaction regulating bone cell functions at the interface. 45S5 has been analyzed in many studies but very little is known concerning the surface reactions and dissolution of the other two materials. We used a-MEM cell culture medium, instead of water, ethanol or simulated body fluid (SBF), which is common in several studies, because the composition of cell culture medium (Table 2) mimics the extra cellular tissue fluid more closely.

α-MEM Component		mM
Inorganic Salts		
Calcium Chloride	$CaCl_2 \cdot 2H_2O$	1.8
Magnesium Sulfate	$MgSO_4$ (anhyd.)	0.814
Potassium Chloride	KCl	5.33
Sodium Bicarbonate	$NaHCO_3$	26.19
Sodium Chloride (ionic medium)	NaCl	117.24
Sodium Phosphate monobasic	$NaH_2PO_4 \cdot 2H_2O$	1.01
Organic		
Amino Acids	21 Amino Acids (Main L-Glutamine)	8.503 (2)
Vitamins	11 Vitamins (Main Ascorbic Acid)	0.324 (0.284)
Other Components	4 Compounds (Main D-Glucose/Dextrose)	6.588 (5.56)

Table 2. α-MEM cell culture medium composition (Gibco®, obtained from Invitrogen. Cat.nr: 22561021).

If analyzing the materials surface reactions using a non-physiological solution, the results will not be relevant when relating to clinical reactions. Furthermore, using a-MEM for incubation of the materials gives us the possibility to use the incubation medium for cell culture experiments which otherwise is not possible.

Our first series of experiments were performed without addition of serum to the incubation medium which means that amino acids but no proteins are present (Mladenovic et al., 2010). In 1991, Kokubo et al., suggested that prediction of an artificial materials bioactivity (i.e. the ability of a material to form a bone-like hydroxyapatite on the surface) is essential and could be evaluated in vitro with simulated body fluid (SBF) (Kokubo & Takadama, 2006). This method has since then been widely used for assessment of bioactivity in vitro. Recently, the use of SBF for testing bioactivity has been questioned and suggestions have been made for improvement of the test in general (Bohner & Lemaitre, 2009). In line with this, Lee et al. (2011) suggested that cell culture medium would be a better option for bioactivity testing of biomaterials. In contrast to Lee et al. (2011) we believe that the presence of serum/proteins is essential when studying the interface and therefore in our second set of experiments we added 10% fetal bovine serum (FBS) to the cell culture medium.

Our two series of experiments correspond to the initial stages of bone regeneration process before cell attachment (Fig. 2.). What interfacial processes can experimentally be observed? What phenomena dominate at the interface? Why do cells find the interface attractive to initiate osteogenesis? We will try to answer some of these questions by applying one of the most powerful surface analysis techniques – X-ray photoelectron spectroscopy (XPS).

4. X-ray Photoelectron Spectroscopy (XPS) with fast-frozen samples

X-Ray Photoelectron Spectroscopy (XPS) is one of the most widely used principal techniques that probe the surface of materials (Ratner & Castner, 2009). The attractiveness of XPS is explained by its relatively simple theoretical background, applicability to virtually any type of samples, very high and sound information output, as well as sufficient availability of commercially manufactured laboratory spectrometers and experienced personal.

XPS experiment is based on the photoelectric effect. When X-rays (usually monochromatic) with known, well-defined, photon energy (hv) strike the sample, photoelectrons from the constituting atoms are emitted. By measuring the kinetic energy (KE) of the emitted electrons, their binding energy (BE) can be calculated in accordance with the energy conservation law:

$$BE = h\nu - KE - \varphi \tag{1}$$

where φ is the spectrometer work function (constant for given instrument).

All electrons with binding energies less than the energy of the X-ray beam (e.g. mono Al K$_\alpha$ 1486.6 eV) are excited and can be counted. The binding energies for core level electrons (1s, 2s, 2p, 3s, 3p, 3d, etc.) are unique for each chemical element in the Periodic Table. Thus, the determination of surface elemental composition is straightforward from the BE values obtained by XPS. Typical XPS spectrum (Fig. 3.) is recorded as a plot of the number of detected electrons (counts per second) on the ordinate and their binding energy (electron-Volts, eV) on the abscissa.

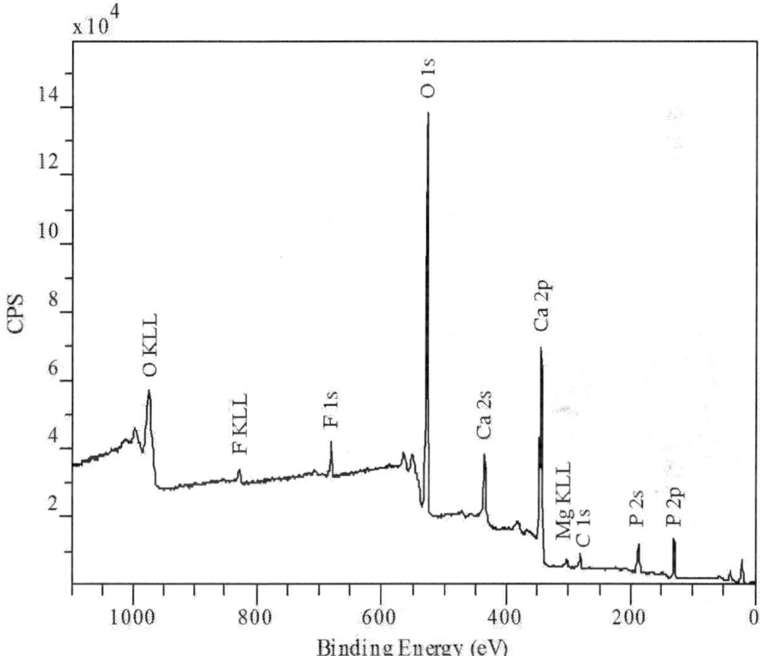

Fig. 3. Survey XPS spectrum of initial (dry) Algipore biomaterial.

X-rays penetrate deep into the sample (tens of microns) and generate photoelectrons over the entire penetration depth. Of interest in XPS are the electrons that do not lose their kinetic energy travelling through the sample. The escape depth of the photoelectrons from solids limits the analysis depth to 2 – 10 nm, providing very high surface sensitivity of the technique. The depth of analysis is dependent on the KE of the measured photoelectron, the sample density and the type of elements present. Typically, the depth of analysis is determined to be 2-3 nm for metals, 3-6 nm for inorganic oxides, and 6-10 nm for organic compounds.

The key feature of XPS making the technique exclusively important for surface chemistry is a direct measurement of a chemical shift. Specific chemical information is obtained on the principle that the binding energies of core electrons of an atom are affected by the valence electrons. Shifts in binding energies of core levels, therefore, occur (and are routinely measured) due to changes in electron density around the atom of a sample. As a result, a change in oxidation state, ligand electronegativity, coordination, protonation, etc. can be immediate experimentally observed. To illustrate the chemical shift effect, high resolution C 1s spectrum showing different functional groups of α-MEM fast-frozen drop without serum is given in Fig. 4. Specific BE values, which have been determined experimentally for different chemical compounds, have been tabulated and can be easily found in handbooks (e.g. Handbook of X-ray Photoelectron Spectroscopy) and databases (e.g. NIST X-ray Photoelectron Spectroscopy Database & The XPS of Polymers Database (Beamson & Briggs, 2001)).

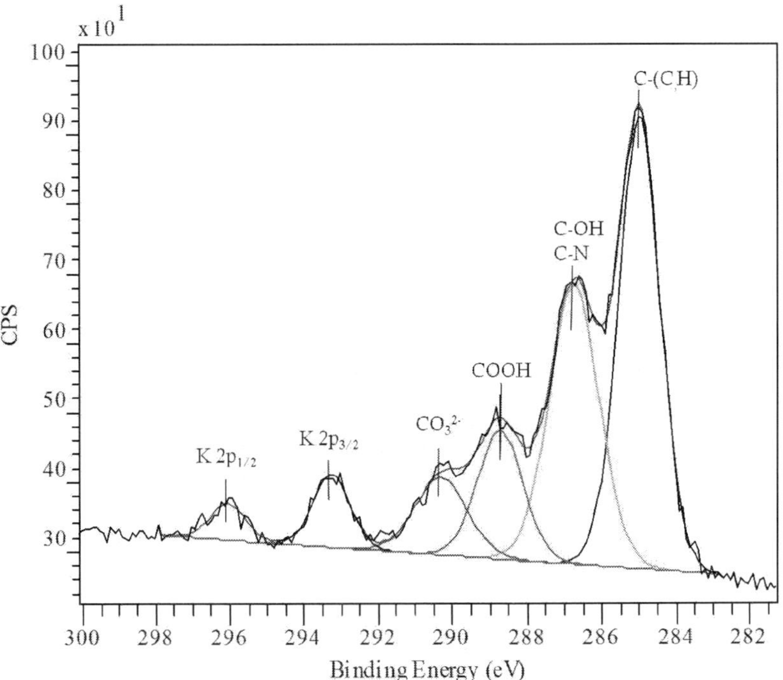

Fig. 4. XPS C 1s spectrum of a-MEM fast-frozen drop without serum.

Intensity of the photoelectron line is proportional to the numbers of atoms at the surface. An error of 10% is typically quoted for routinely determined XPS atomic concentrations using standard relative sensitivity factors. However, the intensities measured from similar samples are reproducible with good precision. Since many surface related problems involve monitoring changes in samples, the precision of XPS makes the techniques quantitatively very powerful. To avoid uncertainties in absolute intensities measurements, atomic ratios calculated from the obtained atomic concentrations are implemented. It is the ratios, which directly relate to the formulas of chemical compounds, that are quantitative values in XPS. Analysis area can be varied in the range of tens μm^2 up to 1 cm^2, therefore small spot analysis and surface mapping are available. Analytical sensitivity of XPS is exceptionally high, and 10% of monolayer (or 0.1 atomic % concentration) being routinely detected and measured in reasonable time with acceptable signal-to-noise spectral ratio.

Modern achievements in XPS applications, including special sample treatment and handling, provide insight to intimate information about chemistry, vertical and lateral distribution of chemical species, and even topography and morphology of a surface. Detailed description of XPS principles, instrumentation, practice, and developing aspects can be found in "XPS Bible" (Briggs & Grant, 2003). Some biomedical (Castner & Ratner, 2002) and bioengineering (McArthur, 2006) applications of XPS were earlier reviewed.

In respect to real solid (biomaterial) - aqueous solution (biological media) interface, there has always been a major disadvantage in XPS that the technique cannot operate in ambient conditions. To detect the photoelectrons, ultra-high vacuum (UHV) is necessary. It is the reason why the technique is traditionally considered to be *ex situ*. Indeed, conventional XPS has very limited application to real wet samples which possess very high vapor pressure under ambient conditions and are not compatible with the UHV environment of electron spectrometer. However, inelastic scattering of photoelectrons on gas phase water molecules can be avoided in two general ways: (i) decrease the distance between the sample and detector or/and (ii) reduce the vapor pressure by lowering the sample temperature or/and reduce the size of evaporating surface. Experimental approaches to real solid-aqueous solution interfaces along these lines were shortly reviewed by Shchukarev (2006a; b), including recent developments for humid conditions (Salmeron & Schlögl, 2008).

Considering conventional laboratory electron spectrometers, only two approaches can be regarded as adequate sample preparation and handling methods which do not significantly alter the real interface. On the basis of well-known "freeze-drying" technological process, a deep freezing (*ex situ*) of hydrated samples followed by ice sublimation (*in situ*) was proposed (Ratner et al, 1978) and developed (Ratner, 1995) aimed towards the application for polymers used in biology and medicine. The freezing of hydrated polymer sample is performed at 160 or 113 K, *ex situ*. The frozen sample is then placed onto precooled sample holder in the spectrometer and pumped down to UHV. XPS spectra taken with the frozen polymer indicate only O 1s photoelectron line corresponding to the water (ice). The sample is then heated *in situ* up to 200 K until C 1s line from the polymer appears, with consecutive XPS measurements performed at 160 K. The final dehydration of the sample is carried out at 303 K, also *in situ*, followed by XPS investigation of dehydrated sample. In spite of important information obtained about the processes at polymer-water interface, the technique is only applicable to pure water or aqueous solutions of very low concentrations. Ice sublimation from natural frozen biological media will "precipitate" all non-volatile solution components, like inorganic salts and organic compounds, thus altering the interface and its real chemical composition. Taken into account the analysis depth of XPS, it is very realistic that only solution components will be seen in the spectra. Moreover, the "freeze-drying" technique has not been widely used because most commercial laboratory instruments are not equipped with accessories necessary to perform controlled ice sublimation, and the experiment is complicated and time consuming (Lukas et al, 1995).

The simplest technique to study aqueous solutions in the form of solid transparent ice was proposed in 1969 (Kramer & Klein, 1969), and developed during 1970's by Burger's group (Burger & Fluck, 1974; Burger et al, 1975; Burger, 1978; Burger et al, 1977). A fast-frozen sample was prepared by direct injection of solution drop inside the spectrometer onto "cold finger" holder precooled by liquid nitrogen. XPS spectra were then acquired also at liquid nitrogen temperature. Freezing rate is so high (15-20 ^{o}C/s) that a frozen drop coming from a pipette nozzle often has a shape of ball or even onion. It is extremely important for the technique that fast-freezing has been proven to preserve the chemical speciation of solutes, and prevents solutions from crystallization.

Recently, the fast-freezing procedure was modified and applied to study the mineral- (Shchukarev & Sjöberg, 2005; Shchukarev, 2006b), bacteria- (Leone et al, 2006; Ramstedt et al, 2011), and biomaterial- (Mladenović et al, 2010) aqueous solution interfaces. Instead of a solution drop, a wet paste (alternatively gel, sediment etc) is applied to the sample holder, usually outside the spectrometer and at room temperature (T). The wet paste is obtained by centrifuging from diluted equilibrated powder (bacterial, colloidal) suspension. If the suspension is sensitive to oxygen, a glove box attached to the entry lock of spectrometer can be easily used. After applying the wet paste, the sample holder is immediately placed onto precooled claw (-170 ^{o}C) of sample transfer rod in the spectrometer entry lock. After a few seconds, the paste becomes visibly frozen and is kept at atmospheric pressure of dry nitrogen for another 30-45 s to be sure that it is cold enough to prevent water losses during the pumping. The pumping and XPS measurements are performed under liquid nitrogen cooling. To investigate the changes of the interface due to water loss, a fast-frozen sample can be left at the analysis position inside the spectrometer overnight without cooling. Slow increase in the sample temperature up to room T causes sublimation of water and other possible volatile species into the vacuum system of the spectrometer. Corresponding changes in composition and chemical speciation at the "dry" surface are monitored and quantified next day by routine XPS measurements.

Cryogenic XPS with fast-frozen samples was shown to be indispensable in the determination of electrolyte interface concentrations, their pH and ionic strength dependences, and particularly the particles surface charge using the atomic ratio of electrolyte counter-ions. As an example, the atomic ratio of Na$^+$ and Cl$^-$ (in the case of NaCl ionic media) at the interface of neutral particles is equal to one. A significant excess of sodium (chloride) ions is observed for negatively (positively) charged surfaces. Practically important for bioceramics, XPS determination of the surface charge is suitable for large particles where traditional zeta-potential measurements are impossible to perform due to immediate sedimentation of the suspension. Finally, XPS measurements repeated the next day at room temperature, with the same sample, have provided even more direct experimental evidences for the important interfacial phenomena, such as specific adsorption, ion pair formation, protonation of amine group as well as remarkable interfacial shifts in protonation constants (Ramstedt et al, 2004; Shimizu et al, 2011). In addition, pH dependent interface ligand exchange reaction, M-OH + Cl$^-$ $\leftrightarrow$ M-Cl + OH$^-$, was found and described; a surface potential drop due to collapse of electrical double layer was observed and measured; and the thickness of interface solution layer about 5 Å was experimentally estimated.

Cryogenic XPS with fast-frozen samples, as a "snapshot" technique, is very promising in describing even dynamic processes at the interface of biomaterial in biological media by changing contact time or/and one of solution parameters (T, concentration, pH, composition, etc.). In this fashion, XPS was used to study dry surfaces of three biomaterials described in previous section as well as their wet pastes obtained from α-MEM solution with and without serum addition. In parallel with XPS measurements, the supernatant is also analyzed, and both sets of experimental data are combined together to obtain chemically self-consistent insights into the phenomena at the interface of bone graft substitute (45S5, Algipore and Bio-Oss) materials in biological (α-MEM) media.

5. Three bone graft material interfaces in biological media

Formation of new bone in close proximity of implanted particles, described in section 3 and shown in Fig 5, indicates that bone formation process is facilitated by the surface of the biomaterial. In the case of a bone graft substitute, it occurs at the interface of the biomaterial which is in equilibrium with surrounding body fluid.

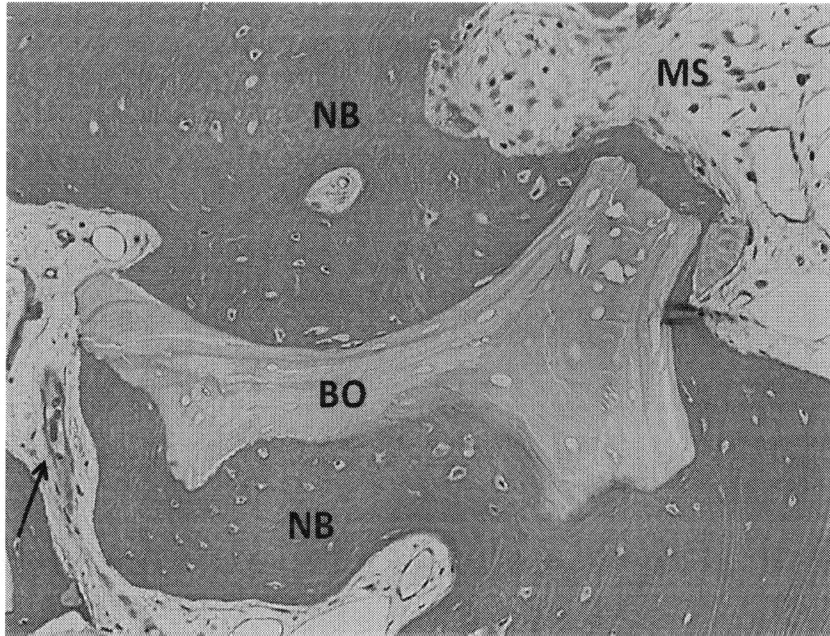

Fig. 5. Light microphotograph of biopsy taken from patient at implant installation, 12 months after bone augmentation with Bio-Oss (without addition of autogenous bone). The histological section shows newly formed bone (NB) in close relation to the Bio-Oss (BO) interface. Marrow space (MS) surrounding NB and BO indicates bone formation growing "out" from the bone graft substitute. Black arrow (↑) points out a blood vesicle. Htx-Eosin staining, original magnification 10x. (Image courtesy of Dr A. Sahlin-Platt).

The equilibrium, in turn, is achieved in the process of initial (dry) particle surface interaction with water and other components of the biological medium. Reference XPS data on initial surface chemical composition of all three bone graft substitutes and both of the biological media (as a fast-frozen drops) are given in Table 3.

Surface composition (Atomic %) of initial biomaterials and α-MEM biological media					
	α-MEM	α-MEM + serum	Algipore	Bio-Oss	45S5
Photoelectron line					
C 1s					
Organic	14.72	33.88	6.79	8.39	12.3
Surface carbonate	0.92	1.67	0.82	1.61	16.78
O 1s	70.48	47.03	56.54	57.66	49.27
N 1s	1.94	7.76	-	-	-
Ca 2p	-	-	19.87	19.49	2.91
P 2p	-	-	12.91	11.59	0.25
Si 2p	-	-	-	-	2.23
F 1s	-	-	2.92	0.82	-
K 2p	0.24	0.33	-	-	-
Na 1s	5.8	4.77	0.61	0.45	16.27
Cl 2p	5.91	4.57	-	-	-

Table 3. Surface composition of initial Algipore, Bio-Oss, Bioactive glass 45S5 and α-MEM biological media.

The chemical compositions of the surface of Bio-Oss and Algipore are very similar to each other, only differing in fluorine content. Except for surface organic contaminations, their surface chemistry is in good agreement with bulk apatite formulation in respect to Ca/P atomic ratio. Conversely, the surface of Bioactive glass particles is covered by sodium bicarbonate (Mladenovic et al., 2010), probably due to increased surface reactivity and prolonged storage of the material in air. Upon the interaction with biological medium, this water soluble surface $NaHCO_3$ phase will dissolve, enriching the medium by additional sodium and carbonate ions and probably causing a momentarily change in the local pH. Due to very low surface area of 45S5 Bioactive glass (Table 1) and the remarkable buffer capacity of the medium with a pH of 7.3, the α-MEM composition will not be noticeably influenced. XPS data, taken from fast-frozen drops of both α-MEM solutions (Table 3), demonstrates a predominance of sodium and chloride in the inorganic part of the medium. It is important to notice that the Na/Cl atomic ratio equal to one (Na/Cl = 5.8/5.91 for pure α-MEM, and 4.77/4.57 for α-MEM with serum). This would indicate a neutral (non-charged) surface of the fast-frozen drops. The amino acids are the dominating contributors to the organic part of pure α-MEM as evident by N 1s spectrum (Fig. 6, a), showing both protonated and neutral amino groups with BE of 402.1 and 400.2 eV, respectively. Serum addition to the biological medium results in the dominance of protein which can be clearly seen by the increased carbon and nitrogen concentrations (Table 3), and also evidenced by the N 1s spectrum (Fig 6, b) being similar to the one observed for bacterial cell walls (Leone et al., 2006).

XPS investigation of fast-frozen biomaterial samples, conditioned in original α-MEM medium (without serum), shows that equilibrium is reached at the interface within only 1 day (Mladenovic, et al., 2010). Corresponding interfacial chemical compositions are given in Table 4.

Surface composition (Atomic %) of biomaterials in original α-MEM after 1 day of equilibration				
Photoelectron line	α-MEM	Algipore	Bio-Oss	45S5
C 1s				
Organic	14.72	35.99	8.33	43.05
Surface carbonate	0.92	2.6	1.5	2.11
O 1s	70.48	34.82	68.56	38.96
N 1s	1.94	3.38	-	2.79
Ca 2p	-	5.7	9.74	0.58
P 2p	-	4.69	7.02	0.16
Si 2p	-	-	-	0.17
F 1s	-	-	-	-
K 2p	0.24	0.46	0.06	0.34
Na 1s	5.8	5.62	3.08	5.04
Cl 2p	5.91	6.76	1.71	6.8
Atomic ratio Na/Cl	0.98 : 1	0.83 : 1	1.8 : 1	0.74 : 1

Table 4. Surface composition of biomaterials in α-MEM after 1 day of equilibration.

The Bio-Oss - α-MEM medium interface does not show any adsorption of organic components from the biological medium therefore similar in composition to the initial (dry) surface, except for the contributions from water and Na^+ and Cl^- ions from the solution. An excess of sodium over chloride, providing the atomic ratio Na/Cl 1.8:1, indicates negative charge of the Bio-Oss particles' surface. Assuming the Bio-Oss is similar to hydroxyapatite (HAP), the isoelectric point of the biomaterial can be expected to have a pH value close to 8.1 (Bengtsson, et al., 2009). Therefore, the surface in α-MEM solution has to be substantially neutral with the atomic ratio Na/Cl close to one. The observed excess of sodium ions may indicate a re-adsorption of the phosphate-related solution species, like HPO_4^{2-}, producing outer-sphere coordination sites for the Na^+. Such occurrences were confirmed for synthetic HAP by the solution analysis and surface complexation modeling (Bengtsson, et al., 2009). Additional adsorption of hydro phosphate ions is indirectly observed by a decrease in the atomic ratio Ca/P at the Bio-Oss - α-MEM medium interface, compared to that at the dry surface.

The interfacial composition of both Algipore and 45S5 is dominated by water, sodium and chloride, and adsorbed organic molecules. The amino acids are the key players in the adsorbed layer, as evident by N 1s spectra (Fig. 6, c and 7, c) which similar to that of the pure α-MEM (Fig. 6, a, and 7, a). As it was noted by Mladenovic et al. (2010), the adsorption of amino acids on the surface of Algipore and 45S5 leads to a partial deprotonation of $-NH_2^+$ group, which can be clearly seen by a decrease in intensity of the N 1s component with BE 402.1 eV. Amine group deprotonation would mean an increased pH value at the interface, compared to that in the solution. The influence of the biomaterial surface on the acid-base equilibria at the interface can be important for the expected attachement of cells. In particular, an increase in interfacial pH from physiological value of 7.4 to 8 significantly enhances both the proliferation and the alkaline phosphatase activity of osteoblasts (Shen, et al., 2011; Liu et al., 2011).

The atomic ratio Na/Cl is lower than 1, and this clearly indicates that the surface of both the Algipore and the Bioglass is positively charged. The positive charge seems to be related to the pronounced adsorption of organic species from the α-MEM solution. Unfortunately, it is difficult to distinguish whether the generation of the surface charge is followed by the formation of electrical double layer, which precedes the adsorption, or vice versa. Since no adsorption occurred at the negatively charged surface of Bio-Oss, the surface charge generation seems to be a first principal phenomenon responsible for the formation of interface's organic (mainly amino acids) layer with specific acid-base properties. We have to emphasize that, had the samples been suspended in SBF, this adsorption would never have been detected, as was shown by Mahmood & Davies, 2000 for the Bioactive glass 45S5 material.

Serum addition to the α-MEM causes significant changes in the interfaces discussed above. Interfacial chemical composition (Table 5) and XPS N 1s spectra, typical for serum (Fig. 6 - 8), demonstrate an immediate adsorption of the protein at the surface of all three biomaterials after only 1 day of equilibration.

Surface composition (Atomic %) of biomaterials in α-MEM with 10% serum after 1 day of equilibration				
Photoelectron line	α-MEM + serum	Algipore	Bio-Oss	45S5
C 1s				
Organic	33.88	52.04	51.0	55.74
Surface carbonate	1.67	2.0	-	2.13
O 1s	47.03	29.42	30.0	26.68
N 1s	7.76	7.17	9.62	5.37
Ca 2p	-	0.73	3.03	0.44
P 2p	-	0.7	3.54	0.21
Si 2p	-	-	-	0.28
F 1s	-	-	-	-
K 2p	0.33	0.32	-	0.29
Na 1s	4.77	2.5	0.6	3.08
Cl 2p	4.57	5.09	0.74	5.76
Atomic ratio Na/Cl	1.04 : 1	0.49 : 1	0.81 : 1	0.53 : 1

Table 5. Surface composition of biomaterials in α-MEM with 10% serum after 1 day of equilibration.

The sum of the atomic concentrations of organic carbon and nitrogen is about 60 atomic %, independent on the bone graft substitute. However, the atomic ratio C/N, calculated for Bio-Oss (5.3:1), is very close to one for the serum-containing α-MEM (4.4:1, Table 5). This may indicate that solution structure of the protein is not noticeably altered upon adsorption. Significant increase in the C/N ratio observed for Algipore (7.3:1) and 45S5 (10.4:1) would mean possible changes in secondary protein structure resulting in the preferential orientation of nitrogen-containing functional groups towards the biomaterial surface and the carbon-enriched polymeric chains towards the solution. The adsorption-induced orientation effect correlates with the different surface charge gained by the biomaterial particles in the original α-MEM. It is important to notice that the adsorption of serum at the surface of Bio-Oss unexpectedly causes the charge reversal at the interface and thereby the positive charge of the particles, as evidenced by the change in the atomic ratio Na/Cl (1.8 → 0.81, Tables 4 and 5). The protein layer formed at the interface could serve as a template for the recruitment of cells, production of bone extracellular matrix, biomineralization, and finally, as an initiator of osteogenesis.

Using cryogenic XPS with fast-frozen samples, two dominating phenomena common for all three biomaterials can clearly be distinguished at the interface with biological media. In the absent of proteins (serum), the particles gain different surface charge. It is the charge that seems to be pivotal for the adsorption of organic molecules from the medium. Significant adsorption of organic species, in particular amino acids, at the positively charged surfaces of Bioactive glass 45S5 and Algipore (completely different in the bulk and surface chemical composition) results in the formation of a thin organic interface layer. Chemically and structurally similar to Algipore, Bio-Oss particles in α-MEM gain negative surface charge which suppresses the adsorption. Even protonated $-NH_3^+$ end members of amino acids are not attracted by the negatively charged surface. An exception to the rule is protein adsorption. Serum adsorbs immediately at the surface of all three biomaterials, independent of their surface charge. Moreover, the protein adsorption at the surface of Bio-Oss results in the charge reversal at the interface and the positive charge of the particles. Serum adsorption cannot be explained by pure electrostatics; an underlying mechanism should include weak hydrophobic/hydrophilic interactions, hydrogen bonding, and change in the protein primary and secondary structures. The protein adsorption at the abiotic materials, independent on their surface charge and chemical composition, seems to be major prerequisite for subsequent cell recognition and biomineralization.

Besides the template formation for bone regeneration, fast adsorption of organic molecules and serum can serve as a dissolution/precipitation barrier, controlling the biomaterial degradation and resorbing processes. In this respect, time dependence of the Ca/P atomic ratios, measured in biological media and at the interface, can indicate the extent of bone graft substitute dissolution and re-adsorption of calcium and phosphate ions. We have monitored the Ca/P atomic ratio in pure α-MEM (Mladenovic, et al., 2010). Since the availability of Ca and P from the medium in our batch experiments is limited, compared to circulating natural body fluids, the results might not be relevant and therefore are not discussed here. Regardless, biomaterial degradation and biomineralization are long-term processes and can be considered as an effect of "secondary order" at the first stages of biomaterial – biological medium interface evolution.

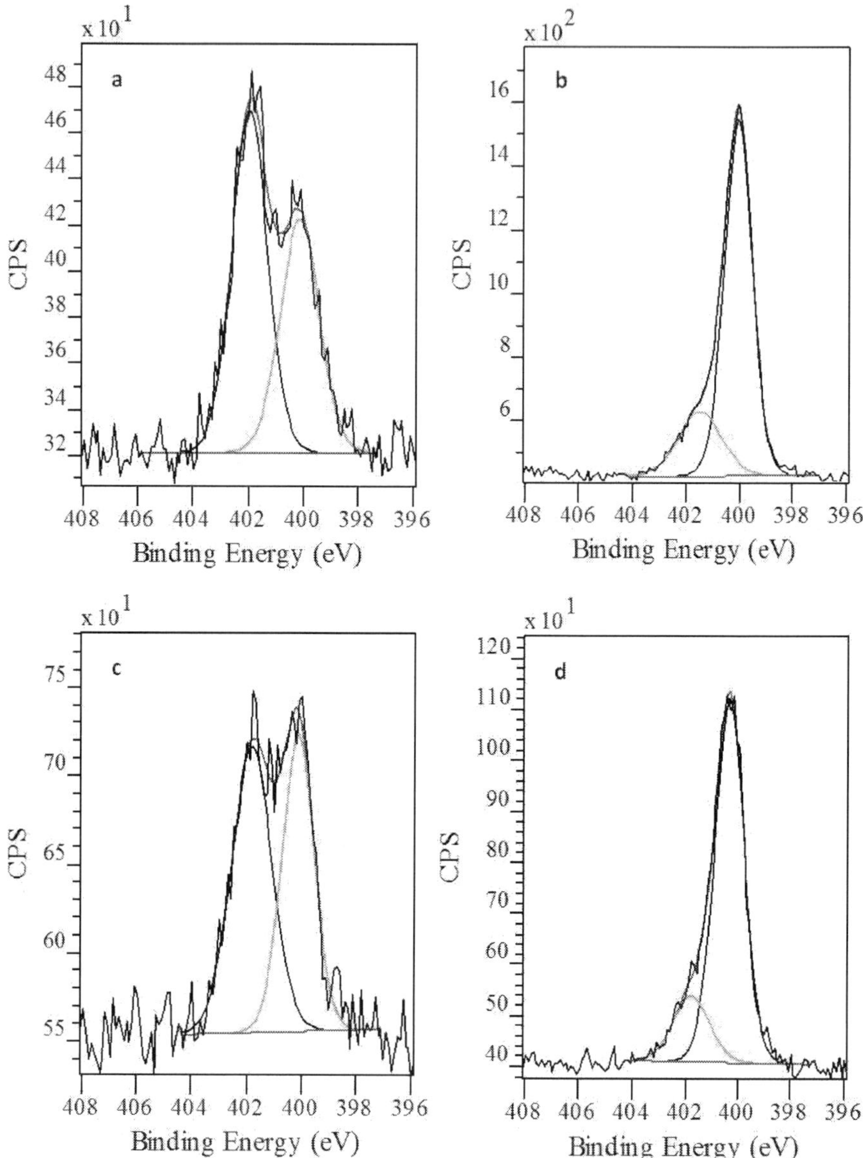

Fig. 6. N 1s spectra of fast-frozen samples: (a) a-MEM without serum, (b) a-MEM with serum, (c) Algipore in a-MEM without serum (1 day), (d) Algipore in a-MEM with serum (1 day).

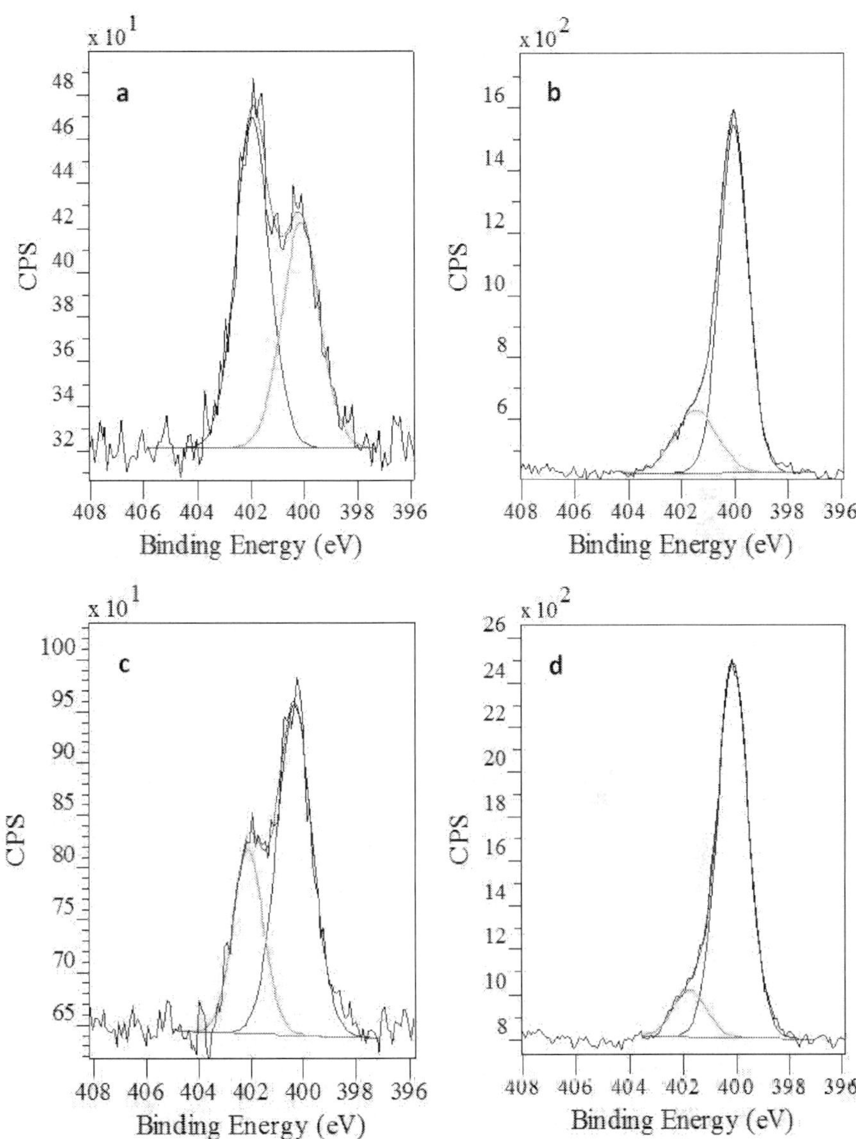

Fig. 7. N 1s spectra of fast-frozen samples: (a) a-MEM without serum, (b) a-MEM with serum, (c) Bioactive glass 45S5 in a-MEM without serum (1 day), (d) Bioactive glass 45S5 in a-MEM with serum (1 day).

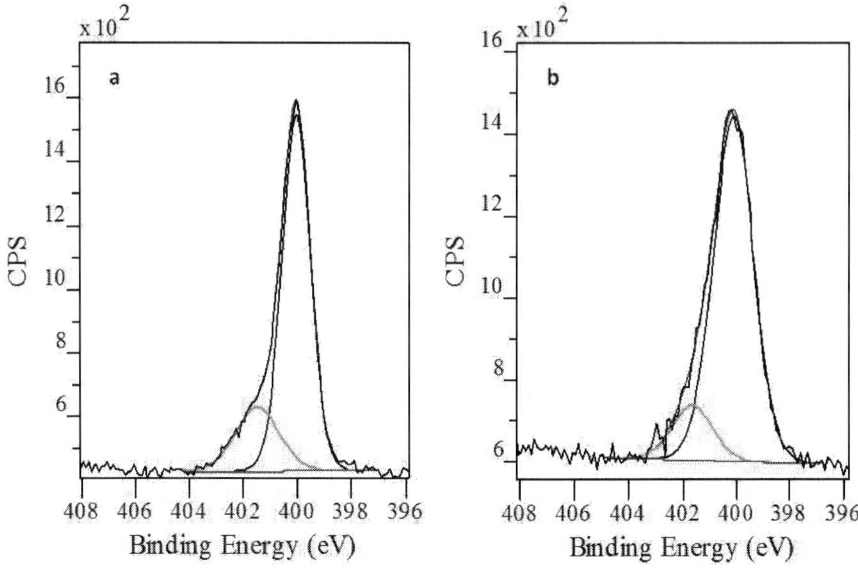

Fig. 8. N 1s spectra of fast-frozen samples: (a) a-MEM with serum, (b) Bio-Oss in a-MEM with serum (1 day).

6. Conclusion

From the cells viewpoint, it is the interface that will determine "to build or not to build" bone. In this respect, cryogenic XPS demonstrates the great potential of fast-freezing technique to study the real biomaterial interface in cell culture medium. Physico-chemical information about the interface, important for cell adherence, such as chemical composition, surface charge, and absorbing molecules can directly be obtained. The next step in this direction would be to acquire nanoscale information on surface structures and adsorption of specific proteins at the biomaterial interface that could have an effect on cellular function. Surface science techniques like Atomic Force Microscopy (*in vivo* and *in vitro*) and Mass Spectrometry (*in vitro*) can be laterally applied to investigate areas occupied by a single cell. The cell surface can be studied in the same manner to elucidate membrane functions responsible for the cell attachment.

The use of bone graft substitutes in clinical practices is still limited by insufficient knowledge of biomaterial interactions with the living tissue. This makes the treatment outcome unpredictable or even unreliable. To improve the knowledge, the great scientific potential existing in biology and material science must be combined thus creating important "interface" between respective scientists. The interactions within this "interface" can result in design of new cost effective biomaterials with improved tissue regenerative potential.

7. Acknowledgment

The authors would like to thank Dr. J.P.L. Kenney for reading the manuscript and suggesting editorial improvements.
The authors would also like to thank Dr. A. Sahlin-Platt for kindly providing the histological light micrograph shown in Fig. 5.
The authors would like to acknowledge that this work was supported by grants from the County Council of Västerbotten, Sweden, and from the Swedish Dental Society.

8. References

Bengtsson, A., Shchukarev, A. Persson, P., & Sjöberg, S. (2009). A Solubility and Surface Complexation Study of a Non-Stoichiometric Hydroxyapatite. *Geochimica Et Cosmochimica Acta,* Vol.73, pp. 257-67

Beamson, G., & Briggs, D. (Eds.).(2001). *The XPS of Polymers Database.* Version 1.0. Surface Spectra Ltd., UK

Bohner, M., & Lemaitre, J. (2009). Can Bioactivity Be Tested in Vitro with SBF Solution?. *Biomaterials,* Vol.30, pp. 2175-9

Briggs, D., & Grant, J.T. (Eds.). (2003). *Surface Analysis by Auger and X-Ray Photoelectron Spectroscopy.* IM Publications, ISBN 9781901019049, Chichester, West Sussex, U.K.

Burger, K. (1978). Charge Correction in XPS/ESCA.Bulk Solvent as Internal Standard in Study of Quick-Frozen Solutions. *Journal of Electron Spectroscopy and Related Phenomena,* Vol.14, pp. 405-10

Burger, K., & Fluck, E. (1974). X-Ray-Photoelectron Spectroscopy (ESCA) Investigations in Coordination Chemistry.1. Solvation of $SbCl_5$ Studied in Quick-Frozen Solutions. *Inorganic & Nuclear Chemistry Letters,* Vol.10, pp. 171-77

Burger, K., Fluck, E., Binder, H., & C. Varhelyi. (1975). X-Ray Photoelectron-Spectroscopy (ESCA) Investigations in Coordination Chemistry.2. Study of Outer Sphere Coordination and Hydrogen Bridge Formation in Cobalt(III) and Nickel(II) Complexes. *Journal of Inorganic & Nuclear Chemistry,* Vol.37, pp. 55-57

Burger, K., Tschimarov, F., & Ebel, H. (1977). XPS-ESCA Applied to Quick-Frozen Solutions.1. Study of Nitrogen-Compounds in Aqueous-Solutions. *Journal of Electron Spectroscopy and Related Phenomena,* Vol.10, pp. 461-65

Castner, D. G., & Ratner, B. D. (2002). Biomedical Surface Science: Foundations to Frontiers. *Surface Science,* Vol.500, pp. 28-60

Diniz, M. G., Soares, G. A., Coelho, M. J., & Fernandes, M. H. (2002). Surface Topography Modulates the Osteogenesis in Human Bone Marrow Cell Cultures Grown on Titanium Samples Prepared by a Combination of Mechanical and Acid Treatments. *J Mater Sci Mater Med,* Vol.13, pp. 421-32

Evans, N. D., Gentleman, E., Chen, X., Roberts, C. J., Polak, J. M., & Stevens, M. M. (2010). Extracellular Matrix Mediated Osteogenic Differentiation of Murine Embryonic Stem Cells. *Biomaterials,* Vol. 31, pp. 3244-52

Ewers, R. (2005). Maxilla Sinus Grafting with Marine Algae Derived Bone Forming Material: A Clinical Report of Long-Term Results. *J Oral Maxillofac Surg,* Vol.63, pp. 1712-23

Feng, X. (2009). Chemical and Biochemical Basis of Cell-Bone Matrix Interaction in Health and Disease. *Curr Chem Biol,* Vol.3, pp. 189-96

Feng, X., & McDonald. J. M. (2011). Disorders of Bone Remodeling. *Annu Rev Pathol*, Vol.6, pp. 121-45.

Geiger, B., Spatz, J. P., & Bershadsky, A. D. (2009). Environmental Sensing through Focal Adhesions. *Nature Reviews Molecular Cell Biology*, Vol.10, pp. 21-33

Gorustovich, A. A., Roether, J. A., & Boccaccini, A. R. (2010). Effect of Bioactive Glasses on Angiogenesis: A Review of in Vitro and in Vivo Evidences. *Tissue Eng Part B Rev,*Vol.16, pp. 199-207

Hamilton, D. W., & Brunette, D. M. (2007). The Effect of Substratum Topography on Osteoblast Adhesion Mediated Signal Transduction and Phosphorylation. *Biomaterials*, Vol.28, pp. 1806-19

Hench, L. L. (2006). The Story of Bioglass. *J Mater Sci Mater Med*, Vol.17, pp. 967-78

Hench, L. L., & Wilson, J. (Eds.). (1993). *An Introduction to Bioceramics (Advanced Series in Bioceramics, Vol.1)*. World Scientific Publishing, ISBN 9789810214005, Singapore

Hing, K. A. (2004). Bone Repair in the Twenty-First Century: Biology, Chemistry or Engineering?. *Philos Transact A Math Phys Eng Sci*, Vol. 362, pp. 2821-50

Hoppe, A., Guldal, N. S. & Boccaccini, A. R. (2011). A Review of the Biological Response to Ionic Dissolution Products from Bioactive Glasses and Glass-Ceramics. *Biomaterials*, Vol.32, pp. 2757-74

Jell, G., & Stevens, M. M. (2006). Gene Activation by Bioactive Glasses. *J Mater Sci Mater Med*, Vol.17, pp. 997-1002

Kokubo, T., & Takadama, H. (2006). How Useful Is SBF in Predicting in Vivo Bone Bioactivity?. *Biomaterials*, Vol.27, pp. 2907-15.

Kramer, L. N., & Klein, M. P. (1969). Solute Investigation by Means of Photoelectron Spectroscopy. *Journal of Chemical Physics*, Vol. 51, pp.3620-1

Kundu, A. K., & Putnam, A. J. (2006). Vitronectin and Collagen I Differentially Regulate Osteogenesis in Mesenchymal Stem Cells. *Biochem Biophys Res Commun*, Vol. 347, pp. 347-57

Lee, J.T.Y., Y. Leng, K.L. Chow, F. Ren, X. Ge, K. Wang, & X. Lu. (2006). Cell Culture Medium as an Alternative to Conventional Simulated Body Fluid. *Acta Biomater*, doi:10.1016/j.actbio.2011.02.034

Leone, L., Loring, J., Sjöberg, S., Persson, P., & Shchukarev, A. Surface Characterization of the Gram-Positive Bacteria Bacillus Subtilis - an XPS Study. *Surface and Interface Analysis*, Vol.38, pp. 202-05

Liu, L., Schlesinger, P. H., Slack, N. M., Friedman, P. A., & Blair, H. C. (2011). High Capacity Na(+) /H(+) Exchange Activity in Mineralizing Osteoblasts. *J Cell Physiol*, Vol. 226, pp. 1702-12

Lu, Z. L., & Zreiqat, H. (2010). The Osteoconductivity of Biomaterials Is Regulated by Bone Morphogenetic Protein 2 Autocrine Loop Involving Alpha 2 Beta 1 Integrin and Mitogen-Activated Protein Kinase/Extracellular Related Kinase Signaling Pathways. *Tissue Engineering Part A*, Vol. 16, pp. 3075-84

Lukas, J., Sodhi, R. N. S., & Sefton, M. V. (1995). An XPS Study of the Surface Reorientation of Statistical Methacrylate Copolymers. *Journal of Colloid and Interface Science*, Vol.174, pp. 421-27

Mahmood, T. A., & Davies, J. E. (2000). Incorporation of Amino Acids within the Surface Reactive Layers of Bioactive Glass in Vitro: An XPS Study. *J Mater Sci Mater Med*, Vol. 11, pp. 19-23

McArthur, S. L. (2006). Applications of XPS in Bioengineering. *Surface and Interface Analysis*, Vol. 38, pp. 1380-85

Milani, R., Ferreira, C. V., Granjeiro, J. M., Paredes-Gamero, E. J., Silva, R. A., Justo, G. Z., Nader, H. B., *et al.* (2010). Phosphoproteome Reveals an Atlas of Protein Signaling Networks During Osteoblast Adhesion. *J Cell Biochem*, Vol. 109, pp. 957-66

Mladenovic, Z., Sahlin-Platt, A., Bengtsson, A., Ransjö, M. & Shchukarev., A. (2010). Surface Characterization of Bone Graft Substitute Materials Conditioned in Cell Culture Medium. *Surface and Interface Analysis,* Vol. 42, pp. 452-56

Mordenfeld, A., Hallman, M., Johansson, C. B., & Albrektsson, T. (2010). Histological and Histomorphometrical Analyses of Biopsies Harvested 11 Years after Maxillary Sinus Floor Augmentation with Deproteinized Bovine and Autogenous Bone. *Clin Oral Implants Res*, Vol. 21, pp. 961-70

Nakahama, K. (2010). Cellular Communications in Bone Homeostasis and Repair. *Cell Mol Life Sci*, Vol.67, pp. 4001-9

"NIST X-ray Photoelectron Spectroscopy Database." NIST Standard Reference Database 20, Version 3.5. Last updated August 27, 2007, http://srdata.nist.gov/xps/

Porter, J. R., Ruckh, T. T., & Popat, K. C. (2009). Bone Tissue Engineering: A Review in Bone Biomimetics and Drug Delivery Strategies. *Biotechnol Prog*, Vol. 25, pp. 1539-60

Raggatt, L. J., & Partridge, N. C. (2010). Cellular and Molecular Mechanisms of Bone Remodeling. *Journal of Biological Chemistry*, Vol.285, pp. 25103-8

Ramstedt, M., Nakao, R., Wai, S. N., Uhlin, B. E., & Boily, J. F. (2011). Monitoring Surface Chemical Changes in the Bacterial Cell Wall Multivariate Analysis of Cryo-X-Ray Photoelectron Spectroscopy Data. *Journal of Biological Chemistry*, Vol.286, pp 12389-12396

Ramstedt, M., Norgren, C., Sheals, J., Shchukarev, A. & Sjöberg, S. (2004). Chemical Speciation of N-(Phosphonomethyl)Glycine in Solution and at Mineral Interfaces. *Surface and Interface Analysis*, Vol.36, pp. 1074-77

Ratner, B. D. (1995). Advances in the Analysis of Surfaces of Biomedical Interest. *Surface and Interface Analysis*, Vol.23, pp. 521-28

Ratner, B.D. & Castner, D.V. (2009). Electron Spectroscopy for Chemical Analysis., In: *Surface Analysis: The Principal Techniques*. Vickerman, J.C., & Gilmore, I.S. pp. 47-109 Wiley, ISBN: 9780470017630, 2nd ed. Chichester, U.K.

Ratner, B. D., Weathersby, P. K., Hoffman, A. S., Kelly, M. A., & Scharpen, L. H. (1978). Radiation-Grafted Hydrogels for Biomaterial Applications as Studied by ESCA Technique. *Journal of Applied Polymer Science*, Vol.22, pp. 643-64

Roach, P., Eglin, D., Rohde, K., & Perry, C. C. (2007). Modern Biomaterials: A Review - Bulk Properties and Implications of Surface Modifications. *J Mater Sci Mater Med*, Vol.18, pp. 1263-77

Salmeron, M., & Schlögl, R. (2008). Ambient Pressure Photoelectron Spectroscopy: A New Tool for Surface Science and Nanotechnology. *Surface Science Reports*, Vol.63, pp. 169-99

Shchukarev, A. (2006a). XPS at Solid-Aqueous Solution Interface. *Advances in Colloid and Interface Science*, Vol.122, pp. 149-57

Shchukarev, A. (2006b). XPS at Solid-Solution Interface: Experimental Approaches. *Surface and Interface Analysis*, Vol.38, pp. (Apr 2006): 682-85

Shchukarev, A., & Sjöberg, S. (2005). XPS with Fast-Frozen Samples: A Renewed Approach to Study the Real Mineral/Solution Interface. *Surface Science,* Vol.584, pp. 106-12

Shekaran, A., & Garcia, A. J. (2011). Nanoscale Engineering of Extracellular Matrix-Mimetic Bioadhesive Surfaces and Implants for Tissue Engineering. *Biochimica Et Biophysica Acta-General Subjects,* Vol.1810, pp. 350-60

Shen, Y. H., Liu, W. C., Lin, K. L., Pan, H. B., Darvell, B. W., Peng, S. L., Wen, C. Y., et al. (2011). Interfacial pH: A Critical Factor for Osteoporotic Bone Regeneration. *Langmuir,* Vol.27, pp. 2701-08

Shimizu, K., Shchukarev, A., & Boily, J-F. (2011). X-ray Photoelectron Spectroscopy of Fast-Frozen Hematite Colloids in Aqueous Solutions. 3. Stabilization of Ammonium Species by Surface (Hydro)oxo Groups. *J. Phys. Chem. C,* Vol.115, pp 6796-6801. doi:10.1021/jp2002035

Stevens, M. M., & George, J. H. (2005). Exploring and Engineering the Cell Surface Interface. *Science,* Vol.310, pp. 1135-8

Tadjoedin, E. S., de Lange, G. L., Holzmann, P. J., Kulper, L. & Burger, E. H. (2000). Histological Observations on Biopsies Harvested Following Sinus Floor Elevation Using a Bioactive Glass Material of Narrow Size Range. *Clin Oral Implants Res,* Vol.11, 334-44

Wang, N., Tytell, J. D., & Ingber, D. E. (2009). Mechanotransduction at a Distance: Mechanically Coupling the Extracellular Matrix with the Nucleus. *Nature Reviews Molecular Cell Biology,* Vol.10, No.1, pp. 75-82

Wilson, C. J., Clegg, R. E., Leavesley, D. I., & Pearcy, M. J. (2005). Mediation of Biomaterial-Cell Interactions by Adsorbed Proteins: A Review. *Tissue Eng,* Vol.11, No.1-2 , pp. 1-18

Yee, K. L., Weaver, V. M., & Hammer, D. A. (2008). Integrin-Mediated Signalling through the Map-Kinase Pathway. *Iet Systems Biology,* Vol.2, No.1, pp. 8-15

Zambuzzi, W. F., Ferreira, C. V., Granjeiro, J. M., & Aoyama, H. (2011). Biological Behavior of Pre- Osteoblasts on Natural Hydroxyapatite: A Study of Signaling Molecules from Attachment to Differentiation. *Journal of Biomedical Materials Research Part A,*Vol. 97A, No.2, pp. 193-200

15

Mechanobiology of Oral Implantable Devices

José Alejandro Guerrero[1], Juan Carlos Vanegas[1], Diego Alexander Garzón[1],
Martín Casale[2] and Higinio Arzate[3]
*[1]Grupo de Modelado Matemático y Métodos Numéricos GNUM-UN,
Departamento de Ingeniería Mecánica y Mecatrónica, Facultad de Ingeniería,
Universidad Nacional de Colombia, Bogotá, Colombia
[2]División de Ortodoncia, Facultad de Odontología, Universidad Nacional de Colombia,
Bogotá, Colombia
[3]Laboratorio de Biología Celular y Tejidos Mineralizados, Universidad
Nacional Autónoma de México, México DF, México
Colombia - México*

1. Introduction

A dental implant is a biomaterial device inserted in the jaw bone to replace the root of a missing tooth Adell et al. (1981); Hansson et al. (1983). With the implant insertion, a junction site between the biomaterial surface and the surrounding bone known as the *bone-dental implant interface* is created Branemark (1983). The formation of a new bone matrix in this interface allows the firm and long lasting connection between the bone and the implant in a process called *osseointegration* Albrektsson & Johansson (2001); Branemark (1983). The success of this contact depends on the restoration of the functional tissues around the implant providing it with mechanical support and anchorage Gapski et al. (2003); Hansson et al. (1983). This tissue restoration is conditioned to cell migration, cell proliferation and cell differentiation phenomena depending on the pathological conditions of the patient, the biological conditions of the host bone, the implant design and surface topography, and the distribution of mechanical loads between the bone and the implant Aukhil (2000); Ellingsen et al. (2006); Sikavitsas et al. (2001). The analysis of both biological and mechanical factors is known as mechanobiology Klein-Nulend et al. (2005); Van der Meulen & Huiskes (2002).

Dental implants are widely used as anchorage devices for restoration and esthetical purposes Branemark (1983). However, in several orthodontic treatments where the anchorage is used, a higher control of treatment time and oral availability is needed Antoszewska et al. (2009); Gapski et al. (2003). In these cases, dental implants are not recommended since specific surgical procedures are required for their implantation usually demanding long lasting recovery times for insertion and loading Antoszewska et al. (2009). Furthermore, most of the orthodontic treatments are based on the temporary and direct application of loads for the movement of teeth Papadopoulos & Tarawneh (2007). This treatments rely on implantable devices capable of transmit immediate loading without the implicit need of being osseointegrated Papavasiliou (1996). Such devices, called mini-implants, reinforce the anchorage of orthodontic devices and speed up treatments given its relative simple insertion procedure and lack of osseointegration Antoszewska et al. (2009); Papadopoulos & Tarawneh (2007); Papavasiliou (1996).

Although a bone-implant interface is also created around the inserted mini-implant, its mechanobiological behavior differs from that of the dental implants since mini-implants are not fully osseointegrated Antoszewska et al. (2009). This means that mechanical considerations as loading conditions, material and surface topography govern tissue recovery and bone formation around mini-implants. Therefore, failure of mini-implants is mostly related to anchorage mechanics as bone thickness, shape design, insertion angle and insertion forces Antoszewska et al. (2009); Gracco et al. (2009); Motoyoshi, Inaba, Ono, Ueno & Shimizu (2009a); Motoyoshi, Okazaki & Shimizu (2009).

The purpose of this chapter is to provide a review of dental implants and mini-implants from a mechanobiological approach. The aim is to present the influence of intrinsic biological and mechanical conditions leading to tissue recovery at the bone-implant interface. External loading, surface treatment and surgical procedures are also addressed as conditions for the success of both kinds of devices. The discussion is supported on a vast literature review of theoretical, clinical and experimental evidence. Self-conducted experiments based on tissue engineering techniques are presented for supporting ideas on the interaction of body components with biomaterials, and for the evaluation of their biocompatibility.

Finally, we suggest several guidelines for the mechanobiological modeling of the bone-implant interface. Although the healing processes at the bone-implant interface have been deeply discussed in the experimental literature, in recent years mathematical models have gain insight for their capacity of reproduce implants behavior. Several models describe the biological events leading to bone healing at the interface and many others describe the mechanical loading environment and the implant surface interactions under the assumption of partially or even fully osseointegrated interfaces. For the modeling guidelines, we present a brief discussion of these models and comment on ways for the formulation of experimental procedures for obtaining numerical parameters used in the models, and for the qualitative and quantitative validation of the numerical results.

2. Dental implants

2.1 The bone-dental implant interface

Teeth are anatomic structures used during chewing. Each tooth has a crown, a neck and a root Lang et al. (2003); Lindhe et al. (2003). The crown is the visible part inside the mouth while the root lies inside the jaw bone. The neck is the boundary between the crown and the root Lindhe et al. (2003). Teeth are positioned in the jaw bone through the so-called *dental alveoli*, a type of sockets formed directly inside the *alveolar bone*.

A dental implant is a biomaterial device surgically inserted in the jaw bone to replace the root of a missing tooth Adell et al. (1981); Hansson et al. (1983). The implant is part of the prosthetic unit that replaces the missing tooth and that compromises the abutment, the joint and the prosthesis that finally replaces the lost crown (Figure 1).

There are several types of dental implants although the most recognized are those with a roughed and screwed body with dimensions ranging from 6.0 mm to 16.0 mm length and 3.5 mm to 5.0 mm diameter, depending of the kind of missing tooth Branemark (1983); Gapski et al. (2003). However, the optimal length and diameter needed for a successful long lasting implantation depends on the anchorage conditions of the host bone, and the biological and mechanical factors associated to bone healing Aukhil (2000); Davies (2003);

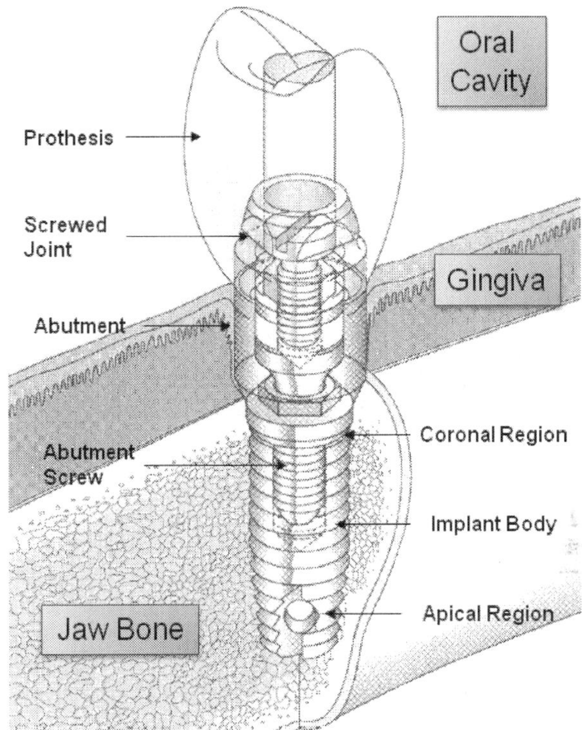

Fig. 1. Detail of a screw-type dental implant. Adapted from Aparicio (2005).

Gapski et al. (2003); Häkkinen et al. (2000); Klein-Nulend et al. (2005); Sikavitsas et al. (2001). There are nowadays different geometries of dental implants available although the most known is the screw-type geometry introduced in the 1960's by PI Branemark Branemark et al. (1969). This type of implant shows high mechanical retention provided by its canalled body, its outstanding ability in transferring compressive forces, and its improved primary stability capability Gapski et al. (2003); Martínez et al. (2002).

Most of the currently commercial dental implants are made of commercially pure titanium Adell et al. (1970) due to its proven biocompatibility, i.e., its acceptance by the living tissues Ellingsen et al. (2006); Ratner et al. (1996). The biocompatibility comprises the absence of corrosion and material wearing that may lead to undesirable inflammatory responses, death of surrounding tissues or thrombus formations by unexpected coagulation effects. It also implies for the living organism not to produce undesirable immunological responses such as an increase in antibody counting, cell mutations or formation of cancer cells Aparicio (2005); Hansson et al. (1983); Ratner et al. (1996).

Besides the implant characteristics, the insertion procedure has been shown to be of importance in the success of the prosthetic unit. The presence of a large number of bacteria inside the mouth demands the injury caused during implant placement to be carefully preserved in order to avoid possible infections leading to implant lost. According to this, the most referenced insertion technique includes the implant coverage with the epithelial tissue originally present at the insertion site Branemark (1983); Branemark et al. (1969); Fragiskos & Alexandridis (2007); Gapski et al. (2003); Leckholm (2003); Lindhe et al. (2003). The use of this technique reduces the wound healing time by the temporally isolation of the implant from an environment full of microorganism as the oral cavity and increases the bone formation at the implant surface reducing the bacterial contamination risk Branemark (1983); Gapski et al. (2003); Hansson et al. (1983). In general terms, this technique known as *two-stages* needs for two surgical interventions to complete the prosthesis placement Branemark et al. (1969). During the first intervention the implant is inserted in the placement site and covered by the epithelial tissue. 4 to 6 week later, a second intervention is carried out to remove the epithelium cover, expose the cortical side of the implant and attach the abutment and the prosthesis Fragiskos & Alexandridis (2007).

However, there is another type of insertion technique in which the implant, the abutment and the prosthesis are placed at the same time during a single surgical intervention. This technique known as *single-stage* avoids the epithelium coverage but reduces the healing time increasing then the patient benefit Heydenrijk et al. (2002). Nevertheless, this technique is less used due to bacterial contamination problems present during wound healing and an increased damage on growing tissues by micromovements caused during the earlier loading of the prosthesis and the implant Gapski et al. (2003); Heydenrijk et al. (2002).

2.2 Osseointegration

Although the evaluation of the anatomic characteristics of the host alveolar bone, the selection of the implant and the use of an adequate insertion protocol are conditions related to the bone-dental implant interface successfully healing, the implant osseointegration depends more on the bone formation directly over the implant surface Albrektsson & Johansson (2001); Branemark (1983). A successful osseointegration requires the action of two previous processes: *osteoinduction* and *osteoconduction* Albrektsson & Johansson (2001). Osteoinduction is the process by which stem cells are somehow stimulated to differentiate at the bone-dental implant interface into osteogenic cells that synthesize bone tissue (Figure 2a). New bone deposition by this cells is known as *osteogenesis* (Figure 2b) Albrektsson & Johansson (2001). There are two kinds of osteogenesis. A first *distant osteogenesis* where bone tissue is formed from the host bone border towards the implant surface Davies (2003), and a second *contact osteogenesis* where bone tissue is formed from the implant surface towards the host bone border (Figure 2b) Davies (2003); Puleo & Nanci (1999).

Contact osteogenesis implies the implant surface to be colonized by the osteogenic cells Davies (2003). This cell colonization or osteoconduction allows the bone growth over a biomaterial surface (Figure 2c) Albrektsson & Johansson (2001); Davies (2003). This process essentially depends on the material biocompatibility and the implant surface characteristics Huang et al. (2005); Wennerberg et al. (2003). Osteoconduction creates a direct contact between the implant and the surrounding growing tissues forming a contact interface that after the complete wound healing process conduces to the implant osseointegration (Figure 2d).

The canalled body of the screw-type dental implant allows it to support stresses and provide stability, while the deep surface irregularities provide the implant surface with a surface patter similar to that left behind by the osteoclasts after bone resorption during bone remodeling Martínez et al. (2002); Stanford & Schneider (2004). This surface pattern allows the osteogenic cells front to synthesize the first new bone line or *cementation line* interlaced with the

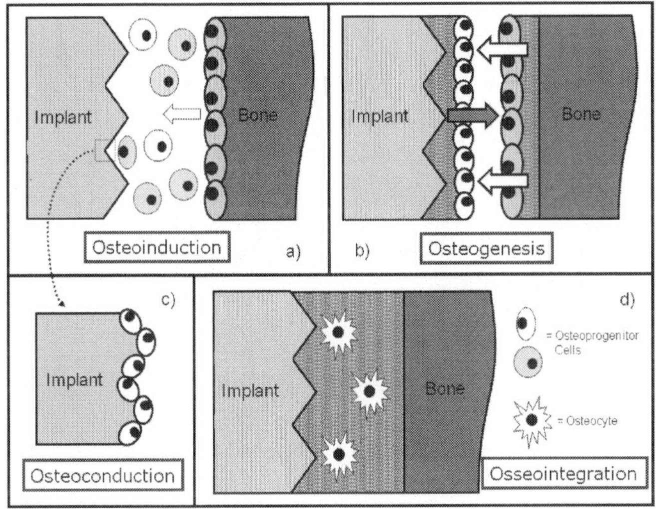

Fig. 2. Osteoinduction (a) is the osteogenic cells differentiation at the interface. Osteogenesis (b) is initiated by the osteogenic cells migrating from the host bone border (direct osteogenesis, white arrows) and by the osteogenic cells colonizing the implant surface (contact osteogenesis, gray arrow). Osteoconduction (c) is the surface colonization by the osteogenic cells. An adequate formation and bone viability (presence of osteocytes) surrounding the implant allows for a successful osseointegration (d) at the bone-dental implant interface.

irregularities, therefore ensuring an adequate new bone formation directly over the implant Davies (2007).

2.3 Mechanobiology of the bone-dental implant interface

Bone-dental implant interface healing consists of four biological stages each one associated with a characteristic biological event Aukhil (2000); Lang et al. (2003) (Figure 3): 1) hematoma formation (bleeding and blood clotting), 2) clot degradation and wound cleansing (fibrinolysis), 3) granulation tissue (fibroplasia and angiogenesis) formation, and 4) new osteoid formation (bone modeling).

Biological wound healing events are activated by bleeding during implant placement Fragiskos & Alexandridis (2007); Heydenrijk et al. (2002). The injured blood vessels initially become constricted and platelets from the bloodstream become activated to form a plug which temporarily stops blood loss Furie & Furie (2005); Minors (2007). Once activated, platelets aggregate and release granules containing several molecules that control the initial activity at the injured area, including platelet-derived growth factor (PDGF) and transforming growth factor beta (TGF-β) Gorkun et al. (1997); Minors (2007).

The temporary plug is later replaced by an hemostatic plug formed by a kinetic reaction between thrombin and fibrinogen, two proteins present in blood Furie & Furie (2005); Mann (2003). Thrombin converts fibrinogen into *fibrin fibers* Gorkun et al. (1997); Minors (2007).

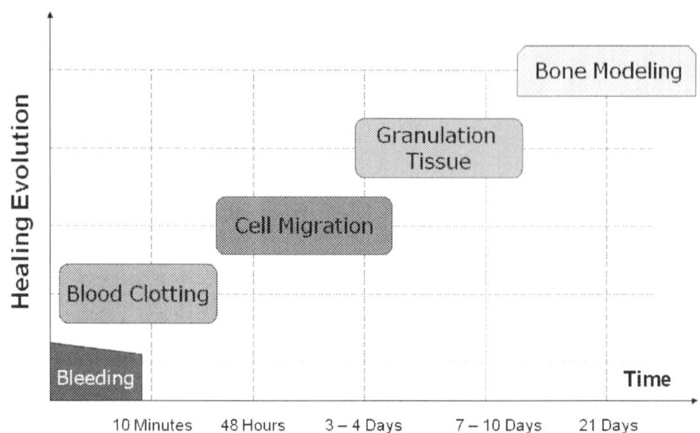

Fig. 3. Biological stages of healing at the bone-dental implant interface Ambard & Swider (2006); Dimitriou et al. (2005); Lang et al. (2003).

These fibers are accumulated to form the *fibrin clot* that completely detains blood flow and also protects tissues left exposed after implantation Aukhil (2000).

Some fibrin fibers are broken down after coagulation to allow the flow of stem cells responsible for restoring the tissues Collen & Lijnen (1991); Pasi (1999). Such degradation is accomplished by *plasmin*, a protein present in blood plasma in its inactive form known as *plasminogen* Aukhil (2000); Collen & Lijnen (1991); Li et al. (2003), and macrophage and neutrophil cleaning activity Davies (2003); Lang et al. (2003). Around the fourth day of healing, a process known as *fibroplasia* begins the replacement of the fibrin clot into a new extracellular matrix known as *granulation tissue* which mainly consist of collagen and new capillaries formed during angiogenesis Aukhil (2000). This new matrix supports the *osteogenic cells* migration Lang et al. (2003), stimulated by several molecules released during blood clotting and clot cleansing, such as PDGF, TGF-β Davies (2003) and fibroblast growth factor (FGF) Dimitriou et al. (2005).

Between the 7-10th day of healing, some of the fibroblasts present in the interface are transformed into *myoblasts* Häkkinen et al. (2000) characterized by smooth muscle α-*actin* cytoplasmatic microfilaments, allowing them to generate contractile forces responsible for wound contraction Aukhil (2000); Davies (2003); Häkkinen et al. (2000). *Osteogenesis* or new bone formation along the vascular structures is started Lang et al. (2003); Meyer & Wiesmann (2006) by day 14 after injury Dimitriou et al. (2005); Lang et al. (2003). Here, granulation tissue is replaced by new collagen fibers that are slowly mineralized to create the new bone matrix Meyer & Wiesmann (2006); Sikavitsas et al. (2001) by contact and direct osteogenesis Davies (2003).

Biological activity regarding wound healing at the bone-dental implant interface concludes with the modeling and subsequent bone remodeling Sikavitsas et al. (2001). Moreover, cell adhesion, cell migration and proliferation on surrounding tissues, and internal and external mechanical loads action modify the new tissue formation profile (Figure 4). Such phenomena may act as follows. First, the adhesion phenomena produced by cells fixation to a substrate Anselme (2000) activate chemical signaling Kasemo (2002) controlling cell proliferation and differentiation, as in platelet aggregation and activation stages Collen & Lijnen (1991).

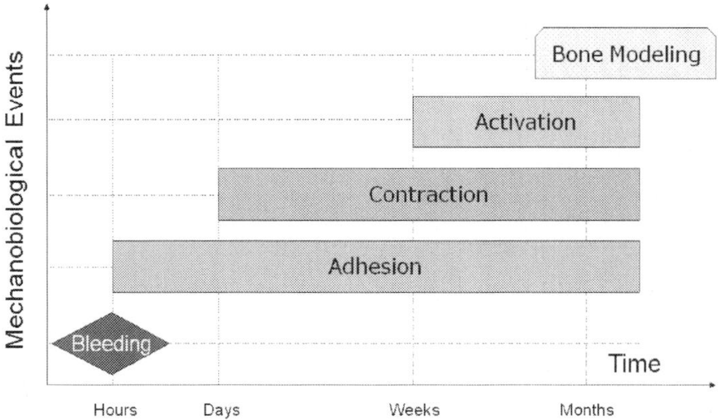

Fig. 4. Mechanical phenomena occurring during the healing of the bone-dental implant interface.

Then, the contraction exerted by cells moving over a substrate Häkkinen et al. (2000) may cause the displacement of the attached fibrin fibers and their detachment from the implant surface Davies (2003; 2007). This kind of event is present during fibroplasia and modeling stages where fibroblasts and osteogenic cells adhere to the fibrin network and begin to move through it in an attempt to colonize the implant surface. Fibers detachment prevents suitable bone formation directly over the implant surface and consequently increases the risk of implant rejection Adell et al. (1981); Branemark (1983).

Finally, the mechanical activation caused by applying an external load may induce the metabolic activity necessary for producing structural changes in the extracellular matrix of the tissues being formed Klein-Nulend et al. (2005); Sikavitsas et al. (2001). A phenomenon of this kind is particularly relevant during *bone mechanotransduction* Burger et al. (2005); Knothe-Tate (2003), which is responsible for controlling functional adaptation to external loads during modeling. Such mechanical adaptation exhibited by bone is widely known as *Wolf's Law* Sikavitsas et al. (2001); Stanford & Schneider (2004).

2.4 Experimental assays

Several studies have been carried out to evaluate the biocompatibility of different materials commonly used for dental implant manufacturing Matsuno et al. (2001); Metikos-Hukovic et al. (2003). It is well known that titanium is the best material of choice due to its high corrosion resistance to the physiological environment and its mechanical stability during the whole healing process Hansson et al. (1983); Watari et al. (2004). It also has been found that niobium, tantalum, zirconium, vanadium, aluminium and molybdenum are the most favourable materials to be used in titanium alloys for biomedical application Niinomi (2003). Although these alloys are non-toxic and highly inert Niinomi (2003); Watari et al. (2004), most of them do not establish a strong connection with the surrounding tissues and often induce the formation of fibrous tissue rather than bone tissue Ellingsen et al. (2006). This fibrous tissue provides inadequate support to the implant avoiding its surface to be colonized by the bony cells Huang et al. (2005). In order to avoid this, dental implant surfaces are modified by surface treatments that create micron and sub-micron scale patterns in the metal surface. These patterns increase the retention of molecules released during the activation of the healing processes at the bone-dental implant interface which also increase the grade of tissues restoration Aukhil (2000); Davies (2007). In addition, the use of inorganic mineral coatings improves the osteogenic capacity of the raw metal surface providing it with an osteoconductive profile that resembles the bone resorption surfaces Davies (2007); Kasemo (2002).

An adequate surface morphology allows the osteogenic cells to adhere and proliferate over the implant surface Kasemo (2002) and to form mineral deposits that become the new bone formation sites Davies (2003); Sikavitsas et al. (2001). Therefore, the strategies that modify the implant surface lead to implants with better osteoinductive and osteoconductive responses and to higher rates of bone deposition that result in successful osseointegration Davies (2007). Here we present a self-conducted experimental analysis perfomed on four different types of dental implant surfaces. The aim is to evaluate cell adhesion and cell proliferation profiles using cell culture techniques. From these assays we obtained initial information about the behavior of each one of the surfaces under a physiological-like environment allowing us to determine which of them had better performance in terms of the osteoinductive and the osteoconductive properties.

For these experimental approach we used 120 Ti-6Al-4V substrates of 15 mm diameter and 2 mm thickness. The substrates were divided into four groups according to their surface morphology: (1) machined, (2) sand-blasted/acid-etching (SBAE) surface, (3) hydroxyapatite-tricalcium phosphate Ha/TCP active surface (TCP), and (4) TCP/acid-etching surface (TCP+acid). All substrates were supplied by MIS Technologies Ltda. (Shlomi, Israel). The substrates were produced by milling and turning machines. Machined surface was achieved by using a cutting and polishing device. SBAE surface roughness and micro geometry was achieved by surface blasting with large particles (300-400 μm size) of Al_2O_3 followed by etching with HCl/H_2SO_4. TCP surface was achieved by blasting HA/TCP particles 200-400 μm size. TCP+acid surface was achieved by blasting HA/TCP particles 200-400 μm size and cleaning with HNO_3. After manufacturing, substrates were sterilized by gamma-radiation and vacuum packed in a clean environment. Before being used in the experiments describe below, substrates were placed in 24-multiwell Costar plates (Costar Corp., Cambridge, MA, USA). The plates were kept under UV radiation for 12 hours and then autoclaved at 120ºC for 6 hours.

Surface morphology characterization of each type of substrate was performed in two ways. First, a Zeiss Stemi SV11 stereozoom microscope with a 4x magnification lens was used. Using a Zeiss AxioCam MRC5 coupled-camera device, macrostructural 2.0 x 1.5 mm field of view images of the four surfaces were obtained. Second, microstructural images at the 10 μm scale were obtained for the four surfaces using a LEICA Stereo Scan 440 scanning electron microscopy (SEM) at 20 KV.

Osteoblastic cementoblastoma-derived cells were derived from a human cementoblastoma through the conventional explant technique and characterized as described elsewhere Arzate et al. (1998; 2002). Ethic considerations were followed as approved by the Internal Review Board of the School of Dentistry of the National University Autonomous of Mexico Arzate et al. (2002). Cells were cultured in 75 cm^2 cell culture flasks containing Dulbecco's modified Eagle's medium (DMEM) supplemented with 10% fetal bovine serum (FBS) and antibiotic solution (100 μg/ml streptomycin and 100 U/ml penicillin) (Sigma Chemical Co., St. Louis, MO, USA). The cells were incubated in a 95% air and 5% CO_2 environment at 37 °C. Cells at the second passage were used for all the experimental procedures.

Cell adhesion assay was performed on three samples of each one of the four types of substrates. Cells were plated and incubated for 24 hours in 500 μl of culture medium at standard conditions, as described above. After incubation, unattached cells were washed off four times with clean water and the remaining cells were fixed and stained with 300 μl of a solution made of 0.1% toluidine blue and 3.5% paraformaldehyde Hayman et al. (1982); Rodil et al. (2003). After 24 hours at room temperature, 100 μl of the supernatant was used for optical absorption reading with an ELISA (Enzyme Linked Immune Sorbent Assay) micro-plate reader at 630 nm. In this technique, the number of attached cells is proportional to the absorbance of the experimental samples Rodil et al. (2003).

Cell proliferation assay was performed on three samples of each one of the four types of substrates. The proliferation of the osteoblastic cementoblastoma-derived cells was determined by the MTT assay. This assay is based on the ability of mitochondrial dehydrogenases to oxidize thiazol blue (MTT), a tetrazolium salt (3-[3, 5+dimethylthiazolyl-2-y] 2, 5-diphenyltetrazolium bromide), to an insoluble blue formazan product Rodil et al. (2003). Cells were plated as in the adhesion assay and incubated for 1, 2, 5, 6 and 7 days. Fresh medium and antibody (500 μl) were added to the cultures every day. After each term, cells were incubated with 60 μl MTT at 37°C for 4 hours. Then, the supernatant was removed and 250 μl of dimethyl sulfoxide (DMSO) was added to each well. After 30 minutes of incubation the absorbance was read at 570 nm. Since the generation of the blue product is proportional to the dehydrogenase activity, a decrease in absorbance at 570 nm provides a direct measurement of the proliferation rate Mosmann (1983).

Surface morphology of the four types of substrates was characterized using stereozoom microscope and SEM. Machined surface exhibited the radial evenly-spaced wave structures created during the cut and polishing manufacturing procedures (Figure 5A). SEM image shows the parallel undulated fashion of these structures at the micron-scale (Figure 5B). For the SBAE substrate, a granular fashion surface was observed (Figure 5C). At the micron-scale, these granules appeared as *peak and valley* surface structures due to the abrasion procedure used during manufacturing (Figure 5D). The stereozoom microscope did not reveal these structures but shows a grain-like surface. TCP surface exhibited micron-surface irregularities with edges and undercuts as in the SBAE surface (Figures 5E and 5F). TCP+acid showed the most dense micron-surface texture (Figures 5G and 5H). A dense grain-like surface was observed through stereozoom. Extensive surface irregularities also with edges and undercuts were observed in SEM.

Cell adhesion and proliferation assays were carried out to evaluate the initial interactions of the osteoblastic-like cells with the biomaterial surface. For the adhesion assay, adherent cells were evaluated 24 hours after plating. The results are shown in Figure 6 and are given in terms of the absorbance measured at 630 nm. There was statistical difference between all results ($p < 0.005$, 95% confidence interval). Adhesion of osteoblasts is favoured in both TCP and TCP+acid surfaces, exceeding in more than 4 and 5 times respectively the cell attachment with respect to the machined surface. This suggests a higher cell interaction due to the presence of edges and undercuts in the biomaterial surface. In counterpart, adhesion to poorly treated surfaces is much lower, revealing the significance of micron and sub-micron structures over the target surface.

Figure 7 shows the results for the proliferation assay carried out after 1, 2, 5, 6 and 7 days of culture. Values are expressed as the absorbance at 570 nm, which is directly proportional to the metabolic activity of the cells and inversely proportional to the toxicity of the material Rodil et al. (2004). As illustrated, all surfaces have a negative proliferation profile between days 1

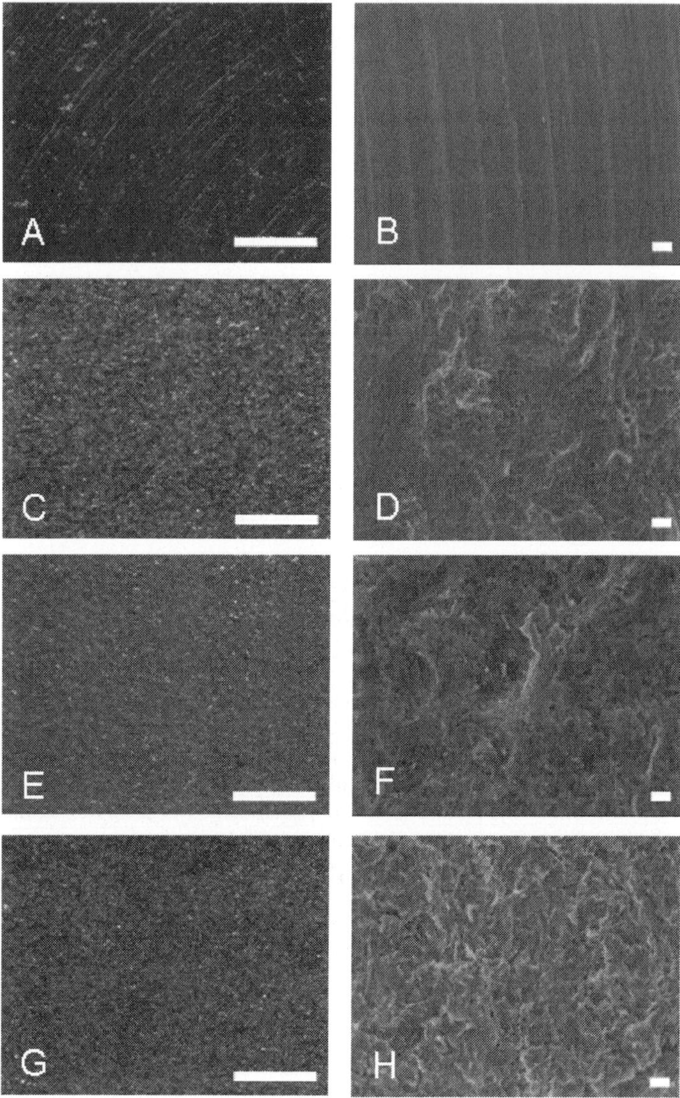

Fig. 5. Stereozoom and SEM micrographs of the four types of surfaces used in this study. Left, stereozoom microscopy (4x, bar = 0.5 mm). Right, SEM microscopy (2.500x, bar = 10 μm). (A) and (B) Machined surface, (C) and (D) SBAE surface, (E) and (F) TCP surface, (G) and (H) TCP+acid surface.

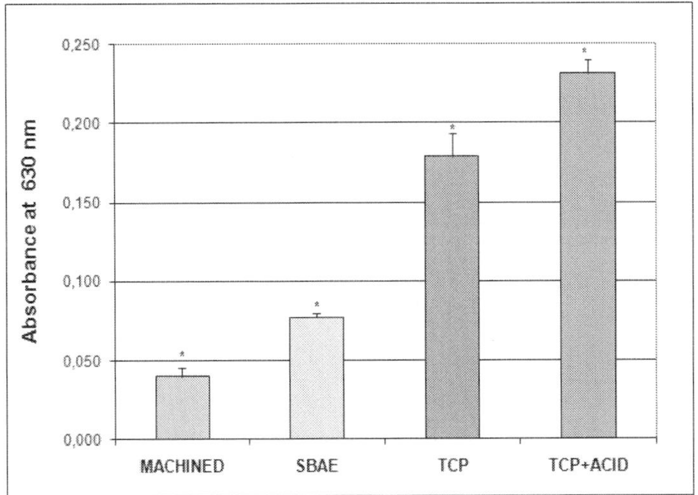

Fig. 6. Cell adhesion results after 24 hours of culture. Error bars = standard error. An asterisk indicates statistical difference between results at $p < 0.005$.

and 2, possibly caused by cell death during initial confluence. However, for machined, SBAE and TCP surfaces between day 2 and 5 proliferation levels increase almost linearly reaching a proliferation peak on day 6, after which the levels slightly decline until a final value on day 7. This proliferation peak at day 6 after culture was found to be higher on the SBAE surface. In contrast, proliferation on the TCP+acid surface shows a slow decrement between day 1 and day 2 together with an exponential-like increment that lasts until day 7. No statistical significant differences between results were found from days 1 to 7.

Concluding results for the adhesion assay showed that TCP and TCP+acid surfaces have better osteoinductive response than machined and SBA surfaces. Cell proliferation assay revealed that after 7 days of culture TCP+acid surface has the lowest proliferation rate, due to its increased surface roughness Aita et al. (2009); Bächle & Kohal (2004). The remarkable adhesion profile of the TCP surface and its considerable high proliferation rate after 7 days of culture were confirmed by the ALP activity results from which TCP surface has better performance and doubles the result obtained for the TCP+acid surface Vanegas et al. (2010). The results obtained suggest that TCP surface promote the formation of mineral deposits, i.e. osteoconduction, in a higher rate compared not only to the TCP+acid surface but also to the machined and SBA surfaces. None of the studied surfaces were toxic to the osteoblastic-like cells since all of them exhibited cell adhesion and proliferation profiles.

2.5 Mathematical modeling

Although at the bone-dental implant interface biological and mechanical factors converge, most of the mathematical models available only consider the mechanical factors obtaining conclusions regarding the long range viability of the implants, the loading distributions and the mechanical behavior of the materials used in the implant manufacture Geng & Tan (2001); Patra et al. (1998). In this type of models the formation of the bone-dental implant

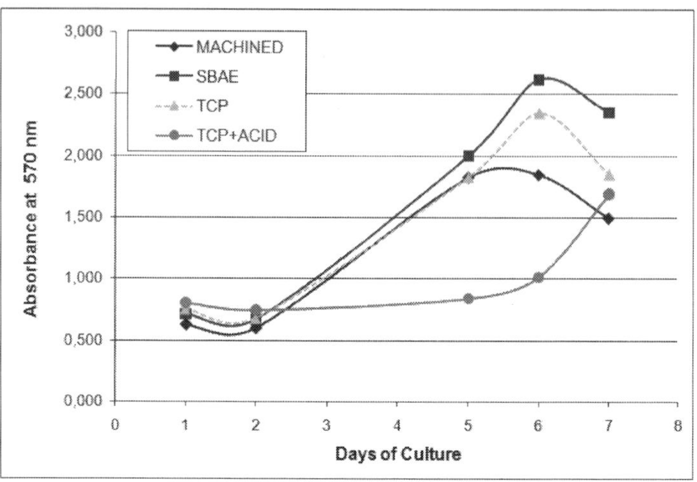

Fig. 7. Cell proliferation after 1, 2, 5, 6 and 7 days of culture.

interface is not considered and the assumed biological starting point is the complete and stable osseointegration Patra et al. (1998). There are also models approaching to the healing biological phenomena at the interface, describing fibrous tissue formation as the results of mechanical variables Huiskes et al. (1997) or by considering the phenomenological behavior of the mechanics involved Isaksson et al. (2008). Nevertheless, there are models with a biological framework that base their descriptions on phase changes at the interface Ambard & Swider (2006) and on variations of the cellular concentrations and extracellular matrix density Bailon-Plaza & van der Meulen (2001); Moreo et al. (2009). In this cases, the equations used include specific terms describing cellular processes as mitosis, proliferation, differentiation and apoptosis, as well as natural biological events leading to the formation, transformation and degradation of the extracellular matrix Bailon-Plaza & van der Meulen (2001); Geris et al. (2008); Moreo et al. (2009).

2.5.1 Mechanical approach

Mathematical modeling of a dental implant with a mechanical approach allows for the evaluation of different implant designs ensuring resistance to certain loading conditions Bonnet et al. (2009); Bülent (2002); Juodzbalys (2005); Kayabasi et al. (2006); Papavasiliou (1996); Poiate et al. (2008). The analysis of the bone-dental implant interface in these models is focused in the stress distribution around the host bone and the implant body, and in the effect of loading in the stability of the alveolar bone Geng & Tan (2001). It is then possible to redesign implants without the complications of physical manufacturing and experimentation saving money and time. Most of these models are computationally implemented through commercial available software that perform a numerical discretization using the finite differences method or the finite elements method Bonnet et al. (2009); Geng et al. (2001); Kayabasi et al. (2006); Poiate et al. (2008).

Most of the authors have reported simulations using geometrical bidimensional and tridimensional models obtaining a graphical sketch of the physical behavior of the interface after loading. This same approach was used in a new numerical simulation of a bone-dental implant interface. Figure 8a shows the compact model of the interface where jaw bone and prosthesis are only visible. A more detail representation is shown in Figure 8b where prostheses were elevated to show three dental implants inserted in the jaw bone. Observe that the jaw bone was modeled as an external trabecular zone covering a cortical zone Davies (2003); Saffar et al. (1997). This model was implemented to performed a comparative study of the biomechanical behavior of individual prosthesis and ferulized prostheses used in oral rehab treatment at the posterior jawbone.

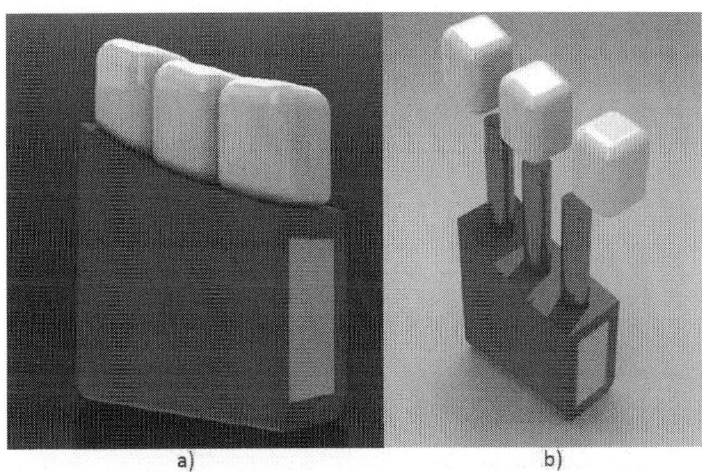

Fig. 8. Sketch of a geometrical model used to simulate a single-crown prosthesis dental-implant condition. a) Compact model. White: Prosthesis. Red: Trabecular bone. Yellow: Cortical bone. b) Detail of the model components. Gray: Dental implants.

We performed a loading conditions analysis at the interface. Shear and longitudinal loads exerted to the prosthesis during chewing induce axial forces and shear momentum resulting in stress gradients at the host bone and the implant body (Geng et al., 2001). Figure 9 shows a simulation of these loading environment on the geometrical mode of Figure 8. Loading conditions applied to the model are similar to those present in the oral cavity during chewing. We simplified these loads as static loads and stress distributions at the interface. Results show a color scale representing the magnitude of the von Misses stress. Maximum value obtained in this simulation was 62.87 MPa. If the yield stress of the dental implant is higher that this value, as is the case of titanium and stainless steel, good performance is expected. Furthermore, surrounding host bone shows an approximated stress value of 17 MPa, that should be compared with the range of functional stress that controls bone deposition and resorption (Geng & Tan, 2001; Rieger et al., 1990). Final conclusion obtained from the numerical results was that best performance is achieved when using single-crown prostheses.

Model simplifications are needed in the mechanical models to manage problems as considering the mechanical differences between cortical and trabecular bone, the complex implant geometries, boundary conditions, computational costs, among others. According to these, a good mechanical model may be formulated based on the following assumptions:

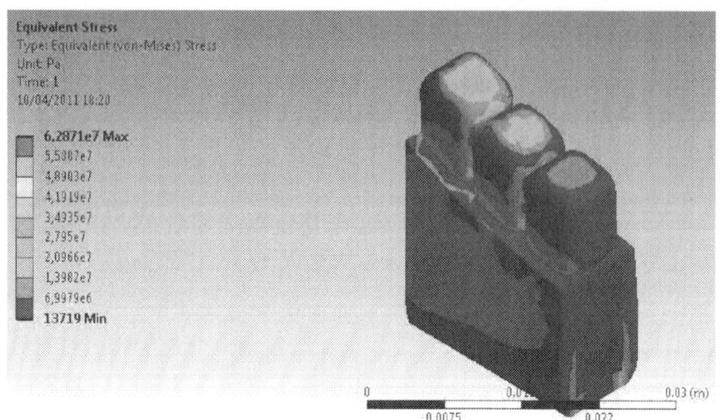

Fig. 9. Stress distribution for the model with single-crown prostheses.

1. For bidimensional studies, bone is modeled using simple rectangular configurations. For tridimensional models jaw is considered as an arch with rectangular sections (Geng et al., 2001). However, advanced medical imaging processing techniques may be useful to better model the geometry of the jaw bone, implants and prosthesis (Bonnet et al., 2009; Geng et al., 2001; Poiate et al., 2008).

2. Material properties for bone and dental implants are mostly considered as isotropic. However, is it well known that bone have orthotropic and anisotropic properties (Huiskes et al., 1997) since stress is not necessarily distributed in a uniform fashion rather than oriented in a preferential direction (Bonnet et al., 2009; Geng et al., 2001; Kayabasi et al., 2006). However, these material assumptions ensure its linearity and homogeneity and allow for an elastic behavior described by the Young and Poisson modules. Nevertheless, a more precise model should account for a better material description, including differences for trabecular and cortical bone, and for other surrounding tissues as muscles and epithelium (Geng et al., 2001).

3. Some assumptions for boundary conditions consider a fixed jaw. However, muscle and ligament function during chewing and functional movement of the temporomandibular joint should be modeled by using additional elements improving model realism and accuracy (Geng et al., 2001).

4. Most models regarding the bone-dental implant interface consider an initial complete osseointegration. However, we have shown that osseointegration is consequence of a biological process involving mechanical conditions. It has been found that osseointegration ratio depends on bone quality, stress distribution during wound healing, implant loading conditions and implant design, among others (Bonnet et al., 2009; Bülent, 2002; Geng et al., 2001; Juodzbalys, 2005; Kayabasi et al., 2006; Papavasiliou, 1996; Poiate et al., 2008). Therefore, a more realistic description of the interface should include additional considerations for the biological phenomena and its temporal evolution, if time evolution starting at implant placement wants to be addressed (Moreo et al., 2009).

2.5.2 Mechanobiological approach

A complete mechanobiological model for the formation of the bone-dental implant interface leading to the implant osseointegration is still not know, although several works have succeeded in the attempt of formulate a mathematical model including the biological and mechanical models related to some of the stages of tissue formation. This is the case of models for cell adhesion and proliferation (DiMilla et al., 1991; Moreo et al., 2008), models for coagulation (Colijn & Mackey, 2005; Vanegas et al., 2010), models for angiogenesis and cell contraction (Mantzaris et al., 2004; Tracqui et al., 2007), and models for bone formation (Amor et al., 2009; Moreo et al., 2009).

From the biological and mechanical reality of the healing and bone formation process at the bone-dental implant interface, and considering the results provided by the abovementioned mathematical models, the following elements should be considered in the formulation of a complete mechanobiological model for the bone-dental implant interface:

1. The biological stages of wound healing at the interface may be assumed as a sequence of events in a time scale divided in minutes, hours, days, weeks and months Ambard & Swider (2006); Aukhil (2000); Dimitriou et al. (2005).

2. The bleeding stage may be simplified as the formation of the fibrin clot by the reaction kinetics between thrombin and fibrinogen (Aukhil, 2000; Minors, 2007).

3. The fibrinolysis stage may be considered as a natural clot degradation term, whereas the fibroplasia and the angiogenesis can be simplified in a single event leading to synthesis of new collagen matrix (Häkkinen et al., 2000).

4. The formation and replacement of the collagen matrix by new bone is related to the presence of an specific concentration of osteogenic cells and the presence of a chemoatractant substance controlling cell migration and proliferation (Davies, 2003; Moreo et al., 2009; Puleo & Nanci, 1999).

5. The adequate bone formation around the dental implant depends on its surface topography and the formation of the cementation line (Davies, 2007; Kasemo, 2002).

6. The adhesion mechanical factors may be considered as part of the cell differentiation process and should be related to the implant surface topography and the implant osteoinduction and osteoconduction properties (Davies, 2007).

7. The contraction and activation mechanical factors are similar at a micro-structural level and therefore may be simplified as the viscoelastic behavior of the fibrillar fibrin matrix compounding the blood clot (Weisel, 2004) and guiding the bone forming process (Saffar et al., 1997; Stanford & Schneider, 2004).

8. The loading effects over the implant may be neglected if the recommended initial healing time of three to six months prior to prosthesis placement is considered (Branemark, 1983; Vanegas et al., 2009).

9. The surface irregularities of the dental implant influence the cell and proliferation profiles, as is shown through the experimental assays.

10. Numerical parameters needed for the mathematical formulation of the model may be obtained through experimental assays. Here, the mathematical formulation should provide enough justification for running the assays.

Most of these elements were used for the formulation of a new mathematical model with a mechanobiological approach (Vanegas et al., 2011). This model includes the biological stages described in Figure 3, some of the mechanical factors described in Figure 4, and the implant surface irregularities. A schematic of this new model is shown in Figure 10.

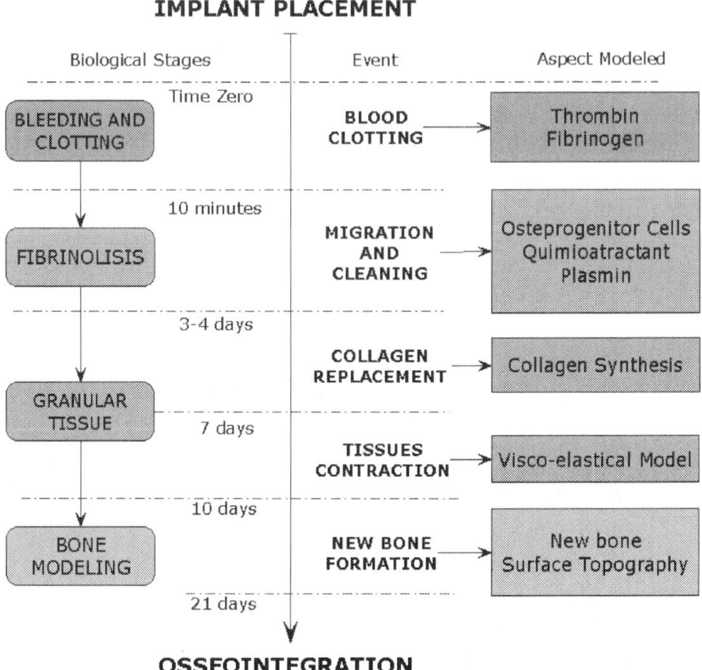

Fig. 10. Schematic of a mechanobiological model of the osseointegration of a dental implant. The sequence of biological stages is shown at the left side and the events and aspects modeled at each stage are shown at the right side. The boxes resume the most important elements used at each event that also represent the simplifications made to the complex chain of biological and mechanical phenomena leading to the dental implant osseointegration.

This model simplifies the biological wound healing process as a sequence of stages each one associated to a series of events. In this way, the bleeding and coagulation stage is simplified as the fibrin blood clot formation by the conversion reaction between thrombin and fibrinogen. During fibroplasia osteogenic cell migration is initiated by the presence of a chemoatractant substance at the same time that the fibrin clot is degraded by the plasmin activity. The new collagen matrix formation by the osteoprogenitor cells simplifies the fibroplasia and angiogenesis processes in a single stage called granular tissue (Aukhil, 2000). The displacement of the osteoprogenitor cells over the collagen matrix causes fibrillar contraction conditioned to the viscoelastic response of the fibers and the collagen mechanical properties. This contraction constitutes the interaction between biological and mechanical factors presented in this mechanobiological model. Finally, the new bone formation process, conditioned to the surface topography of the implant included as a numerical parameter and the adequate formation of the cemented line due to the contact osteogenesis process, leads to the initial osseointegration of the dental implant.

3. Mini-implants

Anchorage is defined as resistance to unwanted tooth movement caused by the reacting force of orthodontic devices Proffit & Fields (1993). Anchorage control is essential in orthodontic biomechanics and is one of the prerequisites for successful orthodontic therapy Chaddad et al. (2008). Traditionally, orthodontic movement of a tooth is anchored by a large group of teeth so as to minimize undesired displacements. Adequate anchorage becomes difficult when teeth are missing or present pathologies like periodontal and endodontic diseases. Several methods have been introduced to provide additional anchorage in orthodontics. Intra-oral and extra-oral auxiliary devices can be used to assist movement, but the effectiveness of these measures depend on patient compliance Egolf et al. (1990).

Conventional dental implants have proven to be successful for orthodontic anchorage because they are suitable for loading and offer absolute anchorage Branemark (1983); Hansson et al. (1983). Since the application of earlier orthodontic forces affect implant osseointegration Adell et al. (1970); Gapski et al. (2003), osseous adaptation mechanisms taking place during orthodontic loading increase bone formation on localized zones in an attempt to counteract the loading effects Klein-Nulend et al. (2005); Wehrbein et al. (1999). However, the larger size of conventional endosseous implants limits their usage as anchorage devices leading to the development of specific orthodontic systems such as plates Chung et al. (2002), onplants Crismani et al. (2008), and mini-implants Kanomi (1997).

Mini-implants are titanium screws with smaller dimensions than dental implants (Figure 11). They are widely used in orthodontic treatments due to their few inherent limitations for the selection of the placement site, they have a simple surgical procedure for insertion and removal, have low cost, cause less trauma to the patient than dental implants and provide easy attachment for additional orthodontic devices Costa et al. (1998). The aim of the mini-implants is to remain stable at the oral cavity during the accomplishment of the orthodontic treatment. Once the treatment is finished, the mini-implant can be easily removed because their osseointegration ratio is only around 13% Zhao et al. (2009). This lower ratio suggest that the mini-implant primary stability is a consequence of a mechanical phenomenon of interaction with the surrounding cortical bone that avoids the need of an initial healing stage prior to orthodontic loading and also allows for an easy final removal Huja et al. (2005). Mini-implants insertion procedure starts with a vertical incision of 3 mm to 4 mm long. Incision borders are separated and a 0.09 mm diameter hole is drilled into the jaw bone. Placement site is cleaned using saline solution to avoid clinical complications. The mini-implant is then inserted leaving at least 2 mm of its distal side exposed in the oral cavity. Finally, orthodontic wire extensions are attached to the mini-implant head in order to include it in the orthodontic treatment Antoszewska et al. (2009).

Contrary to dental implants, mini-implants insertion angle is of paramount importance for the success of the orthodontic treatment Chaddad et al. (2008). It is recognized that a greater insertion angle is useful to increase the screw length inserted inside cortical bone causing an augmented fixation and higher primary stability Costa et al. (1998); Deguchi et al. (2006). Therefore, recommended insertion angles range between 15° and 90° depending of the maxilla dimensions Deguchi et al. (2006). After confirmation of primary stability by

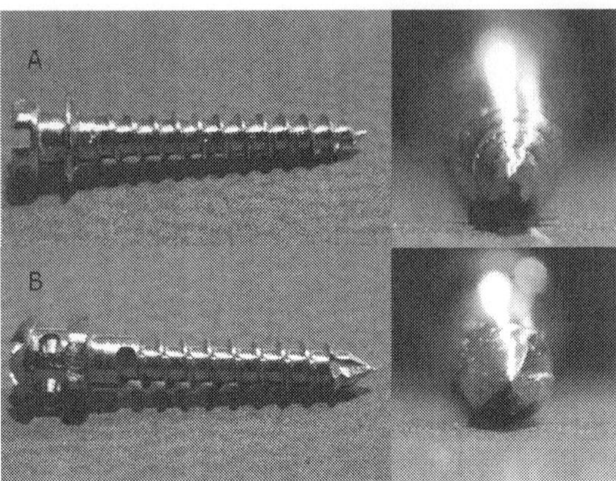

Fig. 11. Different types of mini-implants.(A) Dual-top mini-implant ®(Jeil, Korea). (B) Link orthodontic implant ®(MIS, Israel)

evaluating absence of micromovements, immediate loading is performed with magnitudes ranging from 50 to 200 gr. Melsen & Costa (2002). These loads can be directly or indirectly applied using rubber bands or closed-helical springs after a responsible healing time of two weeks (Antoszewska et al., 2009; Zhao et al., 2009).

Once treatment has started, failure of the mini-implant may occur among other reasons because of low bone density and improper cortical bone thickness at the insertion site Motoyoshi, Inaba, Ono, Ueno & Shimizu (2009b). Experimental tests of these failure factors suggest that minimal cortical bone thickness should be 1 mm (Motoyoshi, Inaba, Ono, Ueno & Shimizu, 2009a). In addition, numerical analyses performed on mini-implant treatments have evaluated these same failure factors. Results from these analyses resume the mechanical conditions required for the use of mini-implants based on a predictive scheme supported on experimental evidence Motoyoshi, Okazaki & Shimizu (2009); Sung et al. (2010).

3.1 Experimental assays

In recent years, there has been an increased concern for better understanding the behavior of mini-implants as anchorage devices in the jaw bone. However, the suitable material, surface treatment, screw design, self-perforating screwing capability, ideal timing for loading and magnitude of loads are still not well defined Chaddad et al. (2008); Seong-Hun et al. (2008). Although the mini-implant stability should be preserved during the entire orthodontic treatment, this is something that not always can be assured since treatment times may vary and in some cases are longer than a year Antoszewska et al. (2009); Costa et al. (1998); Deguchi et al. (2006).

Immediate loading is one of the distinctive characteristics of mini-implants. Since bone healing at the interface is a dynamic process and external mechanical loading induces bone adaptation Klein-Nulend et al. (2005), keeping an unchanging long lasting interface after loading seems somehow unfeasible. Furthermore, it is not clear if the mini-implant may induce higher osseointegration ratios supporting immediate loading or if loads and micromovements induce fibrillar tissue formation at the bone-implant interface that hinder osseintegration Zhang et al. (2010). Recommended mini-implants are made of titanium due to its biocompatibility and bioactivity Ellingsen et al. (2006); Ratner et al. (1996) but have smooth polished surfaces that may explained reduced osseointegration rates Davies (2007). However, a convenient osseointegration ratio may improve mini-implant stability during long lasting orthodontic treatments. Although evidence shows that mini-implants do not osseointegrate, there is no consensus on this matter and further experimental studies are needed to provide more details about the interface behavior Serra et al. (2008).

We have therefore performed a self-conducted experimental assays to analyze samples of bone surrounding mini-implants and evaluate the formation of the bone-implant interface. A total of fifteen 3-months-old male Wistar Rats SPF mean weight 350 gr. were housed with a 12-hour light/dark cycle and fed with a standard pellet diet and tap water at pleasure throughout the experiments Casale & Rivera (2010); Casale & Saavedra (2010). Principles of laboratory animal care and national laws were observed for the present study. Authorization for these experiments was issued by the Ethics Committee of the Dentistry Faculty of the Universidad Nacional de Colombia. Screw-shaped titanium (Ti_6Al_4Va) mini-implants of 10 mm length and 1.6 mm diameter (Link ®, MIS, Israel) were used in the assays. Sample screws were provided by MIS Implants Ltd. Immediate loading of 150 gr. was applied to all the mini-implants. A radiographic image of the insertion site is shown in Figure 12. Histological tissue behavior was evaluated 0, 8, 15, 45 and 120 days after insertion. A control group without immediate loading was also used.

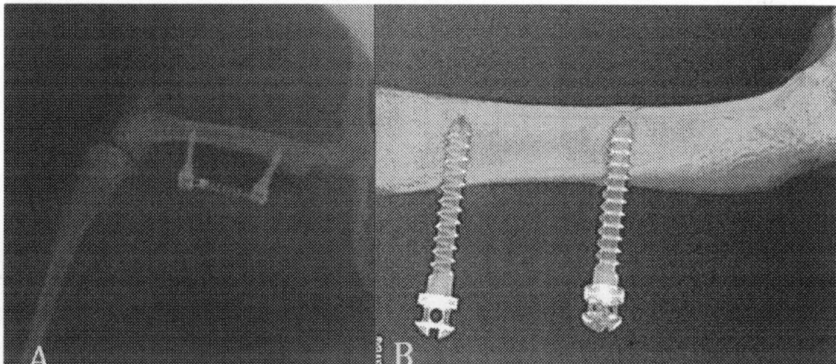

Fig. 12. A) SPF Wistar rat femur radiograph after mini-implant insertion and immediate loading using a nitinol spring (150gr). B) Digital radiograph after sacrifice.

Experimental data analyses (not shown) suggest that immediate loading does not affect the wound healing pattern. Histological observation of loaded samples showed that inflammation is activated between 0 and 15 days after insertion, bone activity under mechanical tension is started after 15 days of insertion and mature bone formation is started at day 45 after insertion. The histological activity in unloaded samples showed suitable bone healing barely starting on day 45 after insertion. These results suggest that wound healing and osseointegration at unloaded samples is similar to that exhibited on dental implants inserted in the cortical region.

Comparative additional experiments for orthodontic immediate loading of 150 gr and 350 gr. showed that higher loading is not feasible in mini-implants because mobility, displacement and screw instability are increased. Conversely, the application of this higher loading on mini-implants unloaded during the entire 45 days experimental time (late loading) showed no mobility or screw displacement. According to this, we provide histological evidence of mini-implant osseointegration at the cortical region Casale & Rivera (2010); Casale & Saavedra (2010) (Figure 13).

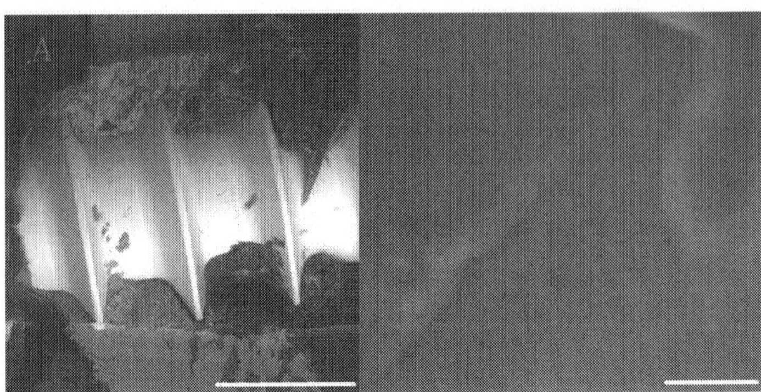

Fig. 13. A) SEM micrograph showing the bone-interface for a mini-implant without immediate loading. Scale bar = 1.0 mm. B) SEM micrograph for the encircled area of A) showing the same interface after 45 days with evidence of osseointegration with cortical bone. Scale bar = 2.0 μm

3.2 Mathematical modeling

Mathematical modeling regarding mini-implants, as applied to dental implants, allows for the description of the bone-implant interface, study factors leading to a successful insertion or a failure in the orthodontic treatment, and redesign mini-implants for better clinical performance. Available mathematical models have been used to study and predict mechanical events occurring during the orthodontic treatments Sung et al. (2010). These studies analyze the influence of mini-implants nearness to adjacent teeth roots, the effect of cortical bone height on treatment success and the consequences of inserting the mini-implant in surrounding poor quality bone Sung et al. (2010).

Since mini-implants are very similar to dental-implants, a good mathematical modeling approach for mini-implants may share conditions of those already mentioned for dental implants modeling. However, specific conditions should be addressed as follows:

1. Biological conditions in mini-implants treatments relate to the primary stability and osseointegration ratio. These same conditions should be accomplished in a good mathematical modeling of dental implants. Since the healing process is the same in dental implants and mini implants, a mathematical model of mini-implants may include the same biological stages as for dental implants. Nevertheless, an additional condition should state that the expected osseointegration ratio is around 13%. This allows for a relative simple adaptation of the models used in dental implants reducing the amount of bone formation around the implant surface, due basically to the absence of surface roughness Papadopoulos & Tarawneh (2007).

2. An additional biological condition for mini-implants successful treatment is the amount of cortical bone tissue in the insertion site. Related studies state that the minimal cortical bone thickness for an suitable functional stability of the device is 1 mm (Motoyoshi, Inaba, Ono, Ueno & Shimizu, 2009a). This condition may be addressed by controlling bone formation

process at the insertion site by an adequate loading stimulation or by increasing the area of the geometrical description (Figure 8).

3. In contrast to the biological approach of the model, the mechanical behavior is different of that regarding dental implants. Although the adhesion and contraction mechanical factors are still present and could be assumed the same as in dental implants, activations mechanical factors controlling the structural behavior of the bone-mini implant interface are quite different. Here, the direction and magnitude of loads, the mini-implant design and body geometry are different to those of dental implants. Therefore, models including these factors should be modified in order to address different types of treatments with specific loading orientations for increase anchorage, different insertions angles and different load transfer profiles.

4. Since mini-implants design have shorter dimensions that those found in dental implants, there must be a biological and mechanical relation of scaling with the surrounding tissues that should be analyzed in order to change the dimensions of the model to the shape profile of the used mini-implant.

4. Conclusions

Here we presented the principal characteristics of dental and mini-implants, the related bone healing process at the bone-implant interface and the mechanical factors involved. We also commented on self-conducted experimental approaches for evaluating the performance of these devices when in contact with living tissues. Finally, we presented a framework for the mathematical modeling of the bone-implant interface in both dental implant and mini-implant cases. Considering that mathematical models are approximations to the real osseointegration process at the bone-implant interface, there are some limitations inherent to the mathematical frameworks. These limitations are the simplification of the mechanobiological bone healing process, the initial conditions for the model, the adequate boundary conditions leading to the appropriate solution, and the model parameters, among many others. However, we should here highlight the prevalence of the latter in obtaining an accurate model.

Although in most of the cases some of the numerical parameter can be estimated from the available literature and others can be estimated from experimental results and previously reported numerical works, the exact value of many of them is an unknown and therefore adjustments are needed to obtain the expected solution. These adjustments may be performed by an iterative solution-based approach, a parameter sensitive analysis or a thorough mathematical analysis of the equations used Amor et al. (2009); Vanegas et al. (2011). Another approach is conducting specific experiments aimed at the quantification of detailed biological or mechanical quantities needed in the model formulation that are otherwise pointless and unjustified. This difficulty to obtain appropriate parameters from experimental evidence is a weakness shared in the models presented here with other works in the area of computational modeling of biological phenomena, and is a circumstance that should be considered before and during the process of model formulation.

Nevertheless these inherent limitations, results obtained from recent models Ambard & Swider (2006); Amor et al. (2009); Moreo et al. (2009); Vanegas et al. (2010; 2011) show that mathematical frameworks are suitable for being used as the methodological basis for the design of predictive tools aimed at the evaluation of the osseointegration ratio in dental and mini-implants considering patient characteristics, anatomy and jaw bone physiology, implant design and surgical procedure used for inserting the implant Fragiskos & Alexandridis (2007); Gapski et al. (2003); Leckholm (2003); Stanford & Schneider (2004). Future applications may deal with developing additional models for the prediction of healing patterns in different types of tissue injury, different implant geometries and surfaces, and for the mechanobiological evaluation of other implantable devices.

5. Acknowledgments

We warmly thank MIS Implants Ltd. and its representative office at Bogota - Colombia for providing the samples used in the experimental assays. We also want to thank Dra. María Cristina Piña from the Materials Research Laboratory at the Universidad Autónoma de México for her support during the dental-implants experimental assays, Dr. Carmen Alicia de Martínez from the Biomimetics Laboratory - Biotechnology Institute at Universidad Nacional de Colombia for providing facilities for the mini-implants experimental assays, and Diana Martínez from the Division of Orthodontics, Faculty of Odontology and Luis Miguel Méndez from the Department of Mechanical and Mechatronics Engineering at Universidad Nacional de Colombia for his valuable comments during the mechanical simulation of dental-implants. JA Guerrero was funded by Colciencias through the program *Jóvenes Investigadores e Innovadores Virgina Gutiérrez de Pineda - año 2009*. JC Vanegas was funded by the Academic Vice-principal Office of Universidad Nacional de Colombia through the program *Becas para Estudiantes Sobresalientes de Posgrado - año 2011*.

6. References

Adell, R., Hansson, B., Branemark, P. & Breine, U. (1970). Intra-osseous anchorage of dental prostheses ii. review of clinical approaches, *Scandinavian Journal of Plastic and Reconstructive Surgery* 4: 19–34.

Adell, R., Lekholm, U., Rockler, B. & Branemark, P. (1981). 15-year study of osseointegrated implants in the treatment of the edentulous jaw, *Journal of Oral Surgery* 10: 387–416.

Aita, H., Hori, N., Takeuchi, M., Suzuki, T., Yamada, M., Anpo, M. & Ogawa, T. (2009). The effect of ultraviolet functionalization of titanium on integration with bone, *Biomaterials* 30: 1015–1025.

Albrektsson, T. & Johansson, C. (2001). Osteoinduction, osteoconduction and osseointegration, *European Spine Journal* 10: S96–S101.

Ambard, D. & Swider, P. (2006). A predictive mechano-biological model of the bone-implant healing, *European Journal of Mechanics and Solids* 25: 927–937.

Amor, N., Geris, L., Vander-Sloten, J. & Van-Oosterwyck, H. (2009). Modelling the early phases of bone regeneration around an endosseous oral implant, *Computer Methods in Biomechanics and Biomedical Engineering* 12(4): 459–468.

Anselme, K. (2000). Osteoblast adhesion on biomaterials, *Biomaterials* 21: 667–681.

Antoszewska, J., Papadopoulos, M., Park, H. & Ludwigd, B. (2009). Five-year experience with orthodontic miniscrew implants: A retrospective investigation of factors influencing success rates, *American Journal of Orthodontics and Dentofacial Orthopedics* 136: 158.e1–158.e10.

Aparicio, C. (2005). *Tratamientos de superficie sobre titanio comercialmente puro para la mejora de la osteointegración de los implantes dentales*, PhD thesis, Universitat Politécnica de Catalunya. Barcelona, España.

Arzate, H., Alvarez-Pérez, M., Aguilar-Mendoza, M. & Alvarez-Fregoso, O. (1998). Human cementum tumor-derived cells have different features from human osteoblastic cells in-vitro, *Journal of Periodontal Research* 33: 249–258.

Arzate, H., Alvarez-Pérez, M., Alvarez-Fregoso, O., Wusterhaus-Chávez, A., Reyes-Gasga, J. & Ximénez-Fyvie, L. (2002). Electron microscopy, micro-analysis and x-ray diffraction characterizationof the mineral-like tissue deposited by human cementum tumor-derived cells, *Journal of Dental Research* 79: 28–34.

Aukhil, I. (2000). Biology of wound healing, *Periodontology 2000* 22: 44–50.

Bächle, M. & Kohal, R. (2004). A systematic review of the influence of different titanium surfaces on proliferation, differentiation and protein synthesis of osteoblast-like mg63 cells, *Clinical Oral Implants Research* 15(6): 683–92.

Bailon-Plaza, A. & van der Meulen, M. (2001). A mathematical framework to study the effects of growth factor influences on fracture healing, *Journal of Theoretical Biology* 212: 191–209.

Bonnet, A., Postaire, M. & Lipinski, P. (2009). Biomechanical study of mandible bone supporting a four-implant retained bridge finite element analysis of the influence of bone anisotropy and foodstuff position, *Elsevier* 31: 806–815.

Branemark, P. (1983). Osseointegration and its experimental background, *Journal of Prosthetic Dentistry* 50(3): 399–410.

Branemark, P., Breine, U., Adell, R., Hansson, O., Lindstöm, J. & Ohlsson, A. (1969). Intra-osseous anchorage of dental prostheses i. experimental studies, *Scandinavian Journal of Plastic and Reconstructive Surgery* 3: 81–100.

Bülent, E. (2002). Numerical analysis of a dental implant system in three Ű dimension, *Advances in engineering software* 33: 109–113.

Burger, E., Klein-Nulend, J. & Mullender, M. (2005). Mechanobiology of bone, *in* H. Petite & R. Quarto (eds), *Engineered Bone*, Landes Bioscience-Eurekah.com, pp. 28–44.

Casale, M. & Rivera, P. (2010). *Evaluacion biomecanica e histologica de implantes ortodonticos de titanio cargados con diferentes brazos de palanca*, Master's thesis, Universidad Nacional de Colombia, Posgrado de Ortodoncia, Facultad de Odontologia.

Casale, M. & Saavedra, M. (2010). *Evaluacion del comportamiento tisular y biomecanico a la carga inmediata y tardia de implantes ortodonticos de titanio*, Master's thesis, Universidad Nacional de Colombia, Posgrado de Ortodoncia, Facultad de Odontologia.

Chaddad, K., Ferreira, A., Geurs, N. & Reddy, M. (2008). Influence of surface characteristics on survival rates of mini-implants, *Angle Orthodontist* 78: 107–113.

Chung, K., Kim, Y., Linton, J. & Lee, Y. (2002). The miniplate with tuve for skeletal anchorage, *Journal of Clinical Orthodontics* 36: 407–412.

Colijn, C. & Mackey, M. (2005). A mathematical model of hematopoiesis i. periodic chronic myelogenous leukaemia, *Journal of Theoretical Biology* 237: 117–132.

Collen, D. & Lijnen, H. (1991). Basic and clinical aspects of fibrinolysis and thrombolysis, *Blood* 78: 3114 3124.

Costa, A., Raffaini, M. & Melsen, B. (1998). Miniscrews as orthodontic anchorage: a preliminary report, *International journal of adult orthodontic and orthognathic surgery* 13: 201–209.

Crismani, A., Bernhart, T., Tangl, S., Celar, A., Fugger, G., Gruber, R., Bantleon, H. & Watzek, G. (2008). Osseointegration and subperiosteal anchoring device in minipig mandible, *American journal of orthodontics and dentofacial orthopedics* 133: 743–747.

Davies, J. (2003). Understanding peri-implant endosseous healing, *Journal of Dental Education* 67(8): 932–949.

Davies, J. (2007). Bone bonding at natural and biomaterial surfaces, *Biomaterials* 28: 5058–5067.

Deguchi, T., Nasu, M., Murakami, K., Yabuuchi, T., Kamioka, H. & Yamamoto, T. (2006). Quantitative evaluation of cortical bone thickness with computed tomography scanning for orthodontic implants, *American Journal of Orthodontics and Dentofacial Orthopedics* 126(6): 721.e7–721.e12.

DiMilla, P., Barbee, K. & Luffenburger, D. (1991). Mathematical model for the effects of adhesion and mechanics on cell migration speed, *Biophysics Journal* 60: 15–37.

Dimitriou, R., Tsiridis, E. & Giannoudis, P. (2005). Current concepts of molecular aspects of bone healing, *Injury: International Journal of the Care of the Injured* 36: 1392–1404.

Egolf, R., BeGole, E. & Upshaw, H. (1990). Factors associated with orthodontic patient compliance with intra-oral elastic and headgear wear, *American Journal of Orthodontics and Dentofacial Orthopedics* 97: 336–348.

Ellingsen, J., Thomsen, P. & Lyngstadaas, P. (2006). Advances in dental implant materials and tissue regeneration, *Periodontology 2000* 41: 136–156.

Fragiskos, F. & Alexandridis, C. (2007). Osseointegrated implants, *in* F. Fragiskos & C. Alexandridis (eds), *Oral Surgery*, Springer Berlin Heidelberg, pp. 337–348.

Furie, B. & Furie, B. (2005). Thrombus formation in vivo, *Journal of Clinical Investigation* 115: 3355–3362.

Gapski, R., Wang, H., Mascarenhas, P. & Lang, N. (2003). Critical review of immediate implant loading, *Clinical Oral Implants Research* 14: 515–527.

Geng, J., Tan, K. & Liu, G. (2001). Application of finite element analysis in implant dentistry: a review of literature, *Journal prosthet dental* 85: 585–598.

Geng, J. & Tan, KBC and, L. G. (2001). Application of finite element analysis in implant dentistry: A review of literature, *Journal of Prosthetic Dentistry* 85: 585–598.

Geris, L., Gerisch, A., Vander-Sloten, J., Weiner, R. & Van-Oosterwyck, H. (2008). Angiogenesis in bone fracture healing: A bioregulatory model, *Journal of Theoretical Biology* 251: 137–158.

Gorkun, O., Veklich, Y., Weisel, J. & Lord, S. (1997). The conversion of fibrinogen to fibrin: Recombinant fibrinogen typifies plasma fibrinogen, *Blood* 89(12): 4407–4414.

Gracco, A., Cirignaco, A., Cozzani, M., Boccaccio, A., Pappalettere, C. & Vitale, G. (2009). Numerical/experimental analysis of the stress field around miniscrews for orthodontic anchorage, *European Journal of Orthodontics* 31: 12–20.

Häkkinen, L., Uitto, V. & Larjava, H. (2000). Cell biology of gingival wound healing, *Periodontology 2000* 24: 127–152.

Hansson, H., Albrektsson, T. & Branemark, P. (1983). Structural aspects of the interface between tissue and titanium implants, *Journal of Prosthetic Dentistry* 50(1): 108–113.

Hayman, E., Engvall, E., A'Hearn, E., Barnes, D., Pierschbacher, M. & Ruoslahti, E. (1982). Cell attachment on replicas of sds polyacrylamide gels reveals two adhesive plasma proteins, *Journal of Cell Biology* 95: 20–23.

Heydenrijk, K., Raghoebar, G., Meijer, H., van der Reijden, W., van Winkelhoff, A. & Stegenga, B. (2002). Two-stage imz implants and iti implants inserted in a single-stage procedure. a prospective comparative study, *Clinical Oral Implants Research* 13: 371–380.

Huang, Y., Xiropaidis, A., Sorensen, R., Albandar, J., Hall, J. & Wikesjo, U. (2005). Bone formation at titanium porous oxide [tiunite] oral implants in type iv bone, *Clinical Oral Implants Research* 16: 105–111.

Huiskes, R., van Driel, W., Prendergast, P. & Soballe, K. (1997). A biomechanical regulatory model for periprosthetic fibrous-tissue differentiation, *Journal of Materials Science: Materials in Medicine* 8: 785–788.

Huja, J., Litsky, A., Beck, F., Johnson, K. & Larsen, P. (2005). Pull-out strength of monocortical screws placed in the maxillae and mandibles of dogs, *American journal of orthodontics and dentofacial orthopedics* 127: 307–313.

Isaksson, H., van Donkelaar, C., Huiskes, R. & Ito, K. (2008). A mechano-regulatory bone-healing model incorporating cell-phenotype specific activity, *Journal of Theoretical Biology* 252: 230–246.

Juodzbalys, G. (2005). Stress distribution in bone: Single Ǔ unit implant prostheses vennered with porcelain or a new composite material, *Implant Dental* 14: 166–175.

Kanomi, R. (1997). Mini-implants for orthodontic anchorage, *Journal of clinical orthodontics* 36: 763–767.

Kasemo, B. (2002). Biological surface science, *Surface Science* 500: 656–677.

Kayabasi, O., Yüzbasioglu, E. & Erzincanli, F. (2006). Static, dynamic and fatigue behaviors of dental implant using finite element method, *Advances in engineering software* 37: 649–658.

Klein-Nulend, J., Bacabac, R. & Mullender, M. (2005). Mechanobiology of bone tissue, *Pathologie Biologie* 53: 576–580.

Knothe-Tate, M. (2003). Whither flows the fluid in bone? an osteocyte's perspective, *Journal of Biomechanics* 36: 1409–1424.

Lang, N., Araujo, M. & Karring, T. (2003). Alveolar bone formation, *in* J. Lindhe, T. Karring & N. Lang (eds), *Clinical Periodontology and Implant Dentistry*, Blackwell Munksgaard, pp. 866–896.

Leckholm, U. (2003). The surgical site, *in* J. Lindhe, T. Karring & N. Lang (eds), *Clinical Periodontology and Implant Dentistry*, Blackwell Munksgaard, pp. 852–865.

Li, W., Chong, S., Huang, E. & Tuan, T. (2003). Plasminogen activator/plasmin system: A major player in wound healing?, *Wound Repair and Regeneration* 11: 239–247.

Lindhe, J., Karring, T. & Araujo, M. (2003). Anatomy of the periodontium, *in* J. Lindhe, T. Karring & N. Lang (eds), *Clinical Periodontology and Implant Dentistry*, Blackwell Munksgaard, pp. 3–49.

Mann, K. (2003). Thrombin formation, *Chest* 124: 4–10.

Mantzaris, N., Webb, S. & Othmer, H. (2004). Mathematical modeling of tumor-induced angiogenesis, *Journal of Mathematical Biology* 49: 111–187.

Martínez, J., Sánchez, C., Trapero, C., Martínez, M. & García, F. (2002). Diseño de los implantes dentales: estado actual, *Avances en Periodoncia* 14(3): 129–136.

Matsuno, H., Yokoyama, A., Watari, F., Uo, M. & Kawasaki, T. (2001). Biocompatibility and osteogenesis of refractory metal implants, titanium,hafnium, niobium, tantalum and rhenium, *Biomaterials* 22: 1253–1262.

Melsen, B. & Costa, A. (2002). Immediate loading of implants used for orthodontic anchorage, *Clinical orthodontics and research* 3: 23–28.

Metikos-Hukovic, M., Kwokal, A. & Piljac, J. (2003). The influence of niobium and vanadium on passivity of titanium-based implants in physiological solution, *Biomaterials* 24: 3765–3775.

Meyer, U. & Wiesmann, H. (2006). *Bone and Cartilage Engineering*, Springer-Verlag, Berlin Heidelberg.

Minors, D. (2007). Haemostasis, blood platelets and coagulation, *Anaesthesia and intensive care medicine* 8(5): 214–216.

Moreo, P., García-Aznar, J. & Doblaré, M. (2008). Modeling mechanosensing and its effect on the migration and proliferation of adherent cells, *Acta Biomaterialia* 4: 613–621.

Moreo, P., García-Aznar, J. & Doblaré, M. (2009). Bone ingrowth on the surface of endosseous implants. part i: Mathematical model, *Journal of Theoretical Biology* 260: 1–12.

Mosmann, T. (1983). Rapid colorimetric assay for cellular growth and survival: Application to proliferation and cytotoxicity assays, *Journal of Immunological Methods* 65: 55–63.

Motoyoshi, M., Inaba, M., Ono, A., Ueno, S. & Shimizu, N. (2009a). The effect of cortical bone thickness on the stability of orthodontic mini-implants and on the stress distribution in surrounding bone, *International journal of oral and maxillofacial surgery* 38: 13–18.

Motoyoshi, M., Inaba, M., Ono, A., Ueno, S. & Shimizu, N. (2009b). The effect of cortical bone thickness on the stability of orthodontic mini-implants and on the stress distribution in surrounding bone, *International journal of oral and maxillofacial surgery* 38(1): 13–18.

Motoyoshi, M Ueno, S., Okazaki, K. & Shimizu, N. (2009). Bone stress for a mini-implant close to the roots of adjacent teeth - 3d finite element analysis, *International journal of oral and maxillofacial surgery* 38: 363–368.

Niinomi, N. (2003). Recent research and development in titanium alloys for biomedical applications and healthcare goods, *Science and Technology of Advanced Materials* 4: 445–454.

Papadopoulos, M. & Tarawneh, F. (2007). The use of miniscrew implants for temporary skeletal anchorage in orthodontics: A comprehensive review, *Oral surgery, oral medicine, oral pathology, oral radiology & endodontics* 103: e6–e15.

Papavasiliou, G. (1996). Three Ű dimensional finite element analysis of stress distribution around single tooth implants as a function of bony support, prosthesis type, and loading during function, *Journal prosthet dental* 76: 633–640.

Pasi, K. (1999). Hemostasis, *in* D. Perry & K. Pasi (eds), *Hemostasis and Thrombosis Protocols*, Humana Press, pp. 3–24.

Patra, A., DePaolo, J., D'Souza, K., Detalla, D. & Meenaghan, M. (1998). Guidelines for analysis and redesign of dental implants, *Implant Dentistry* 7(4): 355–366.

Poiate, I., Vasconcellos, A., Andueza, A., Pola, I. & Poiate, E. (2008). Three dimensional finite element analyses of oral structures by computerized tomography, *Journal of bioscience and bioengineering* 106(6): 606–609.

Proffit, W. & Fields, H. (1993). *Mechanical principles in orthodontic force control. In: Contemporary Orthodontics*, Mosby.

Puleo, D. & Nanci, A. (1999). Understanding and controlling the bone-implant interface, *Biomaterials* 20: 2311–2321.

Ratner, B., Hoffman, A., Shoen, F. & Lemons, J. (1996). *Biomaterials Science: An Introduction to Materials in Science*, Academic Press, San Diego.

Rieger, M., Adams, M. & Kinzel, G. (1990). Finite element analysis of eleven endosseus implants, *Journal of Prosthetic Dentistry* 63(4): 457–465.

Rodil, S., Olivares, R. & Arzate, H. (2004). In vitro cytotoxicity of amorphous carbon films, *Biomedical Materials and Engineering* 15: 101–112.

Rodil, S., Olivares, R., Arzate, H. & Muhl, S. (2003). Properties of carbon films and their biocompatibility using in-vitro tests, *Diamond and Related Materials* 12: 931–937.

Saffar, J., Lasfargues, J. & Cherruau, M. (1997). Alveolar bone and the alveolar process: The socket that is never stable, *Periodontology 2000* 13: 76–90.

Seong-Hun, K., Jae-Hee, C., Kyu-Rhim, C. & Yoon-A, K. (2008). Removal torques values of surface-treated mini-implants after loading, *American journal of orthodontics & dentofacial orthopedics* 134(1): 35–43.

Serra, G., Morais, L., Elias, C., Meyers, M., Andrade, L. & Muller, M. (2008). Sequential bone healing of immediately loaded mini-implants, *American journal of orthodontics & dentofacial orthopedics* 134(1): 44–42.

Sikavitsas, V., Temenoff, J. & Mikos, A. (2001). Biomaterials and bone mechanotransduction, *Biomaterials* 22: 2581–2593.

Stanford, C. & Schneider, G. (2004). Functional behavior of bone around dental implants, *Gerodontology* 21: 71–77.

Sung, S., Jang, G., Chun, Y. & Moond, Y. (2010). Effective en-masse retraction design with orthodontic mini-implant anchorage: A finite element analysis, *American Journal of Orthodontics and Dentofacial Orthopedics* 137: 648–657.

Tracqui, P., Namy, P. & Ohayon, J. (2007). In vitro tubulogenesis of endothelial cells: Analysis of a bifurcation process controlled by a mechanical switch, *in* A. Deutsch, L. Brusch, H. Byme, G. de Vries & H. Herzel (eds), *Mathematical Modeling of Biological Systems*, Vol. I, Springer- Birkhäuser, pp. 47–57.

Van der Meulen, M. & Huiskes, R. (2002). Why mechanobiology? a survey article, *Journal of Biomechanics* 35: 401–414.

Vanegas, J., Landinez, N. & Garzón-Alvarado, D. (2010). Mathematical model of the coagulation at the bone-dental implant interface, *Computers in Biology and Medicine* 40(10): 791–801.

Vanegas, J., Landinez, N., Garzón-Alvarado, D. & Casale, M. (2011). A finite element method approach for the mechanobiological modeling of the osseointegration of a dental implant, *Computer Methods and Programs in Biomedicine* 101: 297–314.

Vanegas, J., Landínez, N. & Garzón, D. (2009). Generalidades de la interfase hueso-implante dental, *Revista Cubana de Investigaciones Biomédicas* 28(3): 130–146.

Vanegas, J., Garzón, D. & Casale, R. (2010). Interacción entre osteoblastos y superficies de titanio: aplicación en implantes dentales, *Revista Cubana de Investigaciones Biomédicas* 29(1): 51–68.

Watari, F., Yokoyama, A., Omori, A., Hirai, T., Kondo, H., Uo, M. & Kawasaki, T. (2004). Biocompatibility of materials and development to functionally graded implantf or biomedical application, *Composites Science and Technology* 64: 893–908.

Welubein, H., Yildirim, M. & Diedrith, P. (1999). Osteodynamics around orthodontically loaded short maxillary implantes, *Jorunal of Orofacial Orthopedics* 60: 409–415.

Weisel, J. (2004). The mechanical properties of fibrin for basic scientists and clinicians, *Biophysical Chemistry* 112: 267–276.

Wennerberg, A., Albrektsson, T. & Lindhe, J. (2003). Surface topography of titanium dental implants, *in* J. Lindhe, T. Karring & N. Lang (eds), *Clinical Periodontology and Implant Dentistry*, Blackwell Munksgaard, pp. 821–828.

Zhang, L., Zhao, Z. & Li, Y. (2010). Osseointegration of orthodontic micro-screw after immediate and early loading, *Angle orthdontists* 80(2): 354–360.

Zhao, L., Xu, Z., Yang, Z., Wei, X., Tang, T. & Zhao, Z. (2009). Orthodontic mini-implant stability in different healing times before loading: A microscopic computerized tomographic and biomechanical analysis, *Oral surgery, oral medicine, oral pathology, oral radiology & endodontics* 108: 196–202.

Part 3

Artificial Tissues Creation and Engineering

16

Dental Pulp Stem Cells and Tissue Engineering Strategies for Clinical Application on Odontoiatric Field

Zavan Barbara et al.[*]
University of Padova
Italy

1. Introduction

Recent advances in tissue engineering have drawn scientists to test the possibility of tooth engineering and regeneration. Tooth regeneration is normally referred to as the regeneration of the entire tooth or root that can be integrated into the jaw bone. This technology is still at its infancy and when it matures, it may be used to restore missing teeth and replace artificial dental implants when the tooth is damaged but still in a reparable condition, regeneration of parts of the tooth structure can prevent or delay the loss of the whole tooth. To engineer and regenerate a whole tooth, the cell source, tissue engineering strategies and specific scaffolds needed to be correct choose.

Indeed, for example, to repair partly lost tooth tissues such as PDL, dentin, and pulp, one or two particular types of dental stem cells may be sufficient to fulfill the need. In light of such considerations, aim of the present chapter is to define the main strategies to isolate dental pulp stem cells, their characterisation and differentiation, tissue enngineering strategies and clinical applications for the creation of artificial tissue useful in odontoiatric field.

2. Dental pulp stem cells

Dental pulp is a well known tissue enrich of adult mesenchymal stem cells: Dental Pulp Stem cells (DPSc). These adult stem cells play an important role in regenerative medicine both for oral and non oral pathoses thanks to their biological properties such as multipotency, high proliferation rates and accessibility (Yamada et al., 2010).

Beyond natural capacity of response to injury, dental pulp stem cells are attractive for their potential to differentiate, in vitro, into several cell types including odontoblasts, neural progenitors, chondrocytes, endothelocytes, adipocytes, smooth muscle cells and osteoblasts. The potential application of dental pulp stem cells and tissue engineering in dentistry are discussed in the present chapter (Sloan & Waddington, 2009).

---

[*] Bressan Eriberto, Sivolella Stefano, Brunello Giulia, Gardin Chiara, Nadia Ferrarese, Ferroni Letizia and Stellini Edoardo

2.1 Tooth anatomy

Each tooth consists of three main parts: the crown, the neck and the root, that we can define with anatomic o clinical criteria. Here follow a brief review of the tooth anatomy (Fig.1) involved on stem cells-tissue engineering field.

2.1.1 Dentin

Dentin is a mesenchymal derived tissue lying between enamel or cementum and dental pulp (pulp chamber and root canal). It is a mineralized connective tissue with an organic matrix. It is made up of 70% inorganic materials (especially hydroxyapatite crystals), 20% organic materials and 10% water by weight. The bulk of organic matrix (85-90%) consists of type I collagen, there is also a minor amount of type V and VI collagen. Noncollagenous molecules of dentin are dentin phosphorines, Gla proteins, acidic glycoproteins, growth-related factors, serum derived proteins, lipids and proteoglycans. Dentin has microscopic channels (0,5-3 µm), called dentinal tubules, radiating outward from pulp cavity to dentino-enamel or dentino-cementum junction. These tubules contain projections of cells secreting dental matrix, known as odontoblasts. The most peripheral aspect of the pulp is lined by the body of these odontoblasts (Yoshiba et al., 2002; Fonzi, 2000).

2.1.2 Pulp

Pulp consists of a loose connective tissue enclosed by rigid predentin and dentin. Along the border between the dentin and the pulp are odontoblasts. The thickness of dentinal layer increases with age due to the deposition of secondary and tertiary dentin, reducing the volume of the pulp chamber and the root canals. The most peripheral aspect of the pulp contains four layers of cells: the odontoblastic layer (the most external one), the cell-free zone, the cell-rich zone and the parietal plexus of nerves. Deep inside is the pulp proper, composed of a great amount of fibroblasts and ECM. Blood vessels and nerves enter the tooth mostly through the apical foramen. Other cells in the pulp include undifferentiated mesenchymal cells, deriving from dental papilla, fibrocytes, macrophages and lymphocytes (Fonzi, 2000; Ferguson, 2002).

2.1.3 Alveolar bone

The bone that supports the teeth is called alveolar bone. It is composed of compact bone and trabecular or spongy bone. The outside wall of the bone is compact bone, such as the thin layer that lines the socket known as lamina dura.

The spongy bone is inside and contains bone marrow. The number and the size of the trabeculae in this bone are determined by the function activity of the organ.

Alveolar bone proper is the part just around the tooth and it gives attachment to the PDL fibres (bundle bone). The alveolar bone proper is also called cribiform plate, due to the presence of perforation for the entry of vessels and nerves.

Bone is made of 65% inorganic material (mainly hydroxyapatite) by weight, 15% water, 20% organic material. The organic matrix is composed of collagen type I (90-95%), Gla proteins, glycoproteins, phosphorines, proteoglycans, growth factors and bone morphogenetic proteins (e.g. osteogenin) (Ferguson, 2002; Fonzi 2000).

2.2 Odontogenesis

During the sixth week of embryogenesis, after the migration of neural crest cells into head and neck mesenchyme, the ectoderm of the first brachial arch begins to proliferate giving rise to the vestibular lamina and the dental lamina (Langman, 2008).

Dental lamina is a band of ectodermic cells growing from the epithelium of the stomodeum into the underlying mesenchyme and giving rise to the enamel organs of teeth, along the horse shoe shaped dental arches.

Several transcription factors are implicated in odontogenesis, including Pax9, Pitx2, Runx2, Msx1, Msx2, Bmp2, Bmp4, Fgf8 and Fgf9 (Zhang et al., 2009; Bei & Maas,1998). The development is commonly divided into the following stages: the bud stage, the cap stage, the bell stage.

The early bell stage of odontogenesis is characterized by epithelial expansion and differentiation into the inner and outer enamel epithelium, stratum intermedium and stellate reticulum.

During the late bell stage, two tooth specific cell types are formed: ameloblasts, which derive from the inner enamel epithelium and produce enamel, and odontoblasts, which differentiate from dental papilla and synthesize dentin.

Dentinogenesis starts before enamel formation with the secretion of an organic matrix in the area directly adjacent to the inner enamel epithelium. Dentin is formed by the production of organic matrix (predentin) and the simultaneous mineralization of this matrix (Hao et al.2009).

After crown formation, root development begins. The cells of the inner and of the outer enamel epithelium become in contact and give rise to the cervical loop at the base of enamel organ (Fonzi, 2000).

The cells of the cervical loop continue to grow away from the crown and become Hertwig's epithelial root sheath. It induce the adjacent cells of dental papilla to differentiate into odontoblasts and produce dentin. Once this structure fragments, the dentin of the root comes in contact with the dental follicle and stimulates the cementoblasts to begin cementum secretion.

The dental follicle also gives rise to the other supporting structure of the tooth: the periodontal ligament and the alveolar bone proper (Luan et al. 2006).

2.3 Dental pulp stem cells

A stem cell is defined as a cell that has the ability to continuously divide to either replicate itself (self-renewing), or produce specialized cells than can differentiate into various other types of cells or tissues (multilineage differentiation). The microenvironment in which stem cells reside is called a stem cell niche and is composed of heterogeneous cell types, extracellular matrix (ECM) and soluble factors to support the maintenance and self-renewal of the stem cells (Yen & Sharp, 2008).

Stem cells can be generally classified in embryonic stem cells (ESCs) and adult stem cells (ASCs). ESCs derive from the early mammalian embryo at the blastocyst stage and have the capability to give rise to all kinds of cells. Thus, ESCs are considered pluripotent. On the contrary, ASCs are just multipotent because their differentiation potential is restricted to certain cell lineages. ASCs reside in several and perhaps most organs and tissues that have already developed. For this reason, ASCs are also referred to as postnatal stem cells.

Mesenchymal stem cells (MSCs) are ASCs with mesodermal and neuroectodermal origin. They are available from many tissues such as bone marrow, adipose tissue, umbilical cord and dental pulp. MSCs are able to differentiate into cells of mesodermal origin like adypocites, chondrocytes or osteocytes, but they can also give rise to representative lineages of the three embryonic layers. For instance, it is well known that MSCs posses an extended degree of plasticity compared to other ASCs populations, including the ability to differentiate in vitro into non-mesodermal cell types such as neurons and astrocytes. MSCs, in addition to their multipotency, are easy to isolate and culture in vitro and they do not apparently represent an ethical issue based on their source of origin.

MSCs were firstly discovered and characterized in bone marrow (Friedenstein et al., 1970). Bone marrow-derived stem cells (BMMSCs) can be easily obtained, greatly expanded in culture and used in cell-mediated therapies and tissue engineering applications. However, the clinical use of BMMSCs is limited by several problems, including a painful recovery often associated to a low number of harvested cells. For these reasons, many researchers begin to investigate alternative tissues for more abundant and accessible sources of MSCs with least invasive collection procedures.

Dental tissue from human third molar represents an easily accessible and often discarded source for MSCs harvesting. The first type of dental stem cell was isolated from the human pulp tissue and termed dental pulp stem cells (DPSCs) (Gronthos et al., 2000). Subsequently, four more types of dental-MSC-like populations were identified: stem cells from exfoliated deciduous teeth (SHED) (Miura et al., 2003), periodontal ligament stem cells (PDLSCs) (Seo et al., 2004), stem cells from apical papilla (SCAP) (Sonoyama et al., 2006), and dental follicle precursor cells (DFPCs) (Morsczeck et al., 2005). Among them, all except SHED are from permanent teeth. All of these stem cells demonstrate multipotentiality and the capability to regenerate multiple dental/periodontal tissues in vitro and in vivo (Huang, 2009).

Dental pulp is the soft connective tissue entrapped within the dental crown. It is divided into four layers. The external layer is made up of odontoblasts producing dentin; the second layer, called "cell free zone", is poor in cells and rich in collagen fibers; the third layer, called "cell rich zone", contains progenitor cells and undifferentiated cells, some of which are considered stem cells. From this layer, undifferentiated cells migrate to various districts where they can differentiate under different stimuli and make new differentiated cells and tissues. The innermost layer is the core of the pulp and comprises the vascular area and nerves (D'Aquino et al., 2009).

Stem cells that reside in dental pulp, called dental pulp stem cells (DPSCs), are considered a population of MSCs, therefore markers that have been used for identifying MSCs are also used for DPSCs. DPSCs result positive to STRO-1, CD13, CD24, CD29, CD44, CD73, CD90, CD105, CD106, CD146, Oct4, Nanog and $\beta2$ integrin, and negative to CD14, CD34, CD45 and HLA-DR. The persistence of negativity for CD45 and positivity for CD34 demonstrates that these cells are not derived from a hematopoietic source, nevertheless they are of mesenchymal origin (D'Aquino et al., 2007). Like all MSCs, DPSCs are also heterogeneous and the various markers listed previously may be expressed by subpopulations of these stem cells (Huang et al., 2009).

DPSCs were firstly identified and isolated by Gronthos and co-workers in 2000, based on their clonogenic abilities and rapid proliferative rates. In addition, they demonstrated that DPSCs can develop into odontoblasts, the cells that form the mineralized matrix of dentin (Gronthos et al., 2000). Moreover, when transplanted in immunocompromised mice, DPSCs mixed with hydroxyapatite/tricalcium phosphate (HA/TCP) form a pulp-dentin-like tissue complex. This complex is composed of a mineralized matrix with tubules lined with odontoblasts, and fibrous tissue containing blood vessels in an arrangement similar to the dentin-pulp complex found in normal human teeth. Interestingly, and in contrast to BMMSCs, DPSCs do not form areas of active hematopoiesis and adipocyte accumulation at the transplantation site. In another study, when DPSCs are seeded onto human dentin surface and implanted into immunocompromised mice, reparative dentin-like structure is deposited on the dentin surface (Batouli et al., 2003). These results together raise the possibility that a protocol for pulp tissue regeneration and new dentin formation for clinical therapeutic purposes could be established.

Later, Laino and co-workers isolated a selected subpopulation of DPSCs called stromal bone producing dental pulp stem cells (SBP-DPSCs), which roughly represent 10% of dental pulp cells (Laino et al., 2005). These cells display a great capability of self-expanding and differentiating in pre-osteoblasts which are able to self-maintain and renew for long time. SBP-DPSCs differentiate into osteoblasts, producing in vitro a living autologous fibrous bone (LAB) tissue. When transplanted into immunocompromised rats, SBP-DPSCs form a lamellar bone containing osteocytes. In this setting, SBP-DPSCs produce bone but not dentin, as shown by mRNA expression of bone markers including osteocalcin, Runx-2, collagen I, alkaline phosphatase, but not dentin sialo phospho protein (DSPP), which is specific for dentin. Moreover, it has been observed that, during their differentiation, about 30% of SBP-DPSCs becomes endothelial cells. These cells are found lining the vessel walls of the newly formed woven bone. In addition, after in vivo transplantation, a complete integration of vessels within bone chips takes place, leading to the formation of a vascularised bone tissue (D'Aquino et al., 2007).

Further characterization revealed that DPSCs also possess adipogenic and neurogenic differentiation capacities by exhibiting adipocyte- and neuronal-like cell morphologies and expressing respective gene markers. In addition, DPSCs were also found to undergo chondrogenic and myogenic differentiation in vitro. The plasticity and multipotential capability of DPSCs can be explained by the fact that dental pulp is made of both ectodermic and mesenchymal components, containing neural crest-derived cells (D'Aquino et al., 2009).

DPSCs can be collected from dental pulp by means of a non-invasive practice that can be performed in the adult during life and in the young after surgical extraction of wisdom teeth. DPSCs can survive for long periods and can be passaged several times. It is possible to obtain more than 80 passages without clear signs of senescence. Furthermore, DPSCs can be cryopreserved and stored for long periods without losing their multipotential differentiation ability (Laino et al., 2005).

3. Role of DPSCs in regenerative medicine

Teeth have a complex structural composition that ensures both hardness and durability. However, this structure is vulnerable to trauma and bacterial infections. When the tooth is damaged but still in a reparable condition, regeneration of parts of the tooth structure can prevent or delay the loss of the whole tooth. This fact is of importance because tooth loss affects not only basic mouth functions but aesthetic appearance and quality of life (Huang, 2009). The regenerative response of teeth to damage and structural degeneration are diverse and compartment-dependent, since teeth, as complex structures, harbour both living (periodontal ligament, cementum, pulp) and acellular (enamel, dentine) tissues. Of all the dental structures, only enamel is incapable of regenerating its original structure, while the remaining tissues possess that capacity in varying degrees, dependent on multiple factors (Inanç & Elçin, 2011).

The dental pulp plays a major role in tooth regeneration after injury, by participating in a process called reparative dentinogenesis. When pulp tissue is exposed as consequence of the loss of the overlaying dentin, direct pulp-capping therapy allows the pulp to form new dentin. It has been observed that the use of various cement-based compounds, such as calcium hydroxide and mineral trioxide aggregate (MTA), promotes the activity of dentinogenesis. Cells that remain in the healthy portion of the pulp migrate to the injured site, proliferate by the growth factors released from surrounding dentin matrix and attach the necrotic layer to form osteodentin. Later, the cells attached to osteodentin differentiate into odontoblasts to produce tubular dentin, thus forming reparative dentine. This early mineralized tissue preserves the pulp integrity and serves as protective barrier upon the injury (Nakashima, 2005).

When the tooth is further damaged, dentin regeneration becomes difficult as it requires a healthy pulp. Thus, bigger traumas or advanced caries are clinically treated with root canal therapy, in which the entire pulp is cleaned out and replaced with a gutta-percha filling. However, living pulp is critical for the maintenance of tooth homeostasis and essential for tooth longevity.

An ideal form of therapy might consist of regenerative approaches in which diseased or necrotic pulp tissues are removed and replaced with regenerated pulp tissues to revitalize the teeth. In particular, the regenerative pulp therapy would reconstitute the normal tissue continuum at the pulp–dentine border by regulating the tissue-specific processes of reparative dentinogenesis. Two types of dental pulp regeneration can be considered based on the clinical situations: in situ regeneration of partial pulp or de novo synthesis of a total pulp replacement (Sun et al., 2010).

Engineering and regeneration of dental pulp tissue still remain a difficult task. A regenerated pulp tissue should be functionally competent: it should be vascularised, contain similar cell density and architecture of ECM to those of natural pulp, be capable of giving rise to new odontoblasts lining against the existing dentin surface, produce new dentin and be innervated. The first step to engineer tissues is to isolate cells with the right phenotype and propagate them in suitable culturing environments. DPSCs can be isolated by two methods: the enzyme-digestion method and the explants outgrowth method. The first method involves the collection of the pulp tissue under sterile conditions, the digestion with appropriate enzymes (collagenase, dispase, trypsin), the seeding in culture dishes containing a special medium supplemented with necessary additives, and then the incubation at 37°C. The second method implies that the extruded pulp tissue is cut into 2 mm^3, and directly incubated in culture dishes containing the essential medium with supplements. A period of two weeks is generally needed to allow a sufficient number of cells to migrate out of the tissue. It has been demonstrated that cells isolated by enzyme-digestion have a higher proliferation rate than those collected by outgrowth (Huang et al., 2006).

Once these cells are grown on a two-dimensional surface, it is possible to transfer them to a three-dimensional scaffold construct. The scaffold provides a 3D environment for cells to attach and grow, therefore mimicking the in vivo condition (Fig.2). An ideal scaffold should be biocompatible, biodegradable, and have adequate physical and mechanical strength. Then, it should be porous to allow placement of cells and effective transport of nutrients, oxygen, waste as well as growth factors. Finally, it should be replaced by regenerative tissue while retaining the shape and form of the final tissue structure (Saber, 2009).

Scaffolds can be fabricated from natural polymers or synthetic materials. The natural polymers have advantages of good biocompatibility and bioactivity. On the contrary, synthetic matrices enable precise control over the physiochemical properties such as degradation rate, porosity, microstructure, and mechanical strength (Sharma & Elisseeff, 2004).

Examples of natural polymers are collagen, gelatin, dextran and fibronectin. Although collagen is a commonly used matrix in which to grow cells in three-dimensions, several cell types are known to cause the contraction of collagen. It has been demonstrated that pulp cells markedly cause the contraction of collagen with a reduction down to ~30%, which might affect pulp tissue regeneration (Huang et al., 2006).

Examples of synthetic polymers are polylactic acid (PLA), polyglycolic acid (PGA) or their co-polymers, poly lactic-co-glycolic acid (PLGA). Recent experiments demonstrate that DPSCs seeded onto PLGA scaffolds regenerate a pulp/dentin-like tissue (Huang, 2009). Other artificial scaffolds are hydrogels, like polyethylene glycol (PEG)-based polymers, or inorganic compounds such as hydroxyapatite (HA), tricalcium phosphate (TCP) and calcium polyphosphate (CPP). These are used to enhance bone conductivity and have proved to be very effective for tissue engineering of DPSCs (Wang et al., 2006).

Apart from DPSCs and an appropriate scaffold, dental pulp regeneration also requires the use of growth factors and ECM molecules that induce specific differentiation pathways and maintain the odontoblast phenotype. It is known that several factors, such as transforming growth factor β (TGFβ), bone morphogenic proteins (BMPs), platelet-derived growth factor (PDGF), fibroblast growth factor (FGF), and vascular endothelial growth factor (VEGF), are secreted by odontoblasts and incorporated within the dentine matrix during dentinogenesis. When these molecules are released from the dentin, they are bioactive and fully capable of inducing cellular responses, as for example those that lead to the generation of reparative dentin and to dental pulp repair (Casagrande et al., 2011).

Dental pulp tissue engineering is an emerging field that can potentially have a major impact on oral health. However, the source of morphogens required for stem cell differentiation into odontoblasts and the scaffold characteristics that are more conducive to odontoblastic differentiation are still unclear. (Demarco 2010) investigated the effect of dentin and scaffold porogen on the differentiation of human dental pulp stem cells (DPSCs) into odontoblasts.Poly-L-lactic acid (PLLA) scaffolds were prepared in pulp chambers of extracted human third molars using salt crystals or gelatin spheres as porogen. DPSCs seeded in tooth slice/scaffolds or control scaffolds (without tooth slice) were either cultured in vitro or implanted subcutaneously in immunodefficient mice.

DPSCs seeded in tooth slice/scaffolds but not in control scaffolds expressed putative odontoblastic markers (DMP-1, DSPP, and MEPE) in vitro and in vivo. DPSCs seeded in tooth/slice scaffolds presented lower proliferation rates than in control scaffolds between 7 and 21 days (p < 0.05). DPSCs seeded in tooth slice/scaffolds and transplanted into mice generated a tissue with morphological characteristics similar to those of human dental pulps. Scaffolds generated with gelatin or salt porogen resulted in similar DPSC proliferation. The porogen type had a relatively modest impact on the expression of the markers of odontoblastic differentiation. Collectively, this work shows that dentin-related morphogens are important for the differentiation of DPSC into odontoblasts and for the engineering of dental pulp-like tissues and suggest that environmental cues influence DPSC behavior and differentiation potential.

Yang et al (Yang 2010) investigated, moreover the in vitro and in vivo behavior of dental pulp stem cells (DPSCs) seeded on electrospun poly(epsilon-caprolactone) (PCL)/gelatin scaffolds with or without the addition of nano-hydroxyapatite (nHA). For the in vitro evaluation, DNA content, alkaline phosphatase (ALP) activity and osteocalcin (OC) measurement showed that the scaffolds supported DPSC adhesion, proliferation, and odontoblastic differentiation. Moreover, the presence of nHA upregulated ALP activity and promoted OC expression. Real-time PCR data confirmed these results. SEM micrographs qualitatively confirmed the proliferation and mineralization characteristics of DPSCs on both scaffolds. Subsequently, both scaffolds seeded with DPSCs were subcutaneously implanted into immunocompromised nude mice. Scaffolds with nHA but without cells were implanted as control. Histological evaluation revealed that all implants were surrounded by a thin fibrous tissue capsule without any adverse effects. The cell/scaffold composites showed obvious in vivo hard tissue formation, but there was no sign of tissue ingrowth. Further, the combination of nHA in scaffolds did upregulate the expression of specific odontogenic genes. In conclusion, the incorporation of nHA in nanofibers indeed enhanced DPSCs differentiation towards an odontoblast-like phenotype in vitro and in vivo.

Galler et (Galler 2008) developped an approach to develop novel regenerative strategies and engineer dental tissues, two dental stem cell lines were combined with peptide-amphiphile (PA) hydrogel scaffolds. PAs self-assemble into three-dimensional networks of nanofibers, and living cells can be encapsulated. Cell-matrix interactions were tailored by incorporation of the cell adhesion sequence RGD and an enzyme-cleavable site. SHED (stem cells from human exfoliated deciduous teeth) and DPSC (dental pulp stem cells) were cultured in PA hydrogels for 4 weeks using different osteogenic supplements. Both cell lines proliferate and differentiate within the hydrogels. Histologic analysis shows degradation of the gels and extracellular matrix production. However, distinct differences between the two cell lines can be observed. SHED show a spindle-shaped morphology, high proliferation rates, and collagen production, resulting in soft tissue formation. In contrast, DPSC reduce proliferation, but exhibit an osteoblast-like phenotype, express osteoblast marker genes, and deposit mineral. Since the hydrogels are easy to handle and can be introduced into small defects, this novel system might be suitable for engineering both soft and mineralized matrices for dental tissue regeneration.

In the future, the success of regenerative endodontic therapy will depend on the ability to yield a functional pulp tissue within cleaned and shaped root canal systems to revitalize teeth. This may be achieved by an in vivo approach, where pulp tissue is regenerated in situ into root canals, or by an ex vivo approach, which implies a de novo engineered pulp relying on the tissue-engineering triad (DPSCs, scaffold, growth factors).

4. Acknowledgment

This research was supported by University of Padova (Italy).

5. References

Batouli, S., Miura, M., Brahim, J., Tsutsui, TW., Fisher, LW., Gronthos, S., Robey, PG. & Shi, S. (2003). Comparison of stem cell- mediated osteogenesis and dentinogenesis. *Journal of Dental Research*, Vol.82, No.12, (December 2003), pp. 976–981, ISSN 0022-0345

Bei, M. & Maas, R. (1998). FGFs and BMP4 induce both Msx1-independent and Msx1-dependent signaling pathways in early tooth development. *Development*. Vol.125, No.21, (November 1998), pp.4325-4333, ISSN 0950-1991

Casagrande, L., Cordeiro, MM., Nör, SA. & Nör, JE., (2011). Dental pulp stem cells in regenerative dentistry. *Odontology*, Vol.99, No.1, (January 2011), pp. 1-7, ISSN 1618-1255

D'Aquino, R., Graziano, A., Sampaolesi, M., Laino, G., Pirozzi, G., De Rosa, A. & Papaccio, G. (2007). Human postnatal dental pulp cells co-differentiate into osteoblasts and endotheliocytes: a pivotal synergy leading to adult bone tissue formation. *Cell Death and Differentation*, Vol. 14, No.6, (March 2007), pp. 1162-71, ISSN 1350-9047

D'Aquino, R., De Rosa, A., Laino, G., Caruso, F., Guida, L., Rullo, R., Checchi, V., Laino, L., Tirino, V. & Papaccio, G. (2009). Human dental pulp stem cells: from biology to clinical applications. *Journal of Experimental Zoology Part B: Molecular and Developmental Evolution*, Vol. 312, No.5, (July 2009), pp. 408-15, ISSN 1552-5007

Demarco FF, Casagrande L, Zhang Z, Dong Z, Tarquinio SB, Zeitlin BD, Shi S, Smith AJ, Nör JE. (2010) Effects of morphogen and scaffold porogen on the differentiation of dental pulp stem cells. J Endod. 2010 Vol 36 (Novembre) 1805-11.

Ferguson, D.B. (2002). *Biologia del Cavo Orale. Istologia, Biochimica, Fisiologia*, Casa Editrice Ambrosiana, ISBN 9788840812045

Fonzi, L. (2000). *Anantomia Funzionale e Clinica dello Splancnocranio*, Edi-Ermes, ISBN 887051238-X, Milano

Friedenstein, AJ., Chailakhjan, RK. & Lalykina, KS. (1970). The development of fibroblast colonies in monolayer cultures of guinea-pig bone marrow and spleen cells. *Cell and Tissue Kinetic*, Vol. 3, No.4, (October 1970) ,pp. 393-403, ISSN 0008-8730

Galler KM, Cavender A, Yuwono V, Dong H, Shi S, Schmalz G, Hartgerink JD, D'Souza RN. (2008) Self-assembling peptide amphiphile nanofibers as a scaffold for dental stem cells. *Tissue Eng Part A*. Vol.12,No.1, pp. 2051-8.

Gronthos, S., Mangani, M., Brahim, J., Robey, PG. & Shi, S. (2000). Postnatal human dental pulp stem cells (DPSCs) in vitro and in vivo. *Proceedings of the National Academy of Sciences of the United States of America*, Vol.97, No.25, (December 2000), pp. 13625-13630, ISSN 0027-8424

Hao, J., Ramachandran, A. & George A. (2009) Temporal and spatial localization of the dentin matrix proteins during dentin biomineralization. *The journal of Histochemistry and Cytochemistry*. Vol.57, No.3, (March 2009), p.227-237 ISSN 0022-1554

Huang, GT., Sonoyama, W., Chen, J. & Park, SH. (2006). In vitro characterization of human dental pulp cells: various isolation methods and culturing environments. *Cell and Tissue Research*, Vol.324, No.2, (May 2006), pp.225-236, ISSN 0302-766X

Huang, GT., (2009). Pulp and dentin tissue engineering and regeneration: current progress. *Regenerative Medicine*, Vol.4, No.5, (September 2009), pp.697-707 ISSN 1746-076X

Huang, GT., Gronthos, S. & Shi, S. (2009). Mesenchymal stem cells derived from dental tissues vs. those from other sources: their biology and role in regenerative medicine. *Journal of Dental Research*, Vol.88, No.9, (September 2009), pp.792-806, ISSN 0022-0345

Inanç, B. & Elçin, YM., (2011). Stem Cells in Tooth Tissue Regeneration-Challenges and Limitations. In: *Stem Cell Reviews*, 18.02.2011, Available from http://www.stemcellgateway.net/ArticlePage.aspx?DOI=10.1007/s12015-011-9237-7

Laino, G., D'Aquino, R., Graziano, A., Lanza, V., Carinci, F., Naro, F., Pirozzi, G., & Papaccio, G. (2005). A new population of human adult dental pulp stem cells: a useful source of living autologous fibrous bone tissue (LAB). *Journal of Bone and Mineral Research*, Vol.20, No.8, (August 2005), pp.1394-1402, ISSN 0884-0431

Langman, (2008). Embriologia Medica di Langman (IV), Elsevier-Masson, ISBN 9788821430459

Luan, X., Ito, Y. & Diekwisch, TGH.(2006). Evolution and Development of Hertwig's Epithelial Root Sheath. *Developmental Dynamics*. Vol. 235, No.5, (May 2006) pp.1167–1180 ISSN 1058-8388

Miura, M., Gronthos, S., Zhao, M., Lu, B., Fisher, LW., Robey, PG. & Shi, S. (2003). SHED: stem cells from human exfoliated deciduous teeth. *Proceedings of the National*

Academy of Sciences of the United Stases of America, Vol.100, No.10, (May 2003), pp.5807-5812, ISSN 0027-8424

Morsczeck, C., Gotz, W., Schierholz, J., Zeilhofer, F., Kuhn, U., Mohl, C., Sippel, C. & Hoffmann, KH. (2005). Isolation of precursor cells (PCs) from human dental follicle of wisdom teeth. *Matrix Biology*, Vol.24, No.2, (April 2005), pp.155-165, ISSN 0945-053X

Nakashima, M. (2005). Bone morphogenetic proteins in dentin regeneration for potential use in endodontic therapy. *Cytokine & Growth Factor Reviews*, Vol.16, No.3, (June 2005), pp.369-376 ISSN 1359-6101

Saber, SE. (2009). Tissue engineering in endodontics. *Journal of Oral Scienc,*. Vol.51, No.4, (December 2005), pp.495-507

Seo, BM., Miura, M., Gronthos, S., Bartold, PM., Batouli, S., Brahim, J., Young, M., Robey, PG., Wang, CY. & Shi, S. (2004). Investigation of multipotent postnatal stem cells from human periodontal ligament. *Lancet*, Vol.364, No.9429, (July 2004), pp.149-155, ISSN 0140-6736

Sharma, B. & Elisseeff, JH. (2004). Engineering structurally organized cartilage and bone tissues. *Annals of Biomedical Engineering*. Vol.32, No.1, (January 2004), pp.148-159, ISSN 0090-6964

Sloan, AJ. & Waddington R. (2009) . Dental pulp stem cells: what, where, how? *International Journal Of Paediatric Dentistry*. Vol. 19, No.1, (January 2009), pp.61-70, ISSN 0960-7439

Sonoyama, W., Liu, Y., Fang, D., Yamaza, T., Seo, BM., Zhang, C., Liu, H., Gronthos, S., Wang, CY., Shi, S. & Wang, S. (2006). Mesenchymal stem cell-mediated functional tooth regeneration in swine. *PLoS One*.Vol.1, (December 2006), pp.e79

Sun, HH., Jin, T., Yu, Q. & Chen, FM. (2011). Biological approaches toward dental pulp regeneration by tissue engineering. *Journal of Tissue Engineering and Regenerative Medicine*. Vol.5, No.4, (April 2011), pp. e1-e16

Wang, FM., Qiu, K., Hu, T., Wan, CX., Zhou, XD. & Gutmann, JL. (2006). Biodegradable porous calcium polyphosphate scaffolds for the three-dimensional culture of dental pulp cells. *International Endodontic Journal*. Vol.39, No.6, (June 2006), pp.477-483, ISSN 0143-2885

Yen, A. & Sharpe, P. (2008). Stem cells and tooth tissue engineering. *Cell and Tissue Research*. Vol.331, No.1, (January 2008), pp.359–372 ISSN 0302-766X

Yamada, Y., Ito, K., Nakamura, S., Ueda, M. & Nagasaka, T. (2010). Promising cell-based therapy for bone regeneration using stem cells from deciduous teeth, dental pulp, and bone marrow. *Cell Transplantation*. [Epub ahead of print], (October 2010)

Yang X, Yang F, Walboomers XF, Bian Z, Fan M, Jansen JA. (2010) The performance of dental pulp stem cells on nanofibrous PCL/gelatin/nHA scaffolds. J Biomed Mater Res A. 2010, Vol 93, No1, pp. 247-57.

Yoshiba, K., Yoshiba, N., Ejiri, S., Iwaku, M. & Ozawa, H. (2002). Odontoblast processes in human dentin revealed by fluorescence labeling and transmission electron microscopy. *Histochemistry and Cell Biology*. Vol.118, No.3,(September 2002), pp.205-212. ISSN 0948-6143

Zhang, Z., Lan, Y., Chai, Y. & Jiang, R. (2009). Antagonistic actions of Msx1 and Osr2 Pattern Mammalian Teeth into Single Row. *Science*. Vol.323, No.5918, (February 2009), pp.1232–1234, ISSN 0036-8075

17

Development of Human Chondrocyte–Based Medicinal Products for Autologous Cell Therapy

Livia Roseti[1], Alessandra Bassi[1], Brunella Grigolo[2] and
Pier Maria Fornasari[1]
[1]PROMETEO Laboratory; [2]RAMSES Laboratory
[1,2]Rizzoli RIT (Research, Innovation & Technology),
Istituto Ortopedico Rizzoli, Bologna,
Italy

1. Introduction

A cell therapy is a clinical treatment including an *ex vivo* cell manipulation step. Such a therapeutical option began more than forty years ago and is now a worldwide reality. Many human cell-based clinical trials have been developed in every medicine's field mostly to cure diseases where conventional treatments are inadequate. Even though there have been few completed trials and some conflicting results on their effectiveness have been reported, the full potential of cell-based treatments remains to be explored and investigated (Park et al., 2008). Moreover, public expectation for such novel therapies, especially for treating incurable and/or rare diseases, remains high. Nowadays each tissue of the human body, including foetal and embryonic ones, can become a reliable source for cell therapy (Mason & Dunnill, 2009). Cells isolated from a specific source can be used also to cure every other tissue of the body and may be administered alone, in combination with biomaterials, scaffolds, cytokines and growth factors or can be genetically manipulated (gene therapy). Cell administrations can be local or systemic, singles or multiples. Treatments may be autologous or allogeneic (from living or cadaver donors). A cell preparation can be crucial for a treatment such as in bone marrow transplantation or otherwise it may be used as an adjuvant to improve clinical results like in regenerative medicine or to slow down the development of several chronic conditions. Cell effect after treatment can be via the ability to differentiate along several lineages or, as recently highlighted for stem cells, also via the capacity to release anti-inflammatory cytokines, growth factors and proteins, collectively known as paracrine factors, which may modulate the host microenvironment by stimulating endogenous stem cells recruitment, differentiation and angiogenesis, thus acting as real drugs (Yagi et al., 2010). *Ex vivo* cell manipulation protocols are different, depending on cell source, type, target, disease and Country regulations. Current European cell therapy laws classify manipulation types according to potentially associated risks. Cutting, grinding, shaping, centrifugation, soaking in antibiotic or antimicrobial solutions, sterilization, irradiation, cell separation, concentration or purification, filtering, lyophilization, freezing, cryopreservation and vitrification are considered "minimal manipulations". On the other hand, cell processing like induction to proliferation, non-homologous use (if cells or tissues are not intended to be used for the same essential functions in the recipient as in the donor) and association with scaffolds or medical devices are defined as "extensive" or " substantial" manipulations. These new kind of extensively manipulated cell-based products are termed "medicinal product for advanced cell therapy" (see Regulation (European Commission [EC]) No 1394/2007 of the European Parliament and of the Council on advanced therapy medicinal products and amending Directive 2001/83/EC and Regulation (EC) No 726/2004). The term "medicinal" is not only a definition, but, of course, holds specific technical and practical consequences entering these products in the "drug world". In fact, pre-clinical and clinical data that are necessary to demonstrate their quality, safety and efficacy should be highly specific. Moreover, it is also mandatory that they have to be manufactured like medicinal products. To this end, it is primarily required to have a suitable environment built as a real pharmaceutical factory and working according to specific rules named Good Manufacturing Practices (GMPs). GMPs are worldwide guidelines for the management of manufacturing and quality control of pharmaceutical products. The Food and Drug Administration [FDA] enforces GMPs in the United States while the European Medicines Agency [EMA] in Europe (The Rules Governing Medicinal Products in the European Union, Volume 4-Guidelines for Good Manufacturing Practices for medicinal products for human and veterinary use, current Edition) where manufacturing must be authorised by the competent Agency of each Member State. Such a structure (also named "cell factory" or "cleanroom facility") allows minimization of any contamination risk by means of standardization and continuous monitoring of specific parameters such as air filtration and ventilation, temperature, relative humidity, differential pressure, number of air particles and bacterial colony forming units. Besides the environmental monitoring, cell culture and reagents must be checked for the presence of bacterial and viral contamination, mycoplasma and endotoxins. Furthermore, standard operative procedures, personnel training and process traceability should be developed and performed. Extensively manipulated cells are generally thought to be elaborate and costly, especially due to GMPs requirements. However they ensure three main characteristics: safety, product consistency and manufacturing quality. Considering that many cell therapies are still in an early experimental phase and that several aspects are not completely understood, these characteristics may represent a real guarantee for safe, standardized and controlled treatments. To date, proposed employments for cell-based medicinal products are quite impressive (Mason & Manzotti, 2010). Fields of interest are musculoskeletal tissue regeneration, autoimmune disorders, myocardial infarction, gastrointestinal diseases, urogenital system disorders, nervous system diseases, wound healing, plastic surgery, organ transplantation, graft versus host disease (GvHD) and diabetes (www.ClinicalTrials.gov). There is tremendous scope for applying cell therapeutics in the musculoskeletal system (Nöth et al., 2010). Cells can be used to repair or regenerate injured or diseased tissues (cartilage, bones, tendons, ligaments, muscles, etc..), or to treat chronic conditions such as Osteoarthritis or Rheumatoid Arthritis. As most cell therapy treatments for musculoskeletal diseases are not life-threatening, safety is a key issue for any clinical application.

1.1 Autologous chondrocyte implantation

Hyaline articular cartilage is a highly specialized tissue derived from mesenchyme during embryonic development. It has the main function to protect the joint by distributing loads and thus preventing potentially damaging stress on the sub-chondral bone. At the same time it provides a low-friction bearing surface to enable free movements. Chondrocytes, the only cell type of mature cartilage, secrete and deposit around themselves a characteristics matrix composed primarily of water, collagens (mainly collagen type II), proteoglycans (mainly aggrecan) and other non-collagenous proteins (Becerra et al., 2010). Articular cartilage lesions are common in the general population and more often anticipated in young and physically active people. Despite the tissue is susceptible to damage, it has limited capacity for regeneration and repair because of poor vascular supply and lack of an undifferentiated cell population capable of migrating and responding to the insult (O'Driscoll, 1998). If left untreated, cartilage injuries may lead to pain and loss of function and predispose individuals to osteoarthritis in later life and eventually to requirements for total joint replacement (arthroplasty). The need of hospital attention is associated with a significant impact on quality of life and represents a huge socioeconomic burden to society. There is no uniform approach to managing cartilage defects (Harris et al., 2010). In younger, active patients "biologic" solutions that prevent or slow down tissue degradation process should be preferred. These procedures can be classified as palliative (arthroscopic debridement), reparative (microfracture) and regenerative such as pertiosteum, perichondrium or osteochondral grafting and Autologous Chondrocyte Implantation (ACI). Conventional treatments by debridement or microfracture produced various outcomes since the resulting repaired tissue is fibrocartilage which lacks the appropriate biochemical and biomechanical properties of normal healthy tissue. Currently, arthroscopic debridement and microfracture are commonly used as first-line treatment for symptomatic focal chondral defects that are relatively small with minimal associated bone loss. Regenerative procedures including ACI can be used as second-line measures to repair chondral or osteochondral defects. ACI therapeutical approach was first used by a Swedish Group (Brittberg et al., 1994) to treat full-thickness chondral defects of the knee and later applied to the ankle. The treatment is now suitable also for other joints such as hip. In the original procedure (first generation technique) small grafts of normal cartilage removed from non-weight bearing areas of the knee were treated in a proper laboratory. Individual chondrocytes were isolated, cultured in specific conditions, and, following a period of cellular division, retrieved for re-implantation. Cell suspension was then injected beneath a periosteal patch harvested from the proximal medial tibia and sutured to the defect. Clinical, radiological and histological results are available at 10 to 20 years after the implantation, suggesting that outcomes remain high, with relatively few complications (Peterson et al., 2010). First generation ACI revealed, however, several disadvantages, such as transplant hypertrophy, calcification, delamination, cell leakage in the articular environment and loss of phenotype due to previous monolayer expansion. Given these shortcomings, recent experimental and clinical research has been directed towards the development of second generation ACI procedures using suitable scaffolds which act as carriers for the implantable cells maintaining at the meantime phenotype stability (Iwasa et al., 2009). These engineered tissues are then cut to the correct size and shape of the defects directly in the operatory room. The scaffolds, which efficiently "mimic" the natural surroundings of cartilage cells, may have different origins (synthetic, natural), characteristics (fibers, gels, sponges, microspheres, etc.) and spatial organization (two- or three-dimensional). Third generation ACI uses three-dimensional (3D) matrices such as hyaluronic acid as scaffolds. The process of implantation in second and third generation techniques can also be performed arthroscopically or with a small incision. Recently, a new technique called 'Characterized

Chondrocyte Implantation' (CCI) utilizes a selected chondrocyte population that expresses a marker profile that can predict cell ability to produce *in vivo* hyaline-like cartilage (Saris et al., 2008). Many Authors in the Literature have suggested ACI effectiveness also for second and third generation techniques and these procedures are now widely diffused and utilized all over the world for cartilage defects repair. However, besides good outcomes evaluations, there is still scepticism about the use of ACI, particularly for its clinical and cost effectiveness in comparison with other traditional treatments and for the steeper learning curve, at least compared with marrow-stimulating techniques (Vasiliadis & Wasiak, 2010). Different Literature revisions highlighted that there is still insufficient evidence to draw conclusions and that further trials with long-term follow up are required in order to clarify ACI clinical benefits..

1.2 Cell manipulation in autologous chondrocyte implantation: from research to cleanroom

Cell manipulation for ACI is crucial step that have to be performed in compliance with GMPs in order to guarantee a safe, standardized and efficacious product to implant. In this perspective the development of a validated and a repeatable process becomes a key issue for this therapy. Chondrocytes can be easily isolated from articular cartilage tissue by enzymatic treatments and then cultured in different conditions like monolayer, bi/three-dimension, chemical or mechanical stimulation or inhibition. Several chondrocyte culture systems that have been developed display a huge number of applications in the research field as attested by worldwide publications. In fact they represent models for cartilage investigation, which is essential for identifying the pathways of both normal development and pathological degeneration and inflammation of the tissue (Roseti et al., 2007). Recently, there has been a great deal of interest in the *ex vivo* development of hyaline cartilage that can be utilized for the regeneration of damaged or diseased tissue. The realization that chondrocytes may act as drugs having therapeutical effects in cartilage regeneration led to new responsibilities and roles for cell culture's laboratory. ACI application required an integration work between clinicians and cell biology experts and it would not have been possible without the support of GMPs specialized laboratories or facilities. Obviously, clinical application of research models presents specific features of quality assurance which must be met. In particular, for cell cultures, a transfer technology step is required or, in other words, research protocols must be translated into GMPs' ones. This involves taking into account not only the peculiar nature of cells and culture's models, but also the mandatory compliance with current GMPs and all the specific cell therapy regulations. This chapter describes the transfer technology utilized to standardize the manufacturing of engineered chondrocyte-based products for applications in ACI. In particular, it focuses on the development and validation of a GMPs' compliant manufacturing process and then of "consistent" chondrocyte-based medicinal products. The GMP facility performing the below described validation processes is located in a public Hospital in Italy. Therefore specific Italian and European rules have been followed.

2. Development and validation of a GMPs compliant chondrocyte culture process suitable for clinical use

Process validation is a pre-requisite to ensure consistent manufacture. Cell processing such as in ACI is a long lasting, articulated process. It comprises three main manipulation phases before final product packaging and release: cell isolation, expansion, and seeding onto biomaterials. For a complete GMPs compliance it is required that each step is validated singularly. One important thing to consider and to perform before starting is a careful evaluation and subsequent choice of high quality raw materials to be used in the process. Once the choice has been taken, the entire validation should be performed using the same materials. Changes are allowed, but a re-validation step is required.

2.1 The choice of raw materials

This is a highly critical step since raw materials, such as culture reagents or plastic wares, become directly in contact with the cells during the process. Moreover other materials, like scaffolds or cell supports, can become an integral part of the medicinal products. Therefore, the choice for such materials should be geared towards products ensuring the highest quality provided by the market at the moment. First of all, the cell producing facility should attest the quality assurance level of each Company/Institution that intends to choose as supplier. This investigation termed "supplier qualification" is mandatory for a fully GMPs compliance and should be performed not only by examining the accreditations provided by the Company itself but also by on-site inspections. Hence, material's full batch documentation/certification should be carefully evaluated to attest fulfilment with specific current regulations. For example, cell culture media sterility should be certified by specific analyses validated in compliance with current European Pharmacopoeia. In particular, it is required that these products are negative for aerobic, anaerobic bacteria, fungi and endotoxins. Documentation control is a delicate and important step, but not sufficient for entering materials into the process. In fact, each declared critical feature should be cross-checked by internal quality controls in the cell factory. Only after passing such internal controls, materials are considered adequate for the process and validation can start. An important point to consider for reagent choice is to avoid zoonosis risks. A recombinant origin should be preferred for enzymes such as Trypsin used for the rapid detachment of adherent cells, like chondrocytes, from the growth substrate and Collagenase type II utilized to digest cartilage and thus isolate cells. In this case users should control documentation also for recombinant source that must be clearly indicated. For animal origin products, like foetal bovine serum (FBS) as supplement in culture, high quality is mandatory. FBS should be free of microbial, mycoplasma and viral contaminations. More importantly, it must be stated and certified its origin from Bovine Spongiform Encephalopathy (BSE) free countries such as Australia and New Zealand. The country of processing should be indicated too, if different. The serum producing process should be described giving evidence to exclude any possibility from contamination with tissues that may harbour the BSE agent, such as the brain, spinal cord and distal ileum. If the final product encloses other components like scaffolds or biomaterials they should be appropriately characterised and evaluated for suitability for the intended use. Nowadays the market displays variability and availability of such materials. For application in cartilage regeneration they have to display several properties: biocompatibility, biodegradability and malleability to fit defects. Moreover they should be viscous and adhesive enough to allow chondrocyte trapping and fixation to the implantation site, respectively. Furthermore they have to guarantee cell viability, growth and phenotype stability (Iwasa et al., 2009). A special comment is earned to cartilage biopsy that is considered a critical raw material since it is the starting point for processing, or the cell source. Even if it is not the focus of this chapter, it is important to underline that cartilage biopsy collection should be validated too in order to minimize microbial contamination, tissue amount and quality (to ensure a sufficient cell number) and possible impurities arising from fragments of the synovial membrane or bony tissue.

2.2 Chondrocyte isolation method

In the last years Researchers have developed several protocols with the aim to isolate chondrocytes from cartilage tissue. In general, they consisted of sequential enzymatic digestions that unbound cells from matrix entrapment. Although effective and repeatable, these methods revealed to be not always suitable for therapeutical applications. In particular, the use of animal origin enzymes and long lasting processing times revealed to be problematic. In fact, GMPs rules require both to avoid animal origin components that can be a source of zoonosis and to shorten product exposition times in order to minimize microbial contamination risks. The method traditionally used by our group for research comprises three sequential digestions including also animal origin reagents (Roseti et al., 2007). We translated this protocol into clinics using recombinant origin enzymes, reducing enzyme number and manipulation times.

2.2.1 Materials and methods

Healthy cartilage samples were harvested from the femoral condyle of three multi-organ donors (age: 23-71 years). To isolate chondrocytes three different methods were used.

A Method. Cartilage samples were minced with a scalpel and carefully washed with cell culture medium Dulbecco's Modified Eagle's Medium-GMP grade (DMEM) (Li StarFish, Carugate, Milano, Italia) supplemented with L-glutamine-GMP grade 4 mM (Li StarFish). The chondrocytes were then isolated by sequential enzymatic digestion: 30 minutes with 880 U/mL hyaluronidase (Sigma, St Louis, MO, USA) (10 mL/gr cartilage); 1h with 26.5 U/mL pronase (Sigma) (10 mL/gr cartilage) and 45 minutes with 740 U/mL collagenase II (Sigma) (20 mL/gr cartilage) at 37 °C, 5% CO_2 and humidified atmosphere. To block collagenase activity, DMEM supplemented with 10% Fetal Bovine Serum-Pharma grade (FBS) (Li StarFish) was added. The isolated chondrocytes were filtered by 100 and 70 μm sterile nylon mesh filters to remove cell raft and matrix debris. The filtrate was then centrifuged and the pellet washed twice. Viable staining assessed cell number and viability.

B Method. After mincing cartilage samples as described above (method A), digestion was performed only with 740 U/mL collagenase II (Sigma) (20 mL/gr cartilage) for 24 h at 37 °C, 5% CO_2 and humidified atmosphere. Isolation was then carried out as already described.

C Method. After mincing cartilage samples as described above (method A), digestion was performed only with 740 U/mL collagenase (Li StarFish) (20 mL/gr cartilage) for 24 h at 37 °C, 5% CO_2 and humidified atmosphere. Isolation was then carried out as already described.

2.2.2 Results

Comparison between the three methods, performed normalizing isolated cell number per cartilage gram, indicated C Method as the most efficient (Figure 1).

2.2.3 Discussion

C methods allowed to avoid animal origin enzymes, to reduce number of reagents (one enzyme instead of three) and manipulation times, giving at the same time the best yield results. Therefore, being the most GMP compliant, it was our choice as isolation method.

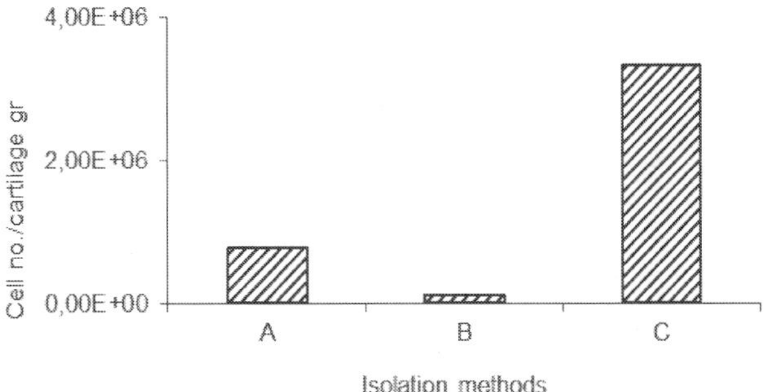

Fig. 1. Comparison of chondrocyte isolation methods.

A Method: sequential incubations with animal origin enzymes; B Method: animal origin collagenase II digestion; C Method: recombinant origin collagenase II digestion. Comparison was performed normalizing cell number per cartilage gram.

2.3 Chondrocyte monolayer expansion

Sera are mixtures of components essential for cell proliferation. They contain growth factors, hormones, molecules promoting cell adhesion and propagation, minerals trace and proteins like transferrin and albumin. FBS is traditionally and successfully used for cell cultures, including chondrocytes. The use of bovine-derived regents in clinical applications carries potential risks of contamination, especially Bovine Spongiform Encephalopathy (BSE) and immune responses. To develop our process we evaluated an alternative solution such as the use of human serum.

2.3.1 Materials and methods

C Method isolated chondrocytes (see section 2.2.1) were cultured in DMEM-GMP grade (Li StarFish) supplemented with L-glutamine-GMP grade 4 mM (Li StarFish) and with 10% FBS-Pharma grade (Li StarFish) or 10% human serum (HS) until passage two (three weeks). At 70% confluence cells were passaged with 1:250 Trypsin-EDTA-GMP grade (Li StarFish) and cell number and viability were assessed by viable staining. A high quality, FBS certified to be free of BSE and microbial, mycoplasma and viral contamination was utilized. Human serum was supplied in a sterile bag containing plasma drawn from donors. Under sterile conditions, plasma was added with calcium gluconate (0.3 ml/10 ml plasma) to induce coagulation process. After about two hours, the obtained serum was aspirated avoiding clots, transferred in a new bag and stored at -20°C.

2.3.2 Results

At passage 2, 10% FBS growth was three-times higher than 10% HS one (Figure 2).

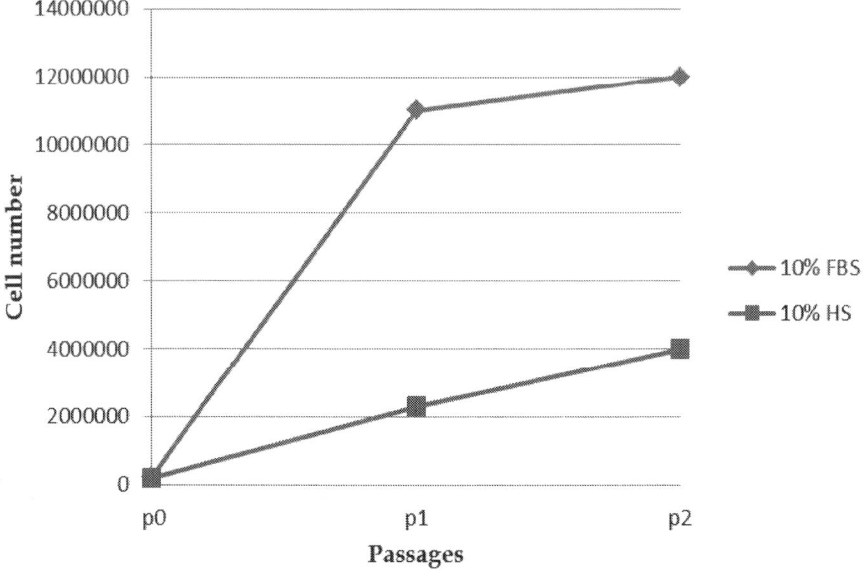

Fig. 2. Comparison of chondrocyte expansion methods. 10% FBS: expansion with 10%FBS; 10% HS: expansion with 10% human serum.

2.3.3 Discussion

The use of FBS for cell therapy is still controversial. Although bovine origin, FBS batches display less variability than the human ones and seem to better enhance cell growth. A reduction batch variability facilitate process standardization thus ensuring products consistency and robustness. Cell growth enhancement allows to shorten manipulation times and to reduce microbial contamination risks as well as progression to senescence. For these reasons our choice was towards FBS. The use of a high quality one (see section 2.1) appeared a good compromise between growth and safety requirements.

2.4 Cartilage engineered cultures

Many studies in the Literature have documented that monolayer expanded chondrocytes lose their phenotype becoming fibroblast-like cells. This de-differentiation process starts in the very first passages and progressively increases with time in culture. However, as attested by a number of study, such a situation can be reverted back when the cells are set in specifically defined conditions. It is as well documented that when de-differentiated chondrocytes are seeded onto a biomaterial this specific configuration is able to allow the re-differentiation process to occur (Roseti et al., 2007). As already evidentiated, new ACI generations utilize different biomaterials as scaffolds onto which expanded chondrocytes can grow and re-acquire their original phenotype. The biomaterial used in our manipulation process is a matrix consisting of collagen without cross-linking or chemical additives (Chondro-Gide®, Geistlich Surgery, Germany). It is sterilized by gamma irradiation and provided in three different formats (2×3 cm², 3×4 cm² and 4×5 cm²). The collagen is extracted from pig and purified to avoid immunological reaction risks. The membrane has a two-dimensional structure: a porous layer allowing cell seeding and culture and compact one acting as a barrier to prevent cell loss in the articular cavity. At the time of implantation the membrane is placed with the porous layer facing cartilage defect and the compact one facing the joint. Pre-clinical studies demonstrated its biocompatibility, low antigenicity, hydrophilicity (due to collagen fiber microstructure) and biodegradability. This membrane has already been used by other groups for ACI (Haddo et al., 2004). Based on their experience and on ours with other biomaterials we started to verify effectiveness of this biomaterial in allowing cell colonization. Due to problems related to FBS for clinical use, we cultured cartilage engineered constructs in medium without this reagent.

2.4.1 Materials and methods

Chondrocyte cultures were carried on in monolayer for three weeks, in the standard conditions described above (C Method for isolation; DMEM with FBS for expansion). After trypsinization pellets were re-suspended in DMEM without FBS. Chondrocytes were then seeded onto collagen membranes at concentration of 1×10^6 cells/cm² surface. Cells were then let to adhere for 15 minutes and finally DMEM without FBS was added to cover the membranes. The loaded scaffolds were incubated at 37° C, 5% CO_2, and humidified atmosphere for 7 days. Cartilage engineered constructs were included in OCT and frozen at -80° C until analysis. Frozen blocks were cut in 25 mµ cryostat slices and slides were thawed and fixed in 4% paraformaldeid (PFA) for 30 minutes at room temperature. The samples were then rehydrated in H_2O for 10 min. Slides were incubated with hematoxylin for 1-2 minutes and, after 4 washes in distilled water, hematoxylin was activated under running water for 10 minutes. Slides were incubated with eosin for 5 minutes and washed four times in distilled water. After dehydration, slides were mounted with Entellan and stored at room temperature. Samples were analysed using a Zeiss Axioscope Microscope (Carl Zeiss, Oberkochen, Germany).

2.4.2 Results

Chondrocytes cultured on collagen-based membranes had a spherical appearance when observed by light microscopy (Figure 3) and were uniformly distributed among collagen fibers in the porous layer.

2.4.3 Discussion

Our data confirmed the ability of the collagen-based bilayer membranes that we intended to use in our process to allow cell colonization. The observed distribution pattern has been already evidentiated in other scaffolds displaying similar composition and structure (De Franceschi et al., 2005). Interestingly, FBS absence did not compromise viability and colonization ability.

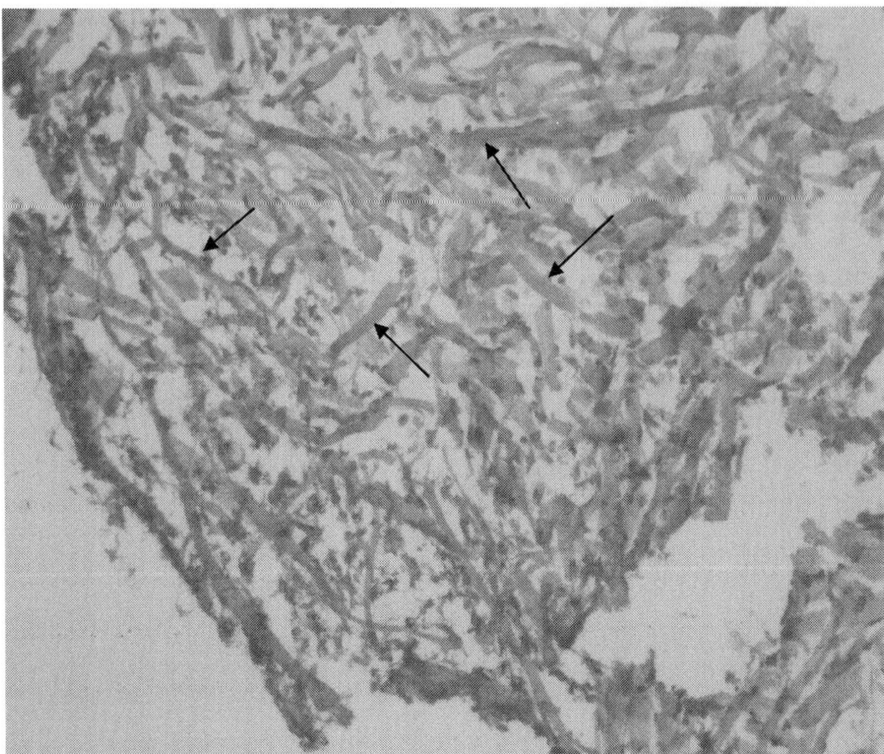

Fig. 3. Hematoxylin and eosin staining of porous layer in cartilage engineered tissues (40X). Spatial cell distribution is homogeneous among collagen fibers. The image is from one representative engineered tissue. Collagen fibers are indicated by black arrows.

3. Final products validation: identification of defined "specifications"

Final products of this GMPs manipulation model consist of engineered tissues derived from the combination of chondrocytes and bi-layer scaffold membranes. Validation is a GMPs requirement aimed at standardizing and thus defining final product peculiar features, technically named "product specifications". They allow the identification of a specific product obtained with a specific process and intended for a specific clinical use. Once defined, specifications cannot be changed. Products displaying even one different feature has to be considered "other" and cannot be released for the intended clinical use. To change product specifications a new re-validation process has to be performed. The first validation to achieve for a cell-based product is sterility. It is aimed at defining absence of aerobic and anaerobic bacteria, fungi, mycoplasma and endotoxins by means of specific analytical methods. It has to be underlined that it also mandatory that these analytical methods have been upstream validated in compliance with current European Pharmacopoeia and GMPs. Besides sterility, that is not the focus of the current chapter, the other specifications needed for cell-based products are: viability, potency, purity, yield and stability. To perform final product validation we utilized the same cells and protocols as for process validation and analyses were performed onto the obtained cartilage engineered tissues.

3.1 Viability

In the Literature there are many methods for viability evaluation. Difficulties for analysis arise when cells are entrapped into a matrix or a scaffold, like in our process. In this case there is a real risk to underestimate the results. The Authors had already faced with this problem when managing other types of chondrocyte-scaffold constructs (Roseti el al., 2007). In that occasions they were able to standardize a feasible method also applied in this model.

3.1.1 Materials and methods

Three engineered tissues -i.e. chondrocytes and collagen-based membranes (see section 2.4.1)- and named Case 1, Case 2 and Case 3 were analysed. Cell viability was determined at the seeding onto the biomaterial (Day 0) and after 7 and 14 days in culture by 3-(4,5-dimetiltiazolo-methyl)-2.5-difeniltetrazolio bromide (MTT) (Sigma)-mitochondrial reduction method, based on Mosmann original protocol (Denizot & Lang, 1986). Briefly, engineered constructs were transferred to 35x10 mm Petri dishes, added with 1 ml of a solution of 1 mg/ml MTT in PBS 1X and incubated for 3 hours. The membranes were then transferred in Eppendorf tubes and added with 1 ml of extraction solution consisting of 0.01N HCl in isopropanol. The membranes, still contained in the Eppendorf tubes were then shaken and centrifuged at 14,000 rpm for 5 min to allow complete solubilisation of formazan. Finally, supernatant absorbance was read at 570 nm using a Beckman spectrophotometer.

3.1.2 Results

MTT testing, which is directly related to chondrocyte activity, showed slight increased values until day 7 for Case 1 and 2 and then a plateau maintained until day 14. Case 3 chondrocytes did not show increased values, but immediately reached a plateau (Figure 4).

3.1.3 Discussion

The engineered constructs showed slight or no chondrocyte growth. These results are in contrast with the ones obtained with other biomaterials (Roseti et al., 2007) that highlighted a cell proliferation increase by time. We believe that this is to be due to FBS deprivation that, however, allowed viability maintenance until day 14. The choice to use FBS only for the monolayer expansion phase where we demonstrated its efficacy in favouring cell growth (see section 2.3) and not for the engineered tissues represents for the Authors a good compromise to guarantee a safer therapy.

3.2 Potency

For chondrocyte-based products potency can be defined as cartilage forming capacity (Reflection paper on *in-vitro* cultured chondrocyte containing products for cartilage repair of the knee. Final. London, 08 April 2010 EMA/CAT/CPWP/568181/2009 Committee for Advanced Therapies [CAT], 2010). Therefore, to investigate if the final products displayed cartilage features we evaluated the presence of collagen type II and aggrecan, that are main recognized markers of this tissue (Becerra et al., 2010).

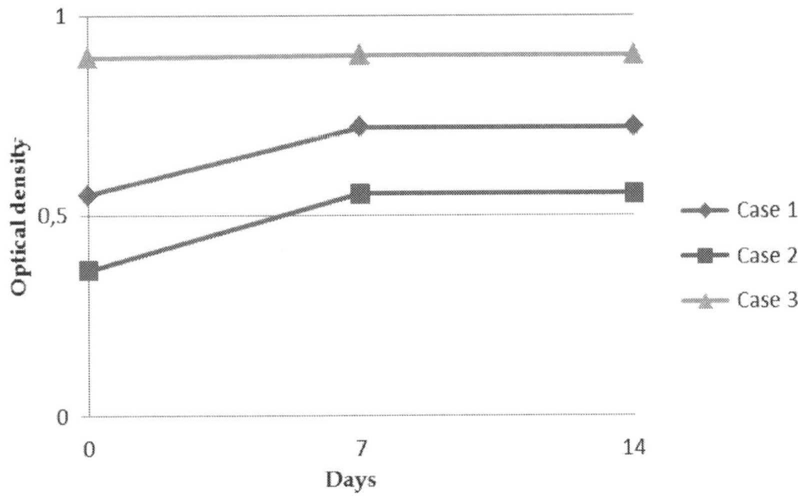

Fig. 4. MTT testing on engineered tissues (chondrocytes and collagen-based membranes) indicated as Case 1, 2 and 3. Samples were analysed after 0, 7 and 14 days in culture. Data are expressed as optical density at 570 nm.

3.2.1 Materials and methods

Tree engineered tissues -i.e. chondrocytes and collagen-based membranes (see section 2.4.1)- were embedded in OCT, snap-frozen in liquid nitrogen, cut into 25 μm sections, air-dried and stored at -20 °C until use. These slides were transferred at room temperature, air-dried for 20 minutes and fixed in 4% PFA at room temperature for 20 minutes. The following primary antibodies were used: mouse anti-human collagen type II monoclonal antibody and mouse anti-human proteoglycans (Chemicon International Temecula, CA, USA). Air-dried fixed samples were rehydrated. Slides for collagen type II determinations were treated with 0.1 % hyaluronidase (Sigma) in Phosphate Buffered Saline (PBS) at 37°C for 5 minutes for epitope unmasking; those for proteoglycans with chondroitinase ABC (Sigma) in Tris-HCl pH=8 for 30 minutes at room temperature. The slides were then incubated with the primary antibodies diluted 1:40 (collagen type II) or 1:50 (proteoglycans) in 0.04M Trizma Base Saline (TBS) pH 7.6 containing 1% BSA and 0.1 % Triton X-100 for 1 hour at room temperature. After washes in PBS with the addition of 1% BSA, the slides were incubated with biotinylated immunoglobulins specific for various animal species (BioGenex, San Ramon, CA, USA) for 20 minutes at room temperature. Then samples were incubated with a phosphatase-labeled streptavidin (BioGenex) for 20 minutes at room temperature, and after washes the reactions were developed using fast red substrate (BioGenex). Negative controls were performed by omitting the primary antibody. Slides were counterstained with hematoxylin and mounted in glycerol gel. All the samples were analysed using a Zeiss Axioscope Microscope (Carl Zeiss).

3.2.2 Results

The engineered tissues revealed to be able to re-express specific cartilage markers. In particular, collagen type II immuno-staining revealed the presence of homogeneously diffused positive cell clusters (Figure 5) while proteoglycans appeared to be distributed in the extra-cellular matrix (Figure 6).

3.2.3 Discussion

This potency validation allowed to define an identity for our products, as specifically required by chondrocyte-based medicinal products guidance. It has to be mentioned that, as evidentiated in the Literature (Saris et al., 2008), one limitation of this system could be the inability to quantify results thus avoiding product good quality ranking. The Authors, on the other hand, suggest that such a quantification could not be really indicative of product quality or positive outcome in patients since there is still not complete evidence of a direct correlation between these features and because of documented patients variability in terms of basal values of cell number and cartilage markers expression.

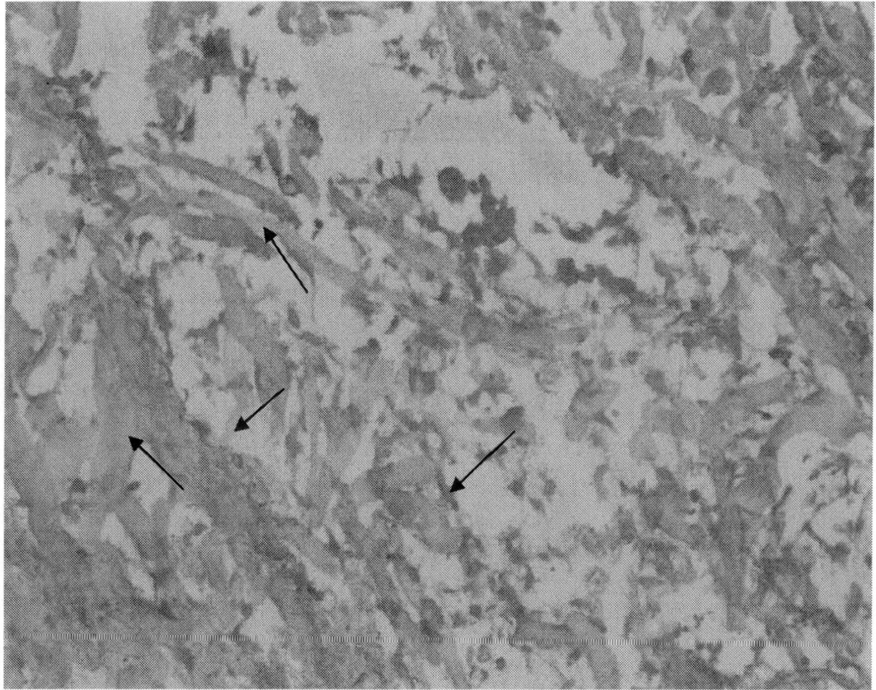

Fig. 5. Collagen type II immuno-staining on engineered constructs (80X).The image from one representative sample shows that the protein is uniformly distributed and mainly located inside the cells or in the peri-cellular matrix. Staining was developed using fast red substrate (red is positive stain). Biomaterial fibers are indicated by arrows.

3.3 Purity

Purity is typically required for drugs. For advanced cell therapy medicinal products it means elimination or decrease of undesired cells. The unique type of cells in cartilage is chondrocyte. However cartilage biopsy should carry possible contaminants arising from fragments of the synovial membrane or from bone. These contaminants could be maintained during culture and thus become a part of the final products. To minimize these risks we standardized biopsy collection (data not shown) and developed a chondrocyte-specific culture method. Nevertheless, to be fully compliant with GMPs, we had to give evidence that our final products were free of cell contaminants. Therefore we analyzed the engineered tissues for the presence of two markers, one typical of fibroblasts (collagen type I) and one of bone phenotype (osteocalcin).

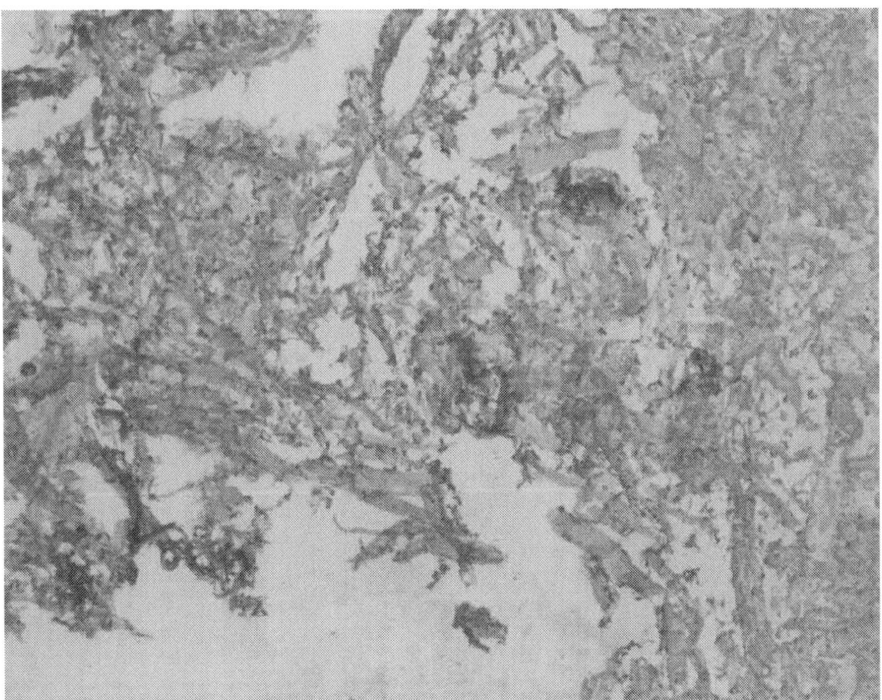

Fig. 6. Proteoglycans immuno-staining on engineered constructs (40X). The image from one representative sample shows that proteoglycans were homogenously distributed throughout the whole extra-cellular matrix. Staining was developed using fast red substrate (red is positive stain).

3.3.1 Materials and methods

The same procedure described for potency validation was utilized (see Section 3.2.1). The following primary antibodies were used: mouse monoclonal anti-human collagen type I

(Chemicon International) diluted 1: 20 and Mouse anti-human osteocalcin (R&D Systems, Inc., Minneapolis, MN, U.S.A.) diluted 1:40. Slides for collagen type I determinations were treated with 0.1 % hyaluronidase (Sigma) in PBS at 37°C for 5 minutes for epitope unmasking.

3.3.2 Results

Engineered constructs were negative for collagen type I (Figure 7,) and osteocalcin (Figure 8) immuno-staining both in the cells and in the newly synthesized matrix.

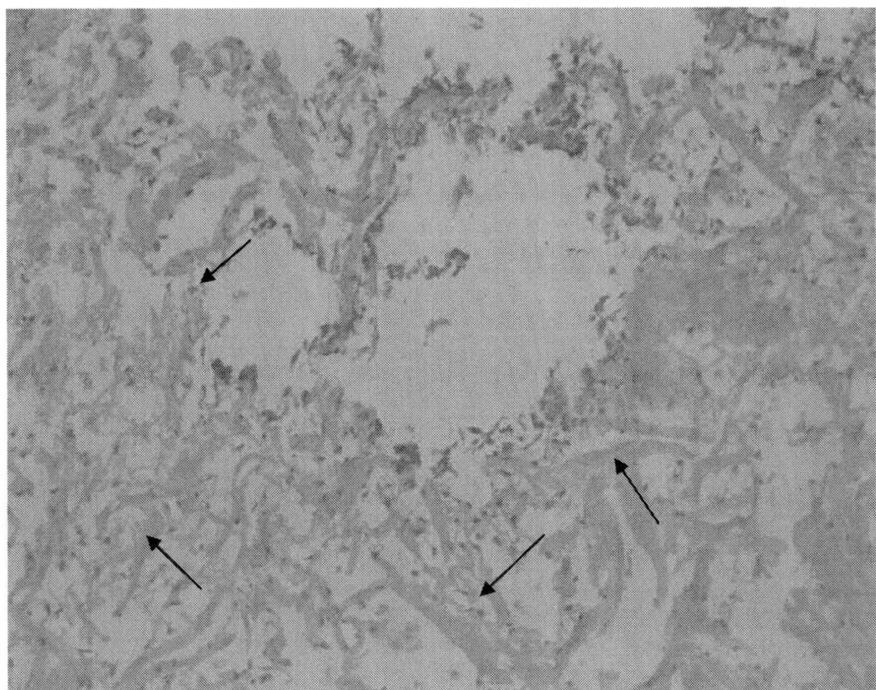

Fig. 7. Collagen type I immuno-staining on engineered constructs (40X). The image from one representative sample shows that the protein is not present (no red areas were observed using fast red substrate). Biomaterial fibers are indicated by arrows.

3.3.3 Discussion

This validation is a further evidence of a defined identity of our products. Collagen type I absence is particularly meaningful: on one hand it indicates that synovial cells or tissues were not present at biopsy harvest and/or were not carried on with cultures; on the other hand it highlightes that chondrocytes within this biomaterial had reverted back to their original phenotype that was lost in monolayer culture.

3.4 Yield

Yield can be defined as the number of cells obtained for each medicinal product. GMPs guidance requires yield validation, but in our experience this can be difficult to perform. In fact, yield is, more than the other specifications, subjected to variables that make quite problematic the required standardization. Intrinsic variability due to patient, cartilage quality or cell growth capacity do not completely depend from operators or process conditions and cannot be totally controlled. When we started our process we had yield validation data allowing a cell seeding onto the biomaterial of 1×10^6 cells per cm^2. Lately, after more than 90 cultures, this initial specification has been replaced by a range of values: $0.1 \div 1.1\times10^6$ cells per cm^2. To accept and apply this new yield scale we had to re-validate the process and the products verifying that each value of the range was able to guarantee the development of viable and cartilage-like engineered tissues.

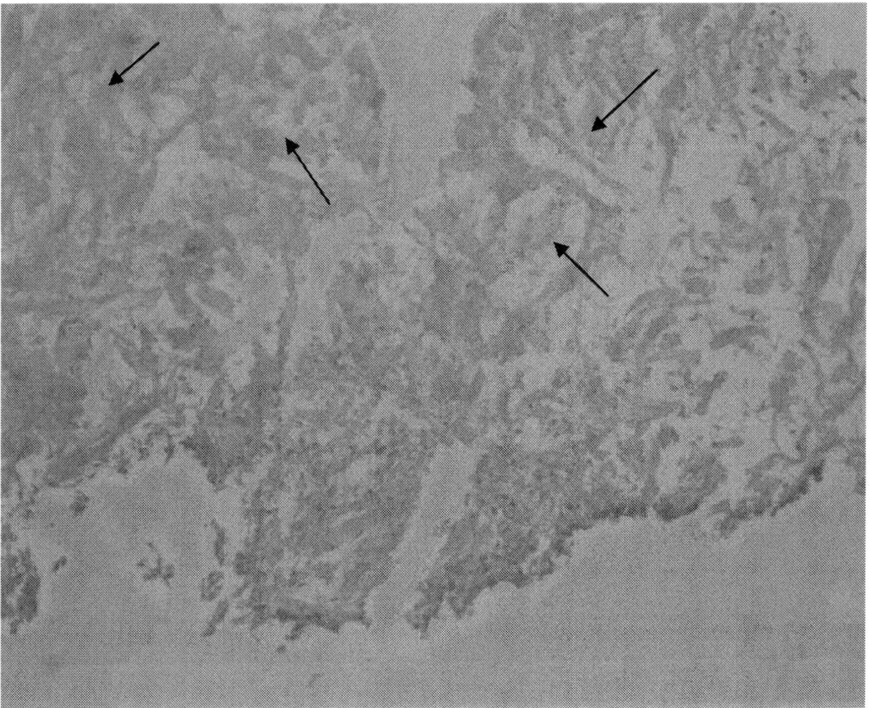

Fig. 8. Osteocalcin immuno-staining on engineered constructs (20X). The image from one representative sample shows that the protein is not present (no red areas were observed using fast red substrate). Biomaterial fibers are indicated by arrows. Biomaterial fibers are indicated by arrows.

3.5 Stability

Process validation ensures the manufacturing of standardized products displaying defined characteristics. But what happens after product release? A critical point is product stability over time: cell manufacturers should find the best storage conditions and times ensuring maintenance of products specifications. To define a product "shelf life" we performed a research in the literature and in the Companies/Institutions already producing this type of cells. Then we evaluated our products for each specification (see section 3) at release and at different times after. As expected, quality products started to decrease from the first day, but we found that after 4 days each specifications was preserved for at least 70% of the initial value. Therefore we chose that period of time as the shelf life of our products.

4. Cryopreservation phase

Chondrocyte manufacturing process may require a cryopreservation step in liquid nitrogen if, for any reason (technical or clinical) the ACI intervention is delayed or if a reservoir of cells is requested by the surgeon for other future treatments in the same patient. In both cases it is required that aliquots of cells are stored as "intermediate products" in proper and defined conditions. A validation step should be performed in order to demonstrate that the freezing/cryopreserving/thawing process does not alter all the cell properties needed for implantation. Considering our experience and literature data we decided to use dimethyl sulfoxide (DMSO) as cryo-protector agent.

4.1 Materials and methods

Three chondrocytes cultures (C Method isolation) at passage 1 were frozen and stored in liquid nitrogen gases. The freezing solution utilized was: DMEM-GMP grade (Li StarFish) supplemented with FBS-Pharma grade 40% and CryoSure-DMSO (Li StarFish) 10%. Samples were then thawed and cells seeded in medium supplemented with 20% FBS. After seven days, cultures returned to usual standard medium conditions (10% FBS). Chondrocytes were then seeded and cultured onto collagen biomaterial as described above. The final products were the checked for viability, yield, potency, purity and stability.

4.2 Results

In the three cultures analyzed the thawing process allowed a survival of 70, 92 and 70% of the cells, respectively. Cells were able to proliferate in monolayer and to reach a number sufficient for the seeding onto the collagen-based biomaterial. The final products revealed to be similar to the ones obtained without the cryopreservation step (data not shown). In particular, cells were viable, and expressing only typical cartilage markers.

4.3 Discussion

We developed a cryopreservation phase that can be included in ACI procedures without altering characteristics of final cartilage constructs.

5. Conclusion

The transfer technology described in this chapter allowed the development and validation of a safe, effective and robust GMPs compliant process articulated in different steps and including a potential stand-by phase in liquid nitrogen. This process results in chondrocyte-based medicinal products (a combination of cells and collagen-based membranes) with defined identity and stability that make them suitable for ACI therapeutical option in patients with articular cartilage damages. The Authors haver three main comments to disclose, based on this experience. The first is about the use of FBS in culture which, besides animal origin related problems, may imply immune responses in patients. We justify our choice to use FBS in monolayer conditions because it allowed a better cell growth standardization. However, since its potentially dangerous action could not be ignored, we decided to avoid FBS presence in the final products. This resolution was supported by validation data that were showing that the engineered tissues were cartilage-like also without this supplement. These results revealed to be in line with the literature and with data obtained by the Authors themselves with other chondrocyte constructs. Only the ability to grow inside the collagen-based biomaterial was importantly reduced, even if viability was consistently maintained. Therefore, considering implications due to FBS presence, our choice could be a good compromise for patient's safety. The second comment is about the validation procedure itself. GMPs Guidance gives strict indication on how to carry on process and product validation thus minimizing contamination risks and variable elements. However such a standardization could be difficult to apply for cells. It is known that cell characteristics in culture are labile and subjected to modulation due non only to culture conditions (times, culture media, supplements an scaffolds) but also to patients (age, gender and pathology). In particular, quantification of some cell properties can become really hard to perform, thus hampering standardization. Therefore validation of a cell-based medicinal product, should mediate between the required "drug rules" and the intrinsic well known cell biological variability that is impossible to eliminate. The last comment is a consideration that cell therapy for cartilage regeneration is under vast exploration and there are now emerging other possibilities as well as improvement in this application. Allogeneic implantation, unexpanded chondrocytes, cell combination with new scaffolds and the use of pre-committed or undifferentiated precursors or mesenchymal stem cells from different sources are some examples of recent advancement in this field. In any case, Country legislation must be applied and our system can be also assumed as a valid model for the compliant GMPs development of other new products suitable for clinical purposes.

6. Acknowledgment

The Authors wish to thank Dr. Giovanna Desando for the technical and scientific support. This work was supported by the grant "Progetto di Medicina Rigenerativa" from "Regione Emilia Romagna" (Delibera di giunta n. 2233/2008).

7. References

Becerra, J., Andrades, J.A., Guerado, E., Zamora-Navas, P., Lo´pez-Puertas, J.M. & Reddi, A. H. (2010). Articular Cartilage: Structure and Regeneration. *Tissue engineering: Part B, Reviews,* Vol.16, No.6, (December 2010), pp. 617-627, ISSN 1937-3368

Brittberg, M., Lindahl, A., Nilsson, A., Ohlsson, C., Isaksson, O. & Peterson, L. (1994). Treatment of deep cartilage defects in the knee with autologous chondrocyte transplantation. *New England Journal of Medicine,* Vol.331, No.14, (October 1994), pp. 617-627, ISSN 0028-4793

De Franceschi, L., Grigolo B., Roseti, L., Facchini, A., Fini, M., Giavaresi, G., Tschon, M. & Giardino, R. (2005). Transplantation of chondrocytes seeded on collagen-based scaffold in cartilage defects in rabbits. *Journal of Biomedical Materials Research Part A,* Vol.75, No.3, (December 2005), pp. 612-622, ISSN 1549-3296

Denizot, F. & Lang, R. (1986). Rapid colorimetric assay for cell growth and survival. Modification to the tetrazolium dye procedure giving improved sensitivity and reliability. *Journal of Immunological Methods,* Vol. 22, No. 89, (May 1986), pp. 271-277, ISSN 0022-1759

Haddo, O., Mahroof, S., Higgs, D., David, L., Pringle, J., Bayliss, M., Cannon, S.R. & Briggs, T.W. (2004). The use of Chondrogide membrane in autologous chondrocyte implantation. *The Knee,* Vol.11, No.1, (February 2004), pp. 51-55, ISSN 0968-0160

Harris, J.D., Siston, R.A., Pan, X., & Flanigan, DC. (2010). Autologous Chondrocyte Implantation: A Systematic Review. *The Journal of Bone and Joint Surgery American,* Vol.92, No.12, (September 2010), pp. 2220-2233, ISSN 0021-9355

Iwasa, J., Engebretsen, L., Shima, Y. & Ochi, M. (2009). Clinical application of scaffolds for cartilage tissue engineering. *Knee Surgery, Sports Traumatology, Arthroscopy,* Vol.17, No.6, (June 2005), pp. 561-577, ISSN 0942-2056

Yagi H., Soto-Gutierre,z A., Parekkadan, B., Kitagawa, Y., Tompkins, R.G., Kobayashi, N. & Yarmush, M.L. (2010). Mesenchymal stem cells: Mechanisms of immunomodulation and homing. *Cell Transplantation.* Vol. 19, No.6, (June 2010), pp. 667-79, ISSN 0963-6897

Mason, C. & Dunnill, P. (2009). Assessing the value of autologous and allogeneic cells for *Regenerative medicine,*Vol.4, No.6, (November 2009), pp. 835-53, ISSN 1746-0751

Mason, C. & Manzotti, E. (2010). Regenerative medicine cell therapies: numbers of units manufactured and patients treated between 1988 and 2010. *Regenerative Medicine,* Vol.5, No.3, (May 2010), pp. 307-313, ISSN 1746-0751

Nöth, U., Rackwitz, L., Steinert, A.F. & Tuan, R.S. (2010). Cell delivery therapeutics for musculoskeletal re generation. *Advanced Drug Delivery Reviews,* Vol.62, No.7-8, (June 2010), pp. 765–783, ISSN 0169-409X

O'Driscoll, S.W. (1998). Current Concepts Review-The healing and regeneration of articular cartilage. The *Journal of Bone and Joint Surgery American,* Vol. 80, No.12, (December 1998), pp. 1795-812, ISSN 0021-9355

Park, D.H., Borlongan, C.V., Eve, D.J. & Sanberg, P.R. (2008). The emerging field of cell and tissue engineering. *Medical Science Monitor,* Vol.14, No.11, (November) 2005), pp. RA206-220, ISSN 1234-1010

Peterson, L., Vasiliadis, H.S., Brittberg, M. & Lindahl, A. (2010). Autologous chondrocyte implantation: a long-term follow-up. *The American Journal of Sports Medicine,* Vol. 38, No. 6, (June 2010), pp. 1117-1124, ISSN 0363-5465

Roseti, L., Facchini, A., De Franceschi, L., Marconi, E., Major, E.O. & Grigolo, B. (2007). Induction of original phenotype of human immortalized chondrocytes: a quantitative gene expression analysis. *International Journal of Molecular Medicine,* Vol. 19, No. 1, (January 2007), pp. 89-96, ISSN 1107-3756

Saris, DB., Vanlauwe, J., Victor, J., Haspl, M., Bohnsack, M., Fortems, Y., Vandekerckhove, B., Almqvist, K.F., Claes, T., Handelberg, F., Lagae, K., van der Bauwhede, J.,

Vandenneucker, H., Yang, K.G., Jelic, M., Verdonk, R., Veulemans, N., Bellemans, J. & Luyten, F.P. (2008). Characterized chondrocyte implantation results in better structural repair when treating symptomatic cartilage defects of the knee in a randomized controlled trial versus microfracture. *The American Journal of Sports Medicine*, Vol. 36, No.2, (February 2008), pp. 235-246, ISSN 0363-5465

Vasiliadis, H.S. & Wasiak, J. (2010). Autologous chondrocyte implantation for full thickness articular cartilage defects of the knee. In: *Cochrane Database of Systematic Reviews*, Bone, Joint and Muscle Trauma Group, Issue 10 Art. No. CD003323, JohnWiley & Sons, Ltd, Retrieved from < http://www.thecochranelibrary.com>

18

Regenerative Medicine for Tendon Regeneration and Repair: The Role of Bioscaffolds and Mechanical Loading

Franco Bassetto, Andrea Volpin and Vincenzo Vindigni
Unit of Plastic and Reconstructive Surgery, University of Padova,
Italy

1. Introduction

Tendons are soft connective tissues, which connect muscle to bone forming a musculo-tendinous unit, whose primary function is to transmit tensile loads generated by muscles to move and enhance joints stability.

Adult tendons have relatively low oxygen and nutrient requirements, low cell density, and poor regenerative capacity.

The biomechanical properties of tendons are mainly attributed to the high degree of organization of the tendon extracellular matrix, primarily composed of collagen type I, arranged in triple-helical molecules bundles that have different dimensions and which are aligned in a parallel manner in a proteoglycan matrix (Fig. 1).

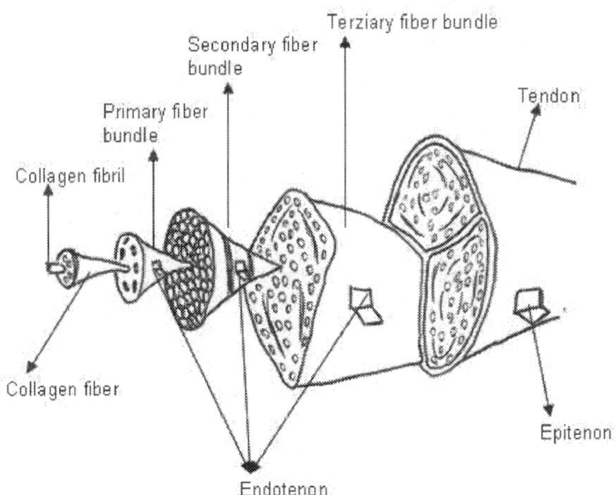

Fig. 1. Hierarchy of tendon structure.

Tendon injuries produce considerable morbidity and affect the quality of life, the disability that they cause may last for several months despite what is considered appropriate management. The basic cell biology of tendons is still not fully understood, and the management of tendon injury poses a considerable challenge for clinicians.

Clinical approaches to tendons rupture often involve surgical repair, which frequently implies working with degenerative, frayed tendon tissue, unable to sustain the rigors of normal activities, and may fail again.

After an injury, the healing process in tendons results in the formation of a fibrotic scar and it is accompanied by an increased risk of further damage. (Longo UG et al. 2010)

The structural, organizational, and mechanical properties of this healed tissue are insufficient as tendons possess a limited capacity to regenerate.(Wong JK et al. 2009) (Woo SL et al.2006)

Adhesion formation after intrasynovial tendon injury poses a major clinical problem. Disruption of the synovial sheath at the time of the injury or surgery allows granulation tissue and fibroblasts from the surrounding tissue to invade the repair site. Exogenous cells predominate over endogenous tenocytes, allowing the surrounding tissue to attach to the repair site, resulting in adhesion formation. (Wong JK et al. 2009) (Woo SL et al.2006)

Despite remodeling, the biochemical and mechanical properties of healed tendon tissue never match those of intact tendon. It is well demonstrated that mechanical loading plays a central role in tenocyte proliferation and differentiation, and that the absence of mechanical stimuli leads to a leak of cellular phenotype. (Wang JH 2006) (Woo SL et al.2006)

While certain tendons can be repaired by suturing the injured tissue back together, some heal poorly in response to this type of surgery, necessitating the use of grafts.(Kim CW & Pedowitz RA 2003)

Unfortunately, finding suitable graft material can be problematic and biological grafts have several drawbacks. Autografts from the patient are only available in limited amounts, they can induce donor site morbidity, while allografts from cadavers may cause a harmful response from the immune system besides also being limited in supply. In both cases, the graft often does not match the strength of the undamaged tissue.(Goulet F et al. 2000) For this reason, obtaining tendinous tissue through tissue engineering approaches becomes a clinical necessity.

Tissue engineering is a multidisciplinary field that involves the application of the principles and methods of engineering and life sciences towards i) the fundamental understanding of structure-function relationships in normal and pathological mammalian tissues and ii) the development of biological substitutes that restore, maintain or improve tissue function.

The goal of tissue engineering is to surpass the limitations of conventional treatments based on organ transplantation and biomaterial implantation. It has the potential to produce a supply of immunologically tolerant, 'artificial' organs and tissue substitutes that can grow within the patient. This should lead to a permanent solution to the damage caused to the organ or tissue without the need for supplementary therapies, thus making it a cost-effective, long-term treatment.

The tissue-engineering approach involves the combination of cells, a support biomaterial construct, and micro-environmental factors to induce differentiation signals into surgically transplantable formats and promote tissue repair, functional restoration, or both.

The earliest clinical application of human cells in tissue engineering, started around 1980, was for skin tissue using fibroblasts, and keratinocytes, on a scaffold. During the last 30 years, many innovative approaches have been proposed to reconstruct different tissues: skin, bone, and cartilage. The field of tendon tissue engineering is relatively unexplored due to the difficulty in *in vitro* preservation of tenocyte phenotype, and only recently has mechanobiology allowed a better understanding of the fundamental role of *in vitro* mechanical stimuli in maintaining the phenotype of tendinous tissues. This chapter analyzes the techniques used so far for the *in vitro* regeneration of tendinous tissues.

2. Scaffolds requirements for tendon tissue engineering

The scaffold should encourage cellular recruitment and tissue ingrowth. Early in the repair process, the scaffold should maintain its mechanical and architectural properties to protect cells and the new, growing tissue from strong forces and early inflammatory events. Subsequently, the scaffold should be gradually reabsorbed allowing a controlled exposure of the regenerating tissue to the local cellular, biochemical and mechanical environment. This will allow the tissue to develop more naturally and function more efficiently.

In order to avoid stress shielding, the scaffold should ideally degrade at the same rate that the new tissue is created. In order to ensure final clinical use, neither the scaffold nor its degradation products should be harmful to the surrounding tissue and they should not result in unresolved inflammation or other deleterious biological responses.

The tendon tissue engineering aims to repair tendon lesions in situ by integrating engineered, living substitutes with their native counterparts in vivo (Fig. 2). For this purpose, scaffolding materials are needed, and these ideally should fulfill the following requirements (Liu Y et al. 2008):

- Biodegradability with adjustable degradation rate.
- Biocompatibility before, during and after degradation.
- Superior mechanical properties and maintenance of mechanical strength during the tissue regeneration process.
- Bio-functionality: the ability to support cell proliferation and differentiation, ECM secretion, and tissue formation.
- Processability: the ability to be processed to form desired constructs of complicated structures and shapes

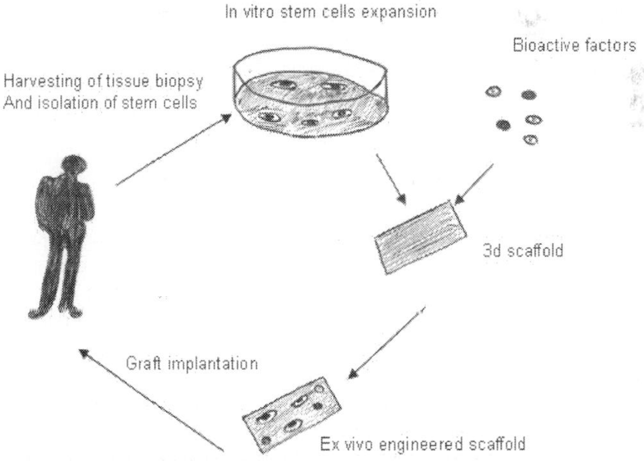

Fig. 2. Overview of tissue engineering approaches employing cell-polymer constrcuts.

In response to these varied criteria, a number of scaffold materials have been examined as scaffolds: porcine small intestine submucosa, (Musahl V at al. 2004) (Rodeo SA et al. 2004) silk fibers,(Altman et GH al. 2002) (Chen J at al. 2003) (Altman GH et al. 2003) semitendinosus tendon (Martinek V 2002), fibronectin/fibrinogen fibres. (Ahmed Z et al. 2000)

In addition to these, historically, three major categories of scaffolding materials have been employed. These are polyesters, polysaccharides, and collagen derivatives.

However, approaches to mimic the native extracellular matrix of tendon have limitetd to their inappropiate mechanical strength or the lack of cell adhesion sites. The use of an acellular graft may prevent the initial cell necrosis observed when autografts and allografts are used wich leads to the deterioration of mechanical strength following implantation.

Natural tissue scaffolds have the advantage of preserved ECM proteins important for cell attachment and the desired mechanical properties.

Natural scaffolds are composed of extracellular matrix proteins that are conserved among different species and wich can act as scaffolds for cell attachment, migration and proliferation.

Natural scaffolds have been decellularized in order to reduce their immunogenicity, a major hurdle to overcome in acellular scaffolds is their capacity for recellularization and regeneration with cellular components in vitro or in vivo, in order to achieve optimal biological and biochemical functions.(Gilbert TW 2006)

2.1 Collagen scaffolds

Collagen derivatives have been intensively investigated for use in tendon tissue engineering applications. Tendon extracellular matrix are mainly composed of type I collagen, so scaffolds based on collagen derivatives are highly biocompatible, then collagen derivatives also exhibit superior bio-functionality: they better support cell adhesion and cell proliferation.

Cells cultured in collagen gels produce extracellular matrix and align longitudinally with the long axis of the tissue equivalent, thereby mimicking cell alignment in ligaments *in vivo*.(Goulet F et al. 2000) (Huang D 1993)

Fibroblasts seeded in collagen gels change their shape and orientation over time (Huang D 1993) (Bell E at al. 1979) (Klebe RJ et al.1989) (Nishiyama T et al. 1993) and these organizational changes have been correlated with cell proliferation, protein synthesis, and matrix morphogenesis. (Ben-Ze'ev A et al. 1980) (Harris AK et al. 1981) (Maciera-Coelho A 1971) Fibroblast-seeded collagen scaffolds have been investigated with regard to their ability to accommodate cell attachment, proliferation, and differentiation. (Goulet F et al. 2000)' (Huang D 1993) (Bellincampi LD et al. 1998) (Dunn MG et al. 1995)

Collagen gel has been reported to augment the quality of tendon repair, but collagen gel does not possess sufficient mechanical strength, it is often accompanied by a high-strength component. For instance, Awad et al. (Awad HA et al. 2003) studied collagen gels in combination with a polyglyconate suture for patellar tendon repair. The biomechanical properties of the resulting tendon tissues were significantly better than those of naturally healed tendons, yet still much inferior to those of uninjured tendons.

Compared with collagen gel, collagen sponges exhibit greater mechanical competence. Given that collagen gels exhibit superior cell-seeding efficiency, a combination of collagen gels with collagen fibres or sponges represents a promising strategy. Juncosa-Melvin et al. (Juncosa-Melvin N et al. 2006) showed that gel–collagen sponge constructs could greatly enhance functional tendogenesis.

Another study, (Gentleman E et al. 2006) provided further evidence that a combination of collagen gels and sponges could bolster development of tendon-like tissue.

Despite its superior bio-functionality and biocompatibility, remaining several limitations to collagen. First, its processability is limited, the degree to which collagen scaffolds can be characterized is restricted. Then, the mechanical strength of collagen scaffolds is much lower than other materials such as the polyesters.

2.2 Polyesters scaffolds

A vast majority of biodegradable polymers for tendon tissue engineering applications are polyesters, such as polyglycolic acid (PGA), polylactic acid (PLA) and their copolymer polylactic-coglycolic acid (PLGA).These polymers are attractive, because their degradation products, glycolic acid and lactic acid, are natural metabolites that are normally present in the human body. Moreover, their good mechanical properties and their processability increase their appeal.

PLGA scaffolds have been reported to improve tendon regeneration considerably. Ouyang et al. (Ouyang, H.W. et al. 2003) found that knitted PLGA scaffolds augmented the tendon healing, both histologically and mechanically. These scaffolds facilitated production of collagen type I and type III fibrils and contributed to the improved mechanical properties.

PGA was also reported as a scaffolding material, Cao et al. (Cao Y.et al. 2002) developed a PGA scaffold that could successfully restore the mechanical capacity of tendons in a hen model. Moreover, Wei et al. (Wei X. et al. 2005) found that woven PGA scaffolds were particularly suitable for tendon tissue engineering because they surpassed the unwoven PGA in mechanical performance and at the same time degraded more slowly.

The cellular responses to these materials, as well as their individual degradation profiles, appear to be very different. Lu et al. (Lu HH et al. 2005) compared scaffolds based on three different materials, PGA, poly-Llactic acid (PLLA) and PLGA. Although the PGA-based scaffolds showed the highest initial strength, they suddenly lost mechanical strength owing to the bulk degradation profile of PGA, and this resulted in a matrix disruption and a loss of integrity. Regarding cellular responses, it was reported that, when using PLLA and PLGA, the morphology of attached cells resembled that of tendons and ligaments, whereas the best cell proliferation was reported for surface-modified PLLA scaffolds.

Despite their advantages, polyesters also suffer from several limitations. First, owing to their hydrophobic nature, poly-a-hydroxyesters do not support a high level of cell adhesion (Wan YQ et al. 2003) which is the initial and crucial step to engineer functional tendons.

Fortunately, this limitation can be overcome by means of surface modification with adhesive agents such as fibronectin (Qin TW et al. 2005)

Second, although degradation products of PGA, PLA and PLGA are natural metabolites, they are also acidic. The presence of these metabolites in large concentrations can therefore give rise to significant systematic or local reactions (Bostman OM et al. 2000)

When the sizes of scaffolds are smaller, the occurrence of such adverse biological reactions is greatly reduced. Therefore, in general, polyesters are more used for repair of smaller defects, which need smaller scaffolds.

2.3 Polysaccharides scaffolds

Polysaccharides have also been applied in the field of soft tissue engineering, and chitosan in particular has been used to regenerate tendons.

Chitosan, a deacetylation product of chitin, is a linear polysaccharide composed of randomly distributed b-(1–4)-linked D-glucosamine (the deacetylated unit) and N-acetyl-D-glucosamine (the acetylated unit), it is hydrophilic and exhibits good cell adhesion and proliferation characteristics (Suh JKF & Matthew, HWT 2000)

Moreover, the N acetylglucosamine present in chitosan is a structural feature that is also found in glycosaminoglycan, which is involved in many specific interactions with growth factors, receptors and adhesion proteins. Chitosan as a glycosaminoglycan analogue might therefore also exhibit similar bio-functionality. Furthermore, chitosan can create highly porous structures that make it especially suitable for a scaffolding material used in tendon tissue engineering (Kumar MNVR et al. 2004)

The bio-functionality of chitosan, such as supporting of cellular attachment and proliferation, and the ability to induce cells to produce ECM has been demonstrated. In a study conducted by Bagnaninchi et al. (Bagnaninchi et al. 2007), porous chitosan scaffolds with microchannels were designed to engineer tendon tissues.

Hyaluronan (HA) is a uniformly repetitive linear GAG composed of disaccharides of glucuronic acid and N-acetylglucosamine: [-b (1,4)- GlcUA- b (1,3)-GlcNAc-]n (Toole BP 2001) It is an essential component of ECM. Anionic hyaluronan interacts with other macromolecules, such as link proteins and proteoglycans, to facilitate tissue morphogenesis, cell migration, differentiation and adhesion (Toole BP 2001), whereas cationic chitosan can elicit electrostatic interactions with anionic glycosaminoglycans and other negatively charged species. (Kumar MNVR et al. 2004)

Hybridization of hyaluronan and chitosan is expected to augment the mechanical properties and bioactivities of tendon tissue engineering scaffolds. Funakoshi et al. (Funakoshi et al. 2005a) demonstrated that scaffolds composed of hybridized chitosan–hyaluronan exhibited enhanced mechanical competence. In another study, Funakoshi et al. (Funakoshi et al. 2005b) reported that the chitosan–hyaluronan scaffold improved the biomechanical properties of the regenerated tendon tissue in the rotator cuff and bolstered production of collagen type I.

Alginate, another type of polysaccharide that can be used for hybridization with chitosan, is an anionic polysaccharide composed of homopolymeric regions of glucuronic acid and mannuronic acid interspersed with mixed sequences (M-G blocks). Because it contains D-glucuronic acid as the main sugar residue in the repeat unit, alginate is often considered to have similar biological activity to glycosaminoglycans. However, owing to its anionic nature, cell adhesion to alginate is often unsatisfactory (Rowley JA et al. 1999) (Genes NG et al. 2004)

Adding cationic chitosan to alginate would augment the bio-functionality of the scaffold because the ionic interactions between alginate and chitosan are expected to facilitate retaining and recruiting of cells and growth factors, as well as cytokines (Madihally SV & Matthew HWT 1999) (Hsu SH et al. 2004)

It was reported that an alginate–chitosan hybrid scaffold showed significantly enhanced cell adhesion to tenocytes. (Majima et al. 2005)

2.4 Decellularized scaffolds

In order to be utilized successfully as a biomaterial, native extracellular matrix must first be decellularized to remove any allogenic cells and to prevent adverse immunological reactions. Native scaffolds are bioactive and promotes cellular proliferation and tissue ingrowth.

An ideal cell removal method would not compromise graft structure and mechanical properties.

Cartmell JS & Dunn MG (Cartmell JS & Dunn MG 2000) compared the effects of three extraction chemicals [t-octyl-phenoxypolyethoxyethanol (Triton X-100), tri(n-butyl)phosphate (TnBP), and sodium dodecyl sulfate (SDS)] on tendon cellularity, structure, nativity, and mechanical properties.

Treatment with 1% SDS for 24 h or 1% TnBP for 48 h resulted in an acellular tendon matrix with retention of near normal structure and mechanical properties, cell removal using SDS and TnBP, suggested these treatments are potentially useful for removing cells from tendon allografts or xenografts without compromising the graft structure or mechanical properties.

In order to function as a living tissue, it is essential that the acellular scaffold is recellularized either in vivo or in vitro prior to implantation, so that remodeling of the scaffold to maintain the correct ultrastructural and physical properties can occur. Recellularization in vitro allows for further conditioning of the graft prior to implantation, and hopefully a more successful outcome.

Several approaches have been developed to reseed scaffolds that are used in tissue engineering, including static culture, pulsatile perfusion and centrifugal force. However, the recellularization in most cases was not homogenous or required large numbers of cells.

Harrison RD Gratzer PF (Harrison RD & Gratzer PF 2005) developed a decellularized bone-anterior cruciate ligament, demonstrating that Triton-X-and TnBP-treated ligaments were more receptive to cellular ingrowth than SDS-treated samples.

Woods T & Gratzer PF (Woods T & Gratzer PF 2005) reported that TnBP treatment slightly decreased the collagen content of the anterior cruciate ligament, but did not alter its mechanical properties.

In a study, Ingram JH et al. (Ingram JH et al. 2007) have decellularized a porcine patella tendon scaffold with hypotonic buffer, 0.1% (w/v) sodium dodecyl sulfate (SDS), then used an ultrasonication treatment in order to produce a microscopically more open porous matrix; cells seeded onto the fascicular scaffolds penetrated throughout the scaffold and remained viable after 3 weeks of culture.

Deeken CR et al. (Deeken CR et al. 2011) decellularized the central tendon of the porcine diaphragm with several treatments but only 1% TnBP was effective in removing cell nuclei while leaving the structure and composition of the tissue intact.

3. Cells

An important prerequisite for current tendon engineering is the successful isolation and selection of functionally active cells, the cells have to retain the ability to proliferate rapidly *in vitro* to provide adequate numbers for *in vivo* implantation.

The most common cell types employed are fibroblasts, tenocytes and mesenchymal stem cells/marrow stromal cells. (Doroski DM et al. 2007)

The main cell type found in tendon tissue is the fibroblast, which is responsible for secreting and maintaining the extracellular matrix. Hence, fibroblasts are the predominant cell type used for tissue engineering applications. (Doroski DM et al. 2007)

Two different fibroblast populations can be found in the tendon: the elongated tenocytes and the ovoid-shaped tenoblasts.(Li F et.al 2008) Elongated tenocytes proliferate well in culture and have optimal morphology in terms of expression of collagen type 1, which is a major component of normal tendons. (Li F et.al 2008) Tendon cells are usually isolated from human tendon samples by tissue dissociation techniques.(Bagnaninchi PO et al. 2007) (Yao L et al 2006) (Cao D et al 2006)

After two or three cell culture passages and before they lose their phenotype, they are seeded into collagen gels or into scaffolds at an appropriate cell density (10 ^{6} cells/mL). (Bagnaninchi PO et al. 2007) (Yao L et al 2006) (Cao D et al 2006)

Anyway the use of tenocytes have some drawbacks such as limited availability of donor sites for cell harvest, the requirement for lengthy *in vitro* culture to expand the number of cells, and donor-site morbidity limit the practicality of this technique. (Hankemeier S et al. 2005) (Awad HA et al. 2000)

Stem cells may represent the ideal source for tendon engineering.There are 2 types of stem cells: embryonic stem cells and adult stem cells, embryonic cells are totipotent, but their practical use may be limited because of ethical issues and concerns regarding cell regulation. Adult stem cells, also known as mesenchymal stem cells, show excellent regenerative capacity, the ability to proliferate rapidly in culture, and the ability to differentiate into a wide variety of cell types.(Gao J and Caplan AI 2003) (Alhadlaq A & Mao JJ 2004) (Bosnakovski D et al. 2005) (Grove JE et al. 2004)

The ability of human marrow derived adult mesenchymal stem cells that have tendinogenic differentiation already has been documented in several studies: mesenchymal stem cells can be stimulated to differentiate into fibroblasts when exposed to mechanical stress, (Ge Z et al. 2005) and their rates of proliferation and collagen excretion have been shown to be higher than those of fibroblasts, so they may be a viable alternative to fibroblasts. (Li F et.al 2008)

The ideal source of autologous stem cells would be one that is easy to obtain, results in minimal patient discomfort, and provides cell numbers substantial enough to obviate extensive expansion in culture.

Studies have shown that raw adipose tissue contains a population of adult stem cells that can differentiate into bone, fat, cartilage, or muscle in vitro.(Lee RH et al. 2004) (Zuk PA et al. 2001) (Lee JA et al. 2003)

These adipose-derived stem cells are easily accessible and unlike marrow are available in large quantities with acceptable morbidity and discomfort associated with their harvest.

The autologous nature of these stem cells together with their putative multipotentiality and ease of procurement may make these cells an excellent choice for many future tendon-engineering strategies and cell-based therapies.(Young RG et al. 1998)

Tissues treated with these cells showed a markedly larger crosssectional area and contained collagen fibers that were better aligned than those in matched controls. (Awad HA et al. 1999)

Compared with their matched controls the MSC-mediated repair tissue showed marked increases in maximum stress, modulus, and strain energy density. Morphometrically,there were no marked differences in microstructure between the experimental and the control sides.

Kryger et al.(Kryger et al. 2007) compared tenocytes and mesenchymal stem cells for use in flexor tendon tissue engineering. They studied four candidate cell types for use in reseeding acellularised tendon constructs. Specifically, they compared epitenon tenocytes, tendon sheath fibroblasts, bone marrow-derived mesenchymal stem cells (BMSCs), and adipoderived mesenchymal stem cells (ASCs) with respect to their in vitro growth characteristics, senescence and collagen production, as well as the viability of reseeded constructs.(Kryger et al. 2007) They also studied the in vitro viability of tendon constructs after reseeding and after *in vivo* implantation in a clinically relevant model of rabbit flexor tendon grafting. Results showed that epitenon tenocytes, tendon sheath cells, bone marrow and adipo-derived stem cells have similar growth characteristics and can be used to successfully reseed acellularized tendon grafts.(Kryger et al. 2007) Constructs using the four cell types were also successfully implanted *in vivo* and showed viability after six weeks following implantation. (Kryger et al. 2007) The most relevant novel finding is that adipo-derived mesenchymal stem cells showed higher proliferation rates at later passages when compared with epitenon tenocytes. adipoderived mesenchymal stem cells have been shown to have multipotency and may be driven toward tenocyte differentiation when seeded into tendon constructs and exposed to the appropriate environment and mechanical forces. (Kryger et al. 2007) As confirmed by immunocytochemistry analysis, these stem cells also produce collagen, suggesting that they would contribute to *in vivo* tendon matrix remodeling. In conclusion, these results suggest that adipoderived mesenchymal stem cells have a practical advantage when compared with epitenon tenocytes and sheath fibroblasts, given that it is easier to harvest large amounts of fat tissue.

4. Local delivery of growth factors and gene therapy

Numerous growth factors that promote soft-tissue regeneration such as platelet-derived growth factor (PDGF), epidermal growth factor, fibroblast growth factor, insulin-like growth factor-I, bone morphogenetic proteins (BMPs) 2 to 7, growth and differentiation factors 5 to 7, and transforming growth factors 1 to 3 have been studied. (Hankemeier S et al. 2005) (Lou J et al. 2001) (Abrahamsson SO et al. 1991) (Rickert M et al. 2001) (Wang XT 2004) (Hsu C & Chang J 2004).

The results of *in vitro* studies have shown that growth factors can promote cell proliferation and protein synthesis (Fu SC et al. 2003) (Jann HW et all 1999)Injured tendons treated with growth factors show improved mechanical properties. (Aspenberg P & Forslund C 1999) (Zhang F et al. 2003) (Chan BP 2000).

Gene therapy in tendon engineering is an attractive new approach to the treatment of tendon lacerations.

The challenge is to define optimal cellular targets and to identify genes that are of therapeutic value and vectors that can deliver these genes with minimal side effects and maximal efficiency and durability.

A variety of gene transfer techniques can be used to maintain local concentrations of growth factor at the repair site by continuous expression of the exogene. (Dai Q et al 2003)

Vectors that enable the uptake and expression of genes into target cells are grouped into viral and nonviral. Viral vectors are viruses that are deprived of their ability to replicate and into which the genetic material can be inserted. These vectors are effective because host cells into which they introduce their genetic material form part of their normal life cycle.(Lou J et al. 1996)

Nonviral vectors such as liposomes are less pathogenic because of the absence of viral proteins, but are less efficient than viral vectors in the transfer of DNA to cells. (Jayankura M et al. 2003)

For effective tissue regeneration it is important to develop methods that will deliver genes to the site of tendon injury. (Lou J 2000) (Hildebrand KA et al. 2004)

Two main strategies for gene transfer using vectors can be envisioned: (1) *in vivo* transfer with a vector that is applied directly to the relevant tissue and (2) removal of cells from the body, transfer of the gene *in vitro*, and after an additional intermediate step that involves culture of the cells, reintroduction into the target site.

Direct gene transfer is less invasive and technically easier than transfer to cells *in vitro* because treatment can be started during the acute phase of injury. One disadvantage is the possibility of nonspecific infection of cells that are adjacent to the injury site. This risk may be complicated further by the fact that owing to the amount of extracellular matrix present and the relative few cells, a vector with high transgenic activity is needed to transfer the gene effectively to enough cells *in vivo*.

Furthermore BMP-12 gene transfer into a complete tendon laceration in a chicken model produced a 2-fold increase in tensile strength and stiffness of repaired tendons. (Lou J et al. 2001)

Recombinant adenovirus-expressing green fluorescent protein or BMP-13 injected into rabbit tendons showed efficient dosedependent transgene expression in all samples at 12 days after injection, although lymphocytic infiltration was noticed at the injection sites with the highest dose of virus, suggesting that injected adenoviral vectors elicit a local inflammatory response.

Although the *ex vivo* indirect technique includes the additional step of preparing cells and maintaining them *in vitro*, it provides a greater margin of safety because modified cells can be tested *in vitro* before administration, and viral DNA that is carried by these cells is not administered directly to the host cells. In addition gene transfer *in vitro* allows the selection of cells that express the trans-gene at high concentrations by using a selectable gene.

A variety of viral vectors such as adenovirus, retrovirus, adeno-associated virus, and liposomes have been evaluated for their ability to deliver genes to tendon, ligament, and meniscal cells. (Gerich TG et al. 1997)

Although adenovirus was the most effective vector in short-term experiments, transgene expression was transient; although the retrovirus gave lower initial transduction efficiencies, the percentage of transduced cells could be increased with a selectable marker gene.

The transfer of PDGF-B DNA to tenocytes increased the expression of PDGF and type I collagen gene expression in cultured tendon cells.Bone-marrow mesenchymal cells transfected with BMP-12 complementary DNA are placed into muscle in nude mice induced a neo-tendinous tissue. (Wang XT 2004)

Furthermore type I collagen synthesis was increased in tendon cells that were transfected with BMP-12 complementary DNA. Platelet-derived growth factor increased flexor tendon cell proliferation *in vitro*. (Thomopoulos S et al. 2005)

The PDGF and insulin-like growth factor-I transduced cells stimulated collagen and DNA synthesis in adjacent tendon cells. (Thomopoulos S et al. 2005)

The polymer that was seeded with tendon cells *in vitro* was used to repair a rotator cuff tear and histological studies showed that the tissue-engineered construct restored tendon with nearly complete repair of the tear. The restoration of normal tendon histology with longitudinally aligned collagen fiber bundles in the experimentally treated animals was shown.

5. Tendon engineering by the application of mechanical load

In vitro tissue development may include the application of mechanical loading to precondition the engineered tissue for the *in vivo* mechanical environment. Mechanical stress plays a significant role in modulating cell behavior and has driven the development of mechanical bioreactors for tissue engineering applications.(Ingber DE 2006) (Barkhausen T et al. 2003) (Brown RA et al. 1998) (Wang JH et al. 2004)

Tendons transmit force from the muscle to the bone and act as a buffer by absorbing external forces to limit muscle damage. Tendons exhibit high mechanical strength, good flexibility, and an optimal level of elasticity to perform their unique role. Tendons are visco-elastic tissues that display stress relaxation and creep. The mechanical behavior of the constituent collagen depends on the number and types of intramolecular and intermolecular bonds.

Experiments have confirmed cell growth and function would be controlled locally through physical distortion of the associated cells or through changes in cytoskeletal tension. Moreover, experimental studies have demonstrated that cultured cells can be switched between different fates including growth, differentiation, apoptosis, directional motility or different stem cell lineages, by modulating cell shape. (Barkhausen T et al. 2003) (Brown RA et al. 1998) (Wang JH et al. 2004) (Schulze-Tanzil G et al. 2004)

Externally applied cyclic strain under in vitro conditions has enormous effects on various functions of tenocytes, such as their metabolism, proliferation, orientation and matrix deposition (Screen HRC et al. 2005) (Yamamoto E et al. 2005)

Kessler et al. (Kessler et al. 2001) demonstrated that collagen fibres and tendon cells can be oriented along the direction of the stress and can upregulate synthesis of tissue inhibitor matrix metalloproteinases-1 and -3 as well as of collagen type I, the main component of tendinous extracellular matrix.

It is known that cyclic strain can affect cell morphology and induce uniaxial cellular alignment. It was observed that cyclic strain stimulation enhanced the cellular alignment and changed the cellular shape.

Other experiments have demonstrated the beneficial effects of motion and mechanical loading on tenocyte function. Repetitive motion increases DNA content and protein synthesis in human tenocytes in culture.(Almekinders LC et al. 1995) (Sharma P and Maffulli N. 2005) Even fifteen minutes of cyclic biaxial mechanical strain applied to human tenocytes, results in improved cellular proliferation.(Sharma P and Maffulli N. 2005) (Zeichen J et al. 2000)

Moreover, *in vitro* cyclic strain allows an increased production of TGF-β, FGF and PDGF by human tendon fibroblasts.(Bagnaninchi PO et al. 2007) (Slutek M et al. 2001)

Cyclic stretching of collagen type I matrix seeded with MSCs for 14 days (8 h/day) resulted in the formation of a tendon-like matrix. (Bagnaninchi PO et al. 2007) (Zeichen J et al. 2000)

Expression of collagen types I and III, fibrinonectin and elastin genes was found to have increased when compared with nonstretched controls in which no ligament matrix was found. (Bagnaninchi PO et al. 2007) (Yang G et al. 2004)

The model reproduces *in vivo* tendon healing by preventing differentiation of tenocytes into fibroblasts.

In animal experiments, mechanical stretching has improved the tensile strength, elastic stiffness, weight and cross-sectional area of tendons.(Sharma P & Maffulli N 2005) (Kannus P et al. 1992) (Kannus P et al. 1997) These effects result from an increase in collagen and extracellular matrix network syntheses by tenocytes. Application of a cyclic load to wounded avian flexor tendons results in the migration of epitendon cells into the wound.(Sharma P and Maffulli N 2005) (Tanaka H et al. 1995)

Also, Qin et al. (Qin et al. 2005) found that cyclic strain promotes cell proliferation, matrix deposition and increased collagen production. In another study, Juncosa-Melvin et al. (Juncosa-Melvin et al. 2007) showed that the application of cyclic strain elevated the gene expression levels of collagen type I. Finally, cyclic strain can enhance mechanical competence of the regenerated tendons (Juncosa-Melvin et al. 2006). The authors of this study found that values for maximum force, linear stiffness, maximum stress, and linear modulus for repaired tendons were close to those of natural patellar tendon. In terms of restoration of key biomechanical parameters, these constructs appear to be the best engineered tendons obtained so far.

Clinical studies have shown the benefit of early mobilization following tendon repair, and several postoperative mobilization protocols have been advocated.(Buckwalter JA 1996) (Chow JA et al 1988) (Elliot D et al. 1994) The precise mechanism by which cells respond to load remains to be elucidated. However, cells must respond to mechanical and chemical signals in a coordinated fashion. For example, intercellular communication by means of gap junctions is necessary to mount mitogenic and matrigenic responses in ex vivo models. (Sharma P & Maffulli N 2005)

Duration, frequencies and amplitude of loading directly influence cellular response and behavior in many other tissues. Understanding the physiological window for these parameters is critical and represents future challenges of research in tendon tissue engineering.

6. References

Longo, UG.; Lamberti, A.; Maffulli, N. & Denaro, V. (Sep 2010). Tissue engineered biological augmentation for tendon healing: a systematic review. Br Med Bull.

Wong, JK.; Lui, YH.; Kapacee, Z.; Kadler, KE.; Ferguson, MW. & McGrouther, DA. (2009). The cellular biology of flexor tendon adhesion formation: an old problem in a new paradigm. Am J Pathol 175: 1938–1951

Woo, SL.; Wu, C.; Dede, O.; Vercillo, F. & Noorani, S. (Sep 2006) Biomechanics and anterior cruciate ligament reconstruction.J Orthop Surg Res. 25;1:2

Kim, CW. & Pedowitz, RA. (2003). Principles of Surgery: Graft Choices and the Biology of Graft Healing in Knee Ligaments: Structure, Function, Injury, and Repair. Ed. Lippincott Williams & Wilkins ISBN-139780781718172. pp 435-55, Philadelphia

Wang, JH. (2006). Mechanobiology of tendon. J Biomech.; 39:1563-82

Goulet, F.; Rancourt, D.; Cloutier, R.; Germain, L.; Poole, AR. & Auger, FA. (2000). Tendons and ligaments. Principles of tissue engineering. 2nd ed. pp 711-22 ISBN 13: 978-0-12-370615-7, *San Diego Academic Press Ltd.*

Liu, Y.; Ramanath, HS. & Wang, DA. (Apr 2008). Tendon tissue engineering using scaffold enhancing strategies. Trends Biotechnol. 26:201-9

Musahl, V.; Abramowitch, S.; Gilbert, T.; Tsuda, E.; Wang, J.; Badylak, S. & Woo, SL. (2004). The use of porcine small intestinal submucosa to enhance the healing of the medial collateral ligament a functional tissue engineering study in rabbits. J Orthop Res; 22:214-20

Rodeo, SA.; Maher, SA. & Hidaka, C. (2004). What's new in orthopaedic research. J Bone J Surg Am; 86:2085-95

Altman, GH.; Horan, RL.; Lu, HH.; Moreau, J.; Martin, I.; Richmond, JC. & Kaplan, DL. (2002) Silk matrix for tissue engineered anterior cruciate ligaments. Biomaterials; 23:4131-41.

Chen, J.; Altman, GH.; Karageorgiou, V.; Horan, R.; Collette, A.; Volloch, V.; Colabro, T. & Kaplan, DL. (2003). Human bone marrow stromal cell and ligament fibroblast responses on RGD-modified silk fibers. J Biomed Mater Res A; 67:559-70.

Altman, GH.; Diaz, F.; Jakuba, C.; Calabro, T.; Horan, RL.; Chen, J.; Lu, H.; Richmond, J. & Kaplan, DL. (2003). Silk-based biomaterials. Biomaterials; 24:401-16.

Martinek, V.; Latterman, C.; Usas, A.; Abramowitch, S.; Woo, SL.; Fu, FH. & Huard, J. (2002). Enhancement of tendon-bone integration of anterior cruciate ligament grafts with bone morphogenetic protein-2 gene transfer: A histological and biomechanical study. J Bone J Surg Am; 84:1123-31.

Ahmed, Z.; Underwood, S. & Brown, RA. (2000). Low concentrations of fibrinogen increase cell migration speed on fibronectin/fibrinogen composite cables. Cell Motil Cytoskeleton; 46: 6-16.

Gilbert, TW.; Sellaro, TL. & Badylak, SF. (Jul 2006). Decellularization of tissues and organs. Biomaterial; 27:3675-83

Huang, D.; Chang, TR.; Aggarwal, A.; Lee, RC. & Ehrlich, HP. (1993). Mechanisms and dynamics of mechanical strengthening in ligament-equivalent fibroblast-populated collagen matrices. Ann Biomed Eng; 3:289-305.

Bell, E.; Ivarsson, B. & Merrill, C. (1979). Production of a tissue-like structure by contraction of collagen lattices by human fibroblasts of different proliferative potential *in vitro* . Proc Nat Acad Sci USA; 3:1274-8.

Klebe, RJ.; Caldwell, H. & Milam, S. (1989). Cells transmit spatial information by orienting collagen fibers. Matrix;6:451-8.

Nishiyama, T.; Tsunenaga, M.; Akutsu, N.; Horii, I.; Nakayama, Y.; Adachi, E.; Yamato, M. & Hayashi, T. (1993). Dissociation of actin microfilament organization from acquisition and maintenance of elongated shape of human dermal fibroblasts in three-dimensional collagen gel. Matrix; 6:447-55.

Ben-Ze'ev, A.; Farmer, SR. & Penman, S. (1980). Protein synthesis requires cell-surface contact while nuclear events respond to cell shape in anchorage-dependent fibroblasts. Cell; 2:365-72.

Harris, AK.; Stopak, D. & Wild, P. (1981). Fibroblast traction as a mechanism for collagen morphogenesis. Nature; 290:249-51.

Maciera-Coelho, A.; Garcia-Giralt, E. & Adrian, M. (1971). Changes in lysosomal associated structures in human fibroblasts kept in resting phase. Proc Soc Exp Biol Med; 2:712-8.

Bellincampi, LD.; Closkey, RF.; Prasad, R.; Zawadsky, JP. & Dunn, MG. (1998). Viability of fibroblast-seeded ligament analogs after autogenous implantation. J Orthop Res; 4:414-20.

Dunn, MG.; Liesch, JB.; Tiku, ML. & Zawadsky, JP. (1995). Development of fibroblast-seeded ligament analogs for ACL reconstruction. J Biomed Mater Res; 11:1363-71.

Awad, HA.; Boivin, GP.; Dressler, MR.; Smith, FN.; Young, RG. & Butler, DL. (2003). Repair of patellar tendon injuries using a cellcollagencomposite. J. Orthop. Res; 21:420–431

Juncosa-Melvin, N.; Boivin, GP.; Gooch, C.; Galloway, MT.; West, JR.; Dunn, MG. & Butler, DL. (2006). The effect of autologous mesenchymal stem cells on the biomechanics and histology of gel-collagen sponge constructs used for rabbit patellar tendon repair. Tissue Eng; 12:369–379

Gentleman, E.; Livesay, GA.; Dee, KC. & Nauman, EA. (2006). Development of ligament-like structural organization and properties in cell-seeded collagen scaffolds in vitro.Ann. Biomed. Eng; 34:726–736

Ouyang, HW.; Goh ,JC.; Thambyah, A.; Teoh, SH. & Lee, EH. (2003). Knitted poly-lactide-co-glycolide scaffold loaded with bone marrow stromal cells in repair and regeneration of rabbit Achilles tendon. Tissue Eng; 9:431–439

Cao, Y.; Liu ,Y.; Liu, W.; Shan, Q.; Buonocore, SD. & Cui, L. (2002). Bridging tendon defects using autologous tenocyte engineered tendon in a hen model. Plast. Reconstr. Surg; 110:1280–1289

Wei, X.; Zhang, PH.; Wang, WZ.; Tan, ZQ.; Cao, DJ. Xu, F.; Cui, L.; Liu, W. & Cao, YL. (2005). Use of polyglycolic acid unwoven and woven fibers for tendon engineering in vitro. Key Eng. Mater; 288–289:7–10

Lu, HH.; Cooper, JA Jr.; Manuel, S.; Freeman, JW.; Attawia, MA.; Ko, FK. & Laurencin, CT. (2005). Anterior cruciate ligament regeneration using braided biodegradable scaffolds: in vitro optimization studies. Biomaterials; 26:4805–4816

Wan, YQ.; Chen, W.; Yang, J.; Bei, J. & Wang, S. (2003). Biodegradable poly(L-lactide)-poly(ethylene glycol) multiblock copolymer: synthesis and evaluation of cell affinity. Biomaterials; 24:2195–2203

Qin, TW.; Yang, ZM.; Wu, ZZ.; Xie, HQ.; Qin, J. & Cai, SX. (2005). Adhesion strength of human tenocytes to extracellular matrix component-modified poly(DL-lactide-coglycolide) substrates. Biomaterials; 26:6635–6642

Bostman, OM. & Pihlajamaki, HK. (2000). Adverse tissue reactions to bioabsorbable fixation devices. Clin. Orthop. Relat. Res; 371:216–227

Suh, JKF. & Matthew, HWT. (2000). Application of chitosan-based polysaccharide biomaterials in cartilage tissue engineering: a review. Biomaterials; 21:2589–2598

Kumar, MNVR.; Muzzarelli, RA.; Muzzarelli, C.; Sashiwa, H. & Domb, AJ. (2004). Chitosan chemistry and pharmaceutical perspectives. Chem. Rev; 104:6017–6084

Toole, BP. (2001). Hyaluronan in morphogenesis. Semin. Cell Dev. Biol; 12:79–87

Funakoshi, T.; Majima, T.; Iwasaki, N.; Yamane, S.; Masuko, T.; Minami, A.; Harada, K.; Tamura, H.; Tokura, S. & Nishimura, S. (2005). Novel chitosan-based hyaluronan hybrid polymer fibers as a scaffold in ligament tissue engineering. J. Biomed. Mater. Res. A; 74:338–346

Funakoshi, T.; Majima, T.; Iwasaki, N.; Suenaga, N.; Sawaguchi, N.; Shimode, K.; Minami, A.; Harada, K. & Nishimura, S. (2005). Application of tissue engineering techniques for rotator cuff regeneration using a chitosan-based hyaluronan hybrid fiber scaffold. Am. J. Sports Med; 33:1193–1201

Rowley, JA.; Madlambayan, G. & Mooney, DJ. (1999). Alginate hydrogels as synthetic extracellular matrix materials. Biomaterials; 20:45–53

Genes, NG.; Rowley, JA.; Mooney, DJ. & Bonassar, LJ. (2004). Effect of substrate mechanics on chondrocyte adhesion to modified alginate surfaces. Arch. Biochem. Biophys; 422:161–167

Madihally, SV. & Matthew, HWT. (1999). Porous chitosan scaffolds for tissue engineering. Biomaterials; 20:1133–1142

Hsu, SH.; Whu, S.; Tsai, CL.; Wu, YH.; Chen, HW. & Hsieh, KH. (2004). Chitosan as scaffold materials: Effects of molecular weight and degree of deacetylation. J. Polym. Res; 11:141–147

Majima T, Funakosi T, Iwasaki N, Yamane ST, Harada K, Nonaka S, Minami A, Nishimura S (2005) Alginate and chitosan polyion complex hybrid fibers for scaffolds in ligament and tendon tissue engineering. J.Orthop. Sci. 10, 302–307

Bagnaninchi, PO.; Yang, Y.; Zghoul, N.; Maffulli, N.; Wang, RK. & Haj, AJ. (2007). Chitosan microchannel scaffolds for tendon tissue engineering characterized using optical coherence tomography. Tissue Eng. 13:323–331

Cartmell, JS. & Dunn MG. (Jan 2000). Effect of chemical treatments on tendon cellularity and mechanical properties. J Biomed Mater Res; 49:134-40

Harrison, RD. & Gratzer, PF. (Dec 2005). Effect of extraction protocols and epidermal growth factor on the cellular repopulation of decellularized anterior cruciate ligament allografts. J Biomed Mater Res A. 75:841-54.

Woods, T. & Gratzer, PF. (Dec 2005). Effectiveness of three extraction techniques in the development of a decellularized bone-anterior cruciate ligament-bone graft. Biomaterials. 26:7339-49

Ingram, JH.; Korossis, S.; Howling, G.; Fisher, J. & Ingham, E. (Jul 2007). The use of ultrasonication to aid recellularization of acellular natural tissue scaffolds for use in anterior cruciate ligament reconstruction. Tissue Eng; 13:1561-72.

Deeken, CR.; White, AK.; Bachman, SL.; Ramshaw, BJ.; Clevel, DS.; Loy, TS. & Grant, SA. (Feb 2011). Method of preparing a decellularized porcine tendon using tributyl phosphate. J Biomed Mater Res B Appl Biomater; 96:199-206

Doroski, DM.; Brink, KS. & Johnna, S. (2007). Techniques for biological characterization of tissue-engineered tendon and ligament. Biomaterials; 28:187-202

Li. F; Li, B.; Wang, QM. & Wang, JH. (2008). Cell shape regulates collagen type I expression in human tendon fibroblasts. Cell Motil Cytoskeleton; 31:1-10.

Yao, L.; Bestwick, CS.; Bestwick, LA.; Maffulli, N. & Aspden, RM. (2006). Phenotipic drift in human tenocyte culture. Tissue Eng; 12:1843-9.

Cao, D.; Liu, W.; Wei, X.; Xu, F.; Cui, L. & Cao, Y. (2006). In vitro tendon engineering with avian tenocytes and poliglicolic acids: A preliminary report. Tissue Eng; 12:1369-77.

Awad, HA.; Butler, DL.; Harris, MT.; Ibrahim, RE.; Wu, Y.; Young, RG.; Kadiyala, S. & Boivin, GP. (2000). In vitro characterization of mesencymal stem cell-seeded collagen scaffolds for tendon repair: Effects of initial seeding density on contraction kinetics. J Biomed Mater Res; 51:233-40

Gao, J. & Caplan, AI. (2003). Mesenchymal stem cells and tissue engineering for orthopaedic surgery. Chir Organi Mov; 88:305–316

Alhadlaq, A. & Mao, JJ. (2004). Mesenchymal stem cells: isolation therapeutics. Stem Cells Dev; 13:436–448.

Bosnakovski, D.; Mizuno, M.; Kim, G.; Takagi, S.; Okumura, M. & Fujinaga, T. (2005). Isolation and multilineage differentiation of bovine bone marrow mesenchymal stem cells. Cell Tissue Res; 319:243–253.

Grove, JE.; Bruscia, E. & Krause, DS. (2004). Plasticity of bone marrow-derived stem cells. Stem Cells; 22:487–500.

Ge, Z.; Gohm, J. & Lee, E. (2005) Selection of cell source for ligament tissue engineering. Cell Transplant; 14:573-83

Lee, RH.; Kim, B.; Choi, I.; Kim, H.; Choi, HS.; Suh, K.; Bae, YC. & Jung, JS. (2004). Characterization and expression analysis of mesenchymal stem cells from human bone marrow and adipose tissue. Cell Physiol Biochem; 14:311–324.

Lee, JA.; Parrett, BM.; Conejero, JA.; Laser, J.; Chen, J.; Kogon, AJ.; Nanda, D.; Grant, RT. & Breitbart, AS. (2003). Biological alchemy: engineering bone and fat from fat-derived stem cells. Ann Plast Surg; 50:610–617

Zuk, PA.; Zhu, M.; Mizuno, H.; Huang, J.; Futrell, JW.; Katz, AJ.; Benhaim, P.; Lorenz, HP. & Hedrick, MH. (2001). Multilineage cells from human adipose tissue: implicationsfor cell-based therapies. Tissue Eng; 7:211–228.

Young, RG.; Butler, DL.; Weber, W.; Caplan, AI.; Gordon, SL. & Fink, DJ. (1998). Use of mesenchymal stem cells in a collagen matrix for Achilles tendon repair. J Orthop Res; 16:406–413.

Awad, HA.; Butler, DL.; Boivin, GP.; Smith, FN.; Malaviya, P.; Huibregtse, B. & Caplan, AI. (1999). Autologous mesenchimal stem cell-mediated repair of tendon. Tissue Eng; 5:267-77

Kryger, GS.; Chong, AK.; Costa, M.; Pham, H.; Bates, SJ. & Chang, J. (2007). A comparison of tenocytes and mesenchymal stem cells for use in flexor tendon tissue engineering. J Hand Surg; 32:597-605

Hankemeier, S.; Keus, M.; Zeichen, J.; Jagodzinski, M.; Barkhausen, T.; Bosch, U.; Krettek, C. & Van Griensven, M. (2005). Modulation of proliferation and differentiation of human bone marrow stromal cells by fibroblast growth factor 2: potential implications for tissue engineering of tendons and ligaments. Tissue Eng; 11:41–49

Lou, J.; Tu, Y.; Burns, M.; Silva, MJ. & Manske, P. (2001). BMP-12 gene transfer augmentation of lacerated tendon repair. J OrthopRes; 19:1199 –1202

Abrahamsson, SO.; Lundborg, G. & Lohmander, LS. (1991). Recombinant human insulin-like growth factor-I stimulates in vitro matrix synthesis and cell proliferation in rabbit flexor tendon. J Orthop Res; 9:495–502.

Rickert ,M.; Jung, M.; Adiyaman, M.; Richter, W. & Simank, HG. (2001). A growth and differentiation factor-5 (GDF-5)-coated suture stimulates tendon healing in an Achilles tendonmodel in rats. Growth Factors; 19:115–126.

Wang, XT.; Liu, PY. & Tang, JB. (2004). Tendon healing in vitro: genetic modification of tenocytes with exogenous PDGF gene and promotion of collagen gene expression. J Hand Surg; 29A:884–890.

Hsu, C. & Chang, J. (2004). Clinical implications of growth factors in flexor tendon wound healing. J Hand Surg; 29A: 551–563

Fu, SC.; Wong, YP.; Chan, BP.; Pau, HM.; Cheuk, YC.; Lee, KM. & Chan, KM. (2003). The roles of bone morphogenetic protein (BMP) 12 in stimulating the proliferation and matrix production of human patellar fibroblats. Life Sci; 72:2965-2974

Jann, HW.; Stein, LE. & Slater, DA. (1999). In vitro effects of epidermal growth factor or insulin-like growth factor on tenoblast migration on absorbable suture material. Vet Surg; 28:268 –278.

Aspenberg, P. & Forslund, C. (1999). Enhanced tendon healing with GDF 5 and 6. Acta Orthop Scand; 70:51–54.

Zhang, F.; Liu, H.; Stile, F.; Lei, MP.; Pang ,Y.; Oswald, TM.; Beck, J.; Dorsett-Martin, W. & Lineaweaver, WC. (2003). Effect of vascular endothelial growth factor on rat Achilles tendon healing. Plast Reconstr Surg;112:1613–1619.

Chan, BP.; Fu, S.; Qin, L.; Lee, K.; Rolf, CG. & Chan, K. (2000). Effects of basic fibroblast growth factor (bFGF) on early stages of tendon healing: a rat patellar tendon model. Acta Orthop Scand; 71:513–518.

Dai, Q.; Manfield, L.; Wang, Y. & Murrell, GA. (2003). Adenovirusmediated gene transfer to healing tendon – enhanced efficiency using a gelatin sponge. J Orthop Res; 21:604–609.

Lou, J.; Manske, PR.; Aoki, M. & Joyce, ME. (1996). Adenovirus-mediated gene transfer into tendon and tendon sheath. J Orthop Res; 14:513–517.

Jayankura, M.; Boggione, C.; Frisen, C.; Boyer, O.; Fouret, P.; Saillant, G. & Klatzmann, D. (2003). In situ gene transfer into animal tendons by injection of naked DNA and electrotransfer. J Gene Med; 5:618–624.

Lou, J. (2000). In vivo gene transfer into tendon by recombinant adenovirus. Clin Orthop; S252–S255.

Hildebrand, KA.; Frank, CB. & Hart, DA. (2004). Gene intervention in ligament and tendon: current status, challenges, future directions. Gene Ther; 11:368 –378.

Lou, J.; Tu, Y.; Burns, M.; Silva, MJ. & Manske, P. (2001). BMP-12 gene transfer augmentation of lacerated tendon repair. J Orthop Res; 19:1199 –1202.

Gerich, TG.; Kang, R.; Fu, FH.; Robbins, PD. & Evans, CH. (1997). Gene transfer to the patellar tendon. Knee Surg Sports Traumatol Arthrosc; 5:118 –123.

Thomopoulos, S.; Harwood, FL.; Silva, MJ.; Amiel, D. & Gelberman, RH. (2005). Effect of several growth factors on canine flexor tendon fibroblast proliferation and collagen synthesis in vitro. J Hand Surg; 30A:441–447.

Ingber, DE. (2006) Mechanical control of tissue morphogenesis during embryological development. Int J Dev Biol; 50:255-66

Barkhausen, T.; van Griensven, M.; Zeichen, J. & Bosch, U. (2003). Modulation of cell functions of human tendon fibroblasts by different repetitive cyclic mechanical stress patterns. Exp Toxicol Pathol; 55:153-8

Brown, RA.; Prajapati, R.; McGrouther, DA.; Yannas, IV. & Eastwood, M. (1998). Tensional homeostasis in dermal fibroblasts: Mechanical responses to mechanical loading in three-dimensional substrates. J Cell Physiol;175:323-32

Wang, JH.; Yang, G.; Li, Z. & Shen, W. (2004). Fibroblast responses to cyclic mechanical stretching depend on cell orientation to the stretching direction. J Biomech; 37:573-6

Schulze-Tanzil, G.; Mobasheri, A.; Clegg, PD.; Sendzik, J.; John, T. & Shakibaei, M. (2004). Cultivation of human tenocytes in high-density culture. Histochem Cell Biol; 122:219-28.

Screen, HR.; Shelton, JC.; Bader, DL. & Lee, DA. (2005). Cyclic tensile strain upregulates collagen synthesis in isolated tendon fascicles. Biochem. Biophys. Res. Commun; 336:424–429

Yamamoto, E.; Kogawa, D.; Tokura, S. & Hayashi, K. (2005). Effects of the frequency and duration ofcyclic stress on the mechanical properties of cultured collagen fascicles from the rabbit patellar tendon. J. Biomech. Eng; 127:1168–1175

Kessler, D.; Dethlefsen, S.; Haase, I.; Plomann, M.; Hirche, F.; Krieg, T. & Eckes, B. (2001). Fibroblasts in mechanically stressed collagen lattices assume a "synthetic" phenotype. J Biol Chem; 276:36575-85.

Sharma, P. & Maffulli, N. (2005). Tendon injury and tendinopathy: Healing and repair. J Bone Joint Surg Am;87:187-202.

Almekinders, LC.; Baynes, AJ. & Bracey, LW. (1995) An *in vitro* investigation into the effects of repetitive motion and nonsteroidal anti inflammatory medication on human tendon fibroblasts. Am J Sports Med; 23:119-23

Zeichen, J.; van Griensven, M. & Bosch U. (2000). The proliferative response of isolated human tendon fibroblasts to cyclic biaxial mechanical strain. Am J Sports Med; 28:888-92

Kannus, P.; Jozsa, L.; Natri, A. & Jarvinen, M. (1997). Effects of training, immobilization and remobilization on tendons. Scand J Med Sci Sports; 7:67-71.

Tanaka, H.; Manske, PR.; Pruitt, DL. & Larson, BJ. (1995). Effect of cyclic tension on lacerated flexor tendons *in vitro* . J Hand Surg Am; 20:467-73.

Qin, TW.; Liang, Z.; Xie, K. & Li, H. (2005). A new construction model of engineered tendonsunder mechanical strain in vitro. Key Eng. Mater. 288–289 (Advanced Biomaterials VI), 19–22

Juncosa-Melvin, N.; Matlin, KS.; Holdcraft, RW.; Nirmalanandhan, VS. & Butler, DL. (2007). Mechanical stimulation increases collagen type I and collagen type III gene expression of stem cell-collagen sponge constructs for patellar tendon repair. Tissue Eng; 13:1219-1226

Juncosa-Melvin, N.; Shearn, JT.; Boivin, GP.; Gooch, C.; Galloway, MT.; West, JR.; Nirmalanandhan, VS.; Bradica, G. & Butler, DL. (2006). Effects of mechanical stimulation on the biomechanics and histology of stem cell-collagen sponge constructs for rabbit patellar tendon repair. Tissue Eng; 12:2291–2300

Buckwalter JA. (1996) Effects of early motion on healing of musculoskeletal tissues. Hand Clin; 12:13-24.

Chow. JA.; Thomes, LJ.; Dovelle, S.; Monsivais, J.; Milnor, WH. & Jackson, JP. (1988). Controlled motion rehabilitation after flexor tendon repair and grafting: A multi-centre study. J Bone Joint Surg Br; 70:591-5

Bagnaninchi, PO.; Yang, Y.; El Hai, AJ. & Maffulli, N. (2007). Tissue engineering for tendon repair. Br J Sports Med; 41:10

Elliot, D.; Moiemen, NS.; Flemming, AF.; Harris, SB. & Foster, AJ. (1994). The rupture rate of acute flexor tendon repairs mobilized by the controlled active motion regimen. J Hand Surg Br; 19:607-12.

19

Xenotransplantation Using Lyophilized Acellular Porcine Cornea with Cells Grown *in vivo* and Stimulated with Substance-P

Jeong Kyu Lee, Seok Hyun Lee and Jae Chan Kim
*Department of Ophthalmology, Chung-Ang University Hospital, Seoul,
Korea*

1. Introduction

Corneal allograft transplantation has a high success rate, but the clinical use of corneal allografts is limited by an insufficient number of human donor corneas (Alldredge & Krachmer, 1981; Sedlakova & Filipec, 2007). Artificial substitutes may serve as an alternative to donor allograft use, but widely accepted substitutes are not currently available (Chen et al., 2001; Griffith et al., 1999; Trinkaus-Randall et al., 1988). Porcine corneas may serve as a reasonable alternative due to the ease of use and the potential for genetic engineering (Auchincloss, 1988; Insler & Lopez, 1991; Ross et al, 1993). In porcine organ xenografts, overcoming xenoantigene expression is of central importance to avoid graft rejection. Gal1α-3Galß1-4GlcNAc (α-gal) on porcine tissues is one of the best known antigens involved in xenograft rejection (Amano et al., 2003; Collins et al., 1995; Good et al., 1992). We previously demonstrated that lyophilization of porcine corneas could eliminate the α-gal antigen by removing antigen-expressing cells, and lyophilized acellular porcine corneas (APCs) survived longer than fresh porcine corneas in pig-to-rat model (Lee et al., 2010). Though decellularization using lyophilization appear favorable in reducing graft rejection, but the early inflammation frequently encountered is another issue which requires resolution. The lyophilization process based on glycerol and surgical manipulation might be responsible for early inflammation, and the delayed healing caused by an acellular substrate graft is another concern.

Recently, several studies have reported that corneal transparency is highly dependent on corneal cells as well as the extracellular matrix (Meek et al., 2003; Mourant et al., 2000). Thus, adequate recellularization of acellular substitute might not only extend graft survival, but enhance optical transparency. Accordingly, we hypothesized that repopulation of lyophilized APCs with cell grown *in vivo* before transplantation can enhance the survival of the graft by reducing the damage caused by inflammation or apoptotic environment. To address this issue, lyophilized APCs were transplanted under the limbus of rabbit corneas in advance for repopulation with cells *in vivo*, then optical transparency and histologic findings were compared with controls over the follow-up period. Furthermore, lamellar keratoplasties were performed to evaluate the usefulness of lyophilized APCs with cells grown *in vivo* in a rabbit model.

2. Materials and methods

The animals in this study were treated according to the ARVO Statement for the Use of Animals in Ophthalmology and Vision Research.

2.1 Preparation of lyophilized acellular porcine cornea

Adult porcine corneas were obtained from a slaughterhouse within 2 hours death, then transported in a 4ᵘC moist chamber to the laboratory. Using sterile techniques, the epithelium of each pig cornea was removed and a 4.0 mm sized stromal button with a thickness of 300 μm was created from the central pig cornea using a microkeratome (Automated Corneal Shaper®; Chiron Vision, Claremont, CA, USA). The corneal button was treated in a mixed solution consisting of 40 μ/ml Dnase and 40 μ/ml Rnase (Sigma-Aldrich, St. Louis, MO, USA) for 30 minutes, followed by distilled water for 2 hours, three freeze-thaw cycles (-196°C liquid nitrogen for 30 minutes, followed by rapid thawing at 37°C for 30 minutes), and centrifuged (15000 ×g, 7 minutes) to removed all of the cellular components. Then, the corneal button was treated in 100 % glycerol (Sigma-Aldrich) at 4 °C for 3 days, stored at -80 °C for 48 hours, then lyophilized using a lyophilizer (SFDSM06; Samwon Freezing Engineering Co., Busan, Korea) at -80 °C for 48 hrs. Finally, the corneal button was irradiated with γ-rays (25 kGy) for sterilization.

2.2 Surgical procedure for *in vivo* recellularization

Forty-eight New Zealand white rabbits of either sex, weighing 2-3 kg were used for this study. Rabbits were divided into 3 groups and anesthetized with an intramuscular injection of mixture of Tiletamine and Zolazepam (Zolatil®, 12.5mg/Kg; Virbac Lab, Carros Cedex, France) and xylazine (Rompun®, 12.5mg/kg, Bayer Korea, Ansan, Korea).

In 16 rabbits, the superior limbal conjunctivae of the left eyes were incised using Westcotts scissors after topical anesthesia, and lyophilized APCs were inserted under the superior conjunctivae. The conjunctivae were closed with 8-0 vicryl. One-half of the implants were treated with substance-P (50 nmol/kg) for 1 hour before grafting. As controls, collagen sheets (CSs) and 8 sheets of bovine amniotic membranes (AMs) were transplanted (4.0 mm diameter) under the superior limbal conjunctivae in 16 rabbits using the same procedure. One-half of the CSs and bovine AMs were also treated with substance-P for comparison. All rabbits received topical levofloxacin (Cravit®, Santen, Osaka, Japan) three times daily until the end of the study.

2.3 Assessment of lyophilized APC with *in vivo* recellularization

The implants were harvested, and transparency was assessed 3day, and 1, 2, and 3 weeks after grafting in 2 rabbits per each time.

2.3.1 Optical property

The harvested lyophilized APC was placed on a numeric panel. A 0-4+ scoring system was devised to describe the transparency semi-quantitatively according to the visibility of a figure through the implants. Scoring was as follows: 0, clear figure image compared with the next numeral; 1+, minimally blurred figure; 2+, half of the blurred figure compared with the next numeral; 3+, intense opacity with the blurred image; and 4+, complete opacification.

2.3.2 Histological examination

The harvested implants were fixed in 4% paraformaldehyde, dehydrated in ethanol, and embedded in paraffin. Cross sections (2-4 μm) were made along the longitudinal axis, and serially-sectioned specimens were stained with hematoxylin and eosin. Sections were examined under light microscopy and photographed for analysis of infiltrating cells on graft. The number of infiltrating mononuclear cells and inflammatory cells per high power field (HPF; 40×objective) was counted from three different regions of each sample for comparison.

2.3.3 Reverse Transcription Polymerase Chain Reaction (RT-PCR)

The harvested implants were examined using RT-PCR to determine the ideal timing of implant removal from recipients. The harvested implants were cut into small pieces by blade, and total RNA was isolated using TRIZOL reagent (Invitrogen., Carlsbad, CA, USA), according to the manufacture's instruction. Briefly, harvested implants were homogenized in 1 ml of TRIZOL reagent, then 200 $\mu\ell$ of chloroform were added. Samples were centrifuged at 12,000 ×g for 15 min at 4 ℃ , and the aqueous phase was transferred to fresh tubes. One ml of isopropanol was added and the mixture was placed at -20 ℃ for 8 hrs. Sedimentation was performed by centrifugation at 12,000 ×g for 15 min at 4 ℃. The resulting RNA pellet was suspended in 75% EtOH and centrifuged at 8,000 ×g for 10 min at 4 ℃. The RNA pellet was dissolved in DEPC-treated RNase-free water. Complementary DNA (cDNA) was made with AccuPower™ RT Premix (Bioneer Co., Daejeon, Korea) in a total volume of 20 $\mu\ell$ containing 250 mM Tris-HCl (pH 8.3), 375 mM KCl, 15 mM $MgCl_2$, 50 mM DTT, 1 mM dNTP, 10 U RNasin, and 20 U RTase. Total RNA (2 μg) and random primer (0.5 μg) were added to the RT Premix. The RT reaction was started at 57 ℃ for 10 min to denature RNA and reverse transcription took place at 42 ℃ for 1 hr, followed by RTase inactivation at 94 ℃ for 5 min. The validity of the RT reaction was determined internally using rabbit glyceraldehyde-3-phosphate dehydogenase (GAPDH) primers. PCR amplification was performed with AccuPower™ PCR Premix (Bioneer Co.) using 5 $\mu\ell$ of the cDNA product in a total volume of 20 $\mu\ell$, containing 10 mM Tris-HCl (pH 9.0), 40 mM KCl, 1.5 mM $MgCl_2$, 1 U DNA polymerase, 1 mM dNTP, and 10 pmole of each specific primers. The reactions were run using a GeneAmp PCR System 2400 (Perkin Elmer Co., Waltham, MA, USA). RT-PCR products were electrophoresed on 1.0% agarose gels in Tris-acetate-EDTA (TAE) buffer, stained with ethidium bromide, and photographed under UV transillumination.

2.4 Biocompatibility tests using lamellar keratoplasty

Five New Zealand white rabbits, weighing 2-3 kg, were used for this study. All rabbits were anesthetized with an intramuscular injection of mixture of Tiletamine and Zolazepam (Zolatil®, 12.5mg/Kg; Virbac Lab) and xylazine (Rompun®, 12.5mg/Kg, Bayer Korea). The superficial corneas of rabbits were excised with a 4-mm punch. Lyophilized APCs treated with substance-P, which were maintained under superior limbal conjunctivae of rabbits for 7 days, were transplanted and fixed with 10-0 nylon (8 sutures) into rabbit corneas. Therapeutic contact lens was applied and tarsorrhaphy was performed. All rabbits received topical levofloxacin (Cravit®; Santen) eye drops three times daily until the end of the study. Therapeutic lens and tarsorrhaphy were removed after complete corneal epithelial regeneration, and the sutures were removed when they were loosened.

Slit lamp biomicroscopic examinations and photographs were done every week for 2 months for changes in transparency and neovascularization. A 0-4+ scoring system was devised to describe the extent of opacification semi-quantitatively (Fantes et al., 1990). Scoring was as follows: 0, totally clear; 0.5+, trace corneal haze seen only by indirect broad tangential illumination; 1+, haze of minimal density seen with difficulty with direct illumination ; 2+, mild haze easily visible with direct focal illumination; 3+, moderate dense opacity that partially obscured the iris details; and 4+, severe dense opacity that obscured completely the details of intraocular structures. A similar scoring system was developed to assess the extent of neovascularization, as follows: 0, no vessels extending toward the graft; 1+, vessels reaching the graft margin; 2+, vessels invading the graft; 3+, many vessels traversing the grafts (Konya et al., 2005; Jeong et al., 2009).

At 4 and 8 weeks, the rabbits were sacrificed, and the corneas were harvested under anesthesia. H&E and vimentin staining were done to observe the stromal cells repopulated in lyophilized APCs.

2.5 Statistical analysis

Data are expressed as the means±standard deviation. Statistical analysis was performed using SPSS (version 17.0 for Windows; SPSS, Inc., Chicago, IL, USA). Intergroup comparisons were analyzed using the Mann-Whitney U test. Statistical significance was set at a p < 0.05.

3. Results

3.1 Assessment of lyophilized APC
3.1.1 Optical transparency

The lyophilized APCs were visually opaque initially. After subconjunctival implantation, the lyophilized APCs were cleared with a transparency of 1+ by POD 7th; subsequently, the APCs gradually became opaque over 21 days. However the CSs and 8 sheets of bovine AMs were opaque, with scores of 3+~4+ over the entire examination point (Fig. 1). There was no difference in visual transparency of each implant whether or not treated with substance-P.

3.1.2 Histologic characterization

H&E staining of lyophilized APCs that were kept subconjunctivally for 7 days showed a number of rounds or spindle-shaped mononuclear cells, suggestive of corneal stroma-like cells, with an occasional inflammatory cell infiltration. Histologic analysis of lyophilized APCs that were kept for 3 days subconjunctivally revealed rare corneal stroma-like cells inside the graft, while APCs that were kept for 14 and 21 days subconjunctivally showed significant inflammatory cell infiltration, including polymorphonuclear leukocytes, and some monocytes (data not shown). Thus, lyophilized APCs that were kept for 7 days subconjunctivally were chosen as the ideal implants with cells grown *in vivo*, and the histologic appearance was compared with controls (Fig. 2).

The number of corneal stroma-like cells in lyophilized APCs and bovine AMs were comparable (average, 16.2±2.1 cells per field for lyophilized APCs vs. 18±4.0 cells per field for bovine AMs; p=0.24), and were significantly greater than CSs (average, 9.6±1.7 cells per field; p<0.05). In implants treated with substance-P, increased infiltration of corneal stroma-like cells was observed, but there was no significant difference following treatment with substance-P (p=0.23).

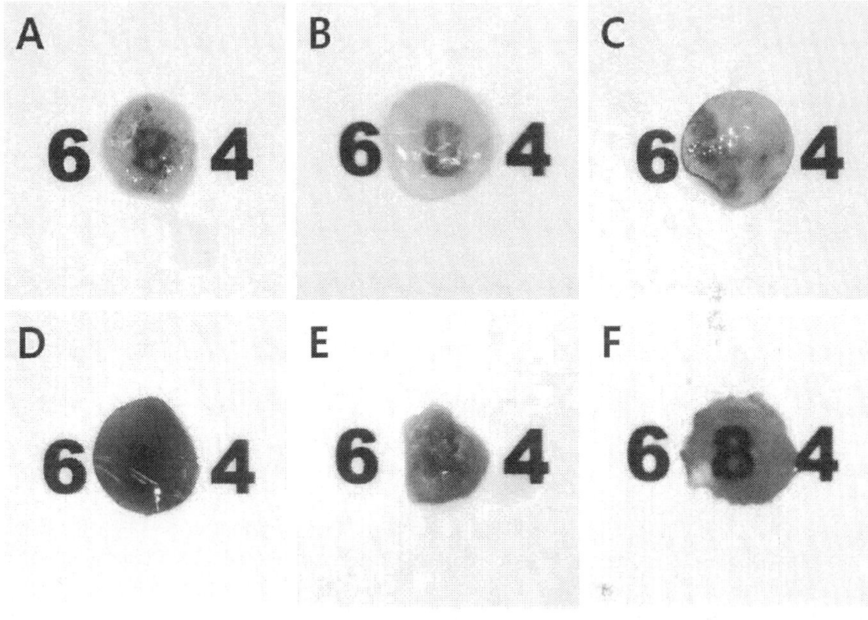

Fig. 1. Optical transparency of subconjunctival implants 7 days postoperatively.
The lyophilized acellular porcine cornea (A) and lyophilized acellular porcine cornea soaked in substance-P (B) showed approximately one-third transparency compared to the surrounding letter. Collagen sheet (C) and collagen sheet soaked in substance-P (D) were visually opaque. Eight sheets of amniotic membrane were also visually opaque (E), but single sheets of amniotic membrane was clear (F).

The average number of inflammatory cell in the lyophilized APC infiltrate was 11.4±1.1 cells per field, which was much less than bovine AM and CS infiltrates (34.8±5.5 cells per field for bovine AMs and 61.8±8.3 cells for CSs). Treatment with substance-P did not have an effect on the inflammatory cell infiltrate in implants.

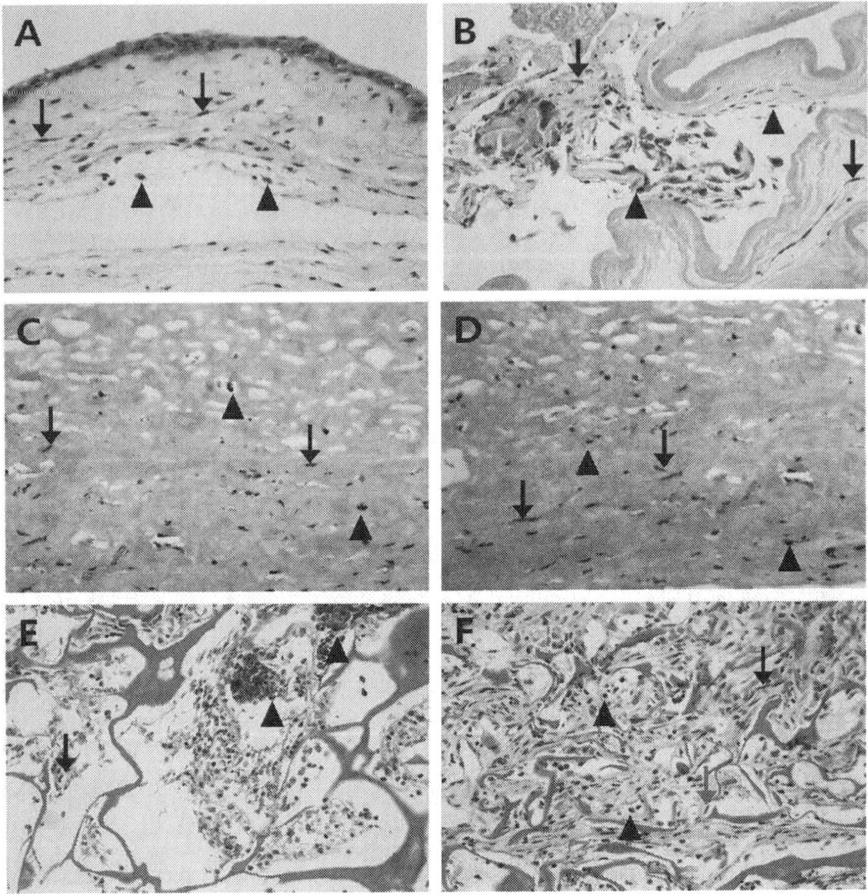

Fig. 2. Hematoxylin-eosin staining of amniotic membranes (A, B), lyophilized acellular porcine corneas (C, D), and collagen sheets (E, F) (B, D, F soaked in substance-P). Amniotic membranes and lyophilized acellular porcine corneas had more keratocytes (black arrow) than collagen sheets. Lyophilized acellular porcine corneas were shown to have less inflammatory cells (black arrowhead) than amniotic membranes and collagen sheets. Keratocytes were more visible in implants treated with substance-P. Original magnification: x400.

3.1.3 Expression of protein markers

The expression of protein markers was examined on implants that were kept for 7 and 14 days after surgery using RT-PCR (Fig. 3). The overall expression of protein markers was more remarkable in implants that were kept for 7 days than 14 days post-operatively, which is in agreement with the histologic results of stroma- like cells 7days after surgery.

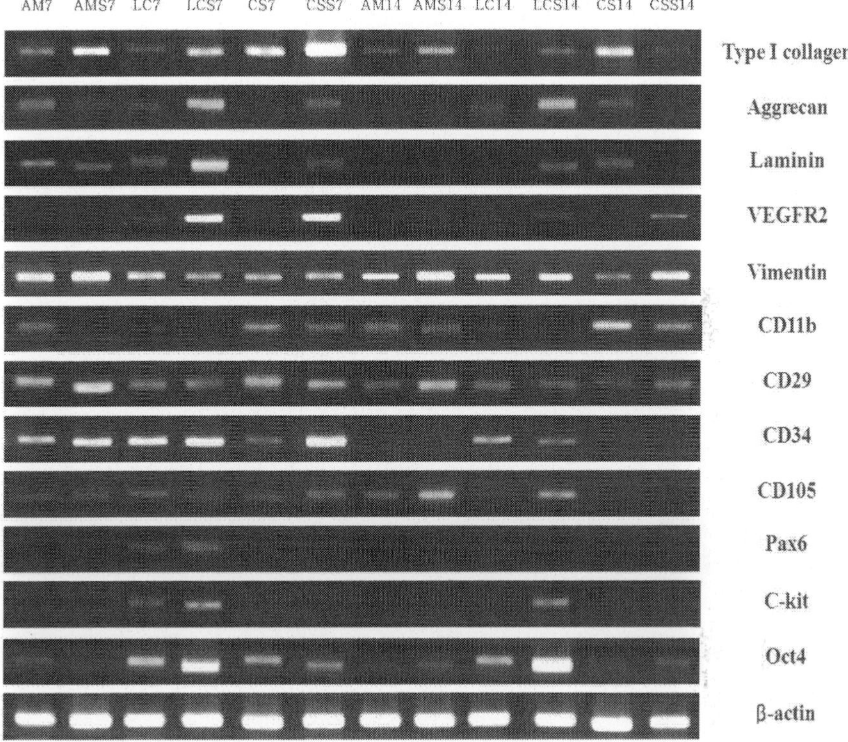

Fig. 3. Expression of various markers in subconjunctival-inserted implants.
Most markers were more highly expressed in the implants 7 days after insertion than implants 14 days after insertion. Also, stem cell markers (c-kit and VEGFR2) and protein markers (aggrecan and laminin) were more expressed in implants treated with substance-P. AM : Amniotic membrane, AMS : Amniotic membrane soaked in substance-P, LS : Lyophilized acellular porcine cornea, LSS : Lyophilized acellular porcine cornea soaked in substance-P, CS : Collagen sheet, CSS : Collagen sheet soaked in substance-P, 7 : Implant which was kept for 7days after subconjunctival insertion, 14 : Implant which was kept for 14days after subconjunctival insertion

The corneal protein markers, aggrecan and laminin, had significantly increased expression in lyophilized APCs, especially APCs treated with substance-P. Among the stem cell markers, CD34, CD29, and CD105 had a similar level of expression between each implant, but c-kit was only expressed in lyophilized APCs. The expression of c-kit and VEGFR2 was more prominent in the lyophilized APCs treated with substance-p. Increased expression of the transcription and differentiation genes (Oct4 and Pax6, respectively) was also observed in lyophilized APCs in comparison with AMs and CSs. The expression of Oct4 and Pax6 was also more distinct in lyophilized APCs treated with substance-P. The expression of type I collagen was higher in CSs compared with AMs or lyophilized APCs.

3.2 Biompatibility of lyophilized APCs treated with substance-P used in lamellar keratoplasty

Although corneal haziness was initially observed and aggravated around the graft in rabbits by 2 weeks, none of the lyophilized APCs showed signs of rejection or severe inflammation. Corneal opacity and neovascularization began to improve 3~4 weeks after surgery, and completely cleared after 6 weeks (Table 1, Fig. 4). The epithelium over the graft was beginning to heal 1 week after surgery, and was usually completed 6 weeks after surgery.

	Opacity scores	Neovacularization scores
1week	1.5±0.8†‡	1.4±0.5†‡
2weeks	2.8±1.0†‡	2.1±0.9†
3weeks	2.4±0.7†‡	2.8±0.7†‡
4weeks	1.4±0.7†‡	1.6±0.5†‡
5weeks	0.4±0.5‡	0.8±0.5†‡
6weeks	0.0±0.0	0.4±0.5‡
7weeks	0.0±0.0	0.0±0.0
8weeks	0.0±0.0	0.0±0.0

† $p < 0.05$ comparison with grade 0 (Mann-Whitney U test)
‡ $p < 0.05$ comparison with previous finding (Mann-Whitney U test)

Table 1. Comparison of corneal opacity and neovascularization in acellular porcine cornea group after surgery.

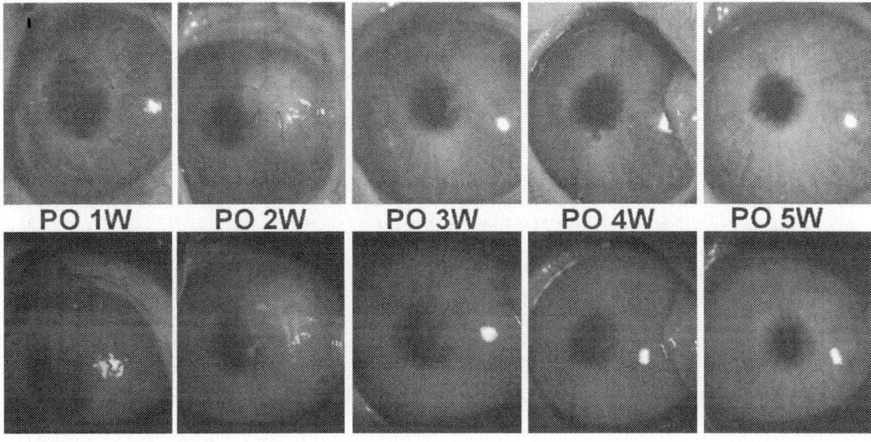

Fig. 4. Results of lamellar keratoplasty using lyophilized acellular porcine corneas.
Corneal opacities and neovascularization appeared 1 week post-operatively, which became
more severe 2 weeks after surgery. Opacity and neovascularization decreased 3~4 weeks
post-operatively, and corneas recovered transparency 5 weeks after surgery.

Based on H&E staining, corneal stromal cells were observed in lyophilized APCs with rare
infiltration of inflammatory cells 1 month after surgery. The lyophilized APCs were well-
integrated into the host tissues, showing indistinct borders with normal rabbit corneal
stroma on histology examination. These findings were more prominent 8 weeks after
surgery. Vimentin staining showed viable stromal cells in grafted lyophilized APCs and
recipient rabbit corneas, and a similar histologic pattern of grafted lyophilized APCs with
recipient rabbit corneas (Fig. 5).

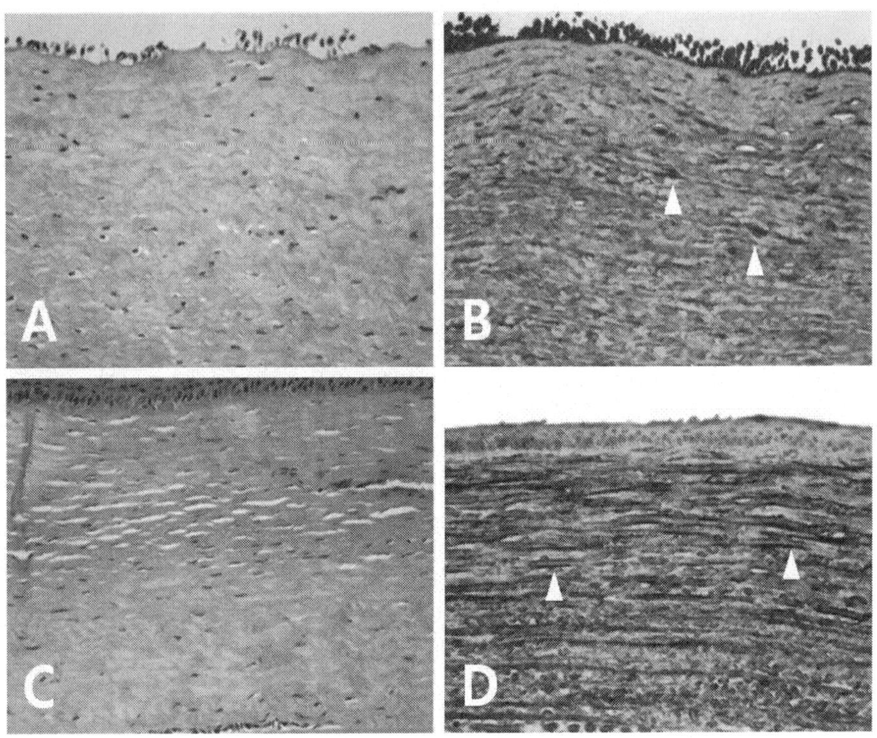

Fig. 5. Hematoxylin-eosin staining (A, C) and vimentin immune-histochemical staining (B, D) of grafted lyophilized acellular porcine cornea.
Corneal stromal cells presented in lyophilized APCs with rare infiltration of inflammatory cells 4 weeks after surgery (A). The lyophilized APCs were well-integrated into rabbit corneas with indistinct borders, and incomplete epithelization was shown. At 8 weeks, the number of corneal stromal cells increased and complete epithelization was observed (C). The staining aspects of vimentin in porcine corneas were very similar to the host cornea (B, D). (original magnification: x400)

4. Discussion

Porcine corneas have a well-organized structure similar to human corneas, and have now been extensively studied as a supplement to corneal collagen (Kampmeier et al., 2000; Xu et al., 2008). With respect to the use of porcine corneas, the most significant limitation is hyperacute or acute rejection, mainly due to xenogenic antigen (Li et al., 1992; Tseng et al., 2005). Several decellularization methods have been reported to decrease the antigenicity by removing stromal cells, but the ideal method for graft survival has not been determined

(Gilbert et al., 2006; Grauss et al, 2005; López-García et al, 2007). In the current study, we used physical methods to decellularize the porcine cornea by a freeze-thaw technique in combination with a centrifuge. In comparison with chemical methods destroying cells and collagen microstructures, this freeze-thaw technique has the advantage in eliminating stromal cells, while minimizing damage to the extracellular matrix (Gulati, 1988; Jackson et al., 1991). Lyophilization makes tissues less immunogenic by removing antigen-expressing cells and keeps the tissue sterilized for a longer period of time (Coombes et al., 2001; Pepose et al., 1991; Rostron et al., 1988; Zavala et al., 1985). We have previously reported the clinical and histological importance of lyophilization in producing less antigenicity and longer survival after xenotransplantation in rabbits (Lee at al., 2010).

Although decellularization is an essential process to minimize immune rejection, but transplantation of acellular lyophilized tissue alone might impede the healing process and increase the risk of infection, especially in immune-privileged tissues, such as the cornea. Efficient tissue regeneration usually needs repopulation of biologic decellularized scaffolds with interstitial cells (Lichtenberg et al., 2006). In the cornea, the role of corneal cells in optical transparency is already known. Accordingly, co-transplantation of acellular scaffolds with viable corneal cells appears to be important for graft survival and optical transparency in the cornea. In this study, we removed the stromal cells, and attempted *in vivo* cultivation rather than direct seeding of cells to increase graft survival and rapid acceptance. *In vivo* cultivation has been used for reconstruction of the ocular surface, and has advantage of maintaining corneal cell characteristics due to cell mitosis *in vivo* (Kim et al., 2008). Direct seeding of cells usually requires *ex vivo* cultivation, which is limited by facilities and equipment, and has a higher risk of infection.

In the present study, we showed that lyophilized APCs recellularized for 1 week had better optical transparency than APCs recellularized for different periods of time. The implants that were kept > 2 weeks had severe fibrosis adherent to the recipient tissue. This result indicates that implants in the limbus appear to serves as a medium for cultivation, and lacking the immune privilege of the eye, were ultimately rejected by the immune response. Nevertheless, it is remarkable that originally opaque APCs had better transparency after recellularization process for 1 week. Histologic examination demonstrated an abundance of stroma-like cells spreading into lyophilized APCs, which appear to be responsible for implants transparency. The expression of aggrecan and laminin in RT-PCR of lyophilized APCs implanted in the limbus of rabbits showed that cells grown in lyophilized APCs had the specific activity of keratocytes. Aggrecan is main glucoprotein of the sclera, and laminin is the main protein of the basement membrane, and thus has an important role in cell proliferation, migration, and adhesion (Doege et al, 1991; Dunlevy & Rada, 2004; Filenius et al., 2001; Yurchenko & Batton, 2009). These findings provide a plausible explanation as to why recellularized APCs in rabbit eyes showed better transparency, and supported the effectiveness of *in vivo* recellularization over the implants. Compared with lyophilized APCs, excessive infiltration of cells (primarily inflammatory cells) were noted on the bovine AMs and CSs. Multiple micropores on bovine AMs and CSs in comparison with the small surface area of lyophilized APC could be responsible for over-infiltration of cells.

Substance-P is known to stimulate migration of bone marrow stem cells, and accelerate the wound healing process (Hong et al., 2009). In this study, we attempted to evaluate the effect of substance-P treatment on implants, and found that expression of mesenchymal stem cell factors (CD 29 and CD105) was not affected by substance-P treatment. However, endothelial stem cell factors (VEGFR2 and c-kit, but not CD34) were highly expressed in lyophilized APCs treated with substance-P. In addition, the increased expression of transcription and differentiation genes (Oct4 and Pax6) and ocular specific proteins (aggrecan and laminin) was also noted in lyophilized APCs treated with substance-P. The transcription factor Oct4 is critical for self-renewal and maintenance of embryonic stem cells, which has a role in controlling cellular phenotype (Zhou et al., 2010). Pax6, located on the short arm of chromosome 11, is known to produce a protein which is very important in ocular development (Ton et al., 1991; Glaser et al., 1992). It is difficult to validate the specific role of mesenchymal and endothelial stem cells during the recellularization process in this study. It is possible that the expression of stem cell factors might contribute to rejection in coordination with higher cellularity and enhanced angiogenesis. Still, higher expression of genes imperative for corneal development in lyophilized APCs treated with substance-P implicated the substantiality and validity of substance-P treatment in xenografts.

To identify the usefulness of our methods in preparing APCs, we transplanted lyophilized APCs with cells grown *in vivo* in rabbit corneas using lamellar keratoplasty. Previous studies using APCs reported at least 8 weeks to accept porcine collagen, suggesting that rejection could be inhibited by removing antigenicity, but failed to decrease the inflammatory reaction caused by the graft itself (Xu et al., 2008; Lin et al., 2008). Our study showed that corneal opacity improved within 4 weeks after the grafting, indicating that corneal stromal cells surviving in APCs function immediately after transplantation. Also, the histologic examination showed well-integrated implants with indistinct borders, and viable corneal stroma-like cells in vimentin staining. It is interesting that corneal haziness becomes aggravated immediately after transplantation, then improved 3~4 weeks after transplantation. The improvement in corneal haziness was accompanied with epithelialization over the implants, indicating the importance of tight junctions in the epithelium that controls the flow of fluid into the cornea.

This study suggests that lyophilized APCs repopulated with recipient allograft cells might be a physiologically functional tissue substitute in xenotransplantation. The rabbit cornea recipients stayed clean after receiving a lyophilized APC. Although, additional experiments are required to clarify the role of substance-P and the recellularization process, we believe that our results might provide a valuable clinical input to tissue-engineered corneal scaffolds using porcine corneas to facilitate the rapid restoration of the ocular surface.

5. Conclusion

The increase of ocular surface disease and shortage of cornea donors need the tissue-engineered corneal equivalent. The lyophilized acellular pig corneal stroma, which is devoid of α-gal epitope, is less antigenic than fresh pig corneal stroma, and might be a useful alternative to corneal tissue. We previously demonstrated that lyophilized APCs survived longer than fresh porcine corneas in pig-to-rat model, but the delayed healing and the risk of infection caused by an acellular substrate graft in immune-privileged corneal tissue require another resolution. In the present study, we investigated the effectiveness of lyophilized APCs with cells grown *in vivo* and stimulated with substance-P. The results showed that lyophilized APCs repopulated with cells grown *in vivo* for 1 week had better optical transparency compared with controls. More infiltrated corneal stromal-like cells observed in lyophilized APC in comparison with controls in histology might explain the better optical transparency, and higher expression of stem cell markers (c-kit and VEGFR) and corneal protein markers (aggrecan and laminin) in lyophilized APCs repopulated with cells grown *in vivo* for 1 week support the importance of recellularization of graft before transplantation. These findings were more remarkable in lyophilized APC treated with substance-P, which implicated the possible role of substance-P in stromal cell maturation in cornea. Lamellar keratoplasty using lyophilized APC containing cells grown *in vivo* and stimulated with substance-P had good graft survival without rejection for 8 weeks. These results might provide a valuable clinical input to xenotransplantation using porcine cornea, and lyophilized APC with cells grown *in vivo* might be useful for ocular surface reconstruction.

6. Acknowledgement

This study was supported by a grant of the Korea Healthcare technology R&D Project, Ministry for Health, Welfare & Family Affairs, Republic of Korea. (A084721)

7. References

Alldredge, OC. & Krachmer, JH. (1981). Clinical types of corneal transplant rejection. Their manifestations, frequency, preoperative correlates and treatment. *Archives of Ophthalmology*, Vol.99, No.4, (April 1981), pp. 599-604, ISSN 0003-9950

Amano, S; Shimomura, N; Kaji, Y; Ishii, K; Yamagami, S & Araie, M. (2003). Antigenicity of porcine cornea as xenograft. *Current Eye Research*, Vol.26, No.6, (June 2003), pp. 313-318, ISSN 0271-3683

Auchincloss, H Jr. (1988). Xenogeneic transplantation. A review. *Transplantation*, Vol.46, No.1, (July 1988), pp. 1-20, ISSN 0041-1337

Chen, KH.; Azar, D. & Joyce, NC. (2001). Transplantation of adult human corneal endothelium ex vivo: a morphologic study. *Cornea*, Vol.20, No.7, (October 2001), pp. 731-737, ISSN 0277-3740

Collins, BH.; Cotterell, AH.; McCurry, KR.; Alvarado, CG.; Magee, JC.; Parker, W. & Platt, JL. (1995). Cardiac xenografts between primate species provide evidence for the importance of the alpha-galactosyl determinant in hyperacute rejection. *Journal of Immunology*, Vol.154, No.10, (May 1995), pp.5500-5510, ISSN 0022-1767

Coombes, AG.; Kirwan, JF. & Rostron, CK. (2001). Deep lamellar keratoplasty with lyophilised tissue in the management of keratoconus. *The British Journal of Ophthalmology*, Vol.85, No.7, (July 2001), pp. 788-791, ISSN 0007-1161

Doege, KJ; Sasaki, M; Kimura, T & Yamada, Y. (1991). Complete coding sequence and deduced primary structure of the human cartilage large aggregating proteoglycan, aggrecan. Human-specific repeats, and additional alternatively spliced forms. *The Journal of Biological Chemistry*, Vol. 266, No.2, (January 1991), pp. 894-902, ISSN 0021-9258

Dunlevy, JR & Rada, JA. (2004). Interaction of lumican with aggrecan in the aging human sclera. *Investigative Ophthalmology & Visual Science*, Vol.45, No.11, (November 2004), pp. 3849-3856, ISSN 0146-0404

Fantes, FE.; Hanna, KD.; Waring, GO 3rd.; Pouliquen, Y.; Thompson, KP. & Savoldelli, M. (1990). Wound healing after excimer laser keratomileusis (photorefractive keratectomy) in monkeys. *Archives of Ophthalmology*, Vol.108, No.5, (May 1990), pp. 665-675, ISSN 0003-9950

Filenius; S; Hormia, M; Rissanen, J; Burgeson, RE; Yamada, Y; Araki-Sasaki, K; Nakamura, M; Virtanen, I & Tervo, T. (2001). Laminin synthesis and the adhesion characteristics of immortalized human corneal epithelial cells to laminin isoforms. *Experimental Eye Research*, Vol.72, No.1, (January 2001), pp. 93-103, ISSN 0014-4835

Gilbert, TW; Sellaro, TL & Badylak, SF. (2006). Decellularization of tissues and organs. *Biomaterials*, Vol.27, No.19, (July, 2006), pp. 3675-3683, ISSN 0142-9612

Glaser, T.; Walton, DS. & Maas, RL. (1992). Genomic structure, evolutionary conservation and aniridia mutations in the human PAX6gene. *Nature genetics*, Vol.3, No.3, (November 1992), pp. 232-239, ISSN 1061-4036

Good, AH.; Cooper, DK.; Malcolm, AJ.; Ippolito, RM.; Koren, E.; Neethling, FA.; Ye, Y.; Zuhdi, N. & Lamontagne, LR. (1992). Identification of carbohydrate structures that bind human antiporcine antibodies: implications for discordant xenografting in humans. *Transplantation proceedings*, Vol.24, No.2, (April 1992), pp. 559-562, ISSN 0041-1345

Grauss, RW; Hazekamp, MG; Oppenhuizen, F; van Munsteren, CJ; Gittenberger-de Groot, AC & DeRuiter, MC. (2005). Histological evaluation of decellularised porcine aortic valves: matrix changes due to different decellularisation methods. *Europian Journal of Cardiothoracic Surgery*, Vol.27, No.4, (April 2005), pp. 566-571, ISSN 1010-7940

Griffith, M.; Osborne, R.; Munger, R.; Xiong, X.; Doillon, CJ.; Laycock, NL.; Hakim, M.; Song, Y. & Watsky, MA. (1999). Functional human corneal equivalents constructed from cell lines. *Science*, Vol.286, No.5447, (December 1999), pp. 2169-2172, ISSN 0193-4511

Gulati, AK. (1988). Evaluation of acellular and cellular nerve grafts in repair of rat peripheral nerve. *Journal of Neurosurgery*, Vol.68, No.1, (January 1988), pp. 117-123, ISSN 0022-3085

Hong, HS.; Lee, J.; Lee, E.; Kwon, YS.; Lee, E.; Ahn, W.; Jiang, MH.; Kim, JC. & Son, Y. (2009). A new role of substance P as an injury-inducible messenger for mobilization of CD29(+) stromal-like cells. *Nature Medicine*, Vol.15, No.4, (April 2009), pp. 425-435, ISSN 1078-8956

Insler, MS & Lopez, JG. (1991). Heterologous transplantation versus enhancement of human corneal endothelium. *Cornea*, Vol.10, No.2, (March 1991), pp. 136-148, ISSN 0277-3740

Jackson, DW.; Grood, ES.; Cohn, BT.; Arnoczky, SP.; Simon, TM. & Cummings, JF. (1991). The effects of in situ freezing on the anterior cruciate ligament. An experimental study in goats. *The Journal of Bone and Joint Surgery. American volume*, Vol.73, No.2, (February 1991), pp. 201-213, ISSN 0021-9355

Jeong, JH.; Chun, YS. & Kim, JC. (2009). The effects of a subtenoncapsular injection of bevacizumab for ocular surface disease with corneal neovascularization. *Journal of the Korean Ophthalmological Society*, Vol.50, No.10, (October 2009), pp. 1475-1482, ISSN 0378-6471

Kampmeier, J.; Radt, B.; Birngruber, R. & Brinkmann, R. (2000). Thermal and biomechanical parameters of porcine cornea. *Cornea*, Vol.19, No.3, (May 2000), pp. 355-363, ISSN 0277-3740

Kim, JT.; Chun, YS.; Song, KY. & Kim, JC. (2008). The effect of in vivo grown corneal epithelium transplantation on persistent epithelial defects with limbal stem cell deficiency. *Journal of Korean Medical Science*, Vol.23, No.3, (June 2008), pp. 502-508, ISSN 1011-8934

Konya, D.; Yildirim, O.; Kurtkaya, O.; Kiliç, K.; Black, PM.; Pamir, MN. & Kiliç, T. (2005). Testing the angiogenic potential of cerebrovascular malformations by use of a rat cornea model: usefulness and novel assessment of changes over time. *Neurosurgery*, Vol.56, No.6, (June 2005), pp. 1339-1345, ISSN 0148-396X

Lee, JK.; Ryu, YH.; Ahn, JI.; Kim, MK.; Lee, TS. & Kim, JC. (2010). The effect of lyophilization on graft acceptance in experimental xenotransplantation using porcine cornea. *Artificial Organs*, Vol.34, No.1, (January 2010), pp. 37-45, ISSN 0160-564X

Li, C; Xu, JT; Kong, FS & Li, JL. (1992). Experimental studies on penetrating heterokeratoplasty with human corneal grafts in monkey eyes. *Cornea*, Vol.11, No.1, (January 1992), pp. 66-72, ISSN 0277-3740

Lichtenberg, A.; Tudorache, I.; Cebotari, S.; Suprunov, M.; Tudorache, G.; Goerler, H.; Park.,JK.; Hilfiker-Kleiner, D.; Ringes-Lichtenberg, S.; Karck, M.; Brandes, G.; Hilfiker, A. & Haverich, A. (2006). Preclinical testing of tissue-engineered heart valves re-endothelialized under simulated physiological conditions. *Circulation*, Vol.114, No.1 suppl, (July 2006), pp. 1559-1565, ISSN 0009-7322

Lin, XC.; Hui, YN.; Wang, YS.; Meng, H.; Zhang, YJ. & Jin, Y. (2008). Lamellar keratoplasty with a graft of lyophilized acellular porcine corneal stroma in the rabbit. *Veterinary Ophthalmology*, Vol.11, No.2, (March 2008), pp. 61-66, ISSN 1463-5216

López-García, JS; Rivas Jara, L; García-Lozano, I & Murube, J. (2007). Histopathologic limbus evolution after alkaline burns. *Cornea*, Vol.26, No.9, (October 2007), pp. 1043-1048, ISSN 0277-3740

Meek, KM.; Leonard, DW.; Connon, CJ.; Dennis, S. & Khan, S. (2003). Transparency, swelling and scarring in the corneal stroma. *Eye (London, England)*, Vol.17, No.8, (November 2003), pp. 927-936, ISSN 0950-222X

Mourant, JR.; Canpolat, M.; Brocker, C.; Esponda-Ramos, O.; Johnson, TM.; Matanock, A.; Stetter, K. & Freyer, JP. (2000). Light scattering from cells: the contribution of the nucleus and the effects of proliferative status. *Journal of Biomedical Optics*, Vol.5, No.2, (April 2000), pp. 131-137, ISSN 1083-3668

Pepose, JS. & Benevento, WJ. (1991). Detection of HLA antigens in human epikeratophakia lenticules. *Cornea*, Vol.10, No.2, (March 1991), pp. 105-109, ISSN 0277-3740

Ross, JR; Howell, DN & Sanfilippo, FP. (1993). Characteristics of Corneal Xenograft Rejection in a Discordant Species Combination. *Investigative Ophthalmology & Visual Science*, Vol.34, No.8, (July 1993), pp. 2469-2476, ISSN 0146-0404

Rostron, CK.; Sandford-Smith, JH. & Morton, DB. (1988). Experimental epikeratophakia using tissue lathed at room temperature. *The British Journal of Ophthalmology*, Vol.72, No.5, (May 1988), pp. 354-360, ISSN 0007-1161

Sedlakova, K & Filipec, M. (2007). Effect of suturing technique on corneal xenograft survival. *Cornea*, Vol.26, No.9, (October 2007), pp. 1111-1114, ISSN 0277-3740

Ton, CC.; Hirvonen, H.; Miwa, H.; Weil, MM.; Monaghan, P.; Jordan, T.; van Heyningen, V.; Hastie, ND.; Meijers-Heijboer, H. & Drechsler, M. (1991). Positional cloning and characterization of a paired box- and homeobox-containing gene from the aniridia region. *Cell*, Vol.67, No.6, (December 1991), pp. 1059-1074, ISSN 0092-8674

Trinkaus-Randall, V.; Capecchi, J.; Newton, A.; Vadasz, A, Leibowitz, H. & Franzblau, C. (1988). Development of a biopolymeric keratoprosthetic material. Evaluation in vitro and in vivo. *Investigative Ophthalmology & Visual Science*, Vol.29, No.3, (March 1988), pp. 393-400, ISSN 0146-0404

Tseng, YL; Kuwaki, K; Dor, FJ; Shimizuk, A; Houser, S; Hisashi, Y; Yamada, K; Robson, SC; Awwad, M; Schuurman, HJ; Sachs, DH & Cooper, DK. Alpha1,3-Galactosyltransferase gene-knockout pig heart transplantation in baboons with survival approaching 6 months. *Transplantation*, Vol.80, No.10, (November 2005), pp. 1493-1500, ISSN 0041-1337

Xu, YG.; Xu, YS.; Huang, C.; Feng, Y.; Li, Y. & Wang, W. (2008). Development of a rabbit corneal equivalent using an acellular corneal matrix of a porcine substrate. *Molecular Vision*, Vol.14, pp. 2180-2189, ISSN 10900535

Yurchenko, P & Batton, BL. (2009). Developmental and pathogenic mechanisms of basement membrane assembly. *Current Pharmaceutical Design*, Vol.15, No.2, (February 2009), pp. 1277-1294, ISSN 1381-6128

Zavala, EY.; Binder, PS.;.Deg, JK. & Baumgartner, SD. (1985). Refractive keratoplasty: lathing and cryopreservation. *The CLAO Journal*, Vol.11, No.2, (April 1985), pp. 155-162, ISSN 0733-8902

Zhou, SY.; Zhang, C.; Baradaran, E. & Chuck, RS. (2010). Human corneal basal epithelial cells express an embryonic stem cell marker OCT4. Current Eye Research, Vol.35, No.11, (November 2010), pp. 978-985, ISSN 0271-3683

20

The Therapeutic Potential of Cell Encapsulation Technology for Drug Delivery in Neurological Disorders

Carlos Spuch and Carmen Navarro
University Hospital of Vigo, Department of Pathology and Neuropathology,Vigo,
Spain

1. Introduction

The brain can be damaged by a wide range of conditions including infections, hypoxia, poisoning, stroke, chronic degenerative disease and acute trauma. Some of the most problematic forms of brain damage are those associated with chronic neurodegenerative diseases or acute brain trauma as a result of contusive or penetrating injury. In these cases, damage results in the loss of specific populations of neurons and the development of defined psychiatric or neurological symptoms. Current treatments for these problems are designed to pharmacologically modify disease symptoms; however, no therapies are yet available that fully restores lost function or slow ongoing neurodegeneration in the brain. Many promising therapies with growth factors has been implicated in brain regeneration, repair and neuroprotection in the central nervous system produced interesting results, such as vascular endothelial growth factor (VEGF) (Spuch et al. 2010) brain derived growth factor (BDNF) (Malik et al., 2010) or nerve growth factor (NGF) (Sharma. 2010). However, the critical problem is the way to deliver, in a continuous and localized manner, and more important is to supply physiological amounts of growth factors into focus damage of the brain tissue. One interesting approach is cell encapsulation, in which engineered somatic cells are protected against immune cell mediated and antibody-mediated rejection through immobilization in a polymer matrix surrounded by a semipermeable membrane. The latter regulates the bidirectional diffusion of nutrients, allowing the controlled and continuous delivery of therapeutic proteins in the absent of immunosupression in the proper concentration and localization.

Encapsulated cells offer enormous potential for the treatment of human disease. Many attempts have been made to prevent the rejection of transplanted cells by the immune system. Cell encapsulation is promising machinery for cell transplantation and new materials and approaches were developed to encapsulate various types of cells to treat a wide range of diseases.

Cell microencapsulation holds promise for the treatment of many diseases by the continuous delivery of therapeutic products. The complexity of many neurological diseases needs the developing of new drugs and especially new pathways to deliver the drug at proper concentration and into the correct localization. One critical problem is the way to deliver, in a continuous and localized manner, physiological amounts of drugs. One promising technology is the developing of new biomaterial components with the capacity of envelope drugs, cells or tissues, being able to distribute the drug therapy, and, at the same time, to be isolated of immune system. The main goal of microcapsules technology is the capacity to release growth factors, peptides, proteins or hormones in a precise location and to keep isolated from immune system attack. This is the critical issue for the long-term efficacy of this biotechnology, due the core of microcapsules is made of cells or tissue that are able to regulate, by themselves, the release of the necessary drug at the implanted tissue, and at the same time, the grafts are isolated from the immune system.

The objective of this chapter is to summarize the recent investigations and news related with cell microencapsulation technology, and the possible therapeutic applications of growth factors-secreting cells on brain impairment in different neurological disorders and brain tumours. We will comment our investigations related last publications and patents in cell microencapsulation and our findings confirming the evidence of a potential therapeutic benefit of growth factors therapy in neuroprotection with VEGF and the last results with BDNF microcapsules implants such as therapeutic value in the treatment and prevention of brain damage.

2. Overview

Each year over 10 million people globally suffer from neurodegenerative diseases. This figure is expected to grow by 20% over the next decade as the aging population increases and lives longer. It is the fourth biggest killer in the developed world after heart, cancer and stroke. There are millions of sufferers worldwide, and can occur at any age but it is more common among the elderly. Many similarities appear which relate these diseases to one another on a sub-cellular level. Discovering these similarities offers hope for therapeutic advances that could ameliorate many diseases simultaneously. Gene defects play a major role in the pathogenesis of degenerative disorders of the nervous system; however a feature observed in the most common neurodegenerative disorders is the dichotomy between familial forms and seemingly non-familial or sporadic forms characterized by the persistent and progressive loss of neuronal subtypes.

The most common neurodegenerative diseases are Alzheimer disease, Parkinson disease, Lewy body dementia, frontotemporal dementia, amyotrophic lateral sclerosis, Huntington disease, and prion diseases. The most widely recognized are Alzheimer's disease and Parkinson's disease, which are among the principal debilitating conditions of the current century. Approximately 24 million people worldwide suffer from dementia, of which 60% is due to Alzheimer's disease occurs in 1% of individuals aged 50 to 70 years old and dramatically increases to 50% of those over 70 years old [Ferri et al, 2005). Alzheimer's disease is typified clinically by learning and memory impairment and pathologically by gross cerebral atrophy, indicative of neuronal loss, with numerous extracellular neuritic amyloid plaques and intracellular neurofibrillary tangles found predominantly in the frontal and temporal lobes, including the hippocampus (Morgan, 2011). The number of cases of dementia was studied by European Community Concerted Action on the Epidemiology and Prevention of Dementia group (EURODEM). The total cost of illness of dementia in the European Union27 in 2008 was estimated to be EUR 160 billions, EUR 22.000 per person with dementia per year. This study also estimated in the United Kingdom that 683.597 people suffered from dementia in 2005, with the total forecasted to increase to 940.110 by 2021 and 1.735.087 by 2051. The economists of United Kingdom calculate in £ billion in care costs and lost of productivity. This terrible estimation will be similar in the rest of European countries (Brayne et al, 2011 and Virues-Ortega, 2011). Only in United States Alzheimer's disease is the sixth leading cause of all death and is the fifth leading cause of death in Americans aged with more than 65 years-old. (Alzheimer's association, 2011). Although other major causes of death have been decrease, deaths because of dementia have been rising dramatically, and the worst is that the true social impact is incalculable.

Few cases of dementia are diagnosed in early stages, as many of the associated symptoms, e.g. memory loss, could be attributed to other conditions such as depression, diabetes, thyroid abnormalities, delirium, alcoholism or simple ageing. This makes diagnosis particularly difficult, such that it may take up to one year or longer for a final diagnosis to be made. Formal testing for dementia requires mental ability tests, such as the Mini Mental State Examination (MMSE), a review of medical history and current medications, an examination of biological markers such as levels of abnormal proteins associated with different dementias, and sometimes imaging scans such as magnetic resonance imaging (MRI) scan to detect changes in the brain. Current treatment options for dementias leave much to be desired. Existing medications, which either prevent the breakdown of neurotransmitters or modulate key receptors in the brain, can temporarily ease some of the cognitive decline associated with the disease, but they do nothing to halt or reverse its progression. And although scientists are developing new therapeutics that target the cause of the different dementias more directly, even these latest experimental drugs might do little to help patients. To make headway, some neuroscientists and neurologist experts now argue that the research community must fundamentally change how it diagnoses the disease and designs clinical trials.

There is still no cost-effective method of identifying people with dementia through population screening. Early diagnosis of dementia is important, allowing those with dementia and their carers to plan better for their future and to start treatments that may slow disease symptoms. Nowadays, pharmaceutical agents that are used to treat brain disorders are usually administered orally, such as donepezil, memantine, rivastigmine, galantamine and tacrine for Alzheimer's disease (Pasic et al, 2011); or levodopa, entacapone, pramipexole and ropinerole for Parkinson´s disease (Morley & Hurtig, 2010). However, most of the ingested drugs does not target the brain in full conditions and is, instead, metabolized totally or partially by the liver. This inefficient utilization of drug may require ingestions of higher drug concentrations that can produce toxic effects such as cardiotoxicity, hepatotoxicity and nephrotoxicity. Also, many therapeutic agents are poorly soluble or insoluble in aqueous solutions. These drugs provide challenges to delivering them orally or parentally, however these compounds can have significant benefits when formulated through nanoparticles or microcapsules technology. More efficient use of the drug can be realized both by eliminating liver metabolism and directly targeting the brain.

Notwithstanding these difficulties, the nanotechnology may provide a solution to overcome the diagnostic and neurotherapeutic challenges for neurodegenerative and neurological diseases. Nanotechnology employs engineered materials or devices with the smallest functional organization on the nanometre scale (1–100 nm) that are able to interact with biological systems at the molecular level. Nanoparticles are able to penetrate the blood brain barrier of *in vitro* and *in vivo* models. Nanotechnology can therefore be used to develop diagnostic tools as well as nano-enabled delivery systems that can bypass the blood brain barrier in order to facilitate conventional and novel neurotherapeutic interventions such as drug therapy, gene therapy, and tissue regeneration. Nanotechnology is currently being used to refine the discovery of biomarkers, molecular diagnostics, drug discovery, and drug delivery, which could be applicable to the management of serious neurodegenerative diseases such as Alzheimer's disease or Parkinson's disease.

The other great promise in the treatment of neurodegenerative diseases is the microencapsulation of cell types secreting bioactive substances locally in the damage area of the brain. The main handicap for the delivery of potentially therapeutic drugs to the brain is hindered by the different blood brain barriers, which restricts the diffusion of drugs from the vasculature to the brain parenchyma. One means of overcoming the blood brain barrier is with cellular implants that produce and deliver therapeutic molecules. Polymer encapsulation, or immunoisolation, provides a means of overcoming the blood brain barrier to deliver therapeutic molecules directly into the central nervous system region of interest. Immunoisolation is based on the observation that xenogeneic cells can be protected from host rejection by encapsulating, or surrounding, them within an immunoisolatory, semi permeable membrane. Cells can be enclosed within a selective, semi permeable membrane barrier that admits oxygen and required nutrients and releases bioactive cell secretions, but restricts passage of larger cytotoxic agents from the host immune defence system. The selective membrane eliminates the need for chronic immunosuppression of the host and allows the implanted cells to be obtained from nonhuman sources.

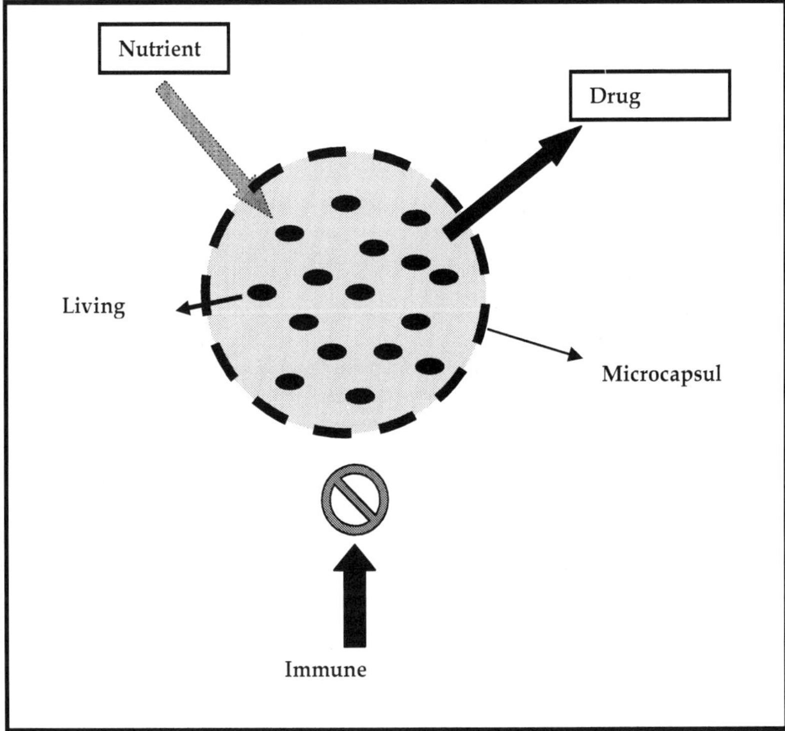

Fig. 1. Scheme of microcapsules containing living cells.

Likewise, a detailed understanding of their genetic and molecular basis will be essential for the development of effective strategies aimed at the early prediction and early prevention/treatment of these devastating diseases. In this review, microencapsulation technology for treating CNS diseases is updated from considerations of device configurations, membrane manufacturing and characterization in different preclinical models of neurodegenerative diseases.

3. Nanotechnology

The most promising aspect of pharmaceuticals and medicine as it relates to nanotechnology is currently drug delivery. Nanotechnology will play a key role developing new diagnostic and therapeutic tools. Nanotechnologies use engineered materials with the smallest functional organization on the nanometre scale in at least one dimension. Some aspects of the material can be manipulated resulting in new functional properties.

Nanoparticles hold tremendous potential as an effective drug delivery system. In this chapter we discuss recent developments in nanotechnology and especially in microencapsulation cell technology for drug delivery, image diagnostics and new therapeutic treatments. To overcome the problems of gene and drug delivery, nanotechnology has gained interest in recent years. Nanosystems with different compositions and biological properties have been extensively investigated for drug and gene delivery applications. To achieve efficient drug delivery it is important to understand the interactions of nanomaterials with the biological environment, targeting cell-surface receptors, drug release, multiple drug administration, stability of therapeutic agents and molecular mechanisms of cell signalling involved in pathobiology of the disease under consideration. Several anti-cancer drugs including paclitaxel, doxorubicin, 5-fluorouracil and dexamethasone have been successfully formulated using nanomaterials. Quantom dots, chitosan, Polylactic/glycolic acid (PLGA) and PLGA-based nanoparticles have also been used for *in vitro* RNAi delivery. Brain cancer, neurological and neurodegenerative diseases are one of the most difficult diseases to detect and treat, mainly because of the difficulty in getting imaging and therapeutic agents past the blood-brain barrier and into the brain. Anti-cancer drugs such as loperamide and doxorubicin bound to nanomaterials have been shown to cross the intact blood-brain barrier and released at therapeutic concentrations in the brain.

4. Process of microencapsulation

The process of microencapsulation has been used for many years, with the introduction of many diverse applications of the basic principle. Microencapsulation is a process by which very tiny droplets or particles of liquid or solid material are surrounded or coated with a continuous film of polymeric material. Most microcapsules have diameters between a few micrometers and a few millimetres. The idea was based on natural subcellular organelles which contain proteins, growth factors or enzymes. The first suggestion was made in 1964 about that encapsulation could be used to replace cells or cell products lost due to genetic defects (Chang, 1964). Although, microencapsulation as a viable procedure to immunoisolate cells for transplantation was introduced more than twenty years ago (Lim & Sum, 1980), it has had a slow progress towards clinical application. Microencapsulation technology holds promise methodology for the treatment of many diseases by the continuous delivery of therapeutic products, for example, due to slow production rates and the appearance of fibrotic overgrowths around the capsules, which can result in endotoxin contamination, e.g., oxygen and nutrient deprivation of the enclosed cells. However, when we talk about how to apply this promising technology in the treatment of neurological and a neurodegenerative disease presents a major problem, how to introduce this therapeutic agent into the brain in a safe an undamaged manner.

The biocompatibility of the microcapsules and their biomaterials components are a critical issue for the long-term efficacy of this technology. Microcapsules are polymers that can carry multiple therapeutic agents such as cells, drugs or proteins. The method of producing and designing these polymers is one of the most important steps in the developing of microencapsulated pharmaceutical products. Polymers are usually formulated using hydrophobic synthetic polymers and copolymers such as PLA and PLGA, polyacrylates and polycaprolactones or natural polymers such as albumin, gelatine, alginate, collagen and chitosan (Orive et al, 2003a). PLA and PGLA are the biomaterials most investigated for microencapsulation for drug delivery (Jain, 2000 and Langer, 1997). The alginate composition and purification, the selection of the polycation, the interactions between the alginates and the polycation, the microcapsule fabrication process, the uniformity of the devices and the implantation procedure are key factors for the correct developing of biocompatible microcapsules.

Microcapsules comprise a wall which surrounds an encapsulated material. Microencapsulation includes bioencapsulation which is more restricted to the entrapment of a biologically active substance generally to improve its performance or enhance its shelf life. Chemical and physical methods for cell immobilization are fairly diverse, and a large number of systems have been created that entrap catalytically active cells in various matrices, such as carrageenan, alginate, polyacrylamide and polyethylene glycol gels, as well as polyurethane foams (Orive et al, 2009). A novel polymeric methodology for microencapsulation based on immobilized living cells has been developed and now is available for different applications. This methodology is based on the formation of gel structures during the process of cooling and subsequent freezing of cell suspensions in polymer solutions.

The wall of the microcapsules should protect the material from external environment and is mainly made by polymers. The polymer used in the encapsulation art has to be biocompatible and biodegradable polymer (Mulqueen et al, 2010). The molecular weight of a polymer is important because influences the biodegradation rate although any desired molecular weight can be used, depending of the properties for the microparticles desired. In certain aspects a high strength polymer is need it to meet strength requirements, or low molecular weight polymers when resorption time is priority. For a diffusional mechanism of bioactive agent release, the polymer should remain intact until all of the drug is released from the polymer and then degrade. In some aspects, the time interval to degrade the polymer within a wanted time can be from about less than one day to months or years. Also, the selection of the polymer can influence the desired lapse time after the implantation. For example, a recent patent disclose a microencapsulation process using epoxy resin is able to change this properties (Cen et al, 2010). An improvement in this process of generation of new polymers was described by a new patent, where new emulsions and solvents it was generated better walls for the microcapsules (Raiche et al, 2010).

After designing the right biodegradable polymers, microencapsulation has permitted controlled release delivery systems. These revolutionary systems allow controlling the rate, duration and distribution of the active drug. With these systems, microparticles sensitive to the biological environment are designed to deliver an active drug in a site-specific way. One of the main advantages of such systems is to protect sensitive drug from drastic environment and to reduce the number of drug administrations for patient (Spuch & Navarro, 2010).

5. Cell microencapsulation implants to treat brain diseases

Drugs agents that are used to treat central nervous system are usually administered orally. However, the penetration of drug into brain decrease exponentially with the distance from cerebrospinal fluid surface, it is necessary to administer high concentrations of drug into cerebrospinal fluid compartment. The ependymal surface is exposed to very high drug concentration, which can have toxic side effects. The treatments with growth factors are very promising; however there are side effects due to the route of administration. It was demonstrated the beneficial effects of different growth factors for the treatment of various brain diseases. For example, it was published the beneficial effects of IGF-I (insulin growth factor-I) associated to a significant increase in brain amyloid-beta complexed to protein carriers such as albumin, apolipoprotein J or transthyrretin, supporting a therapeutic use of IGF-I in neurodegenerative diseases (Carro et al, 2006a, 2006b). Moreover, it was showed the intracerebroventricular (icv) administration of NGF (nerve growth factor) resulted in axonal sprouting and Schwann cell hyperplasia on the ependymal or arachnoids surface (Day-Lollini et al, 1997). The icv administration of fibroblast growth factor (FGF)-2 results in periventricular astrogliosis (Yamada et al, 1991). Also, the icv administration of glial-derived neurotrophic factor (GDNF) in Parkinson's disease, resulted in no penetration of the GDNF into substance nigra or into the caudate putamen nucleus, but did result in high incidence of adverse events (Nutt et al, 2003). Conversely, it is sometime desirable to deliver high concentrations of drug to the meningeal surface of the brain, such as in the treatment of meningeal infiltration of leukemia cells, and this can be achieved with intrathecal drug administration through microcapsules implants.

However, most of these drugs does not target the brain because does not across the blood brain barrier. This inefficient utilization of drug can produce toxic effects in other organs. More efficient use of the drug can be realized both by elimination liver metabolism and directly targeting the brain. Based on these premises, the promising features of microencapsulation technology belong to the direct administration of microencapsulated drug into the brain (Dou et al, 2006).

The microcapsules to be an implantable drug delivery device has to contain a carrier fluid the will dissolve the drug when freed from the capsule, a drug releaser for freeing the microencapsulated drug from the capsule, a reservoir in which the carrier fluid dissolve the drug. An example of the first problems in the microencapsulation technology was the therapeutic use of NGF in the prevention and treatment of many neurodegenerative diseases, such as Alzheimer's disease, degeneration of cholinergic neurons or the natural effects of aging in the brain. The first evidences showed that icv administration of NGF directly or with osmotic pumps prevented the degeneration of these neurons (Cleland et al 2007); however NGF infusion into the brain may be complicated due to stability and degradation in some implants (Schecterson & Bothwell, 2010). The first improvements were made with the NGF microencapsulation, increasing the stability and controlling the release of recombinant NGF, however this type of microcapsules was not enough.

There are various potential micro-implants employed for the treatment of neurological and neurodegenerative disorders, such as polymeric nanoparticles, microcapsules, hydrogels, or liposomes.

5.1 Nanoparticles

The advent of nanotechnology can provide a solution to overcome the future diagnostic and new neurotherapeutic challenges for neurodegenerative diseases as Alzheimer's disease and Parkinson's disease. This technology employs engineered materials with the smallest functional organization on the nanometre scale that are able to interact with biological systems at the molecular level. Nanoparticles are able to penetrate the blood brain barrier of *in vitro* and *in vivo* models disrupting the temporally the barrier and allowing the incorporation the therapeutic agents into the brain (Rempe et al, 2011). One interesting pathway to reach introduce drugs with nanoparticles into the brain can be with previous phagocytise using immune cells that are able to across the blood brain barrier. For example, there is an invention where involves delivering a drug by using macrophages present in the patient's cerebrospinal fluid that are capable to reaching the brain transporting the drug. This particular mode of delivery utilizing macrophages needs of previous uptake by the macrophages of nanoparticles loaded with the drug. The great advantage of this methodology is that these cells are not limited to macrophages, it is possible the use of monocytes, granulocytes, neutrophils, basophils and eosinophils.

Major future and promising uses of nanoparticles can therefore be to develop diagnostic tools. For example, amyloid plaques are one of the pathological hallmarks of Alzheimer's disease, the visualization of amyloid plaques in the brain is important to monitor the progression of this disease and to evaluate the efficacy of therapeutic interventions. Recently, many groups are developing new contrast agents to detect amyloid plaques *in vivo* using ultrasmall superparamagnetic iron oxide nanoparticles, chemically coupled with amyloid-beta (1-42) peptide to detect amyloid deposition (Yang et al, 2011). Further, nanoparticles are currently being used to refine the discovery of biomarkers and molecular diagnostics, which could be applicable to the management of neurodegenerative and neurological diseases (Sahni et al, 2010). Current pharmacotherapies for neurodegenerative diseases that have been successfully encapsulated in nanoparticles are polyphenolic compounds, (EGCG, apolipoprotein E containing curcumin or resveratrol), hormones (estradiol, melatonin, vasoactive intestinal peptide) and amyloid targeted drugs (thioflavin-T and S, coenzyme Q10, amyloid or gold).

5.2 Microcapsules

The technology of cell microencapsulation represents a strategy in which cells that secrete therapeutic products are immobilized and immunoprotected within polymeric and biocompatible devices (Orive et al, 2003b). One of the main advantages of cell microencapsulation is for the treatment of neurological disorders, where some drugs have potential therapeutic possibilities, such as growth factors or peptides, however only at low and constant concentrations. These microcapsules implants are able to secrete only the drug required by the damaged tissue, because the implants with microencapsulated cell are formed by live cells. These immobilized live cells that over-express the drug are able to regulate themselves for the endeavour. One potential impact of this drug delivery approach is that administration of immunosuppressants and implementation of strict immunosuppressive protocols can be reduced or eliminated, therefore the serious risks associated with these drugs can also be avoided.

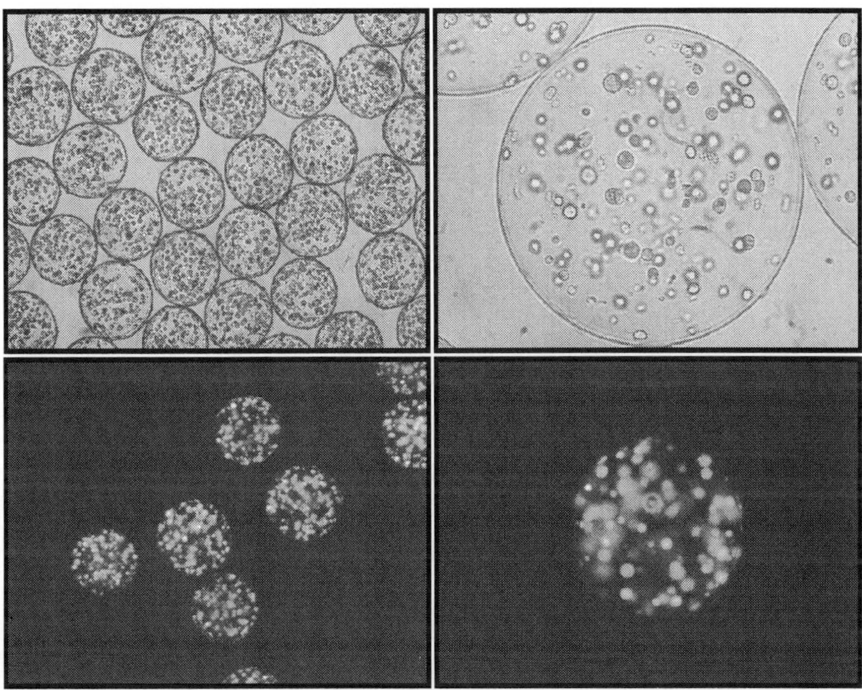

Fig. 2. Light microscopy (upper) and fluorescence microscopy (bottom) of encapsulated cells.

Based on this concept, a wide spectrum of cells and tissues may be immobilized, enhancing the potential applicability of this strategy to the treatment of numerous diseases. The therapeutic use of immortalized lines appropriately modified is employed as a medicinal product in different microcapsules for future treatments in brain disease. Last year, our group has patented one methodology to treat neurodegenerative diseases based in the cell microencapsulation of VEGF (vascular endothelial growth factor) overexpression cells. One goal of theses findings was to develop the therapeutic methodology without alter the blood brain barrier and to reduce the damage of the brain increasing the vascularity at cortical levels and reducing the amyloid-beta deposits (Spuch et al, 2010). Other example using encapsulation procedures are the case of microencapsulation of PC12 cells (dopaminergic cell line) and also embryonic grafts of dopaminergic cells were able to ameliorate behaviours in rat and primates after the implants these microcapsules in experimental parkinsonian models (Cherksey et al, 1996). Recently, based on a previous patent for the treatment of brain tumours with microcapsules, it was published a promising therapy against

Alzheimer's disease with the local and long term administration of CNTF (ciliary neurotrophic factor) using recombinant cells encapsulated with alginate secreting this neurotrophin factor (Orive et al, 2010 & Keunen et al, 2011).

In recent years microencapsulation technology and gene therapy was combined to be use as new therapy to deliver specific substances to target cell in brain tumours. The treatment of brain tumours represents one of the most challenges in oncology. Many anti-cancer drugs developed the last years did not provide effects in brain tumours due to impossibility of cross the blood brain barrier. In particular, the developing of microcapsules loaded with anti-cancer drug implanted into brain allows the treatment of tumours directly in the origin of target cells. Some years ago, it was patented a new system for therapy of malignant brain tumours (Keunen et al, 2011). This methodology used alginate-encapsulated H528 cells releasing antibodies stabilized potentially inhibit a heterogeneous glioma cell population. These microcapsules were implanted into brain and after slow and controlled distribution within all cerebrospinal fluid compartments of the antibodies during 9 weeks, the glioma were significantly reduced. This example can be apply for other potential anti-cancer drugs combined with different producer cells increasing the specificity of the treatment and the potential delivery system for specific brain tumours (Thorsen et al, 2000). Further, new biomaterials are playing an increasingly important role in developing more effective brain tumour treatments. This new biomaterials can also serve as targeted delivery devices for novel therapies including gene therapy, photodynamic therapy, anti-angiogenic and thermotherapy playing key roles in the diagnosis and imaging of brain tumours by revolutionizing both preoperative and intraoperative brain tumour detection.

Monoclonal antibodies have been envisioned as useful agents for human therapeutic and diagnostic applications *in vivo*. Recent results from human clinical trials suggest that this potential is becoming a reality. Attention is now shifting to the development of methods to produce monoclonal antibodies of a quality acceptable for widespread human use and in sufficient quantity to be a commercially viable product. Microencapsulation technology has been demonstrated to be suited to the large-scale production of both human and murine monoclonal antibodies of high purity and activity, for use in applications *in vivo*. It was previously comment the possibility of encapsulate antibodies for the treatment of brain tumours. The same technology using anti-VE-cadherin monoclonal antibodies allowed open a new alternative for the inhibition of angiogenesis and demonstrates the feasibility of using microencapsulated cells as a control-drug delivery system (Orive et al, 2001). Also, it was recently patented the use of human IgM antibodies encapsulated in alginate with demonstrated activity in the treatment of demyelinating diseases as well as other diseases of the central nervous system that are of viral, bacterial or idiopathic origin, including neural dysfunction caused by spinal cord injury (Rodriguez et al, 2009).

Currently, there is ongoing several clinical trial where it is implicated the microencapsulation technology. There is an interesting clinical trial to treat Parkinson's disease with the product named Spheramine. This product, developed by Titan Pharmaceuticals, is currently under safety and efficacy study. This product consists on cultured human retinal pigment epithelial cells on microcarriers. These microcarriers are implanted stereotaxically into both hemispheres of Parkinson's disease patients, and will be evaluated during 24 months (NCT00206687 & NCT00761436). Another two clinical trials are the work from Neurotech Pharmaceuticals to look at the safety and effectiveness of CNTF implants on vision in participants with atrophic macular degeneration (NCT00447954) and retinitis pigmentosa (NCT00447980). The implant is a small capsule that contains human retinal pigment epithelium cells. These cells have been given the ability to make CNTF and release it through the capsule membrane into the surrounding fluid.

5.3 Hydrogels

The microencapsulation technique might to solve different problems with the implantation in several tissues. The brain sometime does not allow the implantation of structures into the brain because there is not enough space. However, one solution to this problem can be the use of hydrogels. In this case, cell clusters are immobilized in hydrogel microspheres. Typically, the semipermeable membranes formed at the microsphere surface, the most common chemical system of the capsule membrane is by ionic or hydrogen bonds between two weak polylectrolytes such as acidic polysaccharides (alginic acid) and cationic polysaccharides (poly-lysine) (Bronich et al, 2006 & Bontha et al, 2006). The entrapment of the cells is obtained by the gelation. These types of microcapsules were developing for new therapies to treat diabetes. However, this technology is more promising for neurological and neurodegenerative diseases. Recently, it was published that the administration of VEGF as a potential neuroprotective strategy following cerebral stroke (Emerich et al, 2010) or Alzheimer's disease (Spuch et al, 2010). VEGF has a short half time life and limited access to the brain parenchyma following systemic administration. Previously, we commented the administration of VEGF in cell microcapsules implants, now we describes the incorporation of VEGF into a sustained release hydrogel delivery system located directly to the site of infarction.

5.4 Liposomes

Liposomes are vesicular structures composed of uni- or multi-lamellar lipid bilayer surrounding internal aqueous compartments. The main advantage of these structures is the relatively large quantities of drug can be incorporated into compartment. However, liposomes structures present various problems related to administration pathway. Orally administration is difficult related to the low pH of the stomach and the presence of bile salts tends to destabilize the liposome complex. Also, liposomes are highly susceptible to destruction via uptake by reticulo-endothelial system of the macrophages. A way of protecting liposomes was studied increasing stable bilayers and regulating the release profile of the liposome.

Between all the applications of this technology, the developing of suitable liposomal carrier to encapsulate neuroactive compounds is very promising. These liposomes are stable enough to carry them to the brain across the blood brain barrier with the appropriate surface characteristics for an effective targeting and for an active membrane transport. It was described the formulation of liposomes with monosialoganglioside allowing the brain uptake of these liposomes and of course making then good candidates as drug delivery system to the brain (Mora et al, 2002).

A novel liposome delivery system was developed for directed transport into olfactory epithelium cells with polyethylene glycol (PEG)ylated (stealth) immunoliposomes directed against human gliofibrillary acidic protein (GFAP). The handicap of theses liposomes are being incapable of penetrating the unimpaired blood brain barrier, nevertheless, may be useful in delivering drugs to glial brain tumours (which continue to express GFAP) or to other pathological loci in the brain with a partially disintegrated blood brain barrier (Chekhonin et al, 2005 & Chekhonin et al, 2008). Furthermore, this transport system mediating liposomes holds promise for the delivery of bioactive substances to olfactory epithelial cells and modulation of their capacity to stimulate axonal regeneration.

Microencapsulation technology has proven to have great potential for providing neuro-therapeutic modalities to limit and reverse the neuropathology of neurodegenerative diseases. Based on this concept our group is working in the establishment of new synapses in the peripheral and central nervous system activating the long-distance retrograde neurotrophin signalling. It is well known that target derived NGF is necessary and sufficient for formation of post-synaptic specialization on dendrites of sympathetic neurons. Based on this concept, we are working with microencapsulation technology to releases NGF with implants outside of brain and induce the formation of new synapses in brain region damaged by ictus and neurodegenerative diseases.

5.5 Exosomes

Recently it has developed a new method for delivering complex drugs directly to the brain. This new way to get effective drugs from blood to the brain it is the possibility to overcome the obstacle utilizing exosomes. New studies demonstrated that exosomes are tiny particles produced naturally by the body and the last investigations adapted them to deliver a gene therapy.

Exosomes are small capsules that are produced by most cells in the body in varying amounts. These natural nanoparticles are thought to be one of the ways cells communicate with each other and the body's immune system. When exosomes break off from the outer walls of cells, they can take various cellular signals and genetic material with them, transporting this material between different cells. The exosomes, injected into the blood, are able to ferry a drug across the normally impermeable blood-brain barrier to the brain where it is needed. Lately, it was published that exosomes endogenous vesicles that transport RNAs and proteins can deliver short interfering RNA to the brain mice. The therapeutic potential of exosome mediated short interfering RNA delivery was demonstrated by the strong messenger RNA and protein knockdown of BACE1, a therapeutic target in Alzheimer's disease, in wild-type mice (Alvarez-Erviti et al, 2011).

Also, it was report that exosomes can deliver anti-inflammatory agents, such as curcumin, to activated myeloid cells in vivo to treat inflammation-related autoimmune/inflammatory diseases and cancers. The specificity of using exosomes as a drug carrier creates opportunities for treatments of many inflammation-related diseases without significant side effects due to innocent bystander or off-target effects (Sun et al, 2010).

6. Conclusions

The natural barrier exists to protect the brain, preventing bacteria from crossing over from the blood, while letting oxygen through. However, this has also produced problems for medicine, as drugs can also be blocked. Currently, less than 5% of drugs are able to cross the barrier; one example is temozolomide, which is the only chemotherapy available for treating brain tumours such as glioblastoma multiforme and progressive anaplastic astrocytoma. These tumours have a poor prognosis and continue to grow, even after treatment with temozolomide. Notwithstanding these difficulties, we expect that cell microencapsulation technology will have a key role in the systemic application of new drugs, such as, growth factors, peptides or hormones. Recent advances in the field have provided novel drugs to fight against neurological or metabolic disorders. Nowadays is impossible to treat correctly many diseases mainly for the localization of damaged tissue or the complexity of tissue affected. The complexity of the disease, and many times, the localization of the tissue damage, difficult the possible treatment, for example, the brain are isolated by the blood brain barrier. Likewise, it is very important a detailed understanding of their genetic basis. Future genetic and cellular studies will be essential for the development of effective strategies aimed at the early prediction and early prevention/treatment of these devastating diseases. Therefore, new therapies for these hard to treat brain diseases are needed urgently alongside brain malfunctions such as Alzheimer's, Parkinson's and more.

Nanoparticles technology is currently being used to refine the discovery of biomarkers, molecular diagnostics, drug discovery, and cell microencapsulation technology for drug delivery, which could be applicable to the management of neurodegenerative diseases. It well demonstrated that the application of neurotrophic factors is able to modulate neuronal survival and synaptic connectivity and it is a promising therapeutic approach for many neurodegenerative diseases. However, it is very difficult to ensure long-term administration into the brain, this technology allow us to use recombinant cells secreting different neurotrophic factors encapsulated in alginate polymers. The implantation of these bioreactors in the damage region of the brain is another handicap to be solved; the correct implantation is associated with the robust improvement of cognitive performances.

One of the most important handicaps for the clinical application of cell microencapsulation technology may represent an improvement of the monitoring and biosafety of encapsulated cells. A significant breakthrough to overcome these problems associated with cell therapy was solved recently by Pedraz's group. They are developing a promising method where demonstrated the simultaneous monitoring and pharmacological control of myoblasts-containing alginate microcapsules. They introduced in the cells the SFG(NES)TGL triple reporter retroviral vector, which contains green fluorescence protein (GFP), firefly luciferase and herpes simplex virus type 1 thymidine-kinase (HSV1-TK). With this reporter they are able to follow up by luminometry if the cell is alive. Also, the treatment with the thymidine-kinase substrate ganciclovir caused death of microencapsulated myoblasts. Hence, they conclude that incorporation of the SFG(NES)TGL vector into microencapsulated cells represents an accurate tool for controlling cell location and viability in a non-invasive way. Moreover, cell death can be induced by administration of ganciclovir, in case therapy needs to be interrupted. This system may represent a step forward in the control and biosafety of cell- and gene-therapy-based microencapsulation protocols (Catena et al, 2010)

The development of novel therapeutic strategies for neurodegenerative and neurological diseases represents one of the biggest unmet medical needs today. The rapid developing of cell microencapsulation technology may provide a solution to overcome these diagnostic and neurotherapeutic challenges for neurodegenerative diseases such as Alzheimer's disease and Parkinson's disease. Although this is a significant and promising result, there are a number of steps to be taken before this new form of drug delivery can be tested in humans in the clinic. Nanotechnology can therefore be used to develop diagnostic tools as well as enabled delivery systems that can bypass the blood brain barrier in order to facilitate conventional and novel neurotherapeutic interventions such as drug therapy, gene therapy, and tissue regeneration.

7. Acknowledgment

We thank Tania Vazquez for editorial assistance; also we are grateful to E. Carro, G. Orive, R.M. Hernandez and J.L. Pedraz for their kind help and collaboration. This work was supported by grants from Xunta de Galicia (INCITE2009, 09CSA051905PR) and "Isidro Parga Pondal" programme.

8. References

Alvarez-Erviti, L.; Seowy, Y.; Yin, H.; Betts, C.; Lakhal, S. & Wood, M. Delivery of siRNA to the mouse brain by systemic injection of targeted exosomes. *Nature Biotechnolology*, in press, (March 2011), ISSN 1087-0156.

Alzheimer's Association. Alzheimer's Association report. 2011 Alzheimer's disease facts and figures. *Alzheimer's & Dementia*, Vol. 7, No.1, (January 2011), pp.208-244, ISSN 1552-5260.

Bontha, S.; Bronich, T. & Kabanov, A. Polymer micelles with cross-linked ionic cores for delivery of anticancer drugs. *Journal of Controlled Release*, Vol. 114, No.2, (August 2006), pp. 163-174, ISSN 0168-3659.

Brayne, C.; Stephan, B. & Mathews, F. A European perspective on population studies of dementia. *Alzheimer's & Dementia*, Vol. 7, No.1, (January 2011), pp. 3-9, ISSN 1552-5260.

Bronich, T.; Bontha, S.; Shlyakhtenko, L.; Bromberg, L.; Hatton, T. & Kabanov, A. Template assisted synthesis of nanogels from pluronic modified poly(acrylic acid). *Journal of Drug Targeting*, Vol. 14, No.6, (July 2006), pp. 357-366, ISSN 1061-186X.

Carro, E.; Trejo, J.; Gerber, A.; Loetscher, H.; Torrado, J.; Metzger, F. & Torres-Aleman, I. Therapeutic actions of insulin-like growth factor I on APP/PS2 mice with severe brain amyloidosis. *Neurobiology of Aging*, Vol. 27, No.9, (September 2006), pp. 1250-1257, ISSN 0197-4580

Carro, E.; Trejo, J.; Spuch, C.; Bohl, D.; Heard, J. & Torres-Aleman, I. Blockade of the insulin like growth factor I receptor in the choroid plexus originates Alzheimer's like neuropathology in rodents: new cues into the human disease? *Neurobiology of Aging*, Vol. 27; No.11, (November 2006), pp. 1618-1631, ISSN 0197-4580.

Catena, R.; Santos, E.; Orive, G.; Hernandez, R.; Pedraz, J. & Calvo, A. Improvement of the monitoring and biosafety of encapsulated cells using the SFGNESTGL triple reporter system. Journal of Controlled Release, Vol. 146, No.1, (August 2010), pp. 93-98, ISSN 0168-3659.

Cen, X.; Hu, Z.; Shen, M. & Wu, K. Microencapsulation expansion type flame retardant and application in epoxy resin composite material thereof. (2010) CN101812186.

Chang, T. Semipermeable microcapsules. *Science*, Vol. 146, No. 3643, (October 1964), pp. 524-525, ISSN 0036-8075.

Cherksey, B.; Sapirstein, V. & Geraci, A. Adrenal chromaffin cells on microcarriers exhibit enhanced long-term functional effects when implanted into the mammalian brain. *Neuroscience*. Vol. 75, No.2, (November 1996), pp. 657-664, ISSN 0306-4552

Day-Lollini, P.; Stewart, G.; Taylor, M., Johnson, R. & Chellman, G. Hyperplastic changes within the leptomeninges of the rat and monkey in response to chronic intracerebroventricular infusion of nerve growth factor. *Experimental Neurology*, Vol. 145, No.1, (May 1997), pp. 24-37, ISSN 0014-4886.

Dou, H.; Destache, C.; Morehead, J.; Mosley, R.; Boska, M.; Kingsley, J.; Gorantla, S.; Poluektova, L.; Nelson, J.; Chaubal, M.; Werling, J.; Kipp, J.; Rabinow, B. & Gendelman, H. Development of a macrophage based nanoparticles platform for antiretroviral drug delivery. *Blood*, Vol.108, No.8, (October, 2006), pp. 2827-2835, ISSN 006-4971.

Emerich, D.; Silva, E.; Ali, O.; Mooney, D.; Bell, W.; Yu, S.;, Kaneko, Y. & Borlogan, C. Injectable VEGF hydrogels produce near complete neurological and anatomical protection following cerebral ischemia in rats. *Cell Transplantion*, Vol. 19, No.9, (April 2010), pp. 1063-1071, ISSN 0963-6897.

Ferri, C.; Prince, M.; Bryne, C.; Brodaty, H.; Fratiglioni, L.; Ganguli, M.; Hall, K.; Hasegawa, K.; Hendrie, H.; Huang, Y.; Jorm, A.; Mathers, C.; Menezes, P.; Rimmer, E.; Scazufca, M. & Alzheimer's Disease International. Global prevalence of dementia: a Delphi consensus study. *Lancet*, Vol. 366, No.9503, (December 2005), pp. 2112-2117, ISSN 0140-6736.

Garcia, P.; Youssef, I.; Utvik, J.; Florent-Berchard, S.; Barthelemy, V.; Malaplate-Armand, C.; Kriem, B.; Stenger, C.; Koziel, V.; Olivier, J.; Escanve, M.; Hanse, M.; Allouche, A.; Desbene, C.; Yen, F.; Bjerkvig, R.; Osetr, T.; Niclou, S. & Pillot T. Ciliary neurotrophic factor cell-based delivery prevents synaptic impairment and improves memory in mouse models of Alzheimer's disease. *Journal of Neuroscience*, Vol.30, No.22, (June 2010), pp. 7516-7527, ISSN 0270-6474.

Jain, R. the manufacturing techniques of various drug loaded biodegradable poly(lactide-co-glicolide) (PLGA) devices. *Biomaterials*, Vol. 21, No. 23, (December 2000), pp. 2475-2490, ISSN 0142-9612.

Keunen, O.; Johansson, M.; Oudin, A.; Sanzey, M.; Rahim, S.; Fack, F.; Thorsen, F.; Taxt, T.; Bartos, M.; Jirik, R.; Miletic, H.; Wang, J.; Stieber, D.; Stuhr, L.; Moen, I.; Rygh, C.; Bjerkvig, R. &, Niclou, S. Anti-VEGF treatment reduces blood supply and increases tumor cell invasion in glioblastoma. *Proceedings of the National Academy of Sciences of the Unites States of America*, Vol. 108, No.9, (March 2011), pp. 3749-3754, ISSN 1091-6490.

Lam, X.; Duenas, E. & Cleland, J. Encapsulation and stabilization of nerve growth factor into poly(lactic-co-glycolitic) acid microspheres. *Journal of Pharmaceutical Sciences*, Vol. 90, No.9, (September 2001), pp. 1356-1365, ISSN 0022-3549.

Langer, R. Tissue emerging: a new field and its challenges. *Pharmaceutical Research*, Vol 14, No.7, (April 1997), pp. 840-841, ISSN 0724-8741.

Lim, F.; & Sun, A. Microencapsulated islets as bioartificial endocrine pancreas. *Science*, Vol. 210, No. 4472, (November 1980), pp. 908-910, ISSN 0036-8075.

Malik, S.; Motamedi, S.; Royo, N.; Lebold, D. & Watson, D. (2011). Identification of potentially neuroprotective genes upregulated by neurotrophin treatment of CA3 neurons in the injured brain. *Journal of Neurotrauma*, Vol. 28, No.3, (March 2011), pp. 415-430, ISSN 0897-7151.

Mora, M.; Sagrista, M.; Trombetta, D.; Bonina, F.; De Pasquale, A. & Saija A. Design and characterization of liposomes containing long chain N-acylPEs for brain delivery: penetration of liposomes incorporating GM1 into the rat brain. *Pharmaceutical Research*, Vol. 19, No.10, (October 2002), pp. 1430-1438, ISSN 0724-8741.

Morgan, D. Immunotherapy for Alzheimer's disease. *Journal of Internal Medicine*, Vol. 269, No.1, (January 2011), pp. 54-63, ISSN 1365-2796.

Morley, J. & Hurtig, H. Current understanding and management of Parkinson's disease: five new things. *Neurology*, Vol.75, No. 18, (November 2010), pp. 9-15, ISSN 0028-3878.

Mulqueen, P.; Taylor, P. & Gittins, D. Microencapsulation, KR20100124284.

Nutt, J.; Burchiel, K.; Comella, C.; Jankovic, J.; Lang, A.; Laws, E.; Lozano, A.; Penn, R.; Simpson, R.; Stacy, M.; Wooten, G. Randomized, double-blind trial of glial cell line derived neurotrophic factor (GDNF) in PD. *Neurology*, Vol. 60, No.1, (January 2003), pp. 69-73, ISSN 0028-3878.

Orive, G.; Hernandez, R.; Gascon, A.; Igartua, M.; Rojas, A. & Pedraz, J. Microencapsulation of an anti-VE-cadherin antibody secreting 1B5 hybridoma cells. *Biotechnolology and Bioengineering*, Vol. 76, No.4, (December 2001), pp. 285-294, ISSN 0006-3592.

Orive, G.; Hernández, R.; Gascon, A.; De Vos, P.; Hortelano, G.; Hunkeler, D.; Lacik, I.; Shapiro, A. & Pedraz, J. Cell encapsulation: promise and progress. *Nature Medicine*, Vol. 9, No.1, (January 2003), pp. 104-107, ISSN 1078-8956.

Orive, G.; Hernandez, R.; Gascon, A.; Dominguez-Gil, A. & Pedraz, J. Drug delivery in biotechnology: present and future. *Current Opinion in Biotechnology*, Vol. 14, No. 6, (December 2003), pp. 659-664, ISSN 0958-1669.

Orive, G.; Anitua, E.; Pedraz, J. & Emerich, D. Biomaterials for promoting brain protection, repair and regeneration. *Nature Reviews Neuroscience*, Vol. 10, No.9, (March 2009), pp. 682-690, ISSN 1471-003X.

Orive, G.; Ali, O.; Anitua, E.; Pedraz, J. & Emerich, D. Biomaterial-based Technologies for brain anti-cancer therapeutics and imaging. *Biochimica et Biophysica Acta*, Vol. 1806, No.1, (August 2010), pp. 96-107, ISSN 0304-419X.

Pasic, M.; Diamandis, E.; McLaurin, J.; Holtzman, D.; Schmitt-Ulms, G. & Quirion, R. Alzheimer's disease: advances in pathogenesis, diagnosis and therapy. *Clinical Chemistry*, Vol.57, No.5, (2011), in press. ISSN 1530-8561.

Raiche, A.; Campbell, J.; Nettles, H. & Womack, A. Microencapsulation process with solvent and salt. (2010). WO2010033776.

Rempe, R.; Cramer, S. & Galla, H. Transport of poly(n-butycyano-acrylate) nanoparticles across the blood brain barrier in vitro and their influence on barrier integrity.

Biochemical Biophysical Research Communications, Vol. 406, No.1, (March 2011), pp. 64-69, ISSN 0006-291X.

Rodriguez, M.; Warrington, A. & Pease, L. Human natural autoantibodies in the treatment of neurologic disease. *Neurology*, Vol. 72, No.14, (April 2009), pp. 1269-1276, ISSN 1015-8618.

Sahni, J.; Doggui, S.; Ali, J.; Baboota, S.; Dao, L. & Ramassamy, C. Neurotherapeutic applications of nanoparticles in Alzheimer's disease. *Journal of Controledl Release*, (December 2010), in press, ISSN 0168-3659.

Schecterson, L. & Bothwell, M. Neurotrophin receptors: old friends with new partners. *Developmental Neurobiology*, Vol. 10, No.5, (April 2010), pp. 332-338, ISSN 1932-8451.

Sharma, N.; Deppmann, C.; Harrington, A.; St Hillaire.; Chen, Z.; Lee, F. & Ginty, D. (2010). Long distance control of synapse assembly by target derived NGF. *Neuron*, Vol.67, No.3, (August 2010), pp. 422-434, ISSN 0896-6273.

Spuch, C. & Navarro, C. The therapeutic potential of microencapsulate implants: patents and clinical trials. *Recent Patents on Endocrine, Metabolic & Immune Drug Discovery*, Vol. 4, No.1, (January 2010), pp. 59-68, ISSN 1872-2148.

Spuch, C.; Antequera, D.; Portero, A.; Orive, G.; Hernandez, R.; Molina, J.; Bermejo-Pareja, F.; Pedraz, J. & Carro, E. The effect of encapsulated VEGF secreting cells on brain amyloid load and behavioural impairment in a mouse model of Alzheimer's disease. *Biomaterials*, Vol.31, No.21, (July 2010), pp. 5608-5618, ISSN 0142-9612.

Sun, D.; Zhuang, X.; Xiang, X.; Liu, Y.; Zhang, S.; Liu, C.; Barnes, S.; Grizzle, W.; Miller, D. & Zhang, H. A novel nanoparticle drug delivery system: the anti-inflammatory activity of curcumin is enhanced when encapsulated in exosomes. *Molecular Therapy*, Vol. 18, No.9, (September 2010), pp. 1606-1614, ISSN 1525-0016.

Thorsen, F.; Read, T.; Lund-Johansen, M.; Bolge, B. & Bjerkvig, R. Alginate encapsulated producers cells: a potential new approach for the treatment of malignant brain tumours. *Cell transplantation*, Vol. 9, No.6, (November 2000), pp. 773-783, ISSN 0963-6897.

Virues-Ortega, J.; de Pedro-Cuesta, J.; Vega, S.; Seijo-Martinez, M.; Saz, P.; Rodriguez, F.; Rodriguez-Laso, A.; Reñe, R.; de las Heras, S.; Mateos, R.; Martinez-martin, P.; Lopez-Pousa, S.; Lobo, A.; Regla, J.; Gascon, J.; Garcia, F.; Fernandez-Martinez F.; Boix, R.; Bermejo-Pareja, F.; Bergareche, A.; Sanchez-Sanchez, F.; de Arce, A.; del Barrio, J. & On behalf of the spanixh epidemiological studies on ageing group. Prevalence and European comparison of dementia in a ≥ 75 years-old composite population in Spain. *Acta Neurologica Scandinavica*, Vol. 123, No. 5, (May 2011), pp. 316-324, ISSN 1600-0404

Yamada, K.; Kinoshita, A.; Kohmura, E.; Sakaguchi, T.; Taguchi, J.; Kataoka, K. & Hayakawa, T. Basic fibroblast growth factor prevents thalamic degeneration after cortical infarction. *Journal of Cerebral Blood Flow & Metabolism*, Vol. 11, No. 3, (May 1991), pp. 472-478, ISSN 0271-678X.

Yang, J.; Zaim Wadghiri, Y.; Minh Hoang, D.; Tsui, W.; Sun, Y.; Chung, E.; Li, Y.; Wang, A.;
    de Leon, M. & Wisniewski, T. Detection of amyloid plaques targeted by USPIO-
    Aβ1-42 in Alzheimer's disease transgenic mice using magnetic resonance
    microimaging, *Neuroimage*, Vol. 55, No.4, (April 2011), pp. 1600-1609, ISSN 1053-
    8119.

Part 4

Biomaterials in Prostheses Production

21

New Developments in Tissue Engineering of Microvascular Prostheses

Vincenzo Vindigni[1], Giovanni Abatangelo[2] and Franco Bassetto[1]
[1]Unit of Plastic and Reconstructive Surgery, University of Padova
[2]Department of Histology, Microbiology, and Medical Biotechnology,
University of Padova
Italy

1. Introduction

Clinical needs to have a ready-to-use small-diameter vascular prostheses are very remarkable and cover different fields of surgery: plastic and reconstructive surgery (microvascular transfer of free flaps), heart surgery (treatment of ischemic heart diseases), vascular surgery (distal revascularization of lower limbs), neurosurgery (substitution of intracranial arteries), paediatric vascular surgery. In particular, there is a substantial need for tissue-engineered, living, autologous replacement materials with the potential for growth in paediatric applications and for substitute small diameter vessel that up to now are defined the "holy grail" of vascular biology. Completely bio-resorbable vascular prostheses with the capacity for induce regeneration and growth of a new vascular segment may overcome the limitations of contemporary artificial prostheses that are nonviable, artificial, or allogenic materials lacking the capacity of growth, repair, and remodelling. These intrinsic properties limit their long-term function, posing the substantial burden of graft failure and related re-operations, particularly on paediatric patient population. Moreover, these synthetic materials are not suitable for the reconstruction of the coronary, carotid, or femoral arteries as well as other small diameter vessels (< 6 mm). Autologous native vessels, i.e., the saphenous vein and mammary artery, are the most currently used material for small-diameter arterial replacement. Immune acceptance is a major advantage offered by this technique of "ready to use" conduits. However, the availability of suitable native replacements is limited when multiple conduits are required, especially in patients with diffuse vascular disease. The need for a prosthetic graft that performs as a small diameter conduit has led investigators to pursue many avenues in vascular biology (Figure 1). There are four main approaches currently being investigated, all of which satisfy an apparent prerequisite to biocompatibilty of a small-diameter graft—that no permanent synthetic materials are used. One approach is acellular, based on implanting decellularized tissues treated to enhance biocompatibility, strength, and cell adhesion/invasion leading to cellularization with host cells. The other three approaches involve implantation of constructs possessing some degree of cellularity. The most recent of these is based on the concept of self-assembly, wherein cells are cultured on tissue culture plastic in medium inducing high ECM synthesis. This leads to sheets of neotissue that are subsequently processed into multilayer tubular form. The other two approaches rely on a polymeric scaffold. One is

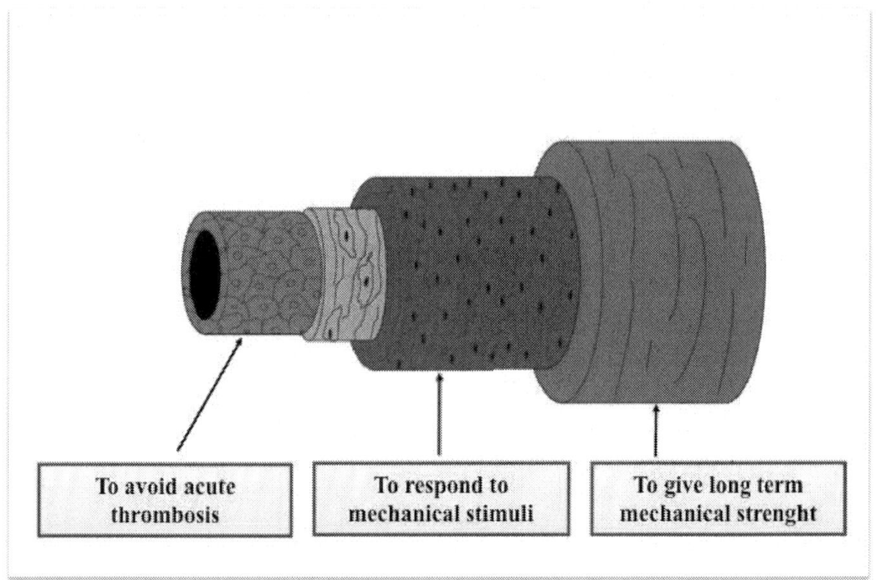

Fig. 1. The ideal micro-vascular prostheses should orchestrate the sequential regeneration of different vascular components: the intimal layer to avoid acute thrombosis, the smooth muscle vascular cell layer to respond to mechanical stimuli and finally the adventitial layer to obtain a long-term vascular patency.

based on forming a tube of a synthetic biodegradable polymer and then seeding the cells (which would not survive the conditions of polymer synthesis), relying on active cell invasion or an applied force to achieve cellularity. The other is based on a tube of a biopolymer, typically a reconstituted type I collagen gel, formed with and compacted by tissue cells, where an appropriately applied mechanical constraint to the compaction yields circumferential alignment of fibrils and cells characteristic of the arterial media. It is this last feature that is most attractive about a biopolymer-based tissue-engineered artery. This follows from two axioms, (i) that native artery function, particularly mechanical function, depends on structure (particularly alignment of the smooth muscle cells and collagen fibers in the medial layer) as much as it depends on composition, and (ii) that the tissue-engineered artery should serve as a functional remodelling template, so that while providing function during the remodelling, the artificial tissue also provides a template for the alignment of the remodelled tissue. To some extent, all these approaches rely on the ability of cells (transplanted or host) to adhere to and migrate within the construct, and to remodel its composition and/or structure. This last point is key, as remodelling confers biocompatibility, in principle, by virtue of complete resorption of the initial scaffold. Of course, the initial scaffold must be replaced by functional cell-derived ECM on the same time scale. Remodelling also determines the ultimate mechanical, transport, and biological properties.

The aim of this chapter is to investigate all the materials commonly studied to create small diameter vascular prostheses. In particular, absorbable biomaterials are also reviewed in depth, for better understanding of the properties of the biomaterials used until now in vascular engineering. A brief analysis of their two main features is necessary: surface properties (haemocompatibility/endothelialisation) and mechanical properties (tensile force/compliance).

2. Material surface

Thrombosis of vascular substitutes is the main mechanism of obliteration and subsequent failure of most vascular conduits. Various methods have been recorded to avoid this phenomenon, such as coatings with antithrombotic drugs, e.g., heparin, hirudin, aspirin, or tissue factor pathway inhibitor (Mooney DJ et al 1996; Kim BS et al 1998; Wake MC et al 1996). There have been attempts to emulate the endothelial cellular surface which, coated with heparan sulphate proteoglycan, produces a negative surface charge which helps to prevent platelet adherence. Some prostheses are therefore coated internally with heparan sulphate, which is quickly degraded, and some materials with an electronegative surface have been created, with uncertain results (Guidoin R et al 1993).

So far, many researchers have described seeding endothelial cells in conduits. A recent study reported a patency rate of 90% in 27 months for ePTFE prostheses used in coronary bypass, after additional incubation with endothelial cells which allowed them to adhere to the material (Laube HR et al 2000). The major limitation of this method is the need for cell cultures and withdrawal of tissue from the patient, and in any case it remains a two-step procedure. A tubular structure of ePTFE is also left in place, with the risk of later infection. Constructs composed entirely of cells (Tissue Engineered Blood Vessels: TEBV) have been devised to overcome these complications (Mooney DJ et al 1996; Kim BS et al 1998; Wake MC et al 1996). Although the method promises amazing results, it is time-consuming and very expensive.

To avoid the cost of cell cultures, many researchers have tried to improve endothelial coverage of prostheses by coating them with endothelial-friendly compounds with good haemocompatibility. For example, e-PTFE prostheses have been coated with perlecan (Zenni GC et al 1993) and endothelial-specific adhesion proteins such as fibrin–gelatin (Kumar TR et al 2002) and hirudin (Salacinski H et al 2002). Fibronectin coating seems to be a successful method, apart from loss of lining at high flow rates. This is why a functional ligand for fibronectin was used, with covalent binding of short peptide sequences (Arg-Gly-Asp, RGD) to improve cell adhesion (Mooney DJ et al 1996; Kim BS et al 1998; Wake MC et al 1996).

Instead of coating prostheses with the above substances, another possibility would be to use absorbable, already biocompatible biomaterials, to make entire prostheses (Kannan RY et al 2005).

In spite of all these experiments, endothelialisation in various types of vascular prostheses has been shown in animals but never satisfactorily in humans. The type of material is not the only essential point in favouring endothelialisation. In order to clarify this, we need to go back a little and recall the physiopathology of endothelialisation, which today takes place in three main ways:
- Trans-anastomatic endothelialisation;
- Transmural endothelialisation;
- Endothelialisation due to 'fall-out' of circulating pluripotent cells.

Therefore, trying to enhance endothelialisation means acting on each of these three modalites of cell growth.

Trans-anastomotic endothelialisation (TAE) appears to be very difficult in humans. Early studies on synthetic prostheses report that they cannot be longer than 0.5 cm, even after prolonged implantation. In spite of a long period of observation, internal endothelialisation has not been observed in humans, except in sites of anastomosis (Berger K et al 1972). Several factors have been observed to influence this, such as species, senescence, anatomic dimensions of the vessel, and prosthetic materials (Zilla P et al 2004), but even in animals TAE is limited (Zhang Z et al 2004).

Study of endothelial cells, both human and canine, compared in vitro, suggest that human cells have a greater potential for migration but a lower capacity for adhesion, which may explain the lack of re-endothelialisation in vivo, when blood flow may obstruct cell adhesion (Dixit P et al 2001).

Instead, the transmural pathway seems to enhance rapid endothelialisation, according to recent studies on materials with sufficient porosity. Pore size takes on importance in these studies, since the prosthesis must be sufficiently large to allow cell growth, but not too large to cause loss of intercellular adhesion (Mooney DJ et al 1996; Kim BS et al 1998; Wake MC et al 1996). Materials with differently sized pores inside and outside the conduit have even been experimented, in order to obtain an ecocompatible surface internally and a colonisable one externally (Mooney DJ et al 1996; Kim BS et al 1998; Wake MC et al 1996).

Pore size also alters the haemocompatibility of biomaterials, as well as their compliance and degradation time. An optimal pore size for vascular engineering has been hypothesised, ranging from 30 to 50 microns. It appears that smaller pores would not allow growth of endothelial cells, and larger ones would cause excessive leakage of blood (Matsuda T et al 1996). Pores in the walls of prosthetic materials can also avoid intimal hyperplasia. It has been hypothesised that a thrombus initially deposited on the walls of the prothesis later organises itself into muscle-like tissue, which then gives rise to intimal hyperplasia. The precocious growth of endothelial tissue would avoid thrombosis and thus the consequent cascade of events leading to intimal hyperplasia (Mooney DJ et al 1996; Kim BS et al 1998; Wake MC et al 1996). Increased pore size causes increased radial compliance of the material. Several studies have shown that vascular implants with fibers organised in a circular fashion do not cause dilation (Mooney DJ et al 1996; Kim BS et al 1998; Wake MC et al 1996). This is local, since cells undergo mechanical stress and are thus conditioned in their spatial orientation.

"Fall-out healing" leads to the formation of endothelial islands, with no connection with the formation of trans-anastomotic or transmural tissue. This is a late phenomenon, not of great importance in materials such as Dacron and e-PTFE grafts, and thus it does not play a central role in present-day prostheses (Zilla PD et al 2007), but it appears to be the mechanism for repairing small vascular lesions (Roberts NM et al 2005). However, recent studies show how this mechanism may be enhanced, by attracting EPC cells to participate (Avci-Adali MG et al 2010).

3. Mechanical properties of biomaterials

Before describing the mechanical properties of vascular replacements, we must digress to describe those of vessels. Arteries are mechanically anisotropic, i.e., their elasticity and resistance (maximum pressure tolerated before bursting) varies according to the direction along which they are measured. The capacity for distension of the arterial wall is generally called compliance (radial elasticity), or the difference in diameter obtained by varying pressures inside the artery itself ($\Delta D/DP$) X 100). The two main structural proteins, collagen and elastin, confer several properties, enabling arteries to distend more at low pressure and become less elastic at high ones. Both properties are extremely important, since inadequate resistance may lead to rupture, and absence of elasticity may disturb flow, leading to thrombosis. In detail, the velocity of propagation (v) of pressure waves depends on the elasticity of vessel walls (E), their thickness (s) and diameter (D), and on the density (P) of blood, as described by the Moens-Korteweg equation, v= Es/ P D. When a section of artery and a replacement conduit have different radial elastic properties, two consequences may ensue. Discontinuity in the speed of propagation of pressure waves leads to turbulence in the area between the natural and artificial vessels which, in turn, leads to local overpressure, the cause of new aneurisms. Apart from the different degree of distension of arterial sections (original artery/prosthesis), this may put greater stress on sutures.

Different compliance has been associated with intimal hyperplasia in an arterial replacement and the artery itself (Stewart SF et al 1992).

Synthetic tissues are far less elastic than arteries, but absorbable materials which could be replaced by the normal vasal extracellular matrix are believed to avoid this problem. Resorption does very often trigger a response similar to that of a foreign body, which leads to the formation of scar tissue which may deprive the original construct of its elasticity (Zilla PD et al 2007).

From the viewpoint of experimental models, elastic components are more or less the same in various species, in spite of changing sizes. Wolinsky and Glagov (1967) have shown that, whereas the total circumferential tension in the vascular wall of the aorta increases 26 times from mouse to pig, defined by Laplace's law as T=PR, tension per lamellar unit (elastic lamina) is similar, being about 1/ 3Nm-1 in various animal aortas, but unfortunately the above study does not compare peripheral vessels. Arterial elasticity causes a reduction in the pressure gradient generated during cardiac systole/distole (Wolinsky et al 1967; Shadwick RE 1999). In peripheral vessels, flow is continuous, thanks to the reservoir role played by the large vessels during cardiac systole. In humans, the ratio between flow gradient and mean flow falls from a value of 6 near the aortic arch to 2 distal to the femoral artery (Wolinsky et al 1967; Shadwick RE 1999). There is therefore a difference between the elasticity of vessels in relation to their size and position within the circulatory system which may influence the design of ideal replacements for them.

According to the above, and in view of the complexity of the cardiocirculatory system and interactions with blood flow, some components are essential for neovessels if they are to guarantee sufficient resistance and compliance and, at the same time, avoid thrombosis and intimal hyperplasia (Mitchell SL et al 2003). At the present time, always in terms of totally absorbable products, there are many approaches involving heterologous tissues, synthetic polymers, biopolymers and totally engineered products (Isenberg BC et al 2006). Every experiment leads to different conclusions, and associations between materials to exploit desired properties while avoiding problems have also been proposed. For example, PGA is a generally stiff product and is frequently associated with other substances to achieve the necessary elasticity (Shinoka T et al 2008).

4. Materials

Many research groups all over the world have approached the problem of developing the ideal prosthesis in a variety of ways. As in all tissue engineering fields, research ranges round the triad Scaffold/Cells/Growth Factor (Figure 2). Scaffolds are ideal biomaterials for conduits, and cells can be seeded and cultivated on them, after preconditioning with various growth factors. Very many materials have been used to make scaffolds, generally subdivided into four main categories: biological materials (allografts, xenografts, and derived products), natural proteins, and both permanent and absorbable synthetic polymers.

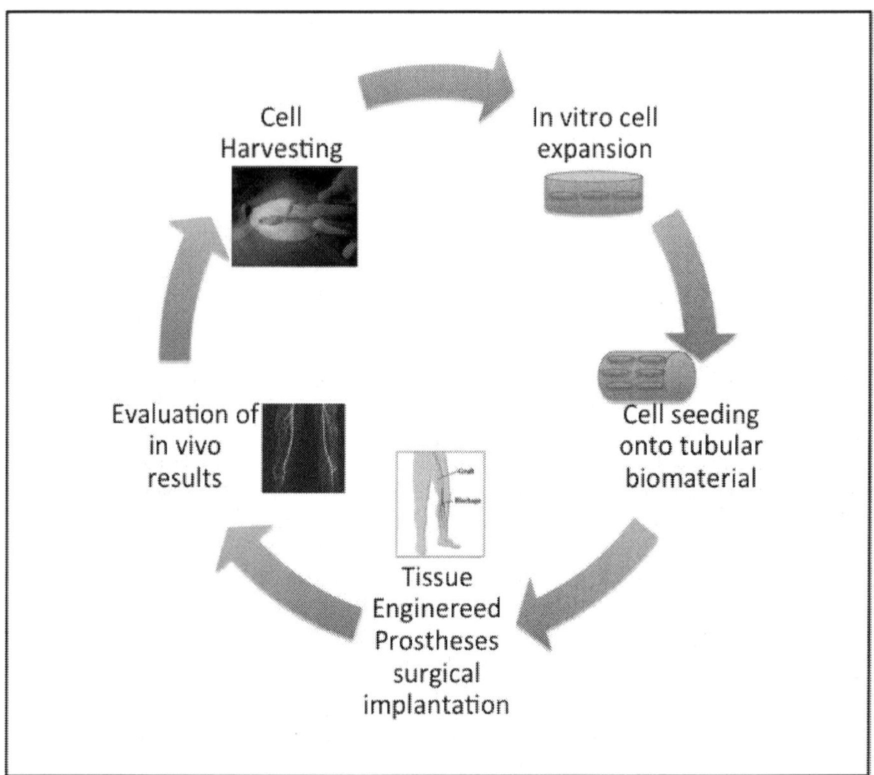

Fig. 2. The traditional vascular tissue engineering approach.

TEBV (Total Engineered Blood Vessels)

Although this review covers biomaterials used as scaffolds in vascular tissue engineering, mention must be made of the methods for reconstructing small diameter vessels which do not use scaffolds. These mass-produced cell laminae are called TEBV (total engineered blood vessels). They have proven mechanical properties comparable with those of the saphenous vein until 8 months after implant. Their anatomic integration is excellent, showing the formation of vasa vasorum histologically in context and the presence of viable endothelium. Results have been so promising that the first studies for clinical application are under way (Mooney DJ et al 1996; Kim BS et al 1998; Wake MC et al 1996). However, TEBV can only be produced in the laboratory over a total period of about 3 months, which does not make them suitable for urgent procedures; they are also extremely expensive. Of the studies examined, only 4 dealt with this method. One significant publication was a multi-centre cohort study illustrating the results of implanting TEBV in access routes in patients under dialysis. Although the cohort was composed of high-risk patients, primary patency of 76% was reported in the first 3 months (Mooney DJ et al 1996; Kim BS et al 1998; Wake MC et al 1996).

Biological materials

The first attempt at vascular replacement was made in 1908 with an allograft in dog, which earned its inventor the Nobel Prize for medicine (Carrel et al 1906). Since then, biological materials have been tested in various species (xeno- and allotransplants) and after various types of preparatory procedures (decellularisation to reduce the immune response, derivation of material for homogenisation, cryoconservation). These biomaterials are widely available and they are of course excellent substrates for cell adhesion. In addition, the processing method can retain all their advantageous mechanical properties (tensile strength, elasticity) (Schmidt CE et al 2000). As the main disadvantages are possible residual antigenicity and infection after implant (Chlupac JE et al 2009), techniques for their decellularisation and sterilisation have been refined. Articles on TEBV published in the last 5 years (biological materials) mainly deal with materials already nautually present as tubular structures in the body (arteries, veins, urethers) and submitted to decellularisation. They are often studied as allo- or xenografts, and enriched with cells (Mooney DJ et al 1996; Kim BS et al 1998; Wake MC et al 1996), bFGF (Mooney DJ et al 1996; Kim BS et al 1998; Wake MC et al 1996), heparin [42] and VEGF (Mooney DJ et al 1996; Kim BS et al 1998; Wake MC et al 1996) to improve patency in the long term. Of special interest for the physiopathology of tissue healing after implant is one study reporting trends after implants of decellularised porcine arteries in rat, concluding that the initial inflammation due to integration in tissues does not interfere with long-term modelling (Mooney DJ et al 1996; Kim BS et al 1998; Wake MC et al 1996). One in vitro study examines the creation of a biotube produced by reaction to a foreign body (Mooney DJ et al 1996; Kim BS et al 1998; Wake MC et al 1996). Other studies examine the use of SIS (Small Intestinal Submucosa), already amply employed in clinical management of wounds. These studies show the good mechanical properties of this biomaterial (Hinds MT et al 2006), but also the poor long-term patency of conduits (Pavcnik D et al 2009).

Permanent materials

To replace large diameter vessels, synthetic materials are now routinely used in clinical practice, but synthetic polymers such as e-PTFE and Dacron have not given good results in small diameter vessels (<6 mm). In view of the enormous number of works on the subject, the problems of permanent synthetic polymers are clearly difficult to overcome. Many works report variable results in terms of long-term vasal patency, with consequent infection of the operation site (Mooney DJ et al 1996; Kim BS et al 1998; Wake MC et al 1996). During experimental studies, variable patency turned out to be considerable. These materials have been implanted in humans, but do not develop an endothelialised surface (Mooney DJ et al 1996; Kim BS et al 1998; Wake MC et al 1996), thus causing platelet adhesion and the development of a fibrin layer which may lead to thrombosis. Later failure may also be due to thrombosis after stenotic occlusion of the vessel consequent upon the development of endothelial hyperplasia. Several methods have been applied in the past to reduce thrombosis after surgery (e.g., anti-thrombotic drugs in their surface or surface ligands (Kidane AG et al 2004). Results were better, although clinically satisfactory criteria were not achieved. In the articles examined – that is, those published in the last 5 years – the main non-degradable biomaterials were PTFE and polyurethanes. A total of 20 articles on PTFE were found, 16 of which described coating surfaces to avoid thrombosis or tissue hyperplasia. This approach is normally followed by endothelialisation of prostheses, and 8 articles described techniques for this (cell cultures). One review (Bordenave L et al 2005) illustrates how this procedure, in time, has moved from one-stage to two-stage techniques although, in spite of discouraging results, not much space was devoted to clinical practice, mainly because of its three most serious limitations: the impossibility of executing these techniques in emergencies; the need for prior withdrawal of cells; and the need for a GLP laboratory to treat human cells.

The remaining approaches for surface coating do not involve cell cultures, and may be interesting as regards future applications of these products in combination with biodegradable materials. Fibrin has been proposed as a coating: results have been either discouraging (Mooney DJ et al 1996; Kim BS et al 1998; Wake MC et al 1996) or encouraging as regards anti-adhesive action in platelets and perlecan (Mooney DJ et al 1996; Kim BS et al 1998; Wake MC et al 1996); one article reported an elastin-like recombining protein [84, 85, 88]. Single peptides like RGD, cyclic RGD (Mooney DJ et al 1996; Kim BS et al 1998; Wake MC et al 1996) have also been proposed, but only in vitro studies are available (Walpoth BH et al 2007; Lord MS et al 2009; Jordan SW et al 2007; Tang C et al 2009; Larsen CC et al 2007).

Polyurethane materials

Polyurethanes are polymers composed of chains of organic units joined by urethane bonds, formed by polymerisation, generally with reagents like monomers containing al least 2 hydroxyl groups (diol), dioisokyanate, and a chain extender. They react to form linear copolymers, segments consisting of alternating stiff and soft segments. The soft segments are derived from polyols such as polyester and the hard ones from isocyanate and chain extenders (e.g., lactic acid/ethylenglycol) (Atala et al 2008). These polymers are biocompatible and highly versatile, since their tensile strength and radial compliance vary according to segment composition (Tiwari et al 2003), stiff segments being responsble for tensile strength and soft segments for elasticity (Kannan et al 2005). Originally produced as permanent biomaterials, they do deteriorate in vivo, due to oxidation and enzymatic and cell-mediated degradation, with the result that their biostability is under revision (Mooney DJ et al 1996; Kim BS et al 1998; Wake MC et al 1996). Oxidation of PU is initiated by oxidase, free radicals and enzymes. The phenomenon of environmental stress cracking is also due to oxidation, a process in which the surface of the biomaterial is coated with proteins which recall adhesion by macrophages, which release oxidising factors (Stokes KR et al 1995). These discoveries were made in the early 1990s and led to the classification of PU as a new category of absorbable materials (Santerre JP et al 2005).

The differing composition of PU segments may lead to products with various degrees of biostability. As regards dioisokyanates/, the aromatic forms are more stable than the aliphatic ones (Fromstein JDW et al 2006; Mooney DJ et al 1996; Kim BS et al 1998; Wake MC et al 1996), although the former were abandoned after it was noted that they release toxic substances with carcinogenic effects (hepatocarcinoma) in laboratory animals (Gunatillake et al 2003). PU have been combined with highly crystalline segments such as polycarbonates and silicon oligomers to inrease their stability (Atala et al 2008). Degradable polyesters such as PLA, PGA and PCL have also been associated as weak segments. Carbonate PU have shown good resistance to hydrolysis and oxidative stress (Salacinski HJ et al 2002). In one in vivo study, such prostheses were implanted in the aorto-iliac segment of dog, where they remained viable for more than 36 months (Mooney DJ et al 1996; Kim BS et al 1998; Wake

MC et al 1996). PU have excellent radial compliance, but are structurally weaker than the others, being more rapidly degraded due to their typical ester bonds (Guan J et al 2002). As well as biodegradability, PU have shown good biocompatibility in in vivo tissue studies (Mooney DJ et al 1996; Kim BS et al 1998; Wake MC et al 1996).

The development of small diameter vascular prostheses in the last 5 years revealed a total of 22 articles on polyurethanes, 14 in vitro, 4 on production of material and and its medical properties, and only 4 in vivo. The cellular compatibility of several PU (associated with other substances) has also been studied according to method of preparation, e.g., the use of porous structures.

Electrospinning has been applied to other materials in the field of vascular engineering (Mooney DJ et al 1996; Kim BS et al 1998; Wake MC et al 1996), and produces small diameter fibers with good tensile strength on the final material. However, this method is difficult to apply when large pores are required. In theory, the immersion-leaching technique (Pan S et al 2005) should not create interconnected pores and thus not be favourable to cell growth, whereas phase inversion (Mooney DJ et al 1996; Kim BS et al 1998; Wake MC et al 1996) may fail in both interconnections and pore size. One in vivo study examined the PU produced by induction of thermal phase separation, but the resulting biomaterial had poor tensile strength and the overall results in terms of implant viability, presence of aneurisms and vessel functionality were mediocre (Mooney DJ et al 1996; Kim BS et al 1998; Wake MC et al 1996).

As regards chemical composition, PU has been combined with silk fibroin, showing better histocompatiblity of pure PU after implant in rat muscular tissue (Wang W et al 2010). Many experiments have also been made on the mixed-composition PU PDMS (polydimethylsiloxane), a silicon-based polymer (Mooney DJ et al 1996; Kim BS et al 1998; Wake MC et al 1996). In this case, PDMS not only increased biostability but also increased haemocompatibility and immunocompatibility (Mooney DJ et al 1996; Kim BS et al 1998; Wake MC et al 1996). In in vivo studies (Mooney DJ et al 1996; Kim BS et al 1998; Wake MC et al 1996) show encouraging long-term viability: in one (Khorasani MT et al 2006), a PEUU/PDMS polymer was created with the spray phase inversion technique in a tubular form with two-phase porosity: a highly porous internal wall (mean interfibrillar distance 40 microns) and an only slightly porous outer one (mean interfibrillar distance 30 microns). It showed good re-endothelialisation 24 months after implant, with remodelling of the vessel parallel with digestion of the material, without aneurismatic dilation or calcification. However, the sample size was very small, and the prostheses showed uniform dilation over time.Another in vivo study in this series used poly(ester urethane)urea (PEUU) combined with a thrombogenic polymer not similar to a phospholipid, poly(2-methacryloyloxyethyl phosphorylcholine-co-methacryloyloxyethyl butylurethane) (PMBU), to create a fibrillar scaffold by electrospinning, with good tensile strength and compliance. In addition, the association with PMBU made the PU less prone to platelet deposition and hypertrophy of muscle cells. The in vivo patency of 1.3-mm conduits implanted in rat aorta after 8 weeks varied from 40% for pure PU to 67% for PU PMBU (Hong Y et al 2009).

Bioresorbable materials

In the last five years, most research groups have concentrated on testing absorbable/ biomaterials to achieve the ideal vascular conduit. As already mentioned, the materials in question may be synthetic or biopolymers already constituting the extracellular matrix.

The most common absorbable biomaterials are polyesters. This category contains poly(α-hydroxylester poly(L-lactic acid) (PLLA), poly(-glycolic acid) (PGA), polylactone polyorthoesters (POE) and polycarbonates. When these materials are implanted in living tissues, their polymeric structure is subject to hydrolysis and the resulting products, such as lactic and glycolic acids, are metabolized. Their safety and biocompatibility are now established (Mooney DJ et al 1996; Kim BS et al 1998; Wake MC et al 1996).

It is generally difficult to examine the use of these biomaterials separately, as they are all linked in the field of vascular tissue engineering. The first to be examined was polyglcolic acid, an asborbable polyester, which has shown good biocompatibility and is chosen for many applications (Mooney DJ et al 1996; Kim BS et al 1998; Wake MC et al 1996).

Polyglycolic acid was the first biodegradable polymer to be used in vascular engineering [84, 85, 88]. It was tested in combination with cultivated bovine muscle cells and then preconditioned in a pulsatile flow bioreactor. After 8 weeks, this construct revealed collagen and had good mechanical properties. Its tensile strength is comparable to that of a vein. However, it also begins to lose its mechanical resistance within 4 weeks of implant and is completely degraded at 6 months. Degradation speed can be controlled by associating it with other polymers such as poly-L-lactic acid (PLLA), polyhydroxyalkanoate (PHA), polycaprolactonecopolylactic acid, and polyethylene glycol (Mooney DJ et al 1996; Kim BS et al 1998; Wake MC et al 1996).

However, the mechanical resistance of these products does not reach the desired levels – an anticipated outcome, as PGA was originally in the form of a non-woven fabric, and thus does not have measurable tensile strength (Sodian R et al 2000). Most studies have therefore concentrated on preconditioning methods to increase resistance, e.g., use of pulsatile flow bireactors, alternative techniques of cell culture, and administration of various growth factors (Niklason LE et al 1997).

Another substantial problem with PGA is its stiffness, which does not confer the elastic properties typical of arterial tissues (Sodian R et al 2000). In this case too, the use of copolymers has improved results. A fibrillar scaffold based on polyglycolic or polylactic acid coated with a 50:50 L-lactate or L-caprolactone (PCLA/PGA or PCLA/PLA) copolymer has been specifically tested for vascular repair, resulting in compliance closer to that of the original vessel with better surgical handling (Watanabe M et al 2001).

One study showed how PGA-based matrices have greater cellularity and production of proteins of extracellular matrices based on PHAV and P4HB. The authors explained this phenomenon as due to the higher porosity of PGA (> 90%), yielding a contact surface greater than that of cells (Sodian R et al 2000).

To support the remodeling process in vivo, a biomaterial that functions only as a temporary absorbable guide, similar to an in vivo "Artery-Bioregeneration Assist Tube" (ABAT), which can promote the sequential and complete regeneration of vascular structures at the implantation site, entirely made of Hyaluronic Acid was used in different in vivo experimental model (Lepidi et al 2006; Pandis L et al 2010; Zavan B et al).

5. Conclusion

Critical reading of researches in the field of microvascular tissue engineering gave the general impression of progress in the search for an ideal replacement for small diameter vessels.

Most studies indicate the use of absorbable biomaterials, in view of their good integration, with the hope of developing autogenous vessels to replace prostheses. However, not one of these products has yet been approved for clinical experimentation, unlike TEBV and products of biological origin. Degradability is one of the characteristics which tend to

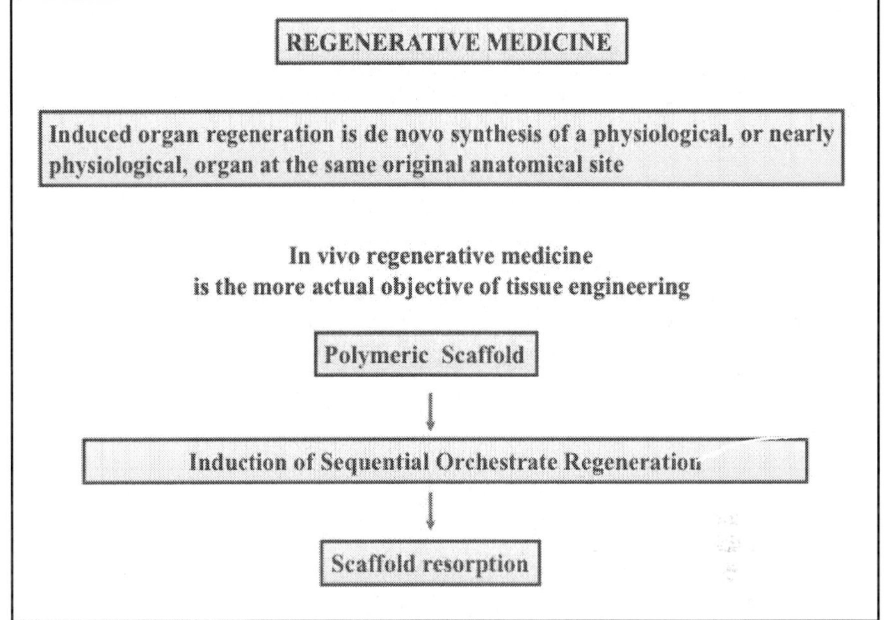

Fig. 3. Regenerative medicine is based on "intelligent" biomaterials able.

dissuade surgeons at the crucial moment of implant. In addition, other variants have been added, making the subject a spiny one. For example, on one hand, the porosity or fibrillar form of these materials not only alters their biostability but also their mechanical characteristics, which are today believed to be essential for implant success. On the other hand, the variability of clinical applications, which may differentiate the desired characteristics of each type of material, requires reflection.

There are many gaps in the examined articles. The first problem, already examined by many authors, is variability in animal models, which hinders direct comparison of results. Homogeneous studies on mechanical studies are also lacking, since so many of them focus on tensile strength, and neglect compliance, which is an essential feature of vessels.

An effective model of an artificial vessel is very far from being achieved, and its development must take into account the context in which it could be applied. Experimental models have already been super-ceded, if we think that the application of a bio-absorbable prosthesis means that cells must be able to reconstruct a new artery and that, in clinical practice, this must be achieved in already damaged arteries.

In elective vascular surgery (e.g., arterial insufficiency in the lower limbs), cellularised replacements are possible, tailored to suit single patients according to their tissue biopsy. However, the procedures are time-consuming and very expensive, requiring dedicated laboratories able to guarantee sterility and suitability for in vivo re-implantation of cell cultures.

As regards urgent procedures, such as revascularisation of all types, the cell culture step should be avoided. The ideal choice would be ready-to-use materials (Figure 3).

6. References

Atala AL et al (2008). Principles of Regenerative Medicine. First ed ed. 2008, Elsevier: Burlington, Massachusetts.

Avci-Adali MG et al (2010). Induction of EPC homing on biofunctionalized vascular grafts for rapid in vivo self-endothelialization--a review of current strategies. Biotechnol Adv;28(1):119-29

Berger K et al (1972). Healing of arterial prostheses in man: its incompleteness. Ann Surg;175(1):118-27

Bordenave et al (2005). In vitro endothelialized ePTFE prostheses: clinical update 20 years after the first realization. Clin Hemorheol Microcirc; 33(3):227-34

Carrel A et al (1906). Uniterminal and biterminal venous transplantations. . Surg Gynecol Obstet;2:266-286

Chlupac JE et al (2009). Blood vessel replacement: 50 years of development and tissue engineering paradigms in vascular surgery. Physiol Res;58 Suppl 2:S119-39

Dixit P et al (2001). Vascular graft endothelialization: comparative analysis of canine and human endothelial cell migration on natural biomaterials. J Biomed Mater Res;56(4):545-55

Fromstein JDW et al (2006). Polyurethane biomaterials. Encyclopedia of Biomaterials and Biomedical Engineering. Eds Marcel Dekker: New York

Guan J et al (2002). Synthesis, characterization, and cytocompatibility of elastomeric, biodegradable poly(ester-urethane)ureas based on poly(caprolactone) and putrescine. J Biomed Mater Res;61(3):493-503

Guidoin R et al (1993). Expanded polytetrafluoroethylene arterial prostheses in humans: histopathological study of 298 surgically excised grafts. Biomaterials;14(9):678-93

Gunatillake PA et al (2003). Biodegradable synthetic polymers for tissue engineering. Eur Cell Mater;5:1-16

Hinds MT et al (2006). Development of a reinforced porcine elastin composite vascular scaffold. J Biomed Mater Res A;77(3):458-69

Hong Y et al (2009). A small diameter, fibrous vascular conduit generated from a poly(ester urethane)urea and phospholipid polymer blend. Biomaterials;30(13):2457-67

Isenberg BC. Small-diameter artificial arteries engineered in vitro. Circ Res;98(1):25-35

Jordan SW et al (2007). The effect of a recombinant elastin-mimetic coating of an ePTFE prosthesis on acute thrombogenicity in a baboon arteriovenous shunt. Biomaterials; 28(6):1191-7

Kannan RY et al (2005) Current status of prosthetic bypass grafts: a review. J Biomed Mater Res B Appl Biomater;74(1):570-81

Khorasani MT et al (2006). Fabrication of microporous polyurethane by spray phase inversion method as small diameter vascular grafts material. J Biomed Mater Res A;77(2):253-60

Kidane AG et al (2004). Anticoagulant and antiplatelet agents: their clinical and device application(s) together with usages to engineer surfaces. Biomacromolecules; 5(3):798-813

Kim BS and Mooney DJ (1998). Engineering smooth muscle tissue with a predefined structure. J Biomed Mater Res;41(2): 322-32

Larsen CC et al (2007). A biomimetic peptide fluorosurfactant polymer for endothelialization of ePTFE with limited platelet adhesion. Biomaterials; 28(24):3537-48

Laube HR et al (2000). Clinical experience with autologous endothelial cell-seeded polytetrafluoroethylene coronary artery bypass grafts. J Thorac Cardiovasc Surg;120(1):134-41

Lord MS et al (2009). The modulation of platelet and endothelial cell adhesion to vascular graft materials by perlecan. Biomaterials;30(28):4898-906

Matsuda T et al (1996). Surface microarchitectural design in biomedical applications: in vitro transmural endothelialization on microporous segmented polyurethane films fabricated using an excimer laser. J Biomed Mater Res;31(2):235-42

Mitchell SL et al (2003). Requirements for growing tissue-engineered vascular grafts. Cardiovasc Pathol;12(2):59-64

Mooney DJ et al (1996). Stabilized polyglycolic acid fibre-based tubes for tissue engineering. Biomaterials; 17(2):115-24

Niklason LE et al (1997). Advances in tissue engineering of blood vessels and other tissues. Transpl Immunol;5(4):303-6

Niklason LE et al (1999). Functional arteries grown in vitro. Science 1999;284(5413):489-93

Pan S et al (2005). Effect of preparation conditions for small-diameter artificial polyurethane vascular graft on microstructure and mechanical properties. Zhongguo Xiu Fu Chong Jian Wai Ke Za Zhi;19(1):64-9

Pandis L et al (2010). Hyaluronan-based scaffold for in vivo regeneration of the rat vena cava: Preliminary results in an animal model. J Biomed Mater Res A;93(4):1289-96

Pavcnik D et al (2009). Angiographic evaluation of carotid artery grafting with prefabricated small-diameter, small-intestinal submucosa grafts in sheep. Cardiovasc Intervent Radiol;32(1):106-13

Roberts NM et al (2005). Progenitor cells in vascular disease. J Cell Mol Med;9(3):583-91

Salacinski H et al (2002). Performance of a polyurethane vascular prosthesis carrying a dipyridamole (Persantin) coating on its lumenal surface. J Biomed Mater Res;61(2): 337-8

Salacinski HJ et al (2002). Thermo-mechanical analysis of a compliant poly(carbonate-urea)urethane after exposure to hydrolytic, oxidative, peroxidative and biological solutions. Biomaterials;23(10):2231-40

Santerre JP et al (2005) Understanding the biodegradation of polyurethanes: from classical implants to tissue engineering materials. Biomaterials;26(35):7457-70

Schmidt CE et al (2000). Acellular vascular tissues: natural biomaterials for tissue repair and tissue engineering. Biomaterials;21(22):2215-31

Shadwick RE. Mechanical design in arteries. J Exp Biol;202(23):3305-13

Shinoka T et al (2008). Tissue-engineered blood vessels in pediatric cardiac surgery. Yale J Biol Med;81(4):161-6

Sodian R et al (2000). Evaluation of biodegradable, three-dimensional matrices for tissue engineering of heart valves. ASAIO J;46(1):107-10

Stewart SF et al (1992). Effects of a vascular graft/natural artery compliance mismatch on pulsatile flow. J Biomech;25(3):297-310

Stokes KR et al (1995). Polyurethane elastomer biostability. J Biomater Appl;9(4):321-54

Tang C et al (2009). Platelet and endothelial adhesion on fluorosurfactant polymers designed for vascular graft modification. J Biomed Mater Res A;88(2):348-58

Tiwari A et al (2003). Improving the patency of vascular bypass grafts: the role of suture materials and surgical techniques on reducing anastomotic compliance mismatch. Eur J Vasc Endovasc Surg;25(4):287-95

Wake MC et al (1996). Fabrication of pliable biodegradable polymer foams to engineer soft tissues. Cell Transplant;5(4):465-73

Walpoth BH et al (2007). Enhanced intimal thickening of expanded polytetrafluoroethylene grafts coated with fibrin or fibrin-releasing vascular endothelial growth factor in the pig carotid artery interposition model. J Thorac Cardiovasc Surg;133(5) 1163-70

Wang W et al (2010). Acute phase reaction of different macromolecule vascular grafts healing in rat muscle]. Sheng Wu Gong Cheng Xue Bao;26(1):79-84

Watanabe M et al (2001). Tissue-engineered vascular autograft: inferior vena cava replacement in a dog model. Tissue Eng;7(4):429-39

Wolinsky H et al. (1967). A lamellar unit of aortic medial structure and function in mammals. Circ Res;20(1):99-111

Zavan B et al (2008). Neoarteries grown in vivo using a tissue-engineered hyaluronan-based scaffold. FASEB J;22(8):2853-61

Zenni GC et al (1993). Modulation of myofibroblast proliferation by vascular prosthesis biomechanics. ASAIO J;39(3):496-500

Zhang Z et al (2004). Pore size, tissue ingrowth, and endothelialization of small-diameter microporous polyurethane vascular prostheses. Biomaterials;25(1):177-87

Zilla PD et al (2007). Human, Prosthetic vascular grafts: wrong models, wrong questions and no healing. Biomaterials;28(34):5009-27

22

Pericardial Processing:
Challenges, Outcomes and Future Prospects

Escande Rémi[1,3], Nizar Khelil[1], Isabelle Di Centa[2], Caroline Roques[1],
Maguette Ba[1,3], Fatima Medjahed-Hamidi[1], Frederic Chaubet[1,3],
Didier Letourneur[1,3], Emmanuel Lansac[1,4] and Anne Meddahi-Pellé[1,3]

1. Introduction

The pericardium is a biological tissue widely used as a biomaterial for tissue engineering applications, including the construction of a variety of bioprostheses such as vascular grafts, patches for abdominal or vaginal wall reparation and, more frequently, heart valves.

However, despite significant advances, some drawbacks have been found in these bioprostheses such as biological matrix deterioration and tissue degeneration associated with calcifications, even though xenopericardium or glutaraldehyde-treated autologous pericardium were used.

In non-autologous pericardial processing, the pericardium must be decellularized in order to remove cellular antigens and procalcific remnants while preserving extracellular matrix integrity. A large variety of decellularization protocols exist, such as chemical, physical or enzymatic methods. Additional cross-linking processing must be carried out to render the tissue non-antigenic and mechanically strong.

So far, almost all bioprosthetic materials made of pericardium, and used in clinical practice, are glutaraldehyde-treated bovine or porcine xenopericardium. However, long-term reports are raising issues concerning their durability, especially highlighting the high risk of calcification. Regarding heart valves, calcification currently represents the major drawback leading to potential failure of the bioprosthesis.

The aim of this review is to present current issues, challenges, outcomes and future prospects of pericardial processing, including decellularization and cross-linking steps. Understanding current issues and improving pericardial processing will allow refining bioprosthesis conception and patients' safety.

2. Characteristics of the pericardium

2.1 Localization and composition

The pericardium is a connective tissue sac surrounding the heart. It is composed by two layers: a deeper layer closely adherent to the heart, the visceral serous pericardium, or

[1]Inserm U698, Bioengineering for Cardiovascular Imaging and Therapy Team, CHU Xavier Bichat, Paris
[2]Ambroise Paré Hospital, vascular surgery department, Boulogne,
[3]Bioengineering for cardiovascular imaging and therapy, Galilée Institut, Paris 13 University
[4]Montsouris Institut, Cardiovascular surgery department, Paris
[1,2,3,4]France

epicardium, and an upper layer: the parietal pericardium. The two layers are separated by the pericardial cavity. The parietal pericardium can be excised and easily tested without causing major complications such as contracture or ischemia (Fomovsky et al., 2010).

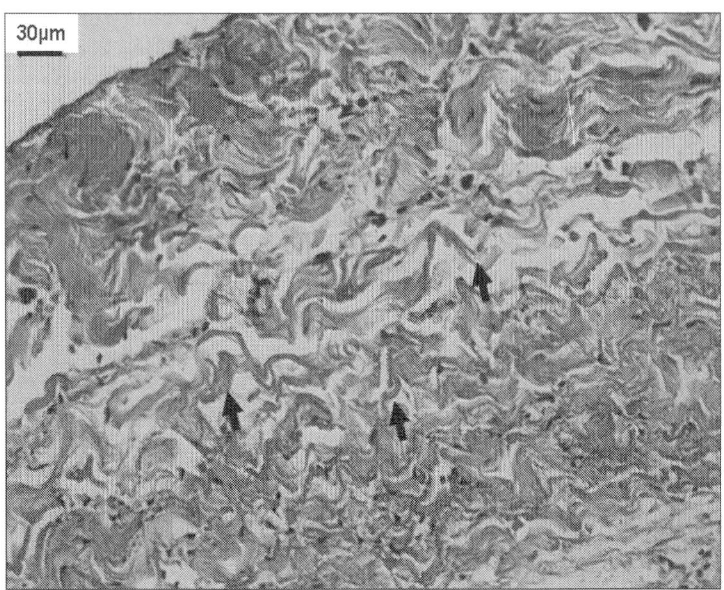

Fig. 1. Histology of ovine pericardium showing the collagen organization (arrows). Hematoxylin-Eosin staining.

The pericardium is composed of a simple squamous epithelium and connective tissue. It is a collagen-rich biological tissue containing mostly type I collagen, as well as glycoproteins and glycosaminoglycans (GAGs) in addition to its constitutive cells (Figure 1). Collagen is structured into different levels of organization ranging from fibrils to laminates, fibers and fiber bundles (Allen & Didio, 1984; Lee & Boughner, 1981). This organization determines the mechanical properties of the pericardial tissue (Sacks, 2003; Liao et al., 2005; Wiegner & Bing, 1981; Lee & Boughner, 1985) and provides an anisotropic and non-linear mechanical behaviour (Zioupos & Barbenel, 1994). Interestingly, depending on the location on the pericardium, the thickness and mechanical properties vary (Hiester & Sacks 1998a, 1998b). Thus, the location of the sample that will be harvested should be carefully selected when designing a tissue engineering protocol.

2.2 Sources of pericardium

Currently marketed heart valve bioprostheses are prepared from bovine or porcine pericardium (Vesely, 2005). Other pericardial tissues from different species have been assessed or are currently used in clinical practice such as equine (DeCarbo et al., 2010; Yamamoto et al., 2009; Sato et al.; 2008.), canine (Lee & Boughner ; 1981; Wiegner & Bing,

1981; 1985), or, even more unusually, ostrich (Maestro et al., 2006) or kangaroo pericardium (Neethling et al., 2000; 2002). However, those exogenous grafts raise several issues, and especially the immune response against the bioprosthesis as well as the viral status of the graft.

Human autologous pericardium is thus an interesting option, presenting several advantages over allografts since it is free of donor-derived pathogens and does not induce any immune response (Mirsadraee et al., 2007), is easily available, easily handled and of low cost. Ultimately, these characteristics allow for shorter and less aggressive pericardial processing before implantation of the bioprosthesis. However, because of intermittent reports of its tendency to retract or become aneurysmal, the general opinion has been negative (Edwards et al., 1969, Bahnson et al., 1970). For cusp tissue replacement or valve tissue replacement, stabilization of pericardium is performed with a solution of 0,2% to 0,6% glutaraldehyde in order to prevent secondary shrinkage (Duran et al., 1998; Al-Halees et al., 1998, 2005; Goetz et al., 2002).

3. Processing of pericardium

As allografts have been the main source for pericardial bioprostheses currently in use, significant processing steps have to be performed prior to clinical use. In particular, as xenogeneic cellular antigens induce an immune response or an immune-mediated rejection of the tissue, decellularization protocols are widely used to reduce the host tissue response (Gilbert et al., 2006.). Once decellularized, the free-cell pericardial tissue is composed of extracellular matrix proteins which are generally conserved among species, and thus can be easily used as a scaffold for the host cell attachment, migration and proliferation (Schmidt & Baier, 2000). This scaffold considerably accelerates tissue regeneration. Overall, tissue decellularization aims at reducing tissue antigenicity and host response while preserving the mechanical integrity, biological activity and composition of the ECM (Simon et al., 2006; Gilbert et al., 2006).

3.1 Extracellular matrix decellularization methods

Most decellularization protocols include a combination of various methods, such as physical, enzymatic or chemical treatments (Gilbert et al., 2006; Crapo et al., 2011). Physical methods can either rely on snap freezing (Jackson et al., 1988; Roberts et al., 1991), mechanical force (Freytes et al., 2004) or mechanical agitation (Schenke-Layland et al., 2003), whereas enzymatic protocols employ nucleases, calcium chelating agents or protease digestion (Teebken et al., 2000; Bader et al., 1998; McFetridge et al., 2004; Gamba et al., 2002). Regarding physical decellularization processes, sonication, based on the use of ultrasounds to disrupt the cell membrane, has been investigated. Such treatment considerably affects the pericardial architecture and full decellularization cannot be achieved. Thus sonication has to be carried out simultaneously with chemical treatments in order to fully decellularize the pericardial tissue and remove cellular debris. However, this combination leads to alterations of the extracellular matrix (ECM) architecture.

For the enzymatic procedure, the main enzyme employed is trypsin, cleaving peptide bonds on the C-side of arginine and lysine and thus allowing separation of the cells from the ECM.

Chemical protocols involve use of alkaline and acid treatments (Freytes et al., 2004), ionic detergents, sodium dodecyl sulfate (SDS), sodium deoxycholate and Triton X-200 (Rieder et al. 2004; Hudson et al., 2004), non-ionic detergents, such as Triton X-100 (Grauss et al., 2003), zwitterionic detergents (Dahl et al., 2003), tri(n-butyl)phosphate (Woods & Gratzer, 2005) as well as hypertonic or hypotonic solutions (Goissis et al., 2000; Woods & Gratzer, 2005; Vyavahareet al., 1997; Dahl et al., 2003). These modalities will either mediate lysis of the cells or solubilization of the cellular components.

Overall, standard decellularization protocols for allografts consist of a multimodal process starting with the lysis of the cell membrane using either ionic solutions or physical treatments. This initial step is then followed by enzymatic treatments to separate any cellular components from the ECM. Subsequently, detergents are used to solubilize the nuclear and cytoplasmic cellular components. At the end of the procedure, all residual cell debris is removed from the remaining ECM. A washing step must also be carried out following the decellularization protocol to remove residual chemicals, thus avoiding any host tissue response (Gilbert et al., 2006). The efficiency of the decellularization protocol and the preservation of the ECM have to be assessed using histological tools.

Concerning pericardial decellularization, several protocols, which have provided interesting results, can be found in the literature. (Courtman et al., 2004; Liang et al., 2004; Wei et al., 2005; Chang et al., 2005; Mendoza-Novelo et al., 2010, Ariganello et al., 2011). Courtman *et al.* proposed the use of a non-ionic detergent, Triton X-100 and an enzymatic extraction process. After this treatment, the acellular matrix was shown to be composed of collagen, elastin and glycosaminoglycans (GAG). Microscopy revealed that all cellular components were removed and that matrix ultrastructure was intact. More recently, Mendoza-Novelo *et al.* compared the surfactant tridecyl alcohol ethoxylate (ATE) and the reversible alkaline swelling (RAS) treatments to Triton X-100 (Mendoza-Novelo et al., 2010). Histological results indicated a significant reduction of cellular antigens with these three decellularization processes. Nevertheless, the native GAG content varied significantly. It decreased from 88.6 ± 0.2% to 62.7 ± 1.1% and 61.6 ± 0.6% for RAS treatment, ATE and Triton X-100 respectively.

On human pericardial tissue, Mirsadraee *et al.* used a protocol employing hypotonic buffer, SDS, protease inhibitors and nuclease solution. Following decellularization, the tissue is decontaminated using a peracetic acid solution (Mirsadraee et al., 2006; 2007). With this process, glycosaminoglycans and structural proteins, such as collagen, remained intact.

Finally, when dealing with autologous pericardium grafting, full decellularization might not be necessary and thus, simpler protocols can be used. For instance, surgeons commonly prepare autologous pericardium for heart valve replacement by mechanical friction. This allows removing sub-pericardial fat before implantation while better preserving the pericardial architecture stability. This mechanical treatment mainly removes superficial cells, thus allowing 50% of viable pericardial cells to remain in the graft (personal data). The preservation of the pericardial architecture as well as part of the pericardial cells, should maintain a better integrity of the graft, while allowing re-cellularization of the superficial layers.

3.2 Effects of decellularization

Depending on the protocol, decellularization may have an impact on the structural and mechanical properties of the treated tissue (Gilbert et al., 2006). According to Zhou *et al.*, decellularization protocols differ significantly in terms of alteration of ECM histoarchitecture (Zhou et al., 2010). For instance, decellularization protocols have a strong impact on the amount of GAGs remaining in a tissue (Badylak et al., 2009; Mendoza-Novelo et al., 2010). Removing GAGs from a tissue leads to adverse effects on pericardial viscoelastic properties. This can be easily understood since water retention is an important function of GAGs in tissues (Lovekamp et al., 2006). Moreover, GAG content plays a key biological role in cellular signaling and communication. Thus, decreasing GAG content leads to an impaired tissue response and repair. Therefore, the decellularization protocol has to be carefully chosen depending on the tissue type as well as the targeted application. Ideally, the process should remove all cellular antigens without compromising the structure and mechanical properties of the tissue.

Liao *et al.* (Liao et al., 2008) investigated the effect of three decellularization protocols on the mechanical and structural properties on porcine aortic valve leaflets. These protocols were based on the use of SDS, Trypsin and Triton X-100. They showed that decellularization resulted in collagen network disruption, and that the ECM pore size varied as a function of the protocol used. For example, leaflets treated with SDS displayed a dense ECM network and small pore sizes, characteristics that may have an impact on the recolonization of interstitial cells.

It has been demonstrated that decellularization of bovine pericardium with SDS causes irreversible denaturation, swelling and a decrease in tensile strength compared to native tissue (Courtman et al. 1994; García-Paéz et al., 2000; Mendoza-Novelo et al., 2009). Because of these deleterious effects on pericardial tissue, non-ionic detergents are preferred for decellularization processes (Mendoza-Novelo et al., 2010). Nevertheless, some issues may be encountered with the use of non-ionic detergents. Indeed, toxic effects (Argese et al., 1994) and estrogenic effects (Soto et al., 1991; Jobling et al., 1993) have been reported after the use of non-ionic detergents such as alkylphenol ethoxylates.

Decellularization mediates alterations of the structural and mechanical properties of the tissue, but this impact varies depending on the protocol used. For instance, Mirsadraee *et al.* (Mirsadraee et al., 2006) did not observe any significant changes using an SDS-based decellularization protocol in the ultimate tensile strength compared to native tissue on human pericardial tissue. They also observed an increased extensibility of the tissue when cut parallel to collagen bundles.

Tissue decellularization reduces the cellular and humoral immune response targeted against the bioprosthesis (Meyer et al., 2005). However, removing cells does not ensure adequate removal of xenoantigens, nor mitigation of the immune response (Goncalves et al., 2005; Kasimir et al., 2006; Simon et al., 2003; Vesely et al., 1995). For this reason, decellularization protocols have turned to antigen removal protocols (Ueda et al., 2006; Kasimir et al., 2005). The presence of cell membrane antigens, such as oligosaccharide beta-Gal has been reported to lead to an immune response that can be prevented by effective decellularization (Badylak et al., 2008). Interestingly, Griffiths *et al.* (Griffiths et al., 2008) used an immunoproteomic approach to study the ability of bovine pericardium to generate a humoral immune response. They identified thirty one putative protein antigens. Some of them, such as albumin, hemoglobin chain A and beta hemoglobin have been identified as xenoantigens. Recently, Ariganello *et al.* provided evidence that decellularized bovine pericardium induced less differentiation of the monocytes to macrophages compared to polydimethylsiloxane or polystyrene surfaces (Ariganello et al., 2010; 2011). Nevertheless, the effects of the host immune response to acellular pericardium remain to be fully characterized. Understanding this phenomenon is necessary to develop new pericardium preparations and thus improve biological scaffold integration and clinical safety (Badylak & Gilbert, 2008).

Overall, no optimal decellularization treatment has been identified so far, but depending on the target tissue as well as the implantation site, the protocol can be adapted to provide the best decellularization efficiency / functional characteristics ratio. Moreover, some additional treatment can be performed following the decellularization step in order to improve the mechanical and biological features of the graft.

4. Pericardial extracellular matrix treatment

The decellularization process will lead to important alterations of the biomaterial. Its mechanical strength will be diminished and after implantation it will undergo rapid resorption. Hence, approximately 60% of the mass of the ECM is degraded and resorbed between one and three months after *in vivo* grafting (Badylak & Gilbert, 2008). It has also been noted that acellular pericardial tissue, mostly made of type I collagen, is highly thrombogenic (Keuren et al., 2004). Finally, preventing calcification of the graft is also a priority to ensure the long-term benefit of the implantation.

To optimize the features of the bioprosthesis before its clinical grafting, several treatments have been developed and are summarized in Table 1.

	Reagents	References
	Acyl azide	(Petite et al., 1990)
	Carbodiimides	(Sung et al., 2003)
	Cyanimide	(Pereira et al., 1990)
	Dye-mediated photooxidation	(Moore et al., 1994)
	Epoxy compound	(Sung et al., 1997)
	Formaldehyde	(Nimni et al, 1988)
	Genepin	(Sung et al., 1999, 2003; Wei et al., 2005)
Cross-linking treatment	Glutaraldehyde	(Huang-Lee et al, 1990; Jayakrishnan et al., 1996; Thubrikar et al., 1983)
	Glutaraldehyde acetals	(Yoshioka et al., 2008)
	Penta-golloyl glucose	(Tedder et al., 2008)
	Phytate	(Grases et al., 2006, 2008)
	Proanthocyanidin	(Han et al., 2003)
	Reuterin	(Chen et al., 2002)
	Tannic acid	(Cwalina et al., 2005; Jastrzebska et al., 2006; Wang et al., 2008)
	Chitosan	(Nogueira et al., 2010)
	RGD polypeptides	(Dong et al., 2009)
Coating treatment	Silk fibroin	(Nogueira et al., 2010)
	Heparin sodium	(Lee et al., 2000)
	Titanium	(Guldner et al., 2009)
	Amino acids	(Jorge-Herrero et al., 1996; Moritz et al., 1991)
	Glycine	(Lee et al., 2010)
	Heparin	(Lee et al., 2000, 2001)
Post-fixative treatment	Hyaluronic acid	(Ohri et al., 2004)
	L-arginine	(Jee et al., 2003)
	L-glutamic	(Grimm et al., 1991; Leukauf et al., 1993)
	Lyophilization	(Santibáñez-Salgado et al., 2010)
	Sulphonated poly(ethylene oxide)	(Lee et al., 2001)

Table 1. Pericardial processing.

4.1 Cross-linking treatment of pericardial tissue

Cross-linking processing must be carried out to render the tissue non-antigenic, mechanically strong and to minimize xenogeneic tissue degradation (Eliezer et al., 2005; Love, 1997). Nevertheless, degradation should not only be considered as a negative phenomenon, as low molecular weight peptides formed during ECM degradation may have a chemo-attractant potential for several cell types (Badylak & Gilbert, 2008). It is thus the degradation rate of the scaffold that should be primarily considered and evaluated. Depending on the application and cells involved, the degradation rate has to be investigated to ensure proper host cell recruitment and tissue remodelling. The pathways of the immune response involved in this process remain to be fully described (Badylak & Gilbert, 2008).

Introducing cross-links between the polypeptide chains of the ECM has been shown to reduce immunogenicity of the pericardium (Mirsadrae et al., 2007) as well as its biodegradability (Taylor et al., 2006) by increasing its resistance to enzymatic degradation.

Until now, glutaraldehyde (GA)-fixed bovine pericardium has been preferred as a substitute to autologous human pericardium. GA was first introduced by Carpentier *et al.* (Carpentier et al., 1969) as a cross-linking reagent to chemically modify the collagen and render the tissue immunologically acceptable in the human host. Fixation was shown to increase stability and strength of the pericardium (Jayakrishnan & Jameela, 1996). GA remains the gold standard as a cross-linking reagent despite its well-known drawbacks. Indeed, GA has been reported to accelerate the calcification process, which considerably limits its application. Calcification is thus the main cause of long-term failure of GA-fixed pericardial valves (Gallo et al., 1985; Grabenwoger et al., 1996). Furthermore, a GA-treated pericardium has a poor ability to regenerate *in vivo* due to the cross-linking of the tissue. Moreover GA residues display cytotoxic effects preventing host cell attachment, migration and proliferation (Huang-Lee et al., 1990).

It is now accepted that GA cross-linking increases tissue stiffness (Thubrikar et al., 1983) with the possibility of tissue buckling (Vesely et al., 1988). Standard use of GA cross-linking leads to a high risk of calcific degeneration as well as tissue fatigue (Grabenwoger et al., 1992). This is mostly due to inflammatory and cytotoxicity changes (Huang Lee et al., 1990), and continuous wear and tear leading to collagen fiber fragmentation.

Besides glutaraldehyde, several cross-linking compounds have been reported in the literature such as genipin (Wei et al., 2005) or epoxy compound (Sung et al., 1997). These alternative methods are used to bridge hydroxylysine residues of different polypeptide chains or amino groups of lysine by oligomeric or monomeric crosslinks (Sung et al., 2003). Because of the adverse effects of cross-linking with glutaraldehyde or other aldehyde treatments such as formaldehyde (Nimni et al., 1988) or dialdehyde starch (Rosenberg, 1978), numerous non-aldehyde treatments have been proposed, such as carbodiimides (Sung et al., 2003), glycerol (Ferrans et al., 1991), glycidal ethers (Thyagarajanet al., 1992) including poly(glycidylether) (Noishiki et al., 1986), acyl azide (Petite et al., 1990), cyanimide (Pereira et al., 1990), genipin (Wei et al., 2005), or dye-mediated photo-oxidation, phytate (Grases et al., 2008).

Genipin, obtained from the fruits of *Gardenia jasminoides* ELLIS (Fujikawa et al., 1987; Tsai et al., 1994), exhibited better results than glutaraldehyde regarding its cytotoxicity (Sung et al., 1999), inflammatory response, ability to prevent calcification and tissue-induced mechanical properties (Wei et al., 2005). Epoxy compound, initially proposed by Noishiki *et al.* (Noishiki et al., 1989), was shown to be less cytotoxic, superior in pliability and to better inhibit calcification than glutaraldehyde (Sung et al., 1997).

Carbodiimides generate amide-type crosslinks via direct cross-linking of the polypeptide chains. Use of carbodiimide cross-linking leads to the activation of the carboxylic acid groups of glutamic or aspartic acid residues to obtain O-acylisourea groups. Hydroxyline residues or free amino groups of lysine generate a nucleophilic attack which allows cross-link formation (Timkovich, 1977). It was noted that adding N-hydroxysuccinimides to carbodiimides considerably increases cross-link number (Olde Damink et al., 1996). In addition, the use of carbodiimides displayed increased stability towards enzymatic degradation on collagen-based tissue such as pericardium (Sung et al., 2003).

Glutaraldehyde acetal cross-linking reagent has been developed with glutaraldehyde in acid ethanolic solution (Yoshioka & Goissis, 2008), protecting free aldehydic reactive groups and minimizing the polymeric formation of glutaraldehyde. This reduces superficial effects with glutaraldehyde cross-linking on pericardial tissue.

Crosslinking of the pericardial tissue with a dye-mediated photo-oxydation process provides chemical, enzymatic and *in vivo* stability as well as biomechanical integrity of the treated tissue (Moore et al., 1994). Penta-golloyl glucose, a collagen-binding polyphenol, stabilizes collagen, preventing its degradation, and allows progressive host cell infiltration as well as ECM remodeling. An *in vivo* study has shown that porcine pericardium does not calcify with such treatment at 6 weeks when implanted subdermally in rats (Tedder et al., 2008). Reuterin, an antimicotic and antibacterial compound obtained from *Lactobacillus reuteri* (Axelsson et al., 1989), has been studied as a crosslinking reagent (Chen et al., 2002). It is a three-carbon aldehyde reacting, as formaldehyde, with free amino groups. Reuterin cross-linked pericardium exhibits comparable results to glutaraldehyde in terms of resistance against enzymatic degradation, denaturation temperature and free amino group content, while decreasing cytotoxic effects (Chen et al., 2002). Tannic acid has been studied on pericardial tissue and was shown to crosslink proteins by creating multiple hydrogen bonds due to its hydroxyl groups (Cwalina et al., 2005; Jastrzebska et al., 2006). It exerts an anti-inflammatory effect, especially on macrophages, as well as an anti-calcification effect on glutaraldehyde-fixed bovine pericardium (Wang et al., 2008). Proanthocyanidin, a natural crosslinking reagent with polyphenolic structures, has the potential to create a stable hydrogen-bonded structure and to increase collagen synthesis, generating nonbiodegradable collagen matrices (Han et al., 2003). Proanthocyanidin-treated pericardial tissues are non-cytotoxic and resist against enzyme digestion, and have been shown to be compatible with cell attachment and proliferation. Phytate has been suggested as an anti-calcification reagent (Grases et al., 2006, 2008) and has achieved promising results, to be validated by further studies. Other amide-type crosslinks, based on the activation of carboxyl groups, have been studied, such as diphenylphosphorylazide or ethyldimethylaminopropyl carbodiimide. It appears, according to Jorge-Herrero *et al.*, that these two chemical treatments are not a good alternative compared to glutaraldehyde. Indeed, pericardial tissues treated with those reagents are less resistant to calcifications and proteolytic attacks (Jorge-Herrero et al., 1999).

Numerous alternative treatments to glutaraldehyde cross-linking have been developed and investigated over the years. However, most of them were mainly evaluated *in vitro* and compared only to glutaraldehyde. A comprehensive comparative study of the different reagents remains to be conducted in terms of benefits regarding the tissue properties as well as their potential toxicity or deleterious effects.

4.2 Coating of the pericardium

Another possible post-decellularization treatment resides in the coating of the bioprosthesis. This procedure should allow improvement of graft integration at the site of implantation as well as decreasing degradation of the pericardial tissue.

Coating bovine pericardium with biopolymeric films, either chitosan or silk fibroin, has been investigated by Nogueira *et al.* (Nogueira et al., 2010). These methods are interesting approaches and both treatments appear to be non-cytotoxic. Nevertheless, chitosan does not allow endothelialisation and silk fibroin-coated bovine pericardium calcifies *in vivo*. Further investigation has to be performed to tackle these major concerns.

In their study, Dong *et al.* suggested treating bovine pericardium with acetic acid coupled with RGD polypeptides (Dong et al., 2009). Acetic acid increases pericardial scaffold pore size and porosity while RGD peptides is meant to improve cell adhesion and growth. Hence, RGD polypeptides have been identified in fibronectin (Pierschbacher & Ruoslahti, 1984), collagen, vitronectin and membrane proteins (Ruoslahti & Pierschbacher, 1987). These sequences have an impact on integrins, which display cell adhesion receptor roles controlling cell signaling pathways.

4.3 Pericardium anti-calcification treatments

The mechanism of calcification on glutaraldehyde-treated pericardium is not well understood because of its complexity. Nevertheless, there is evidence that pericardial tissue residual antigens, free aldhehyde groups of glutaraldehyde and phospholipids are involved in this mechanism.

Thus, circulating antibodies can contribute to pericardial calcification due to a possible immune response. Free aldehyde groups of glutaraldehyde can attract host plasma calcium, increasing tissue calcification. Phospholipids may bind calcium and play an important role in the calcium phosphate crystal formation. Several strategies have been investigated to tackle these major issues.

Suppression of residual antigenicity has been proposed to prevent calcification and it has been shown to be effective. This was performed by fixation treatments using a broad range of high concentrations of glutaraldehyde (Trantina-Yates et al., 2003; Zilla et al., 2000). To remove free aldehyde groups, a large number of amino acids or amino compounds were studied. Post-fixation treatments with amino acids displayed an improved spontaneous endothelialisation *in vivo* of glutaraldehyde-fixed bovine pericardium (Moritz et al., 1991; Jorge-Herrero et al., 1996). The use of L-glutamic acid did reduce residual and unbound aldehyde groups, on glutaraldehyde-fixed bovine pericardium and significantly decreased the risk of calcification (Grimm et al., 1991; Leukauf et al., 1993). Post-treatment with L-arginine also resulted in decreased calcium deposition (Jee et al., 2003). Recently, Lee *et al.* proposed a post-fixation treatment with glycine (Lee et al., 2010). Early results are promising but require further investigation on larger studies.

Alcohol solutions, including ethanol, have been investigated as a treatment to remove tissue phospholipids, thus preventing calcification (Pathak et al., 2004; Vyavahare et al., 1998). Besides, other techniques have been proposed to minimize the side effects of glutaraldehyde residues on GA-treated pericardium. Lyophilization has been shown to decrease aldehyde residues, decreasing the risk of calcification and cytotoxicity (Santibáñez-Salgado et al., 2010).

Moreover, treatments with heparin or sulphonated poly(ethylene oxide) following glutaraldehyde pre-treatment have been proposed (Lee et al., 2000, 2001). Both methods block side effects of GA residues and thus prevent calcification of the pericardium. Finally, a modified adipic dihydrazide hyaluronic acid has been proposed to be grafted on to glutaraldehyde-treated bovine pericardium (Ohri et al., 2004). Calcifications decreased considerably with this post-treatment compared to the control group at two weeks following a subcutaneous implantation in mice.

5. Applications of the pericardium as a biomaterial

So far, the pericardium has been mostly used for cardio-vascular applications, i.e. vascular grafts (Schmidt & Baier, 2000; Chvapil et al., 1970; Matsagas et al., 2006; Menasche et al.,

Pericardium source	Surgical fields	Product	Company
Bovine or porcine	Soft tissue repair Hernia repair Abdominal & thoracic wall defects	-Peripatch® Implantable surgical tissue -TutoMesh®	Neovasc, Maverick Biosciences PTY Limited, Tutogen medical GmbH, RTI Biologics, Med&Care, Biovascular Inc, Novomedics
	Strip reinforcement	-Veritas Peristrips® Dry	Synovis Life Technology
	Orbital repair	-Tutopatch® -Ocugard®	Tutogen medical GmbH, RTI Biologics, Med&Care, Biovascular Inc, Novomedics
	Dural repair	-Lyolem ®r All BP	National tissue Bank Malaysia
	Perivascular Patch	-Peripatch® biologic vascular patch	Neovac
	Cardiac reconstruction and repair	-Peripatch® Implantable Surgical Tissue	Neovasc, Maverick Biosciences PTY Limited
	Heart valve replacement	-PercevalS® aortic valve -Mitroflow® pericardial aortic valve -Freedom solo® -Carpentier-Edwards PERIMOUNT® Magna EaseAortic Heart Valve -Carpentier-Edwards PERIMOUNT® Magna Mitral Ease Heart Valve -Carpentier-Edwards PERIMOUNT® Theon Aortic Heart Valve -Carpentier-Edwards PERIMOUNT® Theon Mitral Replacement System	Sorin group " " Edwards Life Sciences " " "
Equine	Tendon repair	-OrthADAPT®	Synovis Life Technologies Inc
Human	Valvuloplasty Heart valve	-Xeno or (tissue bank) or autologous grafts	Lausberg et al, 2006 Mirsadaee et al, 2006

Table 2. Applications of pericardium as medical devices.

1984; Moon & West; 2008), and heart valves (Ishihara et al., 1981; Schoen & Levy, 1999; Flanagan & Pandit, 2003; Vesely, 2005). Pericardial bioprostheses have also been described for the treatment of acquired cardiac pathologies, including postinfarction septal defects (David et al., 1995), reconstruction of mitral valve annulus (David et al., 1995a, 1995b) or outflow obstruction (Sommers & David, 1997).

Additionally, pericardium has also been used for the construction of bioprostheses in non-cardiac treatments such as patches for vaginal (Lazarou et al., 2005) or abdominal wall reparation (Limpert et al., 2009), dural repair (Cantore et al., 1987) or tracheoplasty (Dunham et al., 1994).

6. Conclusion

For clinical application, pericardial tissue has to be decellularized to prevent an immune responses or immune-mediated rejection of the pericardium. Various decellularization protocols have been largely reviewed here. The choice of the decellularization strategy has an impact on the mechanical properties, the scaffold pore size, the scaffold tissue integration and the development of long-term calcification. All these considerations should be carefully taken into account when designing new pericardial-based biomaterials. Currently, glutaraldehyde is the gold standard for pericardial treatment used in clinical practice. Nevertheless, it has important drawbacks including cytotoxic effects, prevention of host cell attachment, migration and proliferation (Huang-Lee et al., 1990), and a high propensity to calcify. Alternative treatments to replace or complement glutaraldehyde crosslinking of the pericardium have been investigated using other crosslinking reagents, decellularization, lyophilisation or coating with biopolymers (Nogueira et al., 2010). Despite many studies, it is still difficult to know which strategy to adopt regarding pericardial treatment. First, we do not have enough follow-up to permit evaluation of most of these alternatives and treatments. Second, every new treatment proposed is generally compared only to glutaraldehyde. It is thus not possible to classify these treatments by efficiency. Finally, the protocol for an optimal treatment depends largely on the final application targeted. In addition, there have been recent advances in tissue regeneration with the emergence of cell therapy and new pericardial treatments with cellular growth factors promoting recellularization (Chang et al., 2007). However, further improvements need to be achieved to transform these techniques into clinical applications. The use of autologous pericardium in cardiac valvular therapy is also a challenging alternative. Nevertheless, it still currently requires the development of local pericardial treatments aiming to favor the valvular remodelling. The understanding of current issues and the improvement of pericardial processing may have a huge impact for bioprothesis conception and patient safety.

7. References

Al-Halees Z, Gometza B, Duran CM. Aortic valve repair with bovine pericardium. (1998). *Ann Thorac Surg,* Vol. 65, pp. (601-602).

Al-Halees Z, Al Shahid M, Al Sanei A, Sallehuddin A, Duran C. Up to 16 years follow-up of aortic valve reconstruction with pericardium: a stentless readily available cheap valve? (2005) *Eur J Cardiothorac Surg.* Aug; Vol.28, pp.(200-205).

Allen DJ, Didio LJA. (1984). The structure of native human, bovine and porcine parietal pericardium. *Anatomical Record,* Vol. 208, No. 3, pp. (7A-7A).

Argese E, Marcomini A, Bettiol C, Perin G, Miana P. (1994). Submitochondrial particle response to linear alkylbenzene sulfonates, nonylphenol polyethoxylates and their biodegradation derivates. *Environ Toxicol Chem*, Vol. 13, pp. (737-742).

Ariganello MB, Labow RS, Lee JM. (2010). In vitro response of monocyte-derived macrophages to a decellularized pericardial biomaterial. *J Biomed Mater Res*, Vol. 93, No. 1, pp. (280–288).

Ariganello MB, Simionescu DT, Labow RS, Lee JM. (2011). Macrophage differentiation and polarization on a decellularized pericardial biomaterial. *Biomaterials*, Vol. 32, No. 2, pp. (439-449).

Axelsson L, Chung TC, Dobrogosz WJ, Lindgren LE. (1989). Production of a broad spectrum antimicrobial substance by *Lactobacillus reuteri*. *Microbial Ecol Health Disease*, Vol. 2, pp. (131–136).

Bader A, Schilling T, Teebken OE, Brandes G, Herden T, Steinhoff G, Haverich A. (1998). Tissue engineering of heart valves—human endothelial cell seeding of detergent acellularized porcine valves. *Eur J Cardiothorac Surg*, Vol. 14, No. 3, pp. (279–284).

Badylak SF, Freytes DO, Gilbert TW. (2009). Extracellular matrix as a biological scaffold material: structure and function. *Acta Biomater*, Vol. 5, No. 1, pp. (1-13).

Badylak SF, Gilbert TW. (2008). Immune response to biologic scaffold materials. *Seminars in Immunology*, Vol. 20, No. 2, pp. (109–116).

Bahnson HT, Hardesty RL, Baker LD, Brookes D II, Gall DA. (1970) Fabrication and evaluation of tissue leaflets for aortic and mitral valve replacement. *Ann Surg*, Vol.171, pp. (939-947).

Cantore G, Guidetti B, Delfini R. (1987). Neurosurgical use of human dura mater sterilized by gamma rays and stored in alcohol: long-term results. *J Neurosurg*, Vol. 66, No. 1, pp. (93-95).

Carpentier A, Lemaigre G, Robert L, Carpentier S, Dubost C. (1969). Biological factors affecting long-term results in valvular heterografts. *J Thorac Cardiovasc Surg*, Vol. 58, No. 4, pp. (467–483).

Chang Y, Chen SC, Wei HJ, Wu TJ, Liang HC, Lai PH, Yang HH, Sung HW. (2005). Tissue regeneration observed in a porous acellular bovine pericardium used to repair a myocardial defect in the right ventricle of a rat model. *J Thorac Cardiovasc Surg*, Vol. 130, No. 5, pp. (705-711).

Chang Y, Lai PH, Wei HJ, Lin WW, Chen CH, Hwang SM, Chen SC, Sung HW. (2007). Tissue regeneration observed in a basic fibroblast growth factor–loaded porous acellular bovine pericardium populated with mesenchymal stem cells. J Thorac Cardiovasc Surg, Vol. 134, No. 1, pp. (65-73).

Chen CN, Sung HW, Liang HF, Chang WH. (2002). Feasibility study using a natural compound (reuterin) produced by *Lactobacillus reuteri* in sterilizing and crosslinking biological tissues. *J Biomed Mater Res*, Vol. 61, No. 3, pp. (360–369).

Chvapil M, Kronenthal RL, van Winkle Jr W. (1970). Medical and surgical applications of collagen, In: *International review of connective tissue research*, Hall DA, Jackson DS. Vol. 6, pp. (1–61) [Chapter 1], Academic Press, NY.

Courtman DW, Pereira CA, Kashef V, McComb D, Lee JM, Wilson GJ. (1994). Development of a pericardial acellular matrix biomaterial: biochemical and mechanical effects of cell extraction. *J Biomed Mater Res*, Vol. 28, No. 6, pp. (655–666).

Crapo PM, Gilbert TW, Badylak SF. (2011). An overview of tissue and whole organ decellularization processes. *Biomaterials*, Vol. 32, No. 12, pp. (3233-3243).

Cwalina B, Turek A, Nozynski J, Jastrzebska M, Nawrat Z. (2005). Structural changes in pericardium tissue modified with tannic acid. *Int J Artif Organs*, Vol. 28, No.6, pp. (648-653).

Dahl SL, Koh J, Prabhakar V, Niklason LE. (2003). Decellularized native and engineered arterial scaffolds for transplantation. *Cell Transplant*, Vol. 12, No. 6, pp. (659-666).

David TE, Dale L, Sun Z. (1995). Postinfarction ventricular septal rupture: Repair by endocardial patch with infarct exclusion. *J Thorac Cardiovasc Surg*, Vol. 110, No. 5, pp. (1315-1322).

David TE, Feindel CM, Armstrong S, et al. (1995). Reconstruction of the mitral annulus. A ten-year experience. *J Thorac Cardiovas Surg*, Vol. 110, pp. (1323-x1328).

DeCarbo WT, Feldner BM, Hyer CF. (2010). Inflammatory Reaction to Implanted Equine Pericardium Xenograft. *J Foot Ankle Surg*, Vol. 49, No. 2, pp. (155-158).

Dong X, Wei X, Yi W, Gu C, Kang X, Liu Y, Li Q, Yi D. (2009). RGD-modified acellular bovine pericardium as a bioprosthetic scaffold for tissue engineering. *J Mater Sci Mater Med*, Vol. 20, pp. (2327-2336).

Dunham ME, Holinger LD, Backer CL, Mavroudis C. (1994). Management of severe congenital tracheal stenosis. *Ann Otol Rhinol Laryngol*, Vol. 103, pp. (351-356).

Duran CM, Gometza B, Shahid M, Al-Halees Z. Treated bovine and autologous pericardium for aortic valve reconstruction. (1998) *Ann Thorac Surg*. Dec; Vol. 66, pp (S166-199).

Edwards WS. (1969). Aortic valve replacement with autogenous tissue. *Ann Thorac Surg* Vol. 8, pp. (126-32).

Eliezer MA, Lydia MM, Virna VR, Carlos FR. (2005). Mechanics of biomaterials: vascular graft prostheses. *Proceedings of Application of Engineering Mechanics in Medicine*, GED, University of Puerto Rico, Mayaguez, Group A: A1–A25, May 2005.

Ferrans VJ, Milei J, Ishihara T, Storino R. (1991). Structural changes in implanted cardiac valvular bioprostheses constructed of glycerol-treated dura mater. *Eur J Cardiothorac Surg*, Vol. 5, No. 3, pp. (144–154).

Flanagan TC, Pandit A. (2003). Living artificial heart valve alternatives: a review. *Eur Cell Mater*, Vol. 6, pp. (28–45).

Fomovsky GM, Thomopoulos S, Holmes JW. (2010). Contribution of extracellular matrix to the mechanical properties of the heart. *J Mol Cell Cardiol*, Vol. 48, No. 3, pp. (490–496).

Freytes DO, Badylak SF, Webster TJ, Geddes LA, Rundell AE. (2004). Biaxial strength of multilaminated extracellular matrix scaffolds. *Biomaterials*, Vol. 25, No. 12, pp. (2353–2361).

Fujikawa S, Yokota T, Koga K, Kumada J. (1987). The continuous hydrolysis of geniposide to genipin using immobilized β-glucosidase on calcium alginate gel. *J Biotechnol Lett*, Vol. 9, pp. (697–702).

Gallo I, Artinano E, Nistal F. (1985). Four- to seven-year follow-up of patients undergoing Carpentier-Edwards porcine heart valve replacement. *Thorac Cardiovasc Surg*, Vol. 33, No. 6, pp. (347-351).

Gamba PG, Conconi MT, Lo Piccolo R, Zara G, Spinazzi R, Parnigotto PP. (2002). Experimental abdominal wall defect repaired with acellular matrix. *Pediatr Surg Int*, Vol. 18, No. 5-6, pp. (327–331).

García-Paéz JM, Herrero J, Carrera-San Martín A, García-Sestafe JV, Téllez G, Millán I, Salvador J, Cordon A, Castillo-Olivares JL. (2000). The influence of chemical treatment and suture on the elastic behavior of calf pericardium utilized in the construction of cardiac bioprostheses. *J Mater Sci Mater Med*, Vol. 11, No. 5, pp. (273–277).

Gilbert TW, Sellaro TL, Badylak SF. (2006). Decellularization of tissues and organs. Biomaterials, Vol. 27, No. 19, pp. (3675–3683).

Goetz WA, Lim HS, Lansac E, Weber PA, Duran CM. A temporarily stented, autologous pericardial aortic valve prosthesis. (2002) *J Heart Valve Dis*. Sep;11(5):(696-702).

Goissis G, Suzigan S, Parreira DR, Maniglia JV, Braile DM, Raymundo S. (2000). Preparation and characterization of collagen–elastin matrices from blood vessels intended as small diameter vascular grafts. *Artif Organs*, Vol. 24, No. 3, pp. (217–223).

Goncalves AC, Griffiths LG, Anthony RV, Orton EC. (2005). Decellularization of bovine pericardium for tissue-engineering by targeted removal of xenoantigens. *J Heart Valve Dis*, Vol. 14, No. 2, pp. (212–217).

Grabenwoger M, Grimm M, Eybl E, Leukauf C, Muller MM, Plenck Jr H, Böck P. (1992). Decreased tissue reaction to bioprosthetic heart valve material after glutamic acid treatment. A morphological study. *J Biomed Mater Res*, Vol. 26, No. 9, pp. (1231–1240).

Grabenwoger M, Sider J, Fitzal F, Zelenka C, Windberger U, Grimm M, Moritz A, Böck A, Wolner E. (1996). Impact of glutaraldehyde on calcification of pericardial bioprosthetic heart valve material. *Ann Thorac Surg*, Vol. 62, No. 3, pp. (772-777).

Grases F, Sanchis P, Costa-Bauzá A, Bonnin O, Isern B, Perelló J, Prieto RM. (2008). Phytate inhibits bovine pericardium calcification in vitro. *Cardiovasc Pathol*, Vol. 17, No. 3, pp. (139–145).

Grases F, Sanchis P, Perello J, Isern B, Prieto RM, Fernandez-Palomeque C, Fiol M, Bonnin O, Torres JJ. (2006). Phytate (myo-inositol hexakisphosphate) inhibits cardiovascular calcifications in rats. *Front Biosci*, Vol. 11, pp. (136–142).

Grauss RW, Hazekamp MG, van Vliet S, Gittenberger-de Groot AC, DeRuiter MC. (2003). Decellularization of rat aortic valve allografts reduces leaflet destruction and extracellular matrix remodeling. *J Thorac Cardiovasc Surg*, Vol. 126, No. 6, pp. (2003–2010).

Griffiths LG, Choe LH, Reardon KF, Dow SW, Orton EC. (2008). Immunoproteomic identification of bovine pericardium xenoantigens. *Biomaterials*, Vol. 29, No. 26, pp. (3514–3520).

Grimm M, Eybl E, Grabenwoger M, Griesmacher A, Losert U, Bock P, Muller MM, Wolner E. (1991). Biocompatibility of aldehyde-fixed bovine pericardium. An in vitro and in vivo approach toward improvement of bioprosthetic heart valves. *J Thorac Cardiovasc Surg*, Vol 102, No. 2, pp. (195–201).

Guldner NW, Jasmund I, Zimmermann H, Heinlein M, Girndt B, Meier V, Flüß F, Rohde D, Gebert A, Sievers HH. (2009). Detoxification and Endothelialization of Glutaraldehyde-Fixed Bovine Pericardium With Titanium Coating: A New Technology for Cardiovascular Tissue Engineering. *Circulation*, Vol. 119, No. 12, pp. (1653-1660).

Han B, Jaurequi J, Wei Tang B, Nimni ME. (2003). Proanthocyanidin: A natural crosslinking reagent for stabilizing collagen matrices. *J Biomed Mater Res A*, Vol. 65, No. 1, pp. (118–124).

Hiester ED, Sacks MS. (1998a). Optimal bovine pericardial tissue selection sites. I. Fiber architecture and tissue thickness measurements. *J Biomed Mater Res*, Vol. 39, No. 2, pp. (207–214).

Hiester ED, Sacks MS. (1998b). Optimal bovine pericardial tissue selection sites. II. Cartographic analysis. *J Biomed Mater Res*, Vol. 39, No. 2, pp. (215–221).

Huang Lee LLH, Cheung DT, Nimni ME. (1990). Biochemical changes and cytotoxicity associated with the degradation of polymeric glutaraldehyde derived cross-links. *J Biomed Mater Res*, Vol. 24, No. 9, pp. (1185–1201).

Hudson TW, Zawko S, Deister C, Lundy S, Hu CY, Lee K, Schmidt CE. (2004). Optimized acellular nerve graft is immunologically tolerated and supports regeneration. *Tissue Eng*, Vol. 10, No. 11-12, pp. (1641–1651).

Ishihara T, Ferans VJ, Jones M, Boyce SW, Roberts WC. (1981). Structure of bovine parietal pericardium and of unimplanted Ionescu shiley pericardial valvular bioprostheses. *J Thorac Cardiovasc Surg*, Vol. 81, No. 5, pp. (747–757).

Jackson DW, Grood ES, Wilcox P, Butler DL, Simon TM, Holden JP. (1988). The effects of processing techniques on the mechanical properties of bone–anterior cruciate ligament–bone allografts. An experimental study in goats. *Am J Sports Med*, Vol. 16, No. 2, pp. (101–105).

Jastrzebska M, Zalewska-Rejdak J, Wrzalik R, Kocot A, Mroz I, Barwinski B, Turek A, Cwalina B. (2006). Tannic acid-stabilized pericardium tissue: IR spectroscopy, atomic force microscopy, and dielectric spectroscopy investigations. *J Biomed Mater Res A*, Vol. 78, No. 1, pp. (148-156).

Jayakrishnan A, Jameela SR. (1996). Glutaraldehyde as a fixative in bioprostheses and drug delivery matrices. *Biomaterials*, Vol. 17, No. 5, pp. (471–484).

Jee KS, Kim YS, Park KD, Kim YH. (2003). A novel chemical modification of bioprosthetic tissues using L-arginine. *Biomaterials*, Vol. 24, No. 20, pp. (3409–3416).

Jobling S, Sumpter JP. (1993). Detergent components in sewage effluent are weakly estrogenic to fish – an in vitro study using rainbow-trout. *Aquat Toxicol*, Vol. 27, pp. (361–372).

Jorge-Herrero E, Fernandez P, Escudero C, Garcia-Paez JM, Castillo-Olivares JL. (1996). Calcification of pericardial tissue pretreated with different amino acids. *Biomaterials*, Vol. 17, No. 6, pp. (571-575).

Jorge-Herrero E, Fernandez P, Turnay J, Olmo N, Calero P, Garcia R, Freile I, Castillo-Olivares JL. (1999). Influence of different chemical cross-linking treatments on the properties of bovine pericardium and collagen. *Biomaterials*, Vol. 20, No. 6, pp. (539-545).

Kasimir MT, Rieder E, Seebacher G, Nigisch A, Dekan B, Wolner E, Weigel G, Simon P. (2006). Decellularization does not eliminate thrombogenicity and inflammatory stimulation in tissue-engineered porcine heart valves. *J Heart Valve Dis*, Vol. 15, No. 2, pp. (278–286) [discussion 286].

Kasimir MT, Rieder E, Seebacher G, Wolner E, Weigel G, Simon P. (2005). Presence and elimination of the xenoantigen gal (alpha1, 3) gal in tissue-engineered heart valves. *Tissue Eng*, Vol. 11, No. 7-8, pp. (1274–1280).

Keuren JFW, Wielders SJH, Driessen A, Verhoeven M, Hendriks M, Lindhout T. (2004). Covalently-bound heparin makes collagen thromboresistant. *Arterioscler Thromb Vasc Biol*, Vol. 24, No. 3, pp. (613–617).

Lazarou G, Powers K, Pena C, Bruck L, Mikhail MS. (2005). Inflammatory reaction following bovine pericardium graft augmentation for posterior vaginal wall defect repair. *Int Urogynecol J Pelvic Floor Dysfunct*, Vol. 16, No. 3, pp. (242–244).

Lee C, Kim SH, Choi SH, Kim YJ. (2011) High-concentration glutaraldehyde fixation of bovine pericardium in organic solvent and post-fixation glycine treatment: in vitro material assessment and in vivo anticalcification effect. *Eur J Cardiothorac Surg*, Vol. 39, No. 3, pp. (381-387).

Lee JM, Boughner DR. (1981). Tissue mechanics of canine pericardium in different test environment. *Circ Res*, Vol. 49, No. 2, pp. (533–544).

Lee JM, Boughner DR. (1985). Mechanical properties of human pericardium. Differences in viscoelastic response when compared with canine pericardium. *Circ Res*, Vol. 57, No. 3, pp. (475–481).

Lee WK, Park KD, Han DK, Suh H, Park JC, Kim YH. (2000). Heparinized bovine pericardium as a novel cardiovascular bioprosthesis. *Biomaterials*, Vol. 21, No. 22, pp. (2323-2330).

Lee WK, Park KD, Kim YH, Suh H, Park JC, Lee JE, Sun K, Baek MJ, Kim HM, Kim SH. (2001). Improved Calcification Resistance and Biocompatibility of Tissue Patch Grafted with Sulfonated PEO or Heparin after Glutaraldehyde Fixation. *J Biomed Mater Res*, Vol. 58, No. 1, pp. (27–35).

Leukauf C, Szeles C, Salaymeh L, Grimm M, Grabenwoger M, Losert U, Moritz A, Wolner E. (1993). In vitro and in vivo endothelialization of glutaraldehyde treated bovine pericardium. *J Heart Valve Dis*, Vol. 2, No. 2, pp. (230–235).

Liang HC, Chang Y, Hsu CK, Lee MH, Sung HW. (2004). Effects of crosslinking degree of an acellular biological tissue on its tissue regeneration pattern. *Biomaterials*, Vol. 25, No. 17, pp. (3541–3552).

Liao J, Joyce EM, Sacks MS. (2008). Effects of decellularization on the mechanical and structural properties of the porcine aortic valve leaflet. *Biomaterials*, Vol. 29, No. 8, pp. (1065–1074).

Liao J, Yang L, Grashow J, Sacks MS. (2005). Molecular orientation of collagen in intact planar connective tissues under biaxial stretch. *Acta Biomater*, Vol. 1, No. 1, pp. (45–54).

Limpert JN, Desai AR, Kumpf AL, Fallucco MA, Aridge DL. (2009). Repair of abdominal wall defects with bovine pericardium. *Am J Surg*, Vol. 198, pp. (e60–65).

Love JW. (1997). Cardiac prostheses, In: *Principles of tissue engineering*, Lanza R, Langer R, Chick W, pp. (365–378), R.G. Landes Company, New York.

Lovekamp JJ, Simionescu DT, Mercuri JJ, Zubiate B, Sacks MS, Vyavahare NR. (2006). Stability and function of glycosaminoglycans in porcine bioprosthetic heart valves. *Biomaterials*, Vol. 27, No. 8, pp. (1507–1518).

Maestro MM, Turnay J, Olmo N, Fernández P, Suárez D, García Páez JM, Urillo S, Lizarbe MA, Jorge-Herrero E. (2006). Biochemical and mechanical behavior of ostrich pericardium as a new biomaterial. *Acta Biomater*, Vol. 2, No. 2, pp. (213–219).

Matsagas MI, Bali C, Arnaoutoglou E, Papakostas JC, Nassis C, Papadopoulos G, Kappas AM. (2006). Carotid endarterectomy with bovine pericardium patch angioplasty: mid-term results. *Ann Vasc Surg*, Vol. 20, No. 5, pp. (614–619).

McFetridge PS, Daniel JW, Bodamyali T, Horrocks M, Chaudhuri JB. (2004). Preparation of porcine carotid arteries for vascular tissue engineering applications. *J Biomed Mater Res A*, Vol. 70, No. 2, pp. (224–234).

Menasche P, Flaud P, Huc Co A, Piwnica A. (1984). Collagen vascular grafts: a step towards improved compliance in small-caliber by-pass surgery: preliminary report. *Life Support Syst*, Vol. 2, No. 4, pp. (233–237).

Mendoza-Novelo B, Avila EE, Cauich-Rodríguez JV, Jorge-Herrero E, Rojo FJ, Guinea GV, Mata-Mata JL. (2011). Decellularization of pericardial tissue and its impact on tensile viscoelasticity and glycosaminoglycan content. *Acta Biomater*, Vol. 7, No. 3, pp. (1241-1248).

Mendoza-Novelo B, Cauich-Rodriguez JV. (2009). The effect of surfactants, crosslinking agents and L-cysteine on the stabilization and mechanical properties of bovine pericardium. *J Appl Biomater Biomech*, Vol. 7, No. 2, pp. (123–131).

Meyer SR, Nagendran J, Desai LS, Rayat GR, Churchill TA, Anderson CC, Rajotte RV, Lakey JRT, Ross DB. (2005). Decellularization reduces the immune response to aortic valve allografts in the rat. *J Thorac Cardiovasc Surg*, Vol. 130, No. 2, pp. (469-476).

Mirsadraee S, Wilcox HE, Korossis S, Kearney JN, Watterson KG, Fisher J, Ingham E. (2006). Development and characterization of an acellular human pericardial matrix. *Tissue Eng*, Vol. 12, No. 4, pp. (763-773).

Mirsadraee S, Wilcox HE, Watterson KG, Kearney JN, Hunt J, Fisher J, Ingham E. (2007). Biocompatibility of Acellular Human Pericardium. *J Surg Res*, Vol. 143, No. 2, pp. (407–414).

Moon JJ, West JL. (2008). Vascularization of engineered tissues: approaches to promote angiogenesis in biomaterials. *Curr Top Med Chem*, Vol. 8, No. 4, pp. (300–310).

Moore MA, Bohachevsky IK, Cheung DT, Boyan BD, Chen W, Bickers RR, McIlroy BK. (1994). Stabilization of pericardial tissue by dye-mediated photooxidation. *J Biomed Mater Res*, Vol. 28, No. 5, pp. (611–618).

Moritz A, Grimm M, Eybl E, Grabenwoger M, Grabenwoger F, Bock P, Wolner E. (1991). Improved spontaneous endothelialization by postfixation treatment of bovine pericardium. *Eur J Cardiothorac Surg*, Vol. 5, No. 3, pp. (155-159) [discussion 160].

Neethling WM, Cooper S, Van Den Heever JJ, Hough J, Hodge AJ. (2002). Evaluation of kangaroo pericardium as an alternative substitute for reconstructive cardiac surgery. *J Cardiovasc Surg (Torino)*, Vol. 43, No. 3, pp. (301–306).

Neethling WM, Papadimitriou JM, Swarts E, Hodge AJ. (2000). Kangaroo versus porcine aortic valve tissue-valve geometry morphology, tensile strength and calcification potential. *J Cardiovasc Surg (Torino)*, Vol. 41, No. 3, pp. (341–348).

Nimni ME, Cheung D, Strates B, Kodama M, Sheikh K. (1988). Bioprosthesis derived from cross-linked and chemically modified collagenous tissues, In: *Collagen, Vol. III*, Nimni ME, pp. (1-38), CRC Press, Boca Raton, FL.

Nogueira GM, Rodas ACD, Weska RF, Aimoli CG, Higa OZ, Maizato M, Leiner AA, Pitombo RNM, Polakiewicz B, Beppu MM. (2010). Bovine pericardium coated with biopolymeric films as an alternative to prevent calcification: *In vitro* calcification and cytotoxicity results. *Materials Science and Engineering C*, Vol. 30, pp. (575–582).

Noishiki Y, Kodaira K, Furuse M, Miyata T. (1989). Method of preparing antithrombogenic medical materials, U.S. Patent No. 4,806,595.

Noishiki Y, Miyata T, Kodaira K. (1986). Development of a small caliber vascular graft by a new crosslinking method incorporating slow heparin release collagen and natural tissue compliance. *ASAIO Trans*, Vol. 32, No. 1, pp. (114–119).

Ohri R, Hahn SK, Hoffman AS, Stayton PS, Giachelli CM. (2004). Hyaluronic acid grafting mitigates calcification of glutaraldehyde-fixed bovine pericardium. *J Biomed Mater Res A*, Vol. 70, No. 2, pp. (328–334).

Olde Damink LH, Dijkstra PJ, van Luyn MJ, van Wachem PB, Nieuwenhuis P, Feijen J. (1996). Cross-linking of dermal sheep collagen using a water-soluble carbodiimide. *Biomaterials*, Vol. 17, No. 8, pp. (765–773).

Pathak CP, Adams AK, Simpson T, Phillips RE, Moore MA. (2004). Treatment of bioprosthetic heart valve tissue with long chain alcohol solution to lower calcification potential. *J Biomed Mater Res A*, Vol. 69, No. 1, pp. (140–144).

Pereira CA, Lee JM, Haberer S. (1990). Effect of alternative crosslinking methods on the strain rate viscoelastic properties of bovine pericardial bioprosthetic material. *J Biomed Mater Res*, Vol. 24, No. 3, pp. (345–361).

Petite H, Rault I, Huc A, Menasche P, Herbage D. (1990). Use of the acyl azide method for cross-linking collagen-rich tissues such as pericardium. *J Biomed Mater Res*, Vol. 24, No. 2, pp. (179–187).

Pierschbacher MD, Ruoslahti E. (1984). Cell attachment activity of fibronectin can be duplicated by small synthetic fragments of the molecule. *Nature*, Vol. 309, No. 5963, pp. (30–33).

Rieder E, Kasimir MT, Silberhumer G, Seebacher G, Wolner E, Simon P, Weigel G. (2004). Decellularization protocols of porcine heart valves differ importantly in efficiency of cell removal and susceptibility of the matrix to recellularization with human vascular cells. *J Thorac Cardiovasc Surg*, Vol. 127, No. 2, pp. (399–405).

Roberts TS, Drez Jr D, McCarthy W, Paine R. (1991). Anterior cruciate ligament reconstruction using freeze-dried, ethylene oxide-sterilized, bone–patellar tendon–bone allografts. Two year results in thirty-six patients. *Am J Sports Med*, Vol. 19, No. 1, pp. (35–41).

Rosenberg D. (1978). Dialdehyde starch tanned bovine heterografts: Development, In: *Vascular grafts*, Sawyer PN, Kaplitt MJ, pp. (261-270), Appleton Century-Crofts, New York.

Ruoslahti E, Pierschbacher MD. (1987). New perspectives in cell adhesion: RGD and integrins. *Science*, Vol. 238, No. 4826, pp. (491–497).

Sacks MS. (2003). Incorporation of experimentally-derived fiber orientation into a structural constitutive model for planar collagenous tissues. *J Biomech Eng*, Vol. 125, No. 2, pp. (280–287).

Santibáñez-Salgado JA, Olmos-Zúñiga JR, Pérez-López M, Aboitiz-Rivera C, Gaxiola-Gaxiola M, Jasso-Victoria R, Sotres-Vega A, Baltazares-Lipp M, Pérez-Covarrubias D, Villalba-Caloca J. (2010). Lyophilized Glutaraldehyde-Preserved Bovine Pericardium for Experimental Atrial Septal Defect Closure. *Eur Cell Mater*, Vol. 19, pp. (158-165).

Sato H, Suzuki N, Baba T, Ueda T, Mawatari T, Izumiyama O, Morishita K, Hasegawa T. (2008). Repair of ventricular septal perforation with asymmetrical conical patch exclusion. *Ann Thorac Cardiovasc Surg*, Vol. 14, No. 3, pp. (192–195).

Schenke-Layland K, Vasilevski O, Opitz F, Konig K, Riemann I, Halbhuber KJ, Wahlers T, Stock UA. (2003). Impact of decellularization of xenogeneic tissue on extracellular matrix integrity for tissue engineering of heart valves. *J Struct Biol*, Vol. 143, No. 3, pp. (201–208).

Schmidt CE, Baier JM. (2000). Acellular vascular tissues: natural biomaterials for tissue repair and tissue engineering. *Biomaterials*, Vol. 21, No. 22, pp. (2215–2231).

Schoen FJ, Levy RJ. (1999). Tissue heart valves: current challenges and future research perspectives. *J Biomed Mater Res*, Vol. 47, No. 4, pp. (439–465).

Simon P, Kasimir MT, Rieder E, Weigel G. (2006). Tissue Engineering of heart valves-Immunologic and inflammatory challenges of the allograft scaffold. *Progress in Pediatric Cardiology*, Vol. 21, pp. (161–165).

Simon P, Kasimir MT, Seebacher G, Weigel G, Ullrich R, Salzer-Muhar U, Rieder E, Wolner E. (2003). Early failure of the tissue engineered porcine heart valve SYNERGRAFT in pediatric patients. *Eur J Cardiothorac Surg*, Vol. 23, No. 6, pp. (1002–1006) [discussion 1006].

Sommers KE, David TE. (1997). Aortic valve replacement with patch enlargement of the aortic annulus. *Ann Thorac Surg*, Vol. 63, No. 6, pp. (1608-1612).

Soto AM, Justicia H, Wray JM, Sonnenschein C. (1991). P-nonylphenol, an estrogenic xenobiotic released from modified polystyrene. *Environ Health Perspect*, Vol. 92, pp. (167–173).

Sung HW, Chang WH, Ma CY, Lee MH. (2003). Crosslinking of biological tissues using genipin and/or carbodiimide. *J Biomed Mater Res A*, Vol. 64, No. 3, pp. (427–438).

Sung HW, Hsu CS, Wang SP, Hsu HL. (1997). Degradation potential of biological tissues fixed with various fixatives: An *in vitro* study. *J Biomed Mater Res*, Vol. 35, No. 2, pp. (147–155).

Sung HW, Huang RN, Huang LLH, Tsai CC. (1999). *In vitro* evaluation of cytotoxicity of a naturally occurring crosslinking reagent for biological tissue fixation. *J Biomater Sci Polymer*, Vol. 10, pp. (63–78).

Taylor PM, Cass AEG, Yacoub MH. (2006). Extracellular matrix scaffolds for tissue engineering heart valves. *Progress in Pediatric Cardiology*, Vol. 21, pp. (219–225).

Tedder ME, Liao J, Weed B, Stabler C, Zhang H, Simionescu A, Simionescu DT. (2009). Stabilized Collagen Scaffolds for Heart Valve Tissue Engineering. *Tissue Eng Part A*, Vol. 15, No. 6, pp. (1257-1268).

Teebken OE, Bader A, Steinhoff G, Haverich A. (2000). Tissue engineering of vascular grafts: human cell seeding of decellularised porcine matrix. *Eur J Vasc Endovasc Surg*, Vol. 19, No. 4, pp. (381–386).

Thubrikar MJ, Deck JD, Aouad J, Nolan SP. (1983). Role of mechanical stress in calcification of aortic bioprosthetic valves. *J Thorac Cardiovasc Surg*, Vol. 86, No. 1, pp. (115–125).

Thyagarajan K, Nguyen H, Tu R, Lohre J, Guida S, Sagartz JW, Quijano RC. (1992). Preliminary evaluation of calcification potential of Denacolt treated small diameter biological vascular grafts. *Trans Soc Biomater*, Vol. 15, pp. (686-688).

Timkovich R. (1977). Detection of the stable addition of carbodiimide to proteins. *Anal Biochem*, Vol. 79, No. 1-2, pp. (135–143).

Trantina-Yates AE, Human P, Zilla P. (2003). Detoxification on top of enhanced, diamine-extended glutaraldehyde fixation significantly reduces bioprosthetic root calcification in the sheep model. *J Heart Valve Dis*, Vol. 12, No. 1, pp. (93-101) [discussion 100-1].

Tsai TH, Westly J, Lee TF, Chen CF. (1994). Identification and determination of geniposide, genipin, gardenoside, and geniposidic acid from herbs by HPLC/photodiode-array detection. *J Liq Chromatography*, Vol. 17, pp. (2199–2205).

Ueda Y, TorrianniMW, Coppin CM, Iwai S, Sawa Y, Matsuda H. (2006). Antigen clearing from porcine heart valves with preservation of structural integrity. *Int J Artif Organs*, Vol. 29, No. 8, pp. (781–789).

Vesely I, Boughner D, Song T. (1988). Tissue buckling as a mechanism of bioprosthetic valve failure. *Ann Thorac Surg*, Vol. 46, No. 3, pp. (302–308).

Vesely I, Noseworthy R, Pringle G. (1995). The hybrid xenograft/autograft bioprosthetic heart valve: in vivo evaluation of tissue extraction. *Ann Thorac Surg*, Vol. 60, No. Suppl. 2, pp. (S359–364).

Vesely I. (2005). Heart Valve Tissue Engineering. *Circ Res*, Vol. 97, No. 8, pp. (743-755).

Vyavahare N, Hirsch D, Lerner E, Baskin JZ, Schoen FJ, Bianco R, Kruth HS, Zand R, Levy RJ. (1997). Prevention of bioprosthetic heart valve calcification by ethanol preincubation. Efficacy and mechanisms. *Circulation*, Vol. 95, No. 2, pp. (479–488).

Vyavahare NR, Hirsch D, Lerner E, Baskin JZ, Zand R, Schoen FJ, Levy RJ. (1998). Prevention of calcification of glutaraldehyde-crosslinked porcine aortic cusps by ethanol preincubation: Mechanistic studies of protein structure and water–biomaterial relationships. *J Biomed Mater Res*, Vol. 40, No. 4, pp. (577–585).

Wang D, Jiang H, Li J, Zhou JY, Hu SS. (2008). Mitigated calcification of glutaraldehyde-fixed bovine pericardium by tannic acid in rats. *Chin Med J (Engl)*, Vol. 121, No. 17, pp. (1675-1679).

Wei HJ, Liang HC, Lee MH, Huang YC, Chang Y, Sung HW. (2005). Construction of varying porous structures in acellular bovine pericardia as a tissue-engineering extracellular matrix. *Biomaterials*, Vol. 26, No. 14, pp. (1905–1913).

Wiegner AW, Bing OH, Borg TK, Caulfield JB. (1981). Mechanical and structural of canine pericardium. *Circ Res*, Vol. 49, No. 3, pp. (807–814).

Woods T, Gratzer PF. (2005). Effectiveness of three extraction techniques in the development of a decellularized bone-anterior cruciate ligament-bone graft. *Biomaterials*, Vol. 26, No. 35, pp. (7339-7349).

Yamamoto H, Yamamoto F, Ishibashi K, Motokawa M. (2009). In situ replacement with equine pericardial roll grafts for ruptured infected aneurysms of the abdominal aorta. *J Vasc Surg*, Vol. 49, No. 4, pp. (1041–1045).

Yoshioka SA, Goissis G. (2008). Thermal and spectrophotometric studies of new crosslinking method for collagen matrix with glutaraldehyde acetals. *J Mater Sci Mater Med*, Vol. 19, No. 3, pp. (1215–1223).

Zhou J, Fritze O, Schleicher M, Wendel HP, Schenke-Layland K, Harasztosi C, Hu S, Stock UA. (2010). Impact of heart valve decellularization on 3-D ultrastructure, immunogenicity and thrombogenicity. *Biomaterials*, Vol. 31, No. 9, pp. (2549-2554).

Zilla P, Weissenstein C, Human P, Dower T, von Oppell UO. (2000). High glutaraldehyde concentrations mitigate bioprosthetic root calcification in the sheep model. *Ann Thorac Surg*, Vol. 70, No. 6, pp. (2091-2095).

Zioupos P, Barbenel JC. (1994). Mechanics of native bovine pericardium. I. The multiangular behavior of strength and stiffness of the tissue. *Biomaterials*, Vol. 15, No. 5, pp. (366-373).